The Author

Robert Youngson, O.St J., MB, ChB., DTM&H., DO., FRC Ophth, is a medical graduate of Aberdeen University and has postgraduate qualifications from the Royal College of Surgeons of London, the Royal College of Physicians of England, and from the Royal College of Ophthal[illegible]ologists. [illegible] has been a general pr[illegible] medi[illegible] health medical officer and, for a number of years, was responsible for investigating Ministerial enquiries on medical matters for the Army. As a serving officer he qualified as an ophthalmic surgeon and eventually became Consultant Adviser and Head of Ophthalmology in the Army Medical Services, in the rank of Colonel. He has been the chairman of various committees concerned with medical research, clinical instruction, medical libraries, and computerization in medicine. Dr Youngson's many books include the *Collins Dictionary of Medicine*, the *Royal Society of Medicine Health Encyclopedia*, the *Guinness Encyclopedia of Science*, the *Royal Society of Medicine Dictionary of Symptoms* and the NHS Direct *Medical Encyclopedia* on the Internet.

For Elaine

Collins
dictionary *of*
Human Biology

HarperCollins Publishers
Westerhill Road, Bishopbriggs,
Glasgow G64 2QT

www.collins.co.uk

First published 2006

Reprint 10 9 8 7 6 5 4 3 2 1 0

A catalogue record for this book is available from the British Library

ISBN-13 978-0-00-722134-9
ISBN-10 0-00-722134-7

Typeset by Davidson Pre-Press Graphics Ltd, Glasgow
Printed and bound in Great Britain by Clays Ltd, St Ives plc.

Preface

This is not a medical dictionary. But human biology and medicine are so closely inter-related that, inevitably, many medical matters must necessarily be included. The foundations of the medical discipline – anatomy, physiology, biochemistry, genetics, immunology and psychology – are covered here. All of these are fundamental to an understanding of human biology.

Most of the entries go further than simple definitions and are thus more detailed than is usual in a dictionary. Although the dictionary encroaches only peripherally on topics such as signs and symptoms of diseases, pathology, medical and surgical treatment, and prescription drugs, reference to these has been allowed in the hope of enlivening the text and extending the usefulness of the book. It is to be hoped that many students of human biology may, by a perusal of this dictionary, be enabled to build on their knowledge of medical matters. And the slight blurring of the interface between human biology and medicine may perhaps encourage this.

An important feature of the book is the inclusion of a considerable number of detailed key entries written in a style designed to be readily understood by people with no prior knowledge of the topics. Each of these is a fairly full introduction to critically important basic topics in human biology, an understanding of which is essential to the mastery of the subject. Another feature is the weight that has been given to the short biographies of men and women who have made an important contribution to human biology. In many instances these biographies contain useful explanatory matter.

New students of biology and medicine are likely to find that many of their difficulties arise from nothing more than an inadequate grasp of the basic meanings of important terms. Happily, an understanding of the meaning of a few Greek and Latin roots can cast a wide beam of enlightenment. To assist students and others this dictionary provides all the terminological elements – roots, prefixes, suffixes and combining forms – necessary to resolve such difficulties.

It has occasionally been difficult to decide on the precise point at which human biology shades off into other disciplines. When this has happened I have usually decided to interpret the limits of human biology liberally. The topics thus introduced have often seemed to me interesting and helpful in countering undesirable narrowness of perspective.

Robert Youngson
Blandford Forum, Dorset
January 2006

Aa

a- *prefix denoting* not, without or the absence of a quality or object.

AASH adrenal androgen-stimulating hormone.

ab- *prefix denoting* from, outside, away from. From the Latin *ab*, from.

A band the dark band on muscle sarcomere corresponding to the length of the myosin filament.

ABC genes genes that code for the ATP-binding cassette transporter proteins that carry compounds across biological membranes. Fourteen different ABC genes are known to be associated with human genetic diseases.

ABC transporter proteins a large and diverse family of cell membrane transport proteins with two transmembrane domains and two ATP-binding regions within the cytosol. The transmembrane domains provide a controllable pathway for transported substance to pass through the cell membrane into or out of the cell.

ABCA3 gene a gene that codes for a 1704-amino-acid protein found in the limiting membrane of lamellar bodies. Mutations of this gene cause fatal SURFACTANT deficiency in newborn babies.

ABCA4 gene a gene that codes for a transporter protein found in the photoreceptor membrane disks in retinal rod cells.

ABCB1 member 1 of the B subfamily of ABC TRANSPORTER PROTEINS. Also known as MDR1.

ABCG5 genes genes coding for transporter proteins found in the liver and intestines.

ABCG8 genes genes with a function similar to that of ABCG5 GENES.

abdomen the part of the trunk below the chest. The abdominal cavity lies between the DIAPHRAGM, above, and the pelvic floor, below. It contains the LIVER and most of the digestive system, comprising the STOMACH, the DUODENUM, the JEJUNUM, the ILEUM, the CAECUM with the APPENDIX, the COLON, the RECTUM and the ANAL CANAL. Other abdominal organs include the KIDNEYS, the ADRENAL GLANDS, the SPLEEN, the PANCREAS and some large and important blood vessels, such as the AORTA and the inferior VENA CAVA, around which are many chains of LYMPH NODES. The intestines are hung from the back wall of the abdomen by a much-folded membrane, the MESENTERY, and are covered by the OMENTUM. The lower part of the abdomen, the pelvic cavity, contains the BLADDER. In women, the pelvic cavity also contains the womb (UTERUS), the FALLOPIAN TUBES and the OVARIES; in men, the central PROSTATE GLAND and the VAS DEFERENS, on each side. The wall of the abdomen consists of overlapping layers of muscle and sheets of fibrous tissue. The organs and the interior of the walls are covered with PERITONEUM. See also ABDOMINAL REGIONS.

abdominal referring to the ABDOMEN.

abdominal breathing respiration in which most of the work is done by the muscles of the abdominal wall in compressing the abdominal contents and elevating the diaphragm so as to compress the lungs and push out air.

abdominal quadrant any one of the four areas on the front of the abdomen formed by two imaginary lines, one vertical and one horizontal intersecting at the navel. The four quadrants are called upper and lower right and upper and lower left, as seen from the person's point of view.

abdominal region any one of the nine areas into which the surface of the abdomen is divided for descriptive purposes. Centrally, from above down, are the epigastric, umbilical and hypogastric (pubic) regions, and on either side, from above down, are the hypochondriac, lateral and inguinal regions.

abdominoperineal relating to the ABDOMEN and to the PERINEUM.

abdominous having a large belly. Pot-bellied.

abducens nerves the sixth of the 12 pairs of cranial nerves. Each abducens nerve supplies the tiny muscle on the outer side of the eye that moves the eye outwards (the lateral rectus muscle). Also known as abducent nerve.

abducent causing a separation. The word derives from the Latin *ab*, from, and *ducere*, to draw or lead. See also ABDUCTION.

abduction a movement outwards from the mid-line of the body or from the central axis of a limb. The opposite, inward, movement is called ADDUCTION.

abductor muscle a muscle which moves a part away from the mid-line.

aberrant deviating from the normal. The term may be applied to variations in the fine detail of body structure, such as the size and position of small arteries, or to modes of behaviour not generally considered acceptable. See also ABNORMAL.

aberration a deviation from normal. The term derives from the Latin *aberrare*, to wander off. See also ABNORMAL.

-ability *combining form denoting* power or capacity to do something.

abiogenesis the theory of 'spontaneous generation' – the long-discarded notion that living organisms can be formed from non-living matter. Louis Pasteur's work (see PASTEURIZATION) did much to overthrow this idea which was based largely on the observation that maggots often appeared on rotting meat.

abiotic non-living.

abiotrophy a general and inexact term referring to the effects of ageing, or to any degenerative process of unknown cause affecting tissue, especially nerve tissue. As knowledge extends the need for such terms diminishes.

ablactation weaning.

ablation the removal of any part of the body.

abnormal deviating from the observed rule or from the consensus opinion of what is acceptable. ABERRANT. See also ABNORMALITY.

abnormality the condition of not conforming to standard recognized patterns of structure, function, behaviour or phenomenon. In some social contexts, normality is no more than a statistical concept, and may be entirely relative. What is normal in one population or group may be abnormal in another. See also ABERRATION.

abnormal psychology a branch of psychology dealing with disorders of behaviour and mental disturbance, and with certain normal phenomena not clearly understood, such as dreams and altered states of consciousness. See also ABERRATION.

ABO blood groups a system of blood grouping developed from the discoveries of Karl Landsteiner (1868–1943) in 1900. The designations are arbitrary and the four groups are A, B, AB and O. These represent the antigenic differences in the red cells, the ANTIGEN being present on the red cell membranes. Group A, B and AB people have A, B and A and B antigens, respectively, on their red cells. Group O people have no antigens and are known as universal donors, whose blood, other things being equal, may safely be transfused into anyone. Group A people (about 26 per cent in Europe) have antibodies (agglutinins) to B in their serum and must not be given blood with B antigens. Agglutinins cause red cells with the same letter antigens to clump together and to become useless. Group B people (about 6 per cent) have antibodies to A in their serum and must not be given blood with A antigens. Group O people (about 68 per cent) have both A and B antibodies, so must not be given either A or B blood. Group AB people have no ABO blood group antibodies in their serum and are known as universal recipients. See also RHESUS FACTOR DISEASE and KELL BLOOD GROUP SYSTEM.

abreaction a process used in PSYCHOTHERAPY in which repressed thoughts and feelings are brought into consciousness and 'relived'. Abreaction is, it is hoped, followed by CATHARSIS and is most readily achieved when the trouble arises from a recent traumatic event.

absolute refractory period the period during which no stimulus, however strong, is able to evoke a response from an excitable tissue. The absolute refractory period follows immediately after a prior response and is brief.

absolute temperature temperature expressed in the Kelvin scale with absolute zero as 0 kelvin. The magnitude of the kelvin is the same as that of the degree Celsius and any Celsius temperature can be represented as an absolute temperature by degrees C – 273.15. The term 'degree kelvin' is no longer used; absolute temperatures are shown in kelvins (William, Lord Kelvin, British physicist, 1824–1907).

absorption 1 the movement of liquids and of dissolved substances across a membrane, from one compartment of the body to another or into the blood.
2 the assimilation of digested food material into the blood from the small intestine. Compare ADSORPTION. See also DIGESTION.

absorption spectrum the range of wavelengths of light that are absorbed by a pigment, such as the melanion of the skin.

abuse incorrect, improper or excessive use or treatment.

abzyme an antibody that has an enzyme-like (catalytic) action.

acanth-, acantho- *combining form denoting* spine or the prickle cell layer of the skin.

acanthoid resembling a thorn or spiny process.

acapnia absence of CARBON DIOXIDE in the BLOOD or tissues. This cannot occur in life and the term is usually applied imprecisely to a reduced level of CO_2.

acceleration stress the effects on body function and behaviour of periods of exposure to increased gravitational forces ('g') such as are experienced by fighter pilots or astronauts.

accent the speech patterns and pronunciation characteristic of a particular geographic region or social class. Accent sends out strong signals, the reaction to which may be one of reassurance, identification, respect, suspicion, resentment or hostility.

acceptor sites 1 DNA base sequences that bind transcription regulators,
2 Molecules that bind other chemical groups (also known as acceptor molecules).

accessory digestive organs the teeth, tongue, SALIVARY GLANDS, PANCREAS, LIVER and GALL BLADDER.

accessory muscle any muscle whose action reinforces that of any other muscle.

accessory nerve one of the eleventh of the 12 pairs of cranial nerves which arise directly from the BRAIN. The accessory nerve also has a spinal root. Fibres from the root from the brain join the vagus nerve. The accessory nerve supplies many muscles in the PALATE, throat, LARYNX, neck, back and upper chest.

accident-prone unusually liable to suffer accidents. Liability to accidents is probably no more than an effect of carelessness and lack of imaginative foresight or sometimes aggression or non-conformity.

acclimatization physiological adaptation to a new environment or situation. Acclimatization to a hot environment may take as long as three weeks, during which undue exposure to heat, or strenuous exercise, may cause heat exhaustion.

accommodation the automatic process by which the eyes adjust their focus when the gaze is shifted from one point to another at a different distance. Accommodation is achieved by changing the degree of curvature of the internal crystalline lenses of the eyes. In youth, these are naturally elastic and become more curved when the pull on the ligament by which they are suspended is reduced by contraction of the surrounding muscle ring (CILIARY MUSCLE). This allows focusing on near objects. The elasticity of the lenses drops progressively with age so the power of accommodation lessens. Difficulties with near vision in middle age (presbyopia) becomes apparent in all but the near-sighted.

accouchement an old-fashioned term for childbirth or delivery. From the French verb *coucher*, to lie down.

ACE see ANGIOTENSIN CONVERTING ENZYME.

A cells 1 the GLUCAGON-secreting cells of the ISLETS OF LANGERHANS in the PANCREAS. Also known as alpha cells to distinguish from the beta cells that secrete INSULIN. 2 the cells of the adrenal medulla that secrete adrenaline.

acellular devoid of cells. Some connective tissues incorporate acellular areas.

acentric chromosome a chromosome lacking a centromere, being formed from broken pieces of chromosomes. At the next cell division the acentric chromosome will be lost.

acephalous headless.

acet- *combining form denoting* acid. From the Latin *acetum*, vinegar.

acetabulum the socket in the side of the bony pelvis into which the spherical head of the thigh bone (FEMUR) fits. The word is derived from the Latin *acetum*, vinegar and *abulum*, the diminutive of *abrum*, a container. The acetabulum was thought to resemble a small vinegar cup.

acetaldehyde a product of the metabolism of large amounts of alcohol. The reaction is catalyzed in the liver by the enzyme alcohol dehydrogenase. The principal cause of the toxic effects of alcohol.

acetoacetic acid a KETONE body occurring in the urine in uncontrolled diabetes or starvation. See ACETONE BODY.

acetone body a KETONE body. One of the three compounds, acetoacetic acid, 3-hydroxybutanoic acid and acetone, produced when, in the absence of adequate available glucose, fatty acids are used for fuel. Acetone bodies are produced in large quantity in severe untreated diabetes and are the cause of ACIDOSIS and dangerous diabetic COMA.

acetylcholine the acetic acid ester of choline, an important NEUROTRANSMITTER acting at cholinergic synapses to propagate nerve impulses. It occurs in both the brain and the peripheral nervous system and is the neurotransmitter at neuromuscular junctions. It is stored in vesicles until a nerve impulse triggers its discharge across the synapse. It causes depolarization in the target cell. Acetylcholine is released from motor neurons, from all preganglionic neurons of the autonomic nervous system, all postganglionic neurons of the parasympathetic division, and some postganglionic neurons of the sympathetic division. Acetylcholine is inactivated by the enzyme ACETYLCHOLINESTERASE.

acetylcholine receptor an ion channel that opens when acetylcholine binds to it, so converting chemical diffusion into an electrical signal. Also known as the nicotinic acetylcholine receptor.

acetylcholinesterase an ENZYME that rapidly inactivates ACETYLCHOLINE by breaking it down to acetic acid and choline. Also known as cholinesterase.

acetyl coenzyme A, acetyl CoA a small, water-soluble important metabolic agent, formed during oxidation of fatty acids, amino acids and pyruvate, that transfers acetyl groups to citrate in the KREBS CYCLE and to various synthesizing pathways. Usually abbreviated to acetyl CoA.

acetylserotonin a product formed as an intermediate in the synthesis pathway for MELATONIN from SEROTONIN.

ache a persistent dull pain.

achievement motivation the persistent impulse to attain a high standard of performance in any activity.

Achilles tendon the prominent tendon just above the heel by means of which the powerful muscles of the calf are attached to the large heel bone. Contraction of the prominent calf muscles pulls the Achilles tendon upwards so that the ankle is straightened and the heel leaves the ground. The Achilles tendon is essential in walking and running and is easily strained or torn. (Named after Achilles, son of Peleus and Thetis, who, as a baby, was said to have been held by his heel and dipped in the river Styx by his mother to make him invulnerable. He died from a heel wound).

achondroplasia a dominant genetic defect that interferes with the growth of the cartilage at the growing sites at the end of long bones, resulting in a characteristic form of dwarfism. Achondroplasia has no effect other than on growth.

achondroplastic dwarf a person with very short arms and legs as a result of ACHONDROPLASIA. Many circus dwarfs are achondroplastic.

achromatic 1 colourless.
2 failing to take stain.

achromatopsia a rare but severe defect of colour vision in which the world is perceived almost in monochrome.

acid 1 any compound capable of releasing hydrogen ions when dissolved in water. A compound that donates a proton.
2 a solution with a hydrogen ion concentration greater than that of pure water.
3 having a pH of less than 7. pH is the common logarithm of the reciprocal of the hydrogen ion concentration. Acids turn blue litmus red. In the body, the main concentration of acid is in the stomach.

acidaemia a condition of raised blood acidity. The state in which the pH of the blood has fallen below normal.

acid–base balance the effect of mechanisms that operate to ensure that the body fluids remain nearly neutral, being neither significantly acidic nor alkaline. This homeostatic balance is maintained, in spite of the acids produced in metabolic processes, by the controlling action of the kidneys, by breathing out increased or decreased quantities of carbon dioxide as required and by the chemical buffering effect of bicarbonate in the blood.

acid fast of a stained tissue or organism, retaining the stain even in the presence of acid. Acid-fastness is a feature of certain organisms, such as the tubercle bacillus *Mycobacterium tuberculosis*, that have wax and fat in their cell walls. The fast is used as an aid to identification.

acidic hydrolases enzymes that break down molecules in an acidic environment at an optimum pH of about 5. They include acid phosphatases, lipases, proteases, glycosidases and nucleases. Many of the acid hydrolases are held in LYSOSOMES.

acidophilic of cells or tissues that stain readily with acidic dyes.

acidosis an abnormal rise in the acidity of the body fluids. A breakdown of the ACID–BASE BALANCE.

acid-sensitive ion channels voltage-insensitive protein cell-membrane sub-units which are activated and open when the pH falls below about 7.0. Many other calcium-permeable channels are inhibited as pH falls.

acinar pertaining to an ACINUS or acini. Grape-shaped or sac-like.

acinus 1 the part of the air passage of the lung beyond each of the smallest (terminal) bronchioles. There are about 12,000 acini in each lung. The lung ALVEOLI arise from the acini.
2 any terminal sac-like process of a compound gland.

acou- *combining form denoting* hearing. From Greek *akouein*, to hear.

acoustic nerve the short, eighth, cranial nerve connecting each inner ear to the brain and carrying nerve impulses subserving hearing and balance. Also known as the auditory nerve. The acoustic nerve is sometimes the seat of a non-cancerous but dangerous tumour, an ACOUSTIC SCHWANNOMA, which can cause much local damage by compression and which is not readily accessible for removal.

acoustic trauma the often damaging effect of loud noise on the inner ear. The degree of damage is related to the intensity multiplied by the time. A nearby explosion may cause as much damage as years of exposure to noise of lower amplitude. The hair cells in the ORGAN OF CORTI are injured by excessive movement and the higher pitched sounds are the first to be lost.

acquired characteristics bodily features appearing after birth as a result of environmental rather than genetic influences. Acquired characteristics are not passed on to offspring.

acro- *combining form denoting* outermost, at the tip or extremity. From Greek *akros*, an extremity.

acrocentric of a chromosome with the centromere near one end.

acrocephaly dome-headed.

acromegaly a serious disorder resulting from overproduction of growth hormone by the pituitary gland during adult life, after the growing ends of the bones (the epiphyses) have fused and the normal growth process is complete. The condition is usually the result of a benign tumour of the pituitary gland. There is no change in body height, but gradual enlargement of the jaw, tongue, nose, ribs, hands and feet occurs. There is also CUTIS VERTICIS GYRATA. If excessive growth hormone production occurs before the epiphyses have fused the result is gigantism. Acromegaly is treated by removing the cause.

acromioclavicular joint the joint between the outer end of the collar-bone (clavicle) and the ACROMION process on the shoulder-blade (scapula).

acromion the outermost extremity of the spine of the shoulder-blade. The acromion is joined to the outer tip of the collar bone (clavicle) in the acromioclavicular joint.

acrosome a tiny sac or double membrane in the head of a SPERMATOZOON that contains the digestive enzymes needed to break down the protective cell membrane of the egg (OVUM) and allow penetration. Release of these enzymes is called the acrosome reaction.

acrosome stabilizing factor a glycoprotein substance derived from the EPIDIYMUS that inhibits the release of acrosomal enzymes and may thus prevent fertilization.

ACTH adrenocorticotropic HORMONE. This hormone is produced by the pituitary gland, on instructions from the hypothalamus (the part of the brain immediately above the gland) when a stressful situation arises. ACTH is carried by the blood to the adrenal glands and prompts them to secrete the hormone cortisol into the bloodstream.

actin a contractile protein in muscle, found in the thin filaments, to which the myosin cross-bridges bind. Actin filaments are also abundant inside all nucleated cells where they form the cytoskeleton, determining cell shape and, in the case of amoebic cells, cell movement. An actin contractile ring forms around the equator of a dividing cell at the end of MITOSIS and tightens so as to pinch the two daughter cells apart.

actin binding proteins a large collection of different proteins that function by binding to actin as part of the cytoskeleton. They include EF1 in pseudopodia; fascin in stress fibres, microvilli, and acrosomal process; scruin in acrosomal process; villin in microvilli; dematin in the red cell cortical network just under the plasma membrane; fimbrin in microvilli and adhesion plaques; spectrin in cortical networks; dystrophin in muscle cell cortical networks; and filamin in pseudopodia, stress fibres and filopodia.

acting out impulsive, irrational actions that may display previously repressed feelings or emotions or unconscious wishes. The term may be applied to uncontrollable outbursts in children or neurotic adults.

actinic pertaining to radiation from the sun or to the biological effects of sunlight. Public awareness of the dangers of actinic radiation has been growing in recent years and has somewhat diminished enthusiasm for sunbathing.

actino- *combining form denoting* light, having rays, radiational. From the Greek *aktinos*, a ray.

actins filamentary contractile proteins forming the cytoskeleton of cells. Actins contribute to cell shape, cell motility and to some internal cell functions.

action potential the electrical signal propagated in nerve and muscle cells. It consists of a zone of reversal of the normal charge on the membrane so that the outside briefly becomes negative relative to the inside, instead of vice versa. This zone of depolarization, which is caused by the opening of ion channels, then moves along the fibre at a rate very much slower than the speed of normal electrical conduction along a wire. Action potentials operate according to an 'all-or-none' law. They function fully or not at all.

activated cells cells that have changed in response to a stimulus. Examples are macrophages, oocytes and neutrophil polymorphs.

activated macrophage a macrophage that has been rendered more lethal by the action of a cytokine such as interferon-gamma.

activated partial thromboplastin time a measurement of the time taken for the clotting system of the blood to produce the clot protein known as fibrin. The time is generally 25–39 seconds.

activation energy the energy needed to form chemical bonds during a chemical reaction or to break existing ones.

active hyperaemia the increase in blood flow though a tissue or organ occurring during, and as a necessary part of, increased metabolic activity.

active immunity immunity to disease resulting from infection with the disease or immunization with a vaccine. In both cases there is active production of ANTIBODIES.

active site 1 the region of an ENZYME to which the substance being affected binds so as to undergo a catalyzed reaction.
2 the localized part of a protein to which a substrate binds.

active transport the movement of dissolved substances across a membrane in the direction opposite to that of normal diffusion. Active transport operates against gradients of chemical concentration, electrical charge or electrochemical state. It requires the expenditure of energy.

acquired characteristics features of an organism, such as the human being, arising from environmental influences or bodily functioning, rather than from heredity.

acquired immune deficiency syndrome, AIDS a plague of pandemic proportions currently sweeping the world and more than decimating some populations. AIDS differs from all previous infectious diseases in that it deprives the victim of the normal resistance to infection and some kinds of cancer. It is caused by a retrovirus, the human immunodeficiency virus (HIV), that invades a class of T lymphocytes, the helper T cells, killing them and depriving the immune system of an essential component. The incubation period of AIDS – the period from the time of infection to the appearance of the fully established disease – is anything up to 10 years or longer, but cases of accelerated onset have been reported. AIDS features very low resistance to infections of all kinds, and many micro-organisms that do not normally affect the human body produce serious illness. These are called opportunistic infections and treatment exists for all of them. AIDS victims are also prone to develop a form of multiple blood vessel cancer called Kaposi's sarcoma. This is rare except in people with AIDS. The life of AIDS victims can be prolonged almost indefinitely by means of a combination of antiviral drugs.

acu- *combining form denoting* a needle. From the Latin *acus*, a needle.

acuity keenness of sense perception, especially in relation to vision and hearing.

acute short, sharp and quickly over. Acute conditions usually start abruptly, last for a few days and then either settle or become persistent and long-lasting (CHRONIC). From the Latin *acutus*, sharp.

acute phase proteins proteins produced by the liver during injury or infection.

acyl-CoA dehydrogenases enzymes that activate the first stage of the oxidation of fatty acids.

ad- *prefix denoting* to, towards.

ADAM33 an enzyme of the metalloprotease-disintegrin family found mainly on the surface membranes of airway smooth muscle cells.

Adam's apple the popular name for the voice box (LARYNX) at the upper end of the windpipe (TRACHEA). The larynx is larger and more protuberant in men than in women and contains longer vocal cords, which is why men have deeper voices than women.

adaptation 1 adjustment of sensitivity, usually in the direction of reduction, as a result of repeated stimulation.
2 the adjustment of an organism, including man, in part or in whole, to changes in environment or to external stress. Adaptation is an essential feature of all living things and the likelihood of survival often depends on how effectively it operates.

adaptive radiation the evolution of a wide variety of apparently greatly differing species from a single ancestral species or group as a result of wide proliferation and exposure to a range of environmentally different habitats.

addiction dependence for comfort of mind or body on the repeated use of a drug such as nicotine, alcohol or heroin. In some cases, the addiction is physiological – that is, the use of the drug has led to persistent changes in the way the body functions, so that its absence causes physical symptoms (withdrawal symptoms). In others, the dependence is psychological only. Research on rats has indicated that strong conditioned reflexes are often associated with drug addiction and that these make relapses more likely. The nucleus accumbens, linking the dopamine system with the limbic system appears to be importantly involved in addiction.

additive genes two or more genes that act in common to have an effect on the resulting characteristic. This can result in variance in the characteristic. Neither dominance nor interaction is involved.

addressins components in cell membranes that mediate adhesion to particular molecules on the surfaces of other cells. They function to effect cell-to-cell interactions and to help in the distribution of cells within an organism.

adduction a movement towards the centre line of the body. Muscles which adduct are called adductors. The term derives from the Latin *ad*, to and *ducere*, to draw. Compare ABDUCTION.

adenine a purine base. One of the four key biochemical units from which genes are formed in DNA and by which the two helical halves of the DNA molecule are linked together. Adenine pairs with thymine in DNA, but in RNA it pairs with uracil.

adeno- *combining form denoting* gland or glandular.

adenohypophysis the front lobe of the PITUITARY GLAND. This is structurally and functionally distinguishable from the rear lobe (posterior pituitary).

adenoid 1 gland-like.
2 one of the ADENOIDS.

adenoids collections of overgrown gland-like tissue, present on the back wall of the nose above the tonsils in children, which shrivel and disappear in adolescence or early adult life. The adenoids contain LYMPHOCYTES and are part of the body's defence (immune) system.

adenosine a purine NUCLEOSIDE consisting of one molecule of adenine and one molecule of d-ribose. It is formed in the body by the enzymatic breakdown of adenosine triphosphate (ATP). Adenosine is a NEUROTRANSMITTER, a HORMONE, and a NEUROMODULATOR.

adenosine diphosphate (ADP) a compound found in cells, consisting of the NUCLEOSIDE adenosine attached to two phosphate groups which are joined by a high-energy bond.

adenosine monophosphate (AMP) a derivative of ADENOSINE TRIPHOSPHATE (ATP).

adenosine triphosphate (ATP) a compound found in cells, consisting of the NUCLEOSIDE adenosine attached to three molecules of phosphoric acid. Adenosine triphosphate is the main energy-releasing entity of the cell. While it is being formed from adenosine diphosphate (ADP), ATP accepts energy from the breakdown of fuel molecules. During its breakdown to ADP or AMP it donates the energy to cell functions. As the energy source for the entire body, ATP is constantly being formed and broken down. At rest, a human consumes about 40 kg of ATP per day. During strenuous exercise the rate of ATP cycling may reach half a kg per minute.

adenylate cyclase an enzyme that converts ADENOSINE TRIPHOSPHATE (ATP) to cyclic AMP.

adenylate kinase an enzyme that promotes the transfer of a phosphate group from ADENOSINE TRIPHOSPHATE (ATP) to adenosine monophosphate (AMP), producing two molecules of adenosine diphosphate (ADP).

ADH *abbrev. for* ANTIDIURETIC HORMONE.

adherens junction a point of junction between cells at which ACTIN filaments from inside the cells pass across the adjacent cell membranes. A broad, belt-like adherens junction is called an adhesion belt.

adhesins proteins on the outer surfaces of micro-organisms that allow them to bind to cells they are attacking.

adhesion the abnormal healing together of raw tissues within the body that have been deprived of their normal 'non-stick' (epithelial) lining by injury or disease.

adipo- *combining form denoting* fat.

adipocere a wax-like substance, consisting mainly of fatty acids, into which the soft tissues of a dead body, buried in moist earth, are converted.

adipocyte a fat cell. A cell that synthesizes and stores neutral fats (triacylglycerols or TRIGLYCERIDES). Human fat is a liquid at body temperature, so adipocytes are normally filled with oil. Adipocytes develop from adipoblasts, which derive from fibroblasts.

adipose tissue fat cells held together, in large masses, by delicate connective tissue. Adipose tissue is both an insulant and a long-term fuel store, in which food in excess of requirements is converted to neutral fat (see TRIGLYCERIDE) and deposited. The term is derived from the Latin *adeps*, lard.

adipsin an ENZYME secreted by fat cells, and also present in the Schwann cells of nerve fibres, that is thought to regulate fat metabolism. It acts on free lipoproteins and is concerned with fat oxidation. Its levels are markedly reduced in genetically obese mice. The full function of adipsin remains unclear.

adjuvant any substance which, added to an ANTIGEN, non-specifically increases its power to stimulate the production of antibodies (see ANTIBODY).

Adlerian theory a school of psychology originating with the Austrian psychologist Alfred Adler (1870–1937). Adler was an original adherent to Freudian psychoanalysis who defected and formed his own school after coming to reject Freud's ideas of infant sexuality and unconscious motivation. Sex, he thought, was not a significant factor until the development of mature sexual interest – by which time the patterns of personality were already formed. He decided that the child, conscious of its weakness and inferiority, tries to improve its status by self-assertion and the aspiration to grow up and achieve superiority. Compensation for this sense of inferiority can be achieved either by direct effort to overcome it or by efforts directed to achieve success in another field. A major part of motivation was the tendency to avoid situations that emphasize one's sense of inferiority and seek those that diminish it.

Adler used the term 'life-style' to refer to the totality of our perception of ourselves in relation to the world – and especially our goals and how we try to achieve them. A good life-style incorporated a balance between selfish and social impulses and the use of means to overcome inferiority that are not damaging to ourselves or others. Courage, cooperation and a healthy and balanced attitude to sex are required. The neuroses, he believed, were caused by a lack of balance between selfish and social drives.

Adler was a pioneer in child guidance and set up many child guidance clinics in Vienna. He was a successful public lecturer who popularized the phrase 'inferiority complex'. His ideas on how children should be brought up were widely influential, especially in America, where he lectured extensively. There is little to object to in Adler's common sense notions and much to admire. His theory and practice were, however, condemned by most psychoanalysts as naive, superficial and hasty. The untutored public, however, continues to give unacknowledged support to much that he taught.

A-DNA one of the three forms of DNA. This form is found in DNA gels with relatively little water. It is a right-hand helix with about 11 bases per turn. The predominant form is B-DNA.

adnexa adjoining parts of the body. The adnexa of the eyes are the lacrimal glands, the eyelids, and the lacrimal drainage system. The uterine adnexa are the FALLOPIAN TUBES and the OVARIES. The term is derived from the Latin *annexere*, to tie on.

adolescence the period of life between PUBERTY and the achievement of full adult physical maturity around the age of 20.

ADP adenosine diphosphate. See ADENOSINE TRIPHOSPHATE.

adrenal cortex the outer zone of the ADRENAL GLAND that secretes CORTISOL, sex hormones (ANDROGENS) and ALDOSTERONE. See also ADRENAL MEDULLA.

adrenal glands the small internally secreting (ENDOCRINE) organs which sit like triangular caps one on top of each KIDNEY (hence the name). Each adrenal has two distinct parts, the inner core, which produces ADRENALINE, and an outer layer (the cortex) which produces various steroid hormones. Formerly known as suprarenal glands, which, of course, means the same thing.

adrenaline epinephrine, a HORMONE secreted by the inner part of the ADRENAL GLANDS. It is produced when unusual efforts are required. It speeds up the heart, increases the rate and ease of breathing, raises the blood pressure, deflects blood from the digestive system to the muscles, mobilizes the fuel glucose and causes a sense of alertness and excitement. It has been described as the hormone of 'fright, fight and flight'. Also known, especially in USA, as epinephrine.

adrenal medulla the inner part of the ADRENAL GLAND that secretes the hormones ADRENALINE and noradrenaline. The secreting cells are called chromaffin cells. See also ADRENAL CORTEX.

adrenarche the changes in the outer layer of the adrenal gland that occur at puberty, with the production of male sex hormones (androgens) in both sexes that bring about some of the external sexual characteristics and the prompting of libido.

adrenergic having effects similar to those of ADRENALINE. Drugs with adrenaline-like action are called adrenergic. A nerve which releases noradrenaline (a substance closely related to adrenaline) at its endings to pass on its impulses to other nerves, or to muscle fibres, is described as an adrenergic nerve.

adreno- *combining form denoting* the adrenal gland.

adrenocorticotrophic causing stimulation of the outer layer of the adrenal gland and promoting secretion of the adrenal cortical hormones.

adrenoceptor see BETA-ADRENOCEPTOR.

adsorption the process by which a substance, such as a gas or dissolved solid, is attracted to, and adheres to, a surface.

adultery voluntary sexual intercourse between a married person and a person who is not the legal spouse. For good social reasons, adultery has generally been disapproved of. It has ethical, theological and legal implications. In some cultures adultery still carries a death penalty.

adventitia the outer covering or layer of an organ, especially a blood vessel. Also known as tunica externa.

aero- *combining form denoting* air or gas.

aerobic 1 of a process that requires gaseous oxygen.
2 of an organism that is able to live only in the presence of oxygen. Compare anaerobic.
3 relating to aerobics or AEROBIC EXERCISE.

aerobic exercise a system of physical exercising in which the degree of exertion is such that it can be maintained for long periods without undue breathlessness. The object of aerobic exercise is to increase the efficiency of the heart and lungs and their ability to supply oxygen to the tissues.

aerobic respiration a cellular process in which glucose is broken down to produce carbon dioxide, water and energy in the form of 38 molecules of ATP per molecule of glucose. The process includes glycolysis, the citric acid cycle, and the electron transport chain.

aerophagy air swallowing. This is common in people with indigestion whose efforts to bring up wind eventually result in the swallowing of sufficient air to produce a belch.

aerosol a suspension of very small droplets of a liquid or particles of a solid, in air.

aetiology, etiology the study of causes. In common language, the term is used as a synonym for 'cause', especially for the cause of a disease or disorder.

affect mood or emotion. The word is often used to describe the external signs of emotion, as perceived by another person.

affection any disorder or disease of the body.

affective pertaining to mood or emotion.

affective functions the brain functions concerned with the emotions, especially those of fear, pleasure, gratification of all kinds, sexuality and jealousy. Affective functions are centered in the most primitive parts of the brain, especially the limbic area which we have in common with many of the other animals.

afferent directed toward a central organ or part, as in the case of sensory nerves that carry impulses to the spinal cord and brain.

afferent arterioles the small arteries that carry blood to the glomeruli of the nephron in the kidneys.

afferent neuron a nerve cell whose body lies outside the central nervous system and which conveys information to the CNS from sensory receptors at the periphery.

affinity the strength of binding between a receptor, such as an ANTIGEN binding site on an antibody, and a LIGAND, such as an EPITOPE on an antigen.

afterbirth the PLACENTA, the umbilical cord and the ruptured membranes which surrounded the FETUS before birth. These parts are expelled from the womb (UTERUS) within an hour or two of birth.

after-effect a delayed effect of some physiological or psychological stimulus.

after-image a visual negative impression of a bright object or light, which persists for a few seconds after the gaze is shifted or the eyes closed.

afterpotential a short period of hyper-polarization on a nerve cell membrane following an ACTION POTENTIAL.

ageing

The greatly increased interest in the study of ageing in recent decades is a response to our ever-ageing population. A decreasing birth rate and a progressive increase in the expectation of life have unbalanced the populations of the developed countries and the number of the elderly has grown, both in relative and in absolute terms. Life expectancies, at every age, are higher than they have ever been. If the trend continues, the growing dependency of an ever-increasing elderly population on a shrinking young population is likely to become a serious burden. It is for reasons such as this that knowledge of the nature of ageing, and studies into ways in which the elderly can remain independent of help, are urgently required.

The natural life-span in humans appears to be between 100 and 110 years, and there is no reason to suppose that this is increasing. The increasing expectation of life is largely due to advances in medical science which allow more and more people to approach or reach the natural age limit. Some experts dispute the figure for life-span and hold that the human body is programmed to begin to fall apart at about 85. The exponents of this school of opinion point out that, at that age, tiny insults – injuries that would hardly be noticed at 20 – are often enough to cause death. Others question this reasoning and point out that many of the conditions that have traditionally carried away the elderly – cancer, pneumonia, osteoporosis and atherosclerosis – are increasingly susceptible to medical intervention. Studies of mortality rates for people of 85 and over confirm that they have dropped substantially in the last 50 years. Even for people of 100, the number dying in a year has dropped.

Environmental influences

In most people, the phase of bodily development with increasing physical and mental power continues up to the mid-twenties and then undergoes a gradual decline. In some, physical and mental capacity continue to increase until well after that age. Environmental influences and life-style clearly have a major part to play in determining the outcome, and the signs of ageing appear earlier and are more severe in some people than in others. The body is a reactive entity that is strongly affected by many external factors that are clearly related to ageing. Some of the most obvious of these include lack of interest in the world and in health, lack of exercise of body and mind, unsuitable diet, smoking, excessive alcohol intake, use of drugs and undue exposure to sunlight. All of these are known to accelerate the process of ageing.

The progressive reduction in physical and mental capacity deemed to be characteristic of the latter part of life is often merely a reflection of cultural expectations. It would be better to try to emulate the achievements of the exceptional elderly, rather than to conform to cultural stereotypes. Significantly, the achievements of the aged often occasion surprise. There is, however, a substantial literature of examples, in every field of human creativity – art, literature, music and science – of the accomplishments of the old. To an unexpected extent, the body and the mind will deliver the work output required of them, and will adapt in strength and skill, to meet the demand. It is a pity that the common perception of the elderly is that such demands should no longer be made of them. Interest and motivation are essential if ageing is to be retarded. Happily, these facts are gradually becoming understood and the central importance of occupation, throughout life, is becoming appreciated.

The effects of ageing

Some of the commonest effects include: loss of elasticity in the arteries leading to increasing rigidity; thickening of the walls of the arteries; deposition of cholesterol in the arterial walls; widening, elongation and tortuosity of arteries; local ballooning of arteries (aneurysm formation); a gradual rise in the blood pressure; loss of elasticity of the skin leading to wrinkling and sagging; accumulation of skin pigment in patches (lentigenes); loss of bone bulk from reduction in protein and mineral content (osteoporosis); increase in bone brittleness; loss of muscle bulk and power; development of a white ring around the edge of the corneas (arcus senilis); graying and then whitening of the head hair. The deposition of cholesterol in the walls of arteries that occurs with age is a normal, and largely harmless, process and is to be distinguished from the formation of plaques of cholesterol and other materials that is a feature of the disease atherosclerosis.

The emphasis given in the above list to the changes in the arteries is not arbitrary; as the once celebrated French physician

Pierre Jean Georges Cabanis (1757–1808) rightly said: 'A man is as old as his arteries'. This apparent epigram is now known to be all too obviously true. In the Western World, arterial disease, especially atherosclerosis, is by far the commonest cause not only of death but of serious late-life morbidity from strokes, senile dementia, heart attacks and leg gangrene.

In spite of every effort, ageing, or senescence has certain inescapable features or associations. Loss of muscle bulk causes a decrease in power. The lungs lose elasticity and become more prone to local areas of collapse (atelectasis) and to pneumonia, especially after prolonged bodily immobilization. Progressive loss of nerve tissue may lead to such symptoms as unsteadiness, vertigo and reduced powers of memory. Loss of elasticity of the internal lenses of the eyes leads to difficulty in focusing close objects (presbyopia). Gradual denaturing of the protein fibrils of the lenses leads to opacification (cataract). Degenerative changes in the retinas, especially around the most sensitive central zones (age-related macular degeneration) may lead to blindness, as may a gradual failure of the internal fluid drainage mechanisms of the eyes (glaucoma). The immune system declines in efficiency with age so that the susceptibility to infection increases. Various endocrine changes occur with age. The tissues become relatively resistant to insulin so that higher blood sugar levels occur, bringing with them an increased tendency to arterial disease. The output of sex hormones decreases, gradually in men but acutely in menopausal women, and this commonly has serious consequences.

Theories of ageing

Throughout life, DNA undergoes constant change (mutation) as a result of external agencies and as a result of mistakes in replication. Fortunately, because of the double strand structure and the existence of certain repair enzymes, mutations are readily corrected. Theories of ageing include the suggestion that either an increased susceptibility to external mutagenic influences or a decrease in the efficiency of the repair mechanisms, or both occur. So far, however, no evidence has been found of defective repair and the rate of spontaneous mutations from external causes is insufficient to account for the number of changes that occur with ageing. An alternative theory, known as the error catastrophe theory, proposes that errors occurring in replication of DNA and RNA and in protein synthesis gradually and increasingly augment each other until they reach the stage of catastrophic failure at which cells can no longer replicate.

The most generally accepted theory at present is that ageing is regulated by particular genes. Cells in the adult body fall into three groups: those that reproduce continuously, those that reproduce when needed, and those that never reproduce. In the first category are the cells of the skin, intestinal lining and the blood-forming tissues; in the second are cells of the liver; and in the third are the cells of the brain and nervous system and the heart and voluntary muscles. Cells in the first group will readily reproduce in artificial culture.

Forty years ago Leonard Hayflick of the Wistar Institute in Philadelphia demonstrated that artificial cultures of fibroblasts – the main constructional cells of the body – would not continue to divide (replicate) indefinitely, even if provided with ideal conditions and nutrition. The maximum number of population doublings of the cells appeared to be about 50. The most interesting finding in this work, however, was that the number of doublings was apparently determined by the age of the person from whom the cells were taken for culture. Cells from a baby would undergo the full number of divisions; those from an old person only one or two. Likewise, cells from a person with premature senility (progeria) would also divide only a few times. This finding has had a strong influence on gerontological thought and has been taken to indicate that there is a finite limit to human life. Not all, however, agree that one can extrapolate from the culture flask to the human organism.

Telomeres are sections of DNA that form the natural end of a chromosome and are the points that resist union with fragments of other chromosomes. Formation of telomeres involves the shaving-off of some junk DNA at the chromosome end. There is a limit to the number of times that this can occur and this is believed to be the reason for the upper limit in the number of times cells can reproduce. Telomere length is a function of age but there is recent evidence that it may also be an X-linked familial trait.

Premature ageing

There is a rare disorder in which a young person begins to age, gradually over the course of about ten years acquiring all the characteristics of old age – wrinkled skin, white hair, arterial degeneration, high blood pressure, cataracts, increasing susceptibility to disease, and so on – and dying, in early adult life, in a state of apparent senility, usually from coronary artery disease.

This disease is called progeria and may take two forms, distinguished by the age of onset – infantile progeria, which starts in early childhood and in which the whole process of ageing may be complete by the age of 30, and adult progeria, or Werner's syndrome. This usually starts in the 30s or 40s, often with premature baldness or graying of the hair. The skin quickly becomes wrinkled, arterial disease rapidly develops, the sex glands atrophy and diabetes is common. Death occurs often within ten years of onset. The cause of progeria is unknown and there is no effective treatment. The condition, although tragic, is, of course, of intense interest to gerontologists.

ageism discrimination on the grounds of age.

agenesis absence of an organ or part as a result of failure of development in the early stages.

age roles certain characteristic patterns of behaviour expected of people of different ages. Age roles, once of central importance, and marked in their transitional stages by formal rites of passage, have become less prominent in modern industrialized society.

agglutination the clumping and sticking together of normally free cells or bacteria or other small particles so as to form visible aggregates. Agglutination is one of the ways in which ANTIBODIES operate. From the Latin *ad*, to and *glutinare*, to glue.

agglutinin a substance that causes cells or other particles to clump together and, usually, to lose their former properties. 'Warm' agglutinins function at normal body temperatures; 'cold' agglutinins do so at lower temperatures. Agglutinins can cause severe ANAEMIA.

agglutinogen an ANTIGEN that stimulates the production of a substance that induces agglutination (an agglutinin).

aggrecan an extracellular PROTEOGLYCAN that forms exceptionally large aggregates giving cartilage its particular gel-like properties and resistance to deformation.

aggregation factor a protein that forms cross links between the cell membranes of different cells to bind them together to form tissues.

aggression feelings or acts of hostility. Abnormal aggression is often associated with emotional deprivation in childhood, head injury, or brain disease, such as tumour, excessive alcohol intake or the use of drugs such as amphetamines (amfetamines).

agitation a state of mind, usually due to anxiety or tension, which causes obvious restlessness.

agnathia absence of the jaw.

agonal relating to the events occurring in the last moments of life, such as the cessation of breathing or the heartbeat.

agonist 1 a molecule, such as a HORMONE, NEUROTRANSMITTER or drug, that attaches (binds) to a cell receptor site to produce an effect on the cell. An antagonist is a molecule that interferes with or prevents the action of the agonist.
2 a contracting muscle that is opposed by contraction of another associated muscle, the antagonist.

agony 1 intense physical or mental pain.
2 the struggle that sometimes precedes death.

agoraphobia an abnormal fear of open spaces or of being alone or in public places. Agoraphobia may be so severe that the sufferer refuses to leave his or her own home and becomes permanently house-bound. It is the commonest of the phobias. The term derives from the Greek *agora*, an open assembly place or market and *phobia*, fear or horror.

agrin a protein released by motor neurons that brings about the close aggregation of acetylcholine receptors on muscle cells.

AGXT *abbrev. for* alanine glyoxalate aminotransferase, an ENZYME necessary for the metabolism in the liver of the amino acid alanine. The gene that codes for this protein is on the long arm of chromosome number 2.

AID *abbrev. for* ARTIFICIAL INSEMINATION by an anonymous donor. Compare AIH.

AIDS *acronym for* the acquired immune deficiency syndrome.

AIH *abbrev. for* ARTIFICIAL INSEMINATION with semen obtained from the husband or partner.

air the mixture of gases forming the atmosphere of the earth. It consists of about 78% nitrogen, 21% oxygen, 0.1% argon, 0.03% carbon dioxide and smaller proportions of rare gases and ozone. A continuing adequate supply of oxygen is essential to life.

airway 1 the passages from the nose and mouth down to the air sacs in the lungs, by way of which air enters and leaves the body. 2 a curved plastic air tube often used by anaesthetist to prevent the tongue from falling back and obstructing breathing.

akinesia loss of the power of voluntary movement. Paralysis of the motor function.

ala any wing-shaped protrusion in the body, as from a bone. The adjective is alar.

alalia loss of the power of speech.

albinism congenital absence of the normal body colouring pigment MELANIN. This is formed from tyrosine by the action of an enzyme tyrosinase present in cells called melanocytes. There are two kinds of albinism, tyrosinase positive and tyrosinase negative, due to mutations in different GENES that are not ALLELES. Albinism is an AUTOSOMAL RECESSIVE inheritance. Because the two mutations are on unrelated genes, people HETEROZYGOUS for both of the abnormal genes have normal colouring. For the same reason, two albino parents do not necessarily have albino children. Tyrosinase negative people are completely free of pigment, have pure white hair, pink-white skin and blue eyes and usually have visual problems. Tyrosinase positive people are less severely affected. Albinos often have defective eyesight from pigment deficiency in the eyes, and usually suffer from jerky eye movements (nystagmus).

albino a person with a genetic defect causing absence of the normal body pigment, MELANIN, which gives colour to the hair, eyes and skin. The gene responsible is recessive and the parents usually have normal colouring. The term derives from the Latin *albus*, white. See ALBINISM.

albumen 1 a nutritive substance surrounding a developing EMBRYO, as in the white of an egg. 2 ALBUMIN.

albumin a protein, soluble in water, synthesized in the liver and present in the blood PLASMA. Albumin, the most abundant blood protein, concentrates the blood and attracts water, thereby maintaining the circulating blood volume. Compare GLOBULINS.

albumose of a class of albuminous substances formed by the action of digestive enzymes on proteins.

alcohol a colourless volatile liquid obtained by fermenting sugars or starches with yeast and used as a solvent, a skin cleaner and hardener, and as an intoxicating drink. Also known as ethyl alcohol or ethanol.

alcohol forensic tests tests based on alcohol breakdown products used to reveal recent drinking both in the short term (hours) or the longer term (several months). Levels of ethyl glucuronide in urine or blood reveal drinking after 1–5 days; phosphatidyl ethanol levels reveal drinking after one to three weeks; and the combined levels of four fatty acid ethyl esters in hair provide evidence for up to several months after drinking.

alcoholic 1 pertaining to alcohol.
2 containing alcohol.
3 a person who habitually consumes alcoholic drinks to excess or who is addicted to alcohol to the detriment of health. A person suffering from ALCOHOLISM.

alcoholism 1 psychophysiological dependence on ALCOHOL with compulsive consumption of alcohol.
2 damage to the stomach lining, liver, nervous system, heart or voluntary muscles caused by prolonged exposure to high blood levels of alcohol.

alcohology the study of the nature and management of alcohol abuse.

alcohol withdrawal syndrome a group of symptoms and signs that develop within 6–24 hours of taking the last drink in a person suffering from ALCOHOLISM. They include agitation, anxiety, tremors, loss of appetite, nausea, vomiting, sweating, insomnia, disorientation, grand mal seizures and delirium tremens.

aldehyde a product of dehydrogenated (metabolized) alcohol, hence the name. An organic compound containing the group –CHO. Aldehydes cause most of the toxic effects of bibulous overindulgence (hangover).

aldosterone a hormone secreted by the cortex of the adrenal gland that acts on the kidney to decrease the loss of sodium and increase the loss of potassium. It operates in a feed-back mechanism to help to maintain the ionic equilibrium in the body.

aldosteronism the condition caused by excessive secretion of the adrenal HORMONE aldosterone. It features muscle weakness, high blood pressure (HYPERTENSION), ALKALOSIS, excessive urinary output and thirst.

Alexander technique a method of therapy for various physical and psychological disorders said to be caused by faulty posture. The pupils are taught how to break habits of slouching and adopt a new and better bearing, thereby, it is claimed, being relieved of such problems as insomnia, lethargy and chronic ill health.

algesia pain. The term derives from the Greek *algos*, pain.

alicyclic of any organic compound with closed rings of carbon atoms but that still has aliphatic properties.

alienation 1 a state of estrangement from, or inability to relate to, other people, concepts, social norms, or even oneself. Alienation, especially of the latter type, may be a feature of psychiatric disorder, but equally it may result from an accurate perception of the social environment.
2 a feeling that one's thoughts and emotions are under the control of someone else or that others have access to one's mind.

alimentary canal the digestive tract, extending from the mouth to the anus, in which food is converted by enzymes to a form suitable for absorption and through which the processed material passes into the bloodstream. The canal includes the PHARYNX, the OESOPHAGUS, the STOMACH and the small and large INTESTINES.

alimentation the act or process of giving or receiving nourishment.

aliphatic of any organic compound in which all carbon atom chains are arranged linearly rather than in rings, or that have rings that do not have the stability of benzene rings (see AROMATIC). All non-cyclic organic compounds are aliphatic and cyclic aliphatic compounds are called alicyclic compounds.

aliquot 1 a sample quantity of a larger amount of a substance.
2 an amount that is an exact divisor of the whole quantity of a substance.

alisphenoid the greater wing of the SPHENOID BONE.

alkalaemia an increase in the alkalinity of the blood above normal. ALKALOSIS. A rise in the pH of the blood.

alkaline phosphatases ENZYMES with an optimum pH in the alkaline range, that break down phosphate bonds in compounds. They are widely distributed in the body, and are released into the blood in large quantites in a number of conditions including obstructive JAUNDICE and certain bone diseases. This forms the basis of a common biochemical test.

alkaloids a group of bitter tasting plant poisons, many of which are of medical importance. The alkaloids, which include nicotine, strychnine, morphine, codeine, atropine, caffeine and quinine, have powerful actions on the body. Alkaloids are small, complex, nitrogen-containing molecules often produced by plants as an evolutionary survival characteristic in an environment containing herbivorous animals.

alkalosis an abnormal degree of alkalinity of the blood, usually due to loss of acid by prolonged vomiting or to hysterical over-breathing with abnormal loss of carbon dioxide.

allantois a membranous sac that develops from the hindgut in the EMBRYO and takes part in the formation of the umbilical cord and the PLACENTA.

allele short for allelomorph 1 genes that occupy corresponding positions (HOMOLOGOUS loci) on homologous chromosomes. Humans have two, usually identical, alleles for each gene, one on each AUTOSOMAL chromosome of a pair.
2 one of the ways in which a gene, at a particular location on a chromosome, may differ in its DNA sequence from the normal

or from its fellow at the corresponding location on the other chromosome. If different alleles of a gene occur at the corresponding sites on the pair of chromosomes the individual is said to be HETEROZYGOUS for the gene. If the two alleles are abnormal in the same way, the individual is HOMOZYGOUS and the characteristic determined by the gene defect will be present. Heterozygous individuals will show the features of the DOMINANT gene. The other allele is RECESSIVE. The term derives from the Greek *allos*, another.

allelic exclusion a state in a B lymphocyte in which only one of a pair of heterozygous ALLELES can be expressed, the other being inactive. The result is the production of only one of two possible kinds of immunoglobulin. The mechanism is not one of dominance.

allelic series the possible set of alleles to be found at a particular locus.

allelo- *combining form denoting* meaning from one to the other, being in a mutual or reciprocal relation.

allo- *a combining form meaning* 1 different, foreign or pertaining or belonging to a different individual,
2 opposed to.

allogeneic 1 pertaining to the range of genetic differences in different individuals of the same species.
2 of a donated graft from a non-identical donor.

allopatric speciation the formation on new species that can occur when samples of the same species are isolated from each other in geographically and physically different environments.

alloreactive lymphocytes lymphocytes responsible for the strong humoral and cellular responses to non-self antigens.

all-or-none law the property in muscle and nerve fibres of either responding wholly to a stimulus or not at all. The strength of the stimulus must exceed a particular threshold or there will be no response, but when the response occurs it is total. The law applies to individual fibres and a graded response is obtained by a variation in the number of fibres activated.

allosteric protein a protein that changes from one folding conformation to a different shape when another molecule binds to it. The change in conformation alters the activity or properties of the protein.

allozyme one of the variants of an enzyme coded for by different alleles at a particular locus.

alopecia baldness. The commonest form is hereditary and affects males, but baldness may also be caused by old age, disease, chemotherapy or radiation for cancer and treatment with thallium compounds, vitamin A or retinoids.

alpha- *combining form denoting* first. Alpha is the first letter of the Greek alphabet.

alpha-actinin-3 ACTN3, a group of genes one of which, the R allele, codes for the fast-acting muscle protein actinin. The X allele does not produce this protein. Interestingly, 95 percent of champion sprinters have at least one copy of the R allele and 50 percent have two. In controls, 72 percent have one R allele and only 30 percent have two.

alpha-antitrypsin a glycoprotein in the serum which can inactivate protein-splitting enzymes such as TRYPSIN, elastase and collagenase.

alpha cells pancreatic cells in the Islets of Langerhans that secrete the hormone GLUCAGON.

alphafetoprotein a protein synthesized in the fetal liver and intestine and present in fetal blood and in the uterine fluid. Raised levels may indicate that the fetus has spina bifida and further investigation becomes urgent. Alphafetoprotein is also produced by certain malignant tumours and its estimation is sometimes used as a marker for the diagnosis of these tumours or their recurrence.

alpha helix a coiled configuration of a POLYPEPTIDE chain found in many proteins. This is one of the commonest forms of secondary structure in proteins.

alpha motor neurons large motor nerve fibres whose cell bodies are in the front 'horn' of the spinal cord and which innervate voluntary muscles.

alpha rhythm or alpha wave the most prominent waveform found in the

ELECTROENCEPHALOGRAM – the recording electrical activity of the brain. It is characterized by 8–12 oscillations per second when the subject is at rest.

alphatocopherol vitamin E.

Altaic language family a subdivision of the Ural-Altaic family, this group encompasses the Mongolian, Turkic and Manchu-Tungus sub-families. The Altaic languages are spoken over a very wide area. Mongolian extends from China as far west as Afghanistan and the lower Volga; Turkic from Anatolia to the Volga basin; and Manchu-Tungus in the north-east Siberian coast as far south as the Amur river.

alternative hypothesis the possibility (which should always be borne in mind) that an explanation of a phenomenon or result, however apparently obvious, may not be correct. See also NULL HYPOTHESIS.

alternative splicing a process in the expression of some genes in which the initial transcription of RNA can give rise to various related but different messenger RNAs with different combinations of exons that result in a range of related proteins called isoforms. The effect is that the coding capacity of the genome is greater than the number of genes would suggest.

altruism behaviour manifesting unselfish concern for the advantage of others. Much seemingly altruistic behaviour can be shown, on analysis, not to be so, and there are those who hold that altruism is a myth. Most social scientists, however, accept the concept.

alu sequence family a family of DNA sequences in the human genome, some 300 base pairs long, that serves no apparent genetic purpose. The alu sequences are found scattered throughout the chromosomes and occur about 500,000 times, constituting about 5 per cent of human DNA. Copies of the alu sequences occur in slightly varying forms and these variations allow DNA FINGERPRINTING of individuals. Alu sequences are mainly of cytosine/guanine and are specific to species and do not code for any product. They cause visible bands when chromosomes are stained with Gram's stain and can be used to identify human DNA for forensic purposes.

alveolus 1 one of the many million tiny, thin-walled, air sacs in the lungs. 2 a tooth socket in the jaw bone. 3 any small cavity or sac.

Alzheimer's disease a brain disorder that is by far the commonest cause of DEMENTIA. (Alois Alzheimer, German neurologist, 1864–1915).

amaurosis an old-fashioned term for blindness. Amaurosis fugax is a transient form of visual loss suggesting that the affected person is at risk from STROKE and requires urgent medical attention. From the Greek *amaurois*, dark or obscure.

amber codon the three-nucleotide group UAG (uracil, adenine, guanine) that forms a stop CODON marking the point at which the synthesis of a protein ends. Two other codons, UAA and UGA, have the same function. One of these three codons marks the end of every gene.

amber mutation a mutation that creates an AMBER CODON.

ambidexterity the ability to use either hand with the same facility. True ambidexterity, with no bias to one side, is rare and runs in families.

ambisexual having undifferentiated or ambiguous sexual orientation so that attraction to either sex may be experienced. See also BISEXUAL.

amblyopia 1 visual defect resulting from failure, from any cause, to form sharp, central retinal images in early life. In most cases the affected eyes appear structurally normal. 2 visual defect resulting from poisoning (toxic amblyopia) or a deficiency of an essential dietary ingredient. The term is derived from the Greek *ambius*, dull or blunt, and *ops*, an eye.

ambulatory 1 pertaining to walking. 2 capable of walking and not bedridden.

amelo- *combining form denoting* tooth enamel.

amenorrhoea the absence of menstruation. This is normal before puberty and after the menopause. During the reproductive years, the commonest causes are pregnancy and LACTATION, but it can be caused by a number of hormonal and other disorders.

Amerindian language family the group of nearly 1000 languages, many now extinct, that were and are spoken by the indigenous American Indian population in all parts of North and South America. Over 700 Amerindian languages are still spoken.

amide an organic compound containing the group $-CONH_2$.

amine a class of organic compounds derived from ammonia by replacing one or more of the hydrogen atoms by a member of the paraffin series or by an aromatic group. Amines occur widely in the body, and many drugs are amines.

amine hormone any hormone derived from the amino acid tyrosine. Amine hormones include ADRENALINE, NORADRENALINE and thyroid hormones. These are also called catecholamines.

amino acids the basic constituent of PROTEINS. Amino acids can be considered to be the 'alphabet' of letters from which proteins are written. Their properties are determined by their side chains. Body protein breaks down into 20 different amino acids (see below). Some of these can be synthesized by the body but some can not. The latter are known as 'essential amino acids' and must be obtained from protein in the diet. Amino acids group together to form peptides. Linkages between amino acids are called peptide bonds. Dipeptides have two amino acids, polypeptides have many. Polypeptides join to form proteins. The reverse process occurs when proteins are digested. Some amino acids, such as glycine, arginine, aspartic acid and glutamic acid, also perform specific biological functions in addition to helping to form proteins.

The amino acids forming body and other proteins are alanine (Ala), arginine (Arg), asparagine (Asn), aspartic acid (Asp), cysteine (Cys), glutamine (Gln), glutamic acid (Glu), glycine (Gly), histidine (His), isoleucine (Ile), leucine (Leu), lysine (Lys), methionine (Met), phenylalanine (Phe), proline (Pro), serine (Ser), threonine (Thr), tryptophan (Trp), tyrosine (Tyr) and valine (Val).

The German organic chemist Emil fischer (1825–1919) elicited an understanding of amino acids, peptides and proteins which was of fundamental importance in the development of organic chemistry and biochemistry. He was awarded the Nobel Prize in Chemistry in 1902. See also STEREOISOMERS.

aminoacylation the attachment of an amino acid to the acceptor arm of a TRANSFER RNA molecule. This is an essential step in the synthesis of a protein. The attachment is brought about by the appropriate AMINOACYL-tRNA SYNTHETASE.

aminoacyl-tRNA synthetase one of a group of 20 enzymes each of which is specific for the catalyzation of the linkage of a particular AMINO ACID to its own type of TRANSFER RNA during protein synthesis. There are 20 amino acids in body proteins.

amino group a functional end group occurring on all amino acids that participates in the formation of a peptide bond. The amino group has the chemical structure NHO–. Also known as the N-terminal.

aminopeptidases enzymes that split amino acids off, one by one, from the amino (N-terminal) at the end of a POLYPEPTIDE chain. Aminopeptidases occur in the small intestine and participate in the digestive breakdown of proteins to free amino acids.

aminotransferases enzymes present in liver cells which are released into the blood in liver disease, such as HEPATITIS, that damages liver cells. The most important are aspartate aminotransferase (AST) and alanine aminotransferase (ALT). Measurement of the levels of these enzymes in the blood is a valuable test of liver damage.

ammonia a substance produced when AMINO ACIDS are broken down. Ammonia is converted by the liver into urea and excreted in the urine. Urea can be broken down by bacterial enzymes to release ammonia. This may be a cause of nappy rash in babies.

amnesia loss of memory as a result of physical or mental disease or injury, especially head injury. Amnesia for events occurring before the head injury is called retrograde amnesia. Anterograde amnesia is loss of memory for events occurring after the injury.

amniocentesis a method of obtaining early information about the health and genetic constitution of the growing fetus, by taking a sample of the fluid from the womb (AMNIOTIC FLUID) for analysis usually between the 15th and 18th week of pregnancy. Cellular debris in the fluid provides DNA for chromosome analysis and sex determination and the fluid contains specific substances characteristic of various diseases. The risk of fetal loss from this procedure is estimated to be 0.5–1 per cent above the natural level of spontaneous abortion.

amnion one layer of the fluid-filled double membrane surrounding the fetus before birth. The amnion is the inner of two membranes, the other being the chorion. The membranes normally rupture and release the AMNIOTIC FLUID ('breaking of the waters') before the baby is born.

amniotic pertaining to the AMNION.

amniotic fluid a liquid produced by one of the membranes which surround the fetus throughout pregnancy and by fetal urination, in which the fetus floats. The volume of amniotic fluid at full term is usually about 1 litre.

amniotic cavity the space within the AMNION.

amoeba a single-celled microscopic organism of indefinite shape commonly found in water, damp soil and as parasites of other animals. Some amoebae cause disease in humans.

amoeboid resembling an AMOEBA, especially in its method of locomotion. An amoeba is a shapeless, jelly-like, single-celled organism that moves by pushing out a protrusion, a pseudopodium, in the required direction and then flowing into it. Some of the free cells of the immune system are amoeboid.

amorphous 1 of no particular shape or form.
2 lacking distinct crystalline structure.

AMP adenosine monophosphate, an adenosine group with a single phosphate group attached to it by a low-energy bond. AMP is the basic constituent of ADP and ATP (adenosine diphosphate and adenosine triphosphate).

amphetamine amfetamine, a central nervous system (CNS) stimulant drug with few medical uses but commonly abused to obtain a 'high'. Amphetamine use leads to tolerance and sometimes physical dependence. Overdosage causes irritability, tremor, restlessness, insomnia, flushing, nausea and vomiting, irregularity of the pulse, delirium, hallucinations, convulsions and coma. Amphetamine can precipitate a PSYCHOSIS in predisposed people.

amphi- *a prefix meaning* two, both or either.

amphiarthrosis a type of joint allowing very limited movement because the bones are connected by FIBROCARTILAGE. Such joints occur between adjacent vertebrae or between the ribs and the breast bone.

amphipathic of a chemical structure having both hydrophobic and hydrophilic surfaces as in the case of the phospholipid molecules of the cell membrane. Some proteins have amphipathic regions.

amphoteric of a compound that can act as an acid or a base depending on the pH. Amino acids are amphoteric.

amplification the production of extra copies of a DNA sequence. These may be within the chromosomal sequence or outside it.

ampulla 1 a widened (dilated) segment of a gland or small tube.
2 a swelling at the origin of a semicircular canal containing the cupola and hair cells.

ampulla of Vater a small sac-like widening (dilation) at the point of junction of the bile and pancreatic ducts where they enter the DUODENUM.

amputation removal, by surgical operation or injury, or rarely by disease, of part of the body. From the Latin *ambi*, around and *putare*, to prune.

amputee a person who has suffered an AMPUTATION.

amygdala an almond-shaped brain nucleus at the front of the temporal lobe. The amygdala is concerned with memory registration.

amygdalo- *combining form denoting* almond-like, almond-shaped or almond-related.

amygdaloid 1 almond-shaped.
2 relating to the AMYGDALA.

amyl- *combining form denoting* starch.

amylaceous starch-like.

amylase an ENZYME that converts starch to simpler carbohydrates such as disaccharides and small polysaccharides.

amyloid one of a range of proteins deposited in the brain in spongiform encephalopathies such as Creutzfeldt-Jakob disease, in other degenerative brain disorders such as Alzheimer's disease, and in the tissues in a wide range of long-term suppurative disorders. It is a hard, waxy proteinaceous substance in the form of straight, rigid, non-branching fibrils, 10–15 nm in diameter that are insoluble in water and relatively resistant to breakdown by proteolytic enzymes. There is a considerable range of amyloid proteins mainly specific for the different conditions in which amyloid is deposited. From the Latin *amylum*, starch.

amyloid precursor protein a natural brain protein that is coded for on chromosome 21. This protein is related in some way to memory. Point mutations are found in the gene for this protein in people who suffer from a familial form of early-onset Alzheimer's disease. Only a small proportion of cases of Alzheimer's disease, however, are caused in this way. Abnormally folded amyloid protein in the form of plaques and tangles occur in brain cells in Alzheimer's disease but also in chronic suppurative states in other parts of the body.

amylum starch.

amyotonia lack of muscle tone. See AMYOTONIA CONGENITA.

an- *prefix denoting* not, negative. This is the euphonic form of the Greek *a*, not, negative.

ana- *prefix denoting* up, upward, or back, backward.

anabolic 1 pertaining to a chemical reaction in which small molecules, such as amino acids, are combined to form larger molecules, such as proteins.
2 of any substance that increases the rate of metabolism of a cell or organism.
3 of a drug, such as a male sex hormone, that promotes body bulk.

anabolic steroids drugs which promote tissue growth, especially of muscle, by stimulating protein synthesis. Anabolic steroids are synthetic male sex hormones and tend to cause VIRILIZATION. These steroids are sometimes misused by athletes and bodybuilders to gain an unfair advantage. The risks are considerable.

anabolism building up of the tissues. The metabolic process by which the complex biochemical structure of living tissue is synthesized from simple nutritional elements such as sugars, amino acids and fatty acids. Contrast with CATABOLISM, which is the breakdown of complex tissues to simpler, consumable, substances. The term is derived from the Greek *anabole*, to build up or throw up.

anaemia a reduction in the amount of HAEMOGLOBIN in the blood. There are several different kinds of anaemia including simple iron deficiency anaemia, haemolytic anaemia, pernicious anaemia, and aplastic anaemia.

anaemic suffering from anaemia.

anaerobic living and being capable of reproducing in the absence of free oxygen. Only certain very simple organisms, such as some bacteria, are capable of anaerobic existence.

anaerobic respiration the process in which glucose is broken down in the absence of oxygen to produce lactic acid and two molecules of ATP.

anaesthesia loss of the sensations of touch, pressure, pain or temperature in any part of, or in the whole of, the body. This may be due to injury or disease of nerves or brain, or to deliberate medical interference. Drugs are commonly used to effect either general or local anaesthesia.

anaesthetic 1 insensitive.
2 relating to anaesthesia.
3 causing anaesthesia.
4 a drug used to cause unconsciousness or insensitivity to pain.

anaesthetization the induction of anaesthesia.

anal 1 relating to the anus, the controlled terminus of the digestive tract.
2 concerning the putative stage of psycho-sexual development of the child in which pleasure is obtained from sensations associated with the anus. See FREUDIAN THEORY.

anal canal the 5 cm-long terminal portion of the intestine that lies immediately below the RECTUM. The anal canal contains two muscular rings (SPHINCTERS) that can close it tightly and seven or more longitudinal pads of MUCOUS MEMBRANE that contain veins and press together to act as an additional sealing mechanism.

anal-expulsive a term applied by some psychoanalysts to people who show certain personality traits such as ambition, conceit or suspicion. These traits are said to manifest habits, attitudes and values associated with the infantile pleasure in the expulsion of faeces. See also ANAL-RETENTIVE, FREUDIAN THEORY.

analgesic 1 pain-relieving.
2 a pain-relieving drug.

anal intercourse a form of sexual intercourse, usually male homosexual, in which the anus performs the sexual function of the vagina.

anal-retentive a psychoanalytic term indicating traits such as miserliness, obsessiveness, obstinacy and meticulousness. These characteristics are said to originate in values formed during the 'anal stage' of development when particular pleasure was taken in the retention of faeces. See also FREUDIAN THEORY.

anal sphincter the double muscular ring surrounding the anal canal which, in conjunction with the thick, well-vascularized lining of the canal, produces a watertight seal except during defaecation.

anal stage a stage in the psychosexual development, proposed by Sigmund Freud, in which the child's preoccupation is with the anal region and the faeces.

analysis the determination of the constituents of which anything is composed. Compare synthesis. See also PSYCHOANALYSIS.

anaphase a stage in cell division (MITOSIS) in which the separated individual chromosomes migrate to opposite ends of the cell in preparation for the division of the cell into two new individuals.

anaplasia loss of the cellular microscopic features which distinguish one type from another. Anaplastic cells become smaller and simpler in structure and no longer combine to form recognizable tissues. Anaplasia is a common feature of cancer and, in general, the greater the anaplasia the more malignant and dangerous the tumour.

anastomosis a direct surgical connection formed between two tubular structures by stitching or a communication between an artery and a vein without intervening smaller vessels.

anatomical 1 pertaining to the structure of the body (the ANATOMY) or to dissection.
2 structural, as distinct from functional (PHYSIOLOGICAL).

anatomical position an internationally agreed standard position of the body used in descriptions of the position of structures and of directions and motions. The body is supposed to be standing with the head, eyes, palms of the hands and the toes pointing forward and the arms by the sides.

anatomy 1 the structure of the body, or the study of the structure.
2 a textbook or treatise on anatomical science.

ancestral genes DNA sequences in animal species that lived millions of years ago but that, although much modified by mutation and rearrangement are still present in whole or in part in the contemporary human genome. Some distinct and separate human genes share sequences that suggest they may have had a common ancestry.

anchorage dependence a property of cells that can grow and proliferate only if fixed to a substrate. Many cancer cells do not show anchorage dependence and can be grown in a liquid culture.

andro- *combining form denoting* man-like, male, pertaining to man in the sense of a male person.

androgen receptor gene a gene on the X chromosome that codes for the receptors for male sex hormones.

androgens male sex hormones. Androgens are STEROIDS and include testosterone and androsterone. See also ANABOLIC STEROIDS. The term androgen derives from the Greek *andros*, a man and *gennao*, to make.

androgenous pertaining to the birth or preferential production of male offspring.

androgen receptors molecules on the surface of cells that bind androgens and bring about their effects within the cells. They are specific GLYCOPROTEINS coded for by genes on the X chromosome.

androgynous hermaphroditic. Exhibiting both male and female characteristics. From the Greek *andros*, a man and *gune*, a woman.

android 1 a non-biological organism having human characteristics. A synthetic person or humanoid robot.
2 man-like, male-like, as of the pelvis. Compare gynaecoid.

andropause the period, usually occurring between the ages of 45 and 55, during which a man's testosterone levels may fall, leading to a reduction in vigour and sexual drive. The term is rarely used, the etymologically-shaky term 'male MENOPAUSE' being commonly preferred.

androsterone a metabolite of the male sex hormone TESTOSTERONE, excreted in urine of both men and women.

anencephaly absence of the greater part of the brain and of the bones at the rear of the skull. The term derives from the Greek *an*, not and *encephalon*, a brain. See ACEPHALUS.

anergy specific immunological tolerance in which T cells and B cells fail to respond normally by producing an immune response to antigens. The state can be reversed.

aneuploidy an abnormality in the number of CHROMOSOMES by loss or duplication. The number may be smaller or greater than the normal diploid constitution. The loss of a whole chromosome is lethal. A chromosome extra to one of the pairs is called TRISOMY. Trisomy 21, for instance, causes Down's syndrome. DNA aneuploidy refers to abnormal quantities of DNA in a nucleus. See also MOSAICISM.

aneurin vitamin B1, thiamin. This is plentiful in cereals and fresh vegetables, legumes, fruit, meat and milk, but there is very little in refined foods, fat, sugar or alcohol. Deficiency causes BERI-BERI in 1–2 months. Also known as thiamine.

aneurysm a berry-like or diffuse swelling on an artery, usually at or near a branch, and caused by localized damage or weakness to the vessel wall. Aneurysms can also involve the heart wall after a section has been weakened. See also BERRY ANEURYSM. The term derives from the Greek *anaeurusma*, a widening.

angel dust a slang term for the powerful analgesic and anaesthetic drug Phencyclidine commonly abused for recreational purposes. It is also known by the abbreviation PCP. Abuse of this drug can lead to muscle rigidity, convulsions and death.

anger a strong emotion aroused by a sense of wrong, whether justified or not.

angio- *combining form denoting* a blood vessel. From the Greek *angeion*, a vessel.

angioblasts cells in the embryo that give rise to blood capillaries.

angiogenesis the origination and development of new capillary blood vessels. Angiogenesis is necessary so that a growing or enlarging tissue, with its increasing metabolic needs, obtains an adequate blood supply providing oxygen, nutrients and waste drainage. Various angiogenetic factors are secreted by blood-deprived (ischaemic) cells and these operate on the inner lining (endothelium) of existing blood vessels to cause the budding out of new capillaries.

angiogenic growth factors promoters of the production of new small blood vessels. They include VASCULAR ENDOTHELIAL GROWTH FACTOR (VEGF); acidic and basic fibroblast growth factor; transforming growth factor alpha and beta; tumour necrosis growth factor-alpha; platelet-derived growth factor; angiogenin; and interleukin-8.

angiology the study of blood and lymph vessels. From the Greek *angeion*, a vessel and *logos*, a discourse.

angiotensin the vasoconstrictor polypeptide hormone, angiotensin II, which is released by the action of the enzyme renin. Its precursor, angiotensin I, is inactive until acted on by the angiotensin-converting enzyme, mainly in the lungs. Angiotensin II has a powerful effect on raising the blood pressure. It binds to angiotensin receptors and constricts the circular smooth muscle of blood vessel walls. It prompts cells of the adrenal cortex to secrete the hormone ALDOSTERONE. It modulates blood flow through the kidneys and acts directly on the heart muscle. By promoting raised blood pressure it encourages the development of the arterial disease ATHEROSCLEROSIS.

angiotensin-converting enzyme the (ACE) enzyme that converts angiotensin I to the active form angiotensin II. The gene for this enzyme has two alleles, the *I* allele and the

D allele. Research has shown that the *I* allele is associated with significantly better physical performance, endurance and response to physical training than the *D* allele. The difference is especially marked if the *I* allele is present at both loci and compared with people with the *D* allele at both loci.

angiotensinogen a plasma protein acted on by the protease enzyme renin to produce ANGIOTENSIN I.

Ångstrom unit a unit of very small size appropriate to the measurement of molecules and atoms. It is equal to one tenth of a nanometer (0.1 nm).

anhidrosis absence of sweating or inadequate secretion of sweat.

anhydraemia a deficiency of water in the blood.

animism the belief held by many primitive peoples that a spirit resides within every object, controlling its existence and influencing events in the natural world.

anion a negatively charged ion that is attracted to an anode (positively-charged electrode), in electrolysis. Anions are usually shown as the second group in simple inorganic molecules, thus Cl^- is the anion when common salt (NaCl) is dissociated in solution. Na^+ is the CATION.

anis-, aniso- *combining form denoting* unequal.

aniseikonia an ocular defect in which the size or shape of the retinal images are different in the two eyes. This is often the result of ANISOMETROPIA. From the Greek *an*, not *iso*, equal and *eicon*, an image.

aniso- *a prefix meaning* unequal.

anisocoria inequality in the size of the pupils of the eye.

anisocytosis inequality in the size of red blood cells found in many kinds of anaemia but especially in MEGALOBLASTIC anaemia.

anisometropia the condition in which the focus is different in the two eyes. One eye may be normal and the other MYOPIC or HYPERMETROPIC, or one eye may be ASTIGMATIC. From the Greek *an*, not *iso*, equal and *metron*, a measure.

ankle the joint between the lower ends of the TIBIA and FIBULA and the upper surface of the talus bone of the foot. The talus sits on top of the heel bone (calcaneum).

ankylo- *combining form denoting* stiff, fused or fixed.

ankylosis fixation and immobilization of a joint by disease which has so damaged the bearing surfaces that the bone ends have been able to fuse permanently together. Sometimes ankylosis is deliberately performed, as a surgical procedure, to relieve pain. From the Greek *ankylos*, bent.

anlagen a localized cell cluster in an embryo that develops into a particular body part in the mature organism.

annealing the joining-up of complementary single strands of DNA that have been separated by heat, or the joining of parts of separate strands that have complementary base sequences, to form a double helix.

annexins a family of over 20 proteins first described in 1990, that include vascular anticoagulant proteins, placental proteins, placental anticoagulant proteins, lipocortins and endonexins. Annexins have similar structures, usually with four domains of 70 amino acids, and all are capable of binding calcium and phospholipids. They are present in a wide variety of cell types. Annexin II acts to bind plasminogen and TISSUE PLASMINOGEN ACTIVATOR to the endothelial cell surface. Annexin V forms an antithrombotic shield around procoagulant phospholipids.

annular ring-shaped. Mainly used to refer to anything that encircles a hollow tube of organ as in the case of an annular cancer of the colon.

annular ligament a fibrous band that surrounds the ankle joint and the wrist joint, acting to retain in position the ligaments passing across the joint.

annulus any ring-like structure.

annulus fibrosus the tough fibrous outer zone of the intervertebral disc that normally retains the soft inner nucleus pulposus. Weakness of the annulus fibrosus allows prolapse of the pulpy centre in the condition inaccurately known as 'slipped disc'.

anodontia partial or total congenital absence of the teeth.

anogenital pertaining to the anus and to the genitalia.

anomaly anything differing from the normal.

anorectic, anoretic, anorexic featuring or causing loss of appetite.

anorexia loss of appetite, especially as a result of disease. From the Greek *an*, not and *orexis*, appetite.

anosmia loss of the sense of smell. This often results from injury to the delicate fibres of the OLFACTORY NERVE as they pass through the bone above the nose (the cribriform plate).

anovulation failure of the ovaries to produce eggs so that conception is impossible.

anoxaemia an abnormal or extreme reduction in the oxygen content of the blood. Literal anoxaemia is incompatible with life and, as commonly used, the term is incorrect.

anoxia local absence of oxygen, usually as a result of interference with the blood supply. Complete anoxia is rare, the more usual condition being a relative insufficiency, which is known as hypoxia.

ansa an anatomical structure, especially neurological, in the form of a loop or arc. From the Latin *ansa*, the handle of a jug.

antagonist 1 a muscle that acts to oppose the action of another muscle (the agonist). 2 a drug that counteracts or neutralizes the action of another drug. The antonym of antagonist is agonist.

ante- *prefix denoting* before, either in time, order or position. From the Latin *ante*, before.

anteflexion a forward bending of an organ such as the womb. Compare ANTEVERSION.

antegrade, anterograde proceeding in the normal or usual direction. The antonym of antegrade is retrograde.

ante mortem before death.

antenatal before birth. Prenatal.

antepartum before a baby is delivered.

anterior at or towards the front of the body.

anterior commissure a nerve fibre bundle running across the midline of the brain from one hemisphere to the other, just in front of the THIRD VENTRICLE.

anterograde amnesia loss of memory for a variable period following a head injury or an epileptic seizure. The length of the period of AMNESIA following a head injury is usually proportional to the severity. Compare RETROGRADE AMNESIA.

anthropo- *combining form denoting* human or man.

anthropoid 1 resembling man. 2 any member of the suborder of primates *Anthropoidea* that includes monkeys, apes and humans. Humans are apes.

anthropology the science of humankind, and of human cultural differences, from the earliest times to the present. Anthropology is thus a very wide subject, concerned not simply with the less familiar human groups but with every aspect of humankind in a social context. Increasingly, anthropology overlaps the social sciences, but, at the same time, preserves a certain detachment from concern with the more utilitarian aspects of such studies, as befits one of the basic sciences. Cultural anthropology, or ethnology, is a comparative study of cultural systems and includes concern with early archeology, religion, myth, political and economic systems and language. Other branches of cultural anthropology include psychological, legal and urban anthropology. The observation, recording and analysis of anthropological data in the course of 'field work' is called ethnography. Physical anthropology is the study of human evolution, including recent diversification of humans. Social anthropology covers the whole field of humans in their social context.

anthropometry human body measurement and weighing for scientific purposes such as anthropological or nutritional research or as an aid to clinical assessment.

anthropomorphism the conceiving of a deity, lower animal, or other entity in terms of human characteristics or behaviour. Humans, because of their experiential limitations, commonly resorts to an anthropomorphic concept of anything transcendental.

anthropophagous man-eating. Cannibal.

anti- *prefix denoting* against, opposite, counteractive.

anti-amoebic able to destroy or suppress amoebae of medical importance.

antibody a Y-shaped protein molecule, called an IMMUNOGLOBULIN, produced by the B group of lymphocytes in response to the presence of a ANTIGEN. An appropriate B lymphocyte is selected from the existing repertoire. This then produces a clone of PLASMA CELLS each capable of synthesizing large numbers of specific antibodies to combat the infection. The B cells also produce memory cells. Subsequent infection with the same antigen prompts the memory cells to clone plasma cells and produce the correct antibodies without further delay. This is an important way in which an infection leads to subsequent immunity. Antibodies are able to neutralize antigens or render them susceptible to destruction by PHAGOCYTES in the body. The basic structure of an antibody consists of four polypeptide chains linked by disulphide bridges, two larger structures called HEAVY CHAINS and two smaller called LIGHT CHAINS.

antibody-dependent cell-mediated cytotoxicity the killing of a target cell, which has been coated with an antibody, by a MACROPHAGE, neutrophil or natural killer cell that carries the surface receptor that binds to the particular antibody.

anticholinergic antagonistic to the action of acetyl choline or to the parasympathetic or other CHOLINERGIC nerve supply. Acetyl choline stimulates muscle contraction in the intestines and elsewhere and slows the heart. Anticholinergic substances, such as ATROPINE, relieve muscle spasm, dilate the pupils and speed up the heart.

anticholinesterase any substance opposing the action of the enzyme cholinesterase, which breaks down the NEUROTRANSMITTER acetylcholine, releasing the inactive choline for further synthesis to acetylcholine. An anticholinesterase agent thus potentiates the action of acetylcholine, a major neurotransmitter carrying nerve impulses across synapses and from nerves to muscles. Continued action causes serious effects.

anticipation the occurrence of a hereditary disease at progressively earlier ages, and in progressively more severe form, in successive generations. This is a feature of a range of conditions that includes myotonic dystrophy, the fragile X syndrome and Huntington's disease.

anticodon a sequence of three nucleotides in transfer RNA complementary to the three nucleotides in a codon on messenger RNA. The anticodon specifies a particular amino acid which is selected and assembled for protein synthesis.

antidiuretic hormone vasopressin. The hormone released by the rear part of the PITUITARY gland which acts on the distal convoluted tubules of kidneys to control water excretion by promoting reabsorption of water and sodium. In the absence of this hormone, large quantities of urine are produced, as in DIABETES INSIPIDUS.

antiendotoxin antibodies IMMUNOGLOBULIN against the core region that all ENDOTOXINS have in common. Anticore antibody is effective against endotoxin from a wide range of organisms including GRAM NEGATIVE bacteria.

antienzyme a substance that interferes with the action of, or counteracts, an ENZYME.

antigen any molecule recognized by the immune system of the body as signalling 'foreign', and which will provoke the production of a specific ANTIBODY. Antigens include molecules on the surfaces of infective viruses, bacteria and fungi, pollen grains and donor body tissue cells.

antigen/antibody complex the association of an ANTIGEN with its resulting ANTIBODY to form a molecular group which can be damaging to tissues, especially to the lining of blood vessels. More usually called an immune complex. See also IMMUNE COMPLEXES.

antigenicity the power, or degree of power, to act as an ANTIGEN.

antigenic determinant a cluster of EPITOPES.

antigen-presenting cell a cell, such as a macrophage, a B cell or a dendritic cell, that presents processed antigenic peptides and MHC class II molecules to the T cell receptor on CD4 T cells of the immune system.

antihaemophilic globulin the blood coagulation factor, Factor VIII, which is deficient in people suffering from HAEMOPHILIA.

antioxidants substances that inhibit oxidative changes in molecules. Many oxidative changes are destructive and this applies as much to the human body as to non-biological chemistry. Recognition that many of the fundamentally damaging processes in disease are oxidative in nature and result from the action of oxygen FREE RADICALS has raised interest in the possibility of using antioxidants to minimize such damage.

antiparallel of two structures that lie parallel to each other but that run in opposite directions. The best-known example of antiparalellism is that of the two strands in the DNA molecule.

antisense RNA RNA molecules transcribed, not from DNA in the usual manner, but from DNA strands complementary to those that produce normal messenger RNA. Antisense RNA occurs in nature and is inhibitory on gene action. It can be produced synthetically and offers such therapeutic possibilities as turning off viral genes.

antitragus the small projection on the free border of the external ear above the centre of the lobe. The tragus is the more prominent triangular projection on the front of the ear, immediately above the front edge of the lobe.

antitrypsin see ALPHA-ANTITRYPSIN.

antrum a hollow cavity or sinus in a bone. The maxillary antrums (or antra) are the cavities in the cheek bones. The mastoid bones, below and behind the ears, contain the mastoid antrums (antra). From the Latin *antrum*, a cave.

anuria cessation of the production of urine by the kidneys. Anuria is always very serious, unless of brief duration. Uncorrected anuria leads to a build-up of toxic waste material in the blood and eventual death.

anus the short terminal portion of the ALIMENTARY CANAL which contains two SPHINCTERS by means of which the contents of the RECTUM are retained until they can conveniently be discharged as faeces. Between the sphincters are 6–10 vertical columns of mucous membrane containing plexuses of veins. These press together to form watertight seals.

anxiety the natural response to threat or danger, real or perceived and characterized, in its extreme form, by a rapid heart rate, tremulousness, a dry mouth, a feeling of tightness in the chest, sweaty palms, weakness, nausea, bowel hurry with diarrhoea, insomnia, fatigue, headache, and loss of appetite. Anxiety is a response to stress and is a concomitant of a wide spectrum of diseases. But it is also a vital motivating factor causing us to respond constructively to dangers of all kinds and to make greater efforts in all kinds of situations.

aorta the main, and largest ARTERY of the body which springs directly from the lower pumping chamber on the left side of the heart and gives off branches to the heart muscle, the head, arms, trunk, chest and abdominal organs and legs.

aortic body chemoreceptor groups of nerve cells in the arch of the AORTA that are sensitive to, and respond to changes in the concentration of hydrogen ions, oxygen and carbon dioxide in the blood.

aortic valve the three-cusp valve at the origin of the AORTA that allows easy movement of blood from the left VENTRICLE of the heart into the aorta but prevents its backward flow.

apex the tip of an organ with a pointed end. The apex of the heart is at the lower left side and the apex of the lung is at the top. The tooth apex is at the tip of each root.

Apgar score a numerical index used to assess the state of well-being of a new-born baby. The figures 0, 1 or 2 are assigned to each of five variables – the heart rate, the breathing, the muscle tone, the reflex irritability and the skin colour, and added. A normal baby will score 7–10. (Virginia Apgar, American anaesthetist, 1909–74).

aphagia inability to swallow.

apheresis a separating out of a component, usually from the blood. See also PLASMAPHERESIS.

aphonia loss of voice, usually as a result of disorder of the LARYNX or VOCAL CORDS.

aphrodisiac **1** promoting sexual desire or performance.
2 a drug purporting to stimulate sexual interest or excitement or enhance sexual performance. The general medical consensus is that there is no such thing as an aphrodisiac. But any agency, such as Viagra, that can improve confidence in the anticipated performance, can act as an aphrodisiac. From the Greek *Aphrodite*, the goddess of love and beauty.

aplasia failure of the development of an organ or tissue or its congenital absence.

aplastic unable to form new cells or tissue.

apnoea absence of breathing for short periods. This may be a pre-AGONAL effect or may result from forced overbreathing which reduces the blood levels of carbon dioxide and hence a major stimulus to respiration.

apo- *a prefix denoting* separate or derived from.

apocrine pertaining to glands that give off some of their intracellular contents as part of their secretion.

apocrine glands the type of sweat glands found in the hairy parts of the body, especially in the armpits and the groin. Apocrine sweat contains material that is broken down by skin bacteria to substances responsible for unpleasant body odour.

apoenzyme a protein substance which can act as an ENZYME in the presence of a coenzyme.

apoferritin a substance with which absorbed iron may be complexed in the body to form a ferritin store. Ferritin is one of the two forms of iron store in the cells of the body. The other is haemosiderin.

apolipoprotein one of a number of glycoproteins forming part of the surface of LIPOPROTEIN particles in the blood. Apolipoproteins are polar structures which provide structural stability to the lipoprotein and act as receptors that help to determine the fate of the particle. Some act as cofactors for enzymes involved in lipid and lipoprotein metabolism. The level of apolipoprotein-A is genetically induced and a high level is a strong risk factor for premature coronary heart disease. Apolipoprotein-B is the binding site of low-density lipoproteins (LDL) to cellular LDL receptors and is concerned in the movement of CHYLOMICRONS from the intestine. Apolipoprotein-C is present on chylomicrons, high-density lipoproteins and very low-density lipoproteins.

aponeurosis a thin flat sheet of tendinous tissue which covers a muscle or by which broad, flat muscles are connected to bone.

apophyseal joints joints between bony protrusions, as in the spinal column.

apophysis any natural protrusion forming part of a bone, such as a tubercle or tuberosity.

apoptosis cell 'suicide'. A form of programmed cell death, by endonuclease digestion of DNA. Apoptosis is necessary to make way for new cells and occurs constantly in the growing fetus and elsewhere. The p53 gene can induce an organismally-protective apoptosis in cells whose DNA has been dangerously damaged to the point where cancerous change is liable to occur.

appendage a part or organ of the body joined to another part. A protruding part of the body.

appendicular skeleton the bones of the shoulder girdle and arms and of the pelvic girdle and legs. Compare AXIAL SKELETON.

appendix 1 the worm-like structure attached to the CAECUM at the beginning of the large intestine and known as the vermiform appendix.
2 an APPENDAGE.

appetite desire, whether for food, drink, sex, work or anything else that humans can enjoy. Lack of appetite for food is called anorexia, of which a particularly dangerous kind is anorexia nervosa.

aptyalism absence of saliva.

APUD acronyn for amine precursor uptake (and) decarboxylation.

APUD hypothesis the hypothesis that cells of the nervous system and cells that produce peptide hormones (APUD cells) might have a common embryonic origin and that the progenitors migrate to the gastrointestinal tract and the endocrine organs where they differentiate. This idea has been shown to be of only limited application, causing the originator, Professor Anthony Pearse (1916–2003) to suggest that the acronym should stand for 'Anthony Pearse's Ultimate Delusion'.

aqua water.

aqueduct a channel carrying water or other liquid. There are a dozen named aqueducts in the body. The aqueduct of the midbrain (aqueduct of Sylvius) is a narrow channel for cerebrospinal fluid lying between the third and fourth VENTRICLES of the brain.

aqueous humour the watery fluid filling the front chamber of the eye between the back of the CORNEA and the front of the IRIS.

arachidonic acid an unsaturated fatty acid, formed from LINOLENIC acid, and a precursor of prostaglandins and thomboxanes.

arachno- *combining form denoting* a spider, spider-like or resembling a spider's web.

arachnoid the delicate middle layer of the three MENINGES covering the spinal cord and brain, lying between the pia mater and dura mater. Unlike the pia mater, the arachnoid bridges over the grooves (sulci) on the surface of the brain and covers many large blood vessels lying in the sulci. Bleeding from any of these vessels causes a subarachnoid haemorrhage. From the Greek *arachne*, a spider.

arachnoidal granulations projections of the middle (ARACHNOID) layer of the MENINGES through the dura mater into the cerebral veins. The arachnoid granulations are the site of reabsorption of cerebrospinal fluid (CSF) into the blood. Failure of CSF to reach them leads to HYDROCEPHALUS. They are also known as arachnoid villi or Pacchionian bodies.

arachnoid mater see ARACHNOID.

arbo- *combining form denoting* a tree or tree-like.

archenteron the embryonic digestive tract. It is formed at the gastrula stage from the invagination of the blastocoele.

archetypes a term used by the Swiss psychiatrist and mystical thinker Carl Jung (1875–1961) to characterize some of the features of the 'collective unconscious' he believed common to all humankind. Archetypes were, he believed, inherent tendencies to experience and symbolize the many different and important human situations in particular ways. Jung pointed out that all the great mythological and religious systems display these archetypes in common.

arcuate bowed, arched or curved. From the Latin *arcus*, a bow.

arcus senilis a white ring near the outer margin of the CORNEA. This is a normal feature of age and is of no significance. Vision is never affected.

areola the pink or brown area surrounding the nipple of the female breast. It contains tiny protuberances under which are the areolar glands which lubricate the skin to protect it during suckling. From areola, the diminutive of the Latin *area*, a courtyard or space.

areolar tissue loose fibrous connective tissue with a protein matrix.

areflexia absence of tendon jerk reflexes.

arginine vasopressin VASOPRESSIN, one of the two hormones from the rear lobe of the PITUITARY GLAND.

aromatase hypothesis the widely accepted proposition that male sex hormones (androgens) are converted to oestrogens by aromatase enzymes before they can act on certain target cells such as those in the hypothalamus concerned with reproductive functions. In females aromatases convert adrenal androgens to oestrogens in the ovaries, skin and elsewhere.

aromatases a class of enzymes that can act on steroids to produce aromatic rings. All the sex hormones are four-ring structures. The first of the four rings of an oestrogen steroid (the A ring) is aromatic; that of the androgens is not. Aromatases catalyze the desaturation of steroid A rings, so can convert androgens to oestrogens.

aromatic of a class of chemical compounds originally so named because many of them have a fragrant smell derived from benzene. Today, by extension, the term is used to refer to compounds containing one or more structures of the pattern of benzene – a ring of six carbon atoms with alternate single and double bonds. The female sex hormones and many drugs contain aromatic rings. Compounds that contain no rings or rings that are not benzene rings are said to be aliphatic.

arousal the state of heightened awareness and alertness caused by a strong external stimulus such as danger or sexual interest.

arrector pili muscles tiny muscles attached to the hair follicle and dermis that, on contraction, cause the hair to stand on end.

arrest cessation of normal action, especially of the heart.

arrhythmia any abnormality in the regularity of the heart beat. Arrythmia is caused by a defect in the generation or conduction of electrical impulses in the heart.

arterial relating to arteries.

arterio- *combining form denoting* artery.

arteriole a small terminal branch of an artery, intermediate in size between an artery and a CAPILLARY.

artery an elastic, muscular-walled tube carrying blood at high pressure from the heart to any part of the body. From the Greek *arteria*, an air duct. It was once believed that, in life, arteries contained air.

arthritis inflammation in a joint, usually with swelling, redness, pain and restriction of movement.

arthro- *combining form denoting* a joint or articulation.

arthrodesis the fusion of the bones on either side of a joint so that no joint movement is possible. This may occur spontaneously, as a result of disease processes, or may be a deliberate surgical act done to relieve pain and improve function.

arthropathy any disease of a joint.

articular pertaining to a joint.

articular cartilage hyaline cartilage covering the bearing surfaces of bones within a synovial joint.

articulation a joint.

artificial chromosome the result of a biotechnology still at an experimental stage. Human chromosomes, consisting of a centromere of highly repetitive DNA sequences, telomeric DNA, a marker gene and some random sequences of genomic DNA, have been successfully prepared. When these ingredients were introduced into a human cell line they became coated with histone proteins and formed highly coiled structures. They also replicated and passed on copies to daughter cells through more than 200 cell divisions.

artificial insemination a method of achieving pregnancy when normal sexual intercourse is impossible, or when the male partner is sterile. Fresh semen is taken up in a narrow syringe or pipette and injected high into the vagina or into the mouth of the womb. The success rate is high. An example of positive EUGENICS is artificial insemination with sperm from anonymous donors guaranteed to be of high mental or physical calibre. The frozen sperm of certain Nobel Prize winners was stored, for this purpose, by the Repository for Germinal Choice, founded in 1979. See AID, AIH.

artificial intelligence the characteristics of a machine designed to perform some of the perceptive or logical functions of the human organism in a manner appearing to be beyond the merely mechanical. AI is largely a matter of computer programming, in which stored records of past experience are made to modify future responses, but it also encompasses research into humanoid methods of data acquisition, the use of fuzzy logic and of artificial neural networks.

artificial respiration an emergency procedure calculated to save life when normal breathing is absent or insufficient, as in partial drowning, poisoning or head injury. The most effective form of artificial respiration is the mouth-to-mouth method in which the lungs of the victim are repeatedly inflated by blowing into the mouth while pinching the nose.

aryepiglottic pertaining to the ARYTENOID cartilages and to the EPIGLOTTIS.

arytenoid ladle-shaped. Pertaining to the two small cartilages attached to the vocal cords at the back of the LARYNX or to the arytenoid muscles of the larynx. From the Greek *arutaina*, a pitcher or ladle, and *eidos*, like.

ascending aorta the first part of the aorta, extending upwards from its origin in the upper surface of the heart to the aortic arch.

ascending colon the part of the large intestine on the right side, running up from the CAECUM to the bend near the liver.

ascending tracts bundles of nerve fibres in the spinal cord that carry sensory impulses upward to the brain.

ascorbic acid vitamin C. A white, crystalline substance found in citrus fruits, tomatoes, potatoes, and leafy green vegetables. Small doses are needed to prevent the bleeding disease of scurvy and regular large doses are useful as an antioxidant in combatting dangerous FREE RADICALS.

-ase *suffix denoting* an enzyme. In most cases the suffix is added to a term for the substance acted upon or to the biochemical action promoted by the enzyme. Thus a lipase is an enzyme that acts on fats and a reverse transcriptase is an enzyme that promoted the transcription of DNA from RNA (which is the reverse of the usual direction). The suffix originated in the ending of the term diastase, a polysaccharide-splitting enzyme named in 1838.

asepsis the complete absence of all bacteria or other microorganisms capable of causing infection. Asepsis, as distinct from antisepsis, is the concept that made modern surgery possible.

asexual characterizing a simple form of binary, or budding, reproduction in which only a single individual organism is involved. Reproduction without male and female GAMETES.

asparaginase an ENZYME that destroys one of the 20 AMINO ACIDS from which proteins are formed.

aspartate aminotransferase one of the enzymes released into the blood when tissue, such as liver or heart muscle, is damaged. Measurement of the level of such enzymes gives a useful indication of the extent of the damage.

aspartic acid an AMINO ACID which the body can synthesize. It is found in sugar cane and sugar beet and in asparagus.

Asperger's syndrome a condition similar to, but usually less severe than AUTISM, that affects about 1–2 persons in 1000, males more often than females. Affected people, who are normally intelligent, are physically clumsy, have unusual narrow interests and great difficulty in managing social relationships and are often considered simply as eccentric loners. Special training in social skills can be valuable.

aspermia the absence of spermatozoa in the semen or the inability to ejaculate semen. Obstructive aspermia is due to the blockage of the VAS DEFERENS on both sides. This may be congenital or acquired from infection or injury, or from vasectomy.

asphyxia suffocation by interference with the free AIRWAY between the atmosphere and the air sacs in the lungs. Asphyxia is usually the cause of death in drowning, choking, strangling, inhalation of a gas which excludes oxygen, foreign body airway obstruction and swelling of the lining of the LARYNX.

asphyxiation the process of causing, or suffering, ASPHYXIA.

assigned sex the nominal sex or gender assumed or decided upon when a baby has genitals of ambiguous appearance. The assigned sex may or may not correspond to the actual chromosomal sex. Sometimes assigned sex has to be changed later in life and this may lead to TRANSSEXUALISM.

assimilation the process of incorporating nutrient material into cells after digestion and absorption.

association areas areas of the outer layer (cortex) of the brain concerned with the integration of sensory and other data with other aspects of brain function, and the elaboration of them into the complex processes underlying higher mental functions such as language, imagination, judgement and creativity. Thus, damage to the visual association area, while not in any way affecting the primary function of vision, might lead to an inability to recognize or interpret what is seen.

association fibres nerve fibres running just under the surface (CORTEX) of the brain and connecting adjacent parts of the cortex.

astereognosis the inability to recognize the shape of objects by touch.

asthenia lack or loss of strength or energy. From the Greek *asthenes*, weak.

asthenic a slender, lightly muscled physique.

astigmatic of a lens, unable to focus a point image from a point object. Having ASTIGMATISM.

astigmatism an optical error in which objects in the same plane, but of different orientation, are brought to a focus in different planes at the back of the eye. Thus, vertical objects may be seen in focus while horizontal objects may be out of focus. Astigmatism is usually due to a lack of sphericality of the outer surface of the CORNEA, the meridian of maximal curvature being at right angles to that of minimal curvature. An astigmatic lens cannot produce a point image of a point object. The defect can be corrected by cylindrical spectacle lenses. From the Greek *a*, not and *stigma*, a spot or point.

astragalus the talus bone. The upper bone of the foot, on which the tibia rests.

astro- *combining form denoting* a star.

astrocytes star-shaped connective tissue cells of the nervous system that link nerve cells to blood vessels and, by wrapping round brain capillaries, help to form the BLOOD-BRAIN BARRIER. Neurological connective tissue (neuroglial) cells.

asymptomatic free of symptoms or not causing symptoms.

ataraxia tranquillity or peace of mind.

atavism the reappearance of a genetic characteristic after generations of absence. This may be caused by the coincidence of two recessive genes, by recombination, or by mutation. The organism or individual so produced is often called a 'throwback'.

atavistic featuring characteristics not seen for many generations, or of a more primitive evolutionary form of the organism.

ataxia unsteadiness in standing and walking from a disorder of the control mechanisms in the brain, or from inadequate information input to the brain from the skin, muscles and joints. From the Greek *a*, not and *taxis*, order or arrangement.

ataxic 1 showing ATAXIA.
2 a person suffering from ataxia.

atlanto-axial referring to both the ATLAS BONE and the AXIS BONE.

atlas see ATLAS BONE.

atlas bone the uppermost, or first, vertebra of the spinal column. The atlas is unique in having no body or spinous process. In head nodding, the skull moves on the atlas, but in head rotation the atlas locks with the skull and both rotate on the second vertebra, the AXIS.

atomic number a number particular to the atom of each element equal to the number of protons in the nucleus of the atom. Compare ATOMIC WEIGHT.

atomic weight a term now superceded by the term relative atomic mass. This is the ratio of the average mass per atom of the naturally occurring form of an element to one-twelfth of the mass of an atom of carbon-12.

atony sustained abnormal relaxation of muscle. Lack of muscular tone or contractile tendency.

ATP ADENOSINE TRIPHOSPHATE.

ATPase an enzyme that hydrolyses ATP to ADP and inorganic phosphate. See ADENOSINE TRIPHOSPHATE.

atrial natriuretic peptides see ATRIOPEPTIN.

atrio- *combining form denoting* a chamber or ATRIUM.

atriopeptin a hormone stored in the muscle cells of the atria of the HEART and released into the blood when the blood volume increases beyond the optimum. Atriopeptin increases the rate of urine production and salt excretion. The discovery of atriopeptin led to the realization that this was one of a family of small peptides, called the atrial natriuretic peptides (ANPs), with receptors in the heart, lungs, kidneys, bone marrow, thymus and brain. ANPs have amino acid sequences from 28–53 moieties.

atrioventricular pertaining to the association of the upper and lower chambers of the heart – respectively, the atria and the ventricles. It often implies progression from an atrium to a ventricle.

atrioventricular bundle a band of specialized muscle fibres in the heart muscle (myocardium) that conducts the contraction signal from the ATRIOVENTRICULAR NODE to the part of the myocardium that forms the walls of the ventricles.

atrioventricular node a 'junction-box', between the upper and lower chambers of the heart, in the band of specialized heart muscle fibres which conduct the electrical impulses controlling the contraction of the heart. It assists in the correct timing of the contractions, first of the atria and then of the ventricles.

atrioventricular valves the valves lying between the atria and the ventricles of the HEART that ensure movement of the blood from the former to the latter only. They are, on the right side, the tricuspid valve, and on the left, the mitral valve.

atrium one of the thin-walled upper chambers of the heart which receive blood from the veins and pass it down to the lower, more powerful, pumping chambers (the ventricles).

atrophy wasting and loss of substance due to cell degeneration and death. This may be a natural ageing process or it may be due to simple disuse. From the Greek *atrophia*, hunger or want of food.

attention the direction of some of the channels of sensory input to a restricted area of the environment. Since the number of possible sources of information in the environment is so great, attention – which selectively directs and concentrates awareness and controls input – is of the first importance. Attention is seldom continuous for long and is determined mainly by the degree of interest in the source of the information. Attention is closely related to effectiveness in memory storage. It is no accident that people with a wide range of strong interests tend to have well-stocked minds. Motivation, as towards learning or achieving qualification, is a less powerful stimulus to attention than interest. Fortunately, interest grows with knowledge. Attention can be objectively demonstrated by such methods as electro-encephalography or PET scanning, which show special activity in the parts of the brain most employed at the time.

audio- *combining form denoting* hearing.

audiogram a record of the sensitivity, or threshold, of hearing at different frequencies.

audiolingual pertaining to hearing and speaking in the process of learning a language.

audiology the scientific study of hearing and of the medical management of hearing defects.

auditory pertaining to hearing or to the organs of hearing.

auditory association area a region in the cortex of the temporal lobe of the brain where sounds are analyzed and interpreted.

auditory evoked responses subtle electrical changes in the brain waves, demonstrable on the ELECTROENCEPHALOGRAM, which are induced by loud noises. The method provides only a crude estimate of the hearing ability but gives some objective proof of hearing without the cooperation of the subject.

auditory nerve the acoustic or vestibulocochlear nerve. The 8th cranial nerve.

auditory ossicles the chain of three tiny bones in the middle ear which acts as an impedance transformer, efficiently coupling the relatively large low-impedance movement of the ear drum to the smaller, high-impedance movement of the fluid in the cochlea of the inner ear.

aural pertaining to, or perceived by, the ear.

auri- *combining form denoting* the ear.

auricle 1 the pinna, or external ear. 2 an obsolescent term for one of the upper chambers of the heart (atrium).

auricular 1 pertaining to the ear or to hearing. 2 pertaining to an auricle, or atrium, of the heart.

auricularis any of the three small flat muscles attached to the cartilage of the external ear. These are the vestigial remnants of the muscle still highly functional in other mammals, such as cats.

auscultation the act of listening with a stethoscope to the sounds made by the heart, lungs, blood passing through narrowed vessels, the movement of fluid or gas in the abdomen, and so on. The doctor listens for changes in the normal sounds and for new (adventitious) sounds. Heart specialists become skilled in the interpretation of subtle sounds inaudible to the novice. From the Latin *auscultare*, to listen attentively.

australopithecines the earliest known hominids who walked erect and may have known how to use tools. They were apelike primates of the genus *Australopithecus* and related genera, whose remains were found in South and East Africa. Some species are estimated to be over 4.5 million years old.

Australopithecus an extinct human-like genus believed by many to be an evolutionary ancestor of contemporary man. The full title is *Australopithecus africanus* meaning 'southern ape of Africa'.

Austronesian language family a language group, the Malayo-Polynesian family, whose members are spoken half round the world from Easter island to Madagascar, but particularly in Indonesia, the Philippines, Taiwan, Malaysia and Singapore.

aut-, auto- *prefix denoting* self. From the Greek *autos*, self.

autacoid a hormone or other substance produced in an organ and released into the blood to be carried to other parts of the organism to produce various effects.

autoanalysis an attempt to perform PSYCHOANALYSIS upon oneself.

autoantibody an antibody derived from the immune system, which then acts against body tissues or constituents.

autocrine 1 of a substance secreted by a cell and released into the extracellular fluid, that then acts on the cell.

2 of a cell responding to such chemical messengers.

autoeroticism 1 the deliberate arousal of sexual feeling in the absence of a sexual partner.
2 the satisfaction of sexual desire by masturbation.

autogenous originating within the self, as in a vaccine prepared from a person's own bacteria. Compare EXOGENOUS.

autologous derived from the same person, as in the case of transfusions or transplantation.

automatism the quality of acting in a mechanical or involuntary manner.

autonomic nervous system the part of the nervous system controlling involuntary functions, such as the heart beat, the secretion of glands and the contraction of blood vessels. It is subdivided into the SYMPATHETIC and the PARASYMPATHETIC divisions which are, in general, antagonistic and in balance. The term autonomic derives from the Greek *autos*, self, and *nomos*, a law. See Fig. 1 on page 36.

autopsy a postmortem pathological examination done to determine the cause of death or past medical history or to assist in medical research.

autosomal of any chromosome other than the sex chromosome pair.

autosome any one of the ordinary paired CHROMOSOMES other than the sex chromosomes.

autosuggestion a form of self-conditioning involving repeated internal assertion of positive and helpful propositions.

avascular lacking blood vessels.

aversion therapy a form of treatment for addiction or antisocial behaviour in which the undesirable activity is forcibly associated in the mind of the subject with an unpleasant experience. It is not widely used. See also ANTABUSE.

Avery, Oswald Theodore the physician and bacteriologist Oswald Avery (1877–1955) studied at Colgate University where he graduated in medicine in 1904. Most of his working life was spent at the Rockefeller Institute Hospital, New York, as a bacteriologist. In 1932 he began to study how bacteria could change from non-virulent to virulent forms and discovered that a substance could be extracted from killed virulent bacteria that transformed non-virulent living strains to the dangerous form. When he and his colleagues purified this substance in 1944, it was found to be DNA. This was a discovery of the greatest importance because prior to that time the material that transmitted the hereditary elements was believed to be a protein. No one was then interested in DNA, but from that time on the focus was on the nucleic acids. Less than ten years later, Watson and Crick determined the structure of the DNA molecule and started a revolution in biology.

AV node see ATRIOVENTRICULAR NODE.

AVP *abbrev. for* arginine vasopressin, a hormone secreted by the rear lobe of the pituitary gland, usually called vasopressin. It is also known as the ANTIDIURETIC HORMONE.

axial skeleton the skull and spine (vertebral column).

axial musculature the muscles connecting the head to the spine and those lying along the long axis of the spine.

axilla the armpit.

axillary pertaining to all the structures lying in the AXILLA, such as the axillary lymph nodes.

axis 1 (Bone) the second of the vertebrae of the spine, upon which the skull and first vertebra (ATLAS) can rotate. The axis bone has a short, stout vertical peg called the odontoid process around which the atlas can rotate.
2 an imaginary central line of a part or of the body.
3 the meridian in a cylindrical lens that possesses no optical power. The curve of maximal power is at right angles to the axis.

axon the long fibre-like process of a nerve cell which, bundled together with many thousands of other axons, forms the anatomical structure known as a nerve. The axon conducts nerve impulses away from the nerve body.

azoospermia absence of spermatozoa from the seminal fluid, a cause of male sterility. Sperm may still be being produced in the testes.

azygous occurring singly rather than in pairs. From the Greek *azugos*, unyoked.

azygous vein one of the three unpaired veins of the ABDOMEN and THORAX.

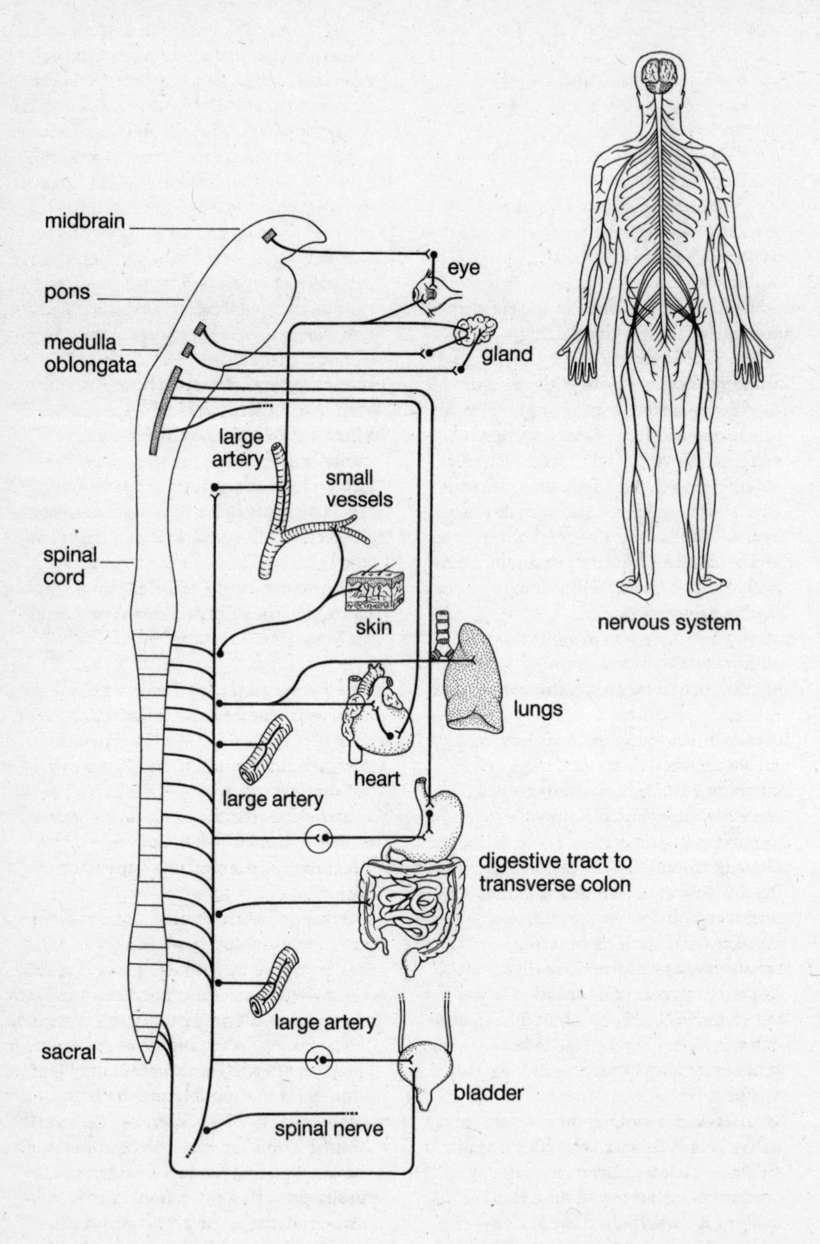

Fig. 1 **Autonomic nervous system**

Bb

backbone the vertebral column or spine.

back mutation the return of a gene to its original nucleotide sequence after a mutation.

baldness premature hair loss. Male baldness is common and is thought to be due to an AUTOSOMAL gene that behaves in a DOMINANT manner in males and in a RECESSIVE MANNER in females. This is called a sex-influenced trait. See also ALOPECIA.

banding patterns transverse stripes visible on stained chromosomes, that vary with different kinds of stain.

bariatric pertaining to obesity and weight control.

baroreceptor a nerve ending which produces an output when there is a change in ambient pressure. Also known as baroceptor.

Barr body a condensed clump of CHROMATIN occurring in the nucleus of cells in normal females and corresponding to an inactive X chromosome. The Barr body also occurs in males with two or more X chromosomes an addition to the Y chromosome. See also X-INACTIVATION. (Murray Llewellyn Barr, Canadian anatomist, 1908–95).

barren sterile. Incapable of producing offspring.

Bartholin's glands mucus-secreting glands lying between the back part of the vaginal orifice and the lesser lips (labia minora) on either side. They secrete under the influence of sexual excitement and facilitate sexual intercourse. (Carpar Secundus Bartholin, 1655–1738, Danish surgeon).

basal pertaining to, situated at, or forming, an anatomical base of any kind.

basal cell a cell of the single-cell-thick lowest layer of epithelia, such as the epidermis of the skin, from which all the more superficial layers are derived by MITOSIS.

basal ganglia the discrete, grey nerve cell masses lying deep in the lower part of the brain within the white matter. They consist, on each side, of the caudate nucleus, the putamen and the globus pallidus. They receive numerous connections from the outer layer (cortex) of the CEREBRUM, above, and from the CEREBELLUM, behind. They are concerned with the control of motor function.

basal metabolic rate the rate at which energy is used by a person at rest. The BMR is measured in terms of the heat given off in a given time.

basal metabolism the minimum amount of energy needed to maintain the vital functions – heart beat, respiration and digestion – in a person at complete rest.

base a chemical compound that combines with an acid to form a salt and water. A substance that releases hydroxyl ions (OH^-) in solution. The term applied in genetics to one of the four nitrogenous bases of DNA and RNA. In DNA the bases are adenine, thymine, guanine and cytosine. Guanine and adenine are PURINES and cytosine and thymine are PYRIMIDINES. In RNA, the pyrimidine base uracil replaces thymine. (See also BASE PAIR).

base analogue any compound structurally similar to a DNA BASE that can substitute for a base and thus cause a mutation.

base complementarity the pairing relationship shown by the four bases in DNA, such that adenine always links with thymine and guanine always links with cytosine.

base deletion one of the types of genetic mutations. In this case a single BASE PAIR is missing thereby causing a FRAME-SHIFT MUTATION.

base insertion one of the types of genetic mutations. In this case an additional BASE PAIR is inserted into the nucleoside sequence thereby causing a FRAME-SHIFT MUTATION.

base pair two linked molecules, one a PURINE the other a PYRIMIDINE, that lie across the two strands of the DNA double helix. The bases are linked by easily-broken hydrogen bonds and the linkage occurs only in a particular, complementary, way – adenine with thymine and guanine with cytosine. This is the essence of DNA replication, which starts with the separation of a length of the two strands at bonds. In RNA, uracil replaces thymine and adenine links with it. Distance along a DNA sequence is measured as the number of base pairs.

basement membrane a thin lamina, consisting of collagen, glycosaminoglycans, fibronectins and other substances, on which one or more layers of cells, especially epithelial cells, rest. Basal membranes are double-layered, the outer layer being secreted by the epithelial cells and the inner layer by connective tissue.

basilar relating to, or situated at or near the bottom (base) of an anatomical structure, especially the base of the skull or of the brain.

basilar artery an important artery formed from the junction of the two vertebral arteries that run up through the side processes of the vertebrae of the neck. The basilar artery runs up in a groove on the front surface of the PONS to supply most of the BRAINSTEM and CEREBELLUM and then joins the arterial circle, the circle of Willis, on the base of the brain.

basilar membrane the membrane in the COCHLEA of the inner ear which supports the ORGAN OF CORTI – the mechanism by which sound vibrations are converted to nerve impulses. The basilar membrane vibrates sympathetically in different parts under the influence of vibrations of different frequencies causing the stimulation of different groups of hair cells.

basilic BASILAR.

basophil having an affinity for alkali. The term is used conveniently to refer to the group of blood white cells (leukocytes) whose internal granules take up an alkaline stain. The granules in basophils are mainly histamine and it is the release of this powerful chemical that causes most of the trouble in allergy. Basophils closely resemble tissue MAST CELLS.

basophilic staining readily with basic dyes.

BAT *abbrev. for* brown adipose tissue.

B cells B LYMPHOCYTES, one of the two main classes of lymphocytes, white cells found in the blood, lymph nodes and tissues which, with other cells, form the immune system of the body. B lymphocytes form CLONES of plasma cells which manufacture antibodies (IMMUNOGLOBULINS).

Bateson, William One of the most notable pioneers of genetics – a term he invented – was the English scientist William Bateson (1861–1926). Educated at Rugby School and St John's College, Cambridge, from which he graduated in 1883, Bateson was the first to hold a chair in genetics, being appointed professor in that discipline at Cambridge in 1908. He became director of the newly instituted John Innes Horticultural Institution in Surrey in 1910 and remained there for the rest of his life. He was also Fullerian Professor of Physiology at the Royal Institution, London, from 1912–14.

During a two-year spell of embryological research in the USA Bateson found evidence that chordate animals had evolved from echinoderms. This sparked his interest in evolution, an interest that persisted throughout his life. After several years of study of the fauna in salt lakes of Europe, Egypt and central Asia, Bateson proposed a theory of evolutionary discontinuity which helped to explain the extended process of evolution. He suggested that instead of being a process of gradual change, evolution occurred in a series of discontinuous leaps. Bateson was an enthusiastic supporter of Gregor Mendel's work on heredity and repeated some of Mendel's breeding experiments. He proved that certain traits are consistently inherited together. This linkage was later shown to be the effect of genes being located close together on the same chromosome.

B-DNA the predominant DNA configuration with a right-handed double helix and one full turn every 10th base. See also A-DNA.

Beadle, George Wells The American scientist George Wells Beadle (1903–89) graduated from the University of Nebraska in 1926 and in 1931 joined the laboratory of Thomas Hunt Morgan at the California Institute of Technology, where he studied the genetics of the fruit fly *Drosophila melanogaster*. He showed that the eye colour of *Drosophila* was the result of chemical reactions under the control of genes. Wishing to study the genetics of an even simpler organism than the fruit fly, Beadle worked on the effects of X-radiation on colonies of the bread mould *Neurospora crassa*. He showed that the mutations caused by the X rays resulted in the production of changed enzymes. This work led him to propose and prove that each gene coded for a single enzyme, an extremely important advance that opened up wide new avenues of research and led to the award in 1958 of the Nobel Prize which he shared with Edward L Tatum and Joshua Lederberg. From 1937–46 Beadle was Professor of Biology at Stanford University, California, and he was Professor of the Division of Biology at the California Institute of Technology from 1945–61. In 1968 he became Director of the American Medical Association's Institute for Biomedical Research.

bearing down the process of assisting in the expulsion of the baby in childbirth by holding the breath and tightening the abdominal muscles so as to compress the abdominal contents in the manner of constipated attempts at defaecation.

behaviourism the psychological school that holds that information about the mind can be reliably derived only from observation of behaviour and not from reports of conscious experience.

belching bringing up gas (eructation), which is usually air, swallowed during the attempts to achieve the relief of a belch or during greedy eating. It is uncommon for gases to be formed in the stomach or intestines as a result of dyspeptic conditions.

belly a common name for the ABDOMEN.

belly button an informal term for the navel (UMBILICUS).

bereavement serious loss, usually of that of a beloved person, but also of any valued thing, including health and wealth. Bereavement gives rise to a characteristic pattern of psychological reaction involving various recognizable stages, known as mourning. The strength of the reaction varies with the perceived value of what is lost.

Berkeley, George Bishop George Berkeley (1685–1753) was a leading exponent of philosophical idealism – the principle that only minds have real existence and that the material world only seems to exist. Berkeley analysed the processes of perception and pointed out, for instance, that the colour of objects could not be said to have any real existence because perception of colour necessarily involved the action of light, eyes and minds. Berkeley dealt similarly with the other properties of matter showing that our perception or conception of them depended wholly on the existence of the mind. These views still carry much weight and it is now clear, from the convergence of physics and physiology, that whatever may be the true nature of the physical world, it is not as it seems.

Berkeley took an extreme position and held that *all* the qualities of matter depend on the mind. The criticism – that common sense experience tells us that objects have continuing existence – he met triumphantly by simply affirming that the whole universe exists by being perceived in the mind of God. This argument has been found by some to be an adequate proof of the existence of God. Others, such as Monsignor Ronald Knox (1888–1957), have been more sceptical.

Bertillon system a long obsolete system of criminal identification, developed by the French anthropometrist Alphonse Bertillon

(1853–1914) in which frontal and profile photographs and extensive body measurements were taken for purposes of identification.

bestiality sexual intercourse with an animal other than a human.

beta the second letter of the Greek alphabet, often used to denote the order in a sequence.

beta-adrenergic pertaining to the class of receptors for ADRENALINE and NORADRENALINE known as the beta receptors.

beta-adrenoceptor one of the many the receptor sites at which noradrenaline and other hormones act to cause muscle to contract or relax. Beta-adrenoceptors occur in blood vessels, in the heart, in the bronchi, in the intestines, in the bladder, in the womb and elsewhere. The effect of the hormones at these sites can be prevented by beta-blocker drugs.

beta carotene a precursor of vitamin A.

beta cells 1 the cells in the islets of Langerhans in the PANCREAS that produce insulin.
2 the basophil cells of the front lobe of the PITUITARY GLAND.

beta-endorphin one of the body's own substances with morphine-like actions. It is part of the precursor molecule of ACTH.

beta rhythm the electrical brain wave on the ELECTROENCEPHALOGRAM associated with a state of alertness and having a frequency of 18–30 Hz.

Betz cells the large pyramidal cells of the motor cortex of the brain. The medical student joke that equates 'abetzia' with lack of intellectual capacity is based on inaccurate knowledge of the function of these cells. (Vladimir Aleksandrovich Betz, 1834–94, Ukrainian anatomist).

bezoar a ball of hair and other material forming in the stomach or intestine and rare in the psychologically normal. In more gullible times bezoars have been valued for their magical properties.

biceps muscle the prominent and powerful muscle on the front of the upper arm which bends the elbow and rotates the forearm outwards, as in using a screwdriver.

bicipital pertaining to the biceps.

bicornuate having two horns or horn-like projections, as in bicornuate uterus.

bicuspid having two cusps, or projections, as on the biting surface (crown) of a PREMOLAR tooth. One of the valves in the heart, the mitral valve, is bicuspid.

bicuspid valve the valve between the atrium and the ventricle on the left side of the heart, so called because it has two leaflets. Compare TRICUSPID VALVE.

B.I.D. *abbrev. for* brought in dead.

bifid divided into two parts. Forked or cleft.

bifocal having two different foci. In bifocal spectacles each lens has a lower segment of greater power for convenience in reading.

bifurcation a fork or double prong. Bifurcations are very common in blood vessels and in the bronchial 'tree' of the lungs. At a bifurcation the sum of the cross-sectional area of the two branches usually exceeds that of the parent branch. Since this happens many times, there is a progressive increase in the unit volume of the system.

bigamy purporting to marry when already legally married. Bigamy is usually for purposes of sexual access by deception and some cases may thus be regarded as a kind of rape.

bigeminy the occurrence of events in pairs. The term is applied especially to coupled heart beats, suggesting heart disease.

bilateral involving or affecting both sides. In the case of paired organs, bilateral means affecting both of them.

bile the dark greenish-brown fluid secreted by the LIVER, stored and concentrated in the GALL BLADDER, and ejected into the DUODENUM to assist in the absorption of fats. Bile contains bile salts which help to emulsify fats, bile pigments derived from the breakdown of red blood cells, cholesterol, lecithin and traces of various minerals and metals.

bile acids cholic and chenodeoxycholic acids. These are produced in the liver from cholesterol, linked with glycine or taurine to form BILE SALTS and passed into the small intestine in the bile.

bile duct the narrow tube which carries BILE from the liver to the bowel. The bile collecting tubules in the liver join up to form a main tube called the hepatic duct.

Just under the liver, this gives off a branch, the CYSTIC DUCT, to the gall bladder. The duct continues down, as the 'common bile duct' to run into the DUODENUM.

bile salts the sodium salts found in bile. Sodium taurocholate and sodium glycocholate. These salts act as emulsifying agents to assist in the absorption of dietary fats.

bilirubin a coloured substance in bile derived from the breakdown of haemoglobin in effete red blood cells at the end of their 120 day life. Bilirubin is conjugated with glucuronic acid in the liver and excreted in the bile, giving the stools their characteristic colour. When it cannot escape freely into the bowel it accumulates in the blood, staining the skin to cause JAUNDICE. The stools become pale and the urine dark. Conjugated bilirubin is water-soluble.

binaural 1 related to two ears. 2 hearing with both ears.

Binet-Simon scales a series of intelligence tests for children. (Alfred Binet, 1857–1911, French psychologist).

binge drinking the practice of drinking excessive amounts of alcohol regularly. A binge has been defined as a pattern of drinking that brings the blood alcohol level to 80 mg per 100 ml (0.08 per cent) or more. A 2003 British Government report indicates that binge drinking in the UK has risen markedly in recent years, especially among young people and that the problem is markedly more severe in the UK than in other European countries. Nearly 20 per cent of the total alcohol taken is consumed by underage drinkers.

binocular pertaining to both eyes or to the simultaneous use of both eyes.

binocular vision simultaneous perception with both eyes.

binomial nomenclature a system for naming animals and plants using two Latin names, of which the first represents the genus and the second the species, as in *Homo sapiens*. By convention, the generic (genus) term is always capitalized but the specific (species) never. The system was originated by the Swedish botanist Carolus Linnaeus (1707–78) and is often known as Linnaean classification.

binovular of twins, derived from two separate eggs and thus non-identical. Compare UNIOVULAR.

bio- *combining form signifying* life. From the Greek *bios*, life.

biochemistry the study of the chemical processes going on in living organisms, especially humans. Biochemistry is concerned, among other things, with the acceleration of biochemical processes by ENZYMES; with the chemical messengers of the body (HORMONES); with communication between cells at cell membranes; with the chemical processes which govern cell survival and reproduction; with the production of energy in cells; and with the processes of digestion of food and the way in which the resulting chemical substances are utilized for energy and structural purposes.

bioengineering see BIOLOGICAL ENGINEERING.

bioethics the study of the ethical and moral questions arising from the growing possible application of biological and genetic knowledge, especially in BIOLOGICAL ENGINEERING.

biofeedback the provision, usually in 'real time', of information to a person about the levels of activity of normally unconscious bodily processes. This is done in the hope that some control or adjustment may be exercised. The information is provided in the form of a moving meter needle, a changing sound, a light of varying brightness, or any other form of display. There is evidence that biofeedback methods can lower blood pressure, but only by a small amount. It is probably valuable as an aid to learning how to relax. Most of the popular claims for biofeedback cannot be substantiated.

biogenesis the recognition that complex living organisms arise only from other living organisms and do not originate by spontaneous generation, as was once believed.

bioinformatics the branch of information science concerned with large databases of biochemical or pharmaceutical information.

biological engineering a range of techniques in which biological substances are used, often at an industrial level, for practical purposes. The scope of biological engineering is widening rapidly and includes the extensive use of natural ENZYMES and the application of GENETIC ENGINEERING.

biological marker biomarker, a substance, physiological characteristic, gene, etc. that indicates, or may indicate, the presence of disease, a physiological abnormality or a psychological condition.

biological warfare the use of micro-organisms capable of spreading and causing epidemics of disease, for military purposes.

biological weapons micro-organisms and the means of their deliberate dissemination for the purpose of killing or disabling members of military or civilian populations for warlike purposes.

biology the science and study of living organisms and life processes.

biome 1 a group of ECOSYSTEMs of similar climate, latitude and altitude.
2 an extensive ecological community, usually featuring a dominant vegetation.

biomechanical engineering the applications of the principles of mechanical engineering to improve the results of surgical treatment, especially by the design of prosthetic parts and the use of new materials. Engineers have answers to many surgeon's problems; close cooperation is essential.

biomedical engineering the cooperative investigation by engineers and doctors of the effective application of all branches of engineering so as to broaden the scope of medicine. Electronics, robotics, hydraulics, rheology, materials science and software engineering are among the more fruitful branches from which medicine has benefited.

biometrics the statistical study of biological data.

biomicroscopy examination of living structure using a microscope. Biomicroscopy is mainly employed, in a clinical context, by ophthalmologists.

bionic relating to a living or life-like system enhanced by, or constructed from, electronic or mechanical components.

bionics biological principles applied to the design of engineering systems, especially electronic systems.

biophysics the physics of biological processes and systems.

biopsy a small sample of tissue, taken for microscopic examination, so that the nature of a disease process can be determined.

biorhythms

Periodicity is very much a feature of the human organism which is locked into a number of biorhythms, especially the diurnal 24-hour, or circadian, clock rhythm. Other biorhythms include the periodic patterns of brain waves revealed by the electroencephalogram, the roughly monthly periodicity of the menstrual cycle in women, and the cyclical release of hormones in pulses at intervals of minutes or hours. These rhythms are controlled by biological clocks – self-sustaining timing mechanisms that control the occurrence or varying intensity of physiological events. These clocks are essentially free-running but some, if not all, of them are maintained at a fixed frequency by external synchronization. At the present stage of knowledge, most is known about physiological events with a circadian periodicity. The process of synchronization of the circadian clocks is now becoming clear.

It should not cause surprise that human functioning is regulated by clocks with a periodicity linked to natural phenomena. Throughout the course of evolution, natural events, especially the sequence of day and night, must have had a great effect on organisms. These effects are demonstrable in many species today. The fiddler crab, for instance, even under laboratory conditions, shows an activity cycle of 24 hours, 50 minutes – the length of the lunar day. This activity is synchronous with low tides caused by the gravitational effect of the Moon. Many other similar examples can be cited.

Circadian rhythms in humans

Many important physiological processes have a circadian periodicity. Production of the natural adrenal hormone cortisol rises at the time of waking, peaking at about 09.00, and is at its lowest about midnight. This is the result of an increased output of a stimulating hormone (ACTH) from the pituitary gland, prompted by the suprachiasmatic nucleus (see below). This periodicity can be lost as a result of stress, depression and heart failure. The circadian rhythm thus has a link with the immune system or at least with our resistance to infection, the outcome of which can vary with the time of exposure. The peak of cortisol production in the morning reduces the efficacy of the immune system at that time. Resistance to the effects of drugs is similarly contingent on the time of dosage.

Thyroid stimulating hormone from the pituitary is also released in accordance with the circadian rhythm. The peak output is at around 23.00 hours and the lowest output around 11.00 hours. Thyroid hormone influences the rate of activity of almost all body cells. Free calcium and phosphate in the body vary in a circadian manner. Endorphin and sex hormone production is circadian. Blood pressure varies characteristically over the 24 hour cycle, as do blood sugar levels, the secretion of acid by the stomach (maximal at 22.00 and minimal at 09.00 in fasting subjects), the tension in the muscles of the air tubes (bronchi) of the lungs (greatest at 04.00 and least at 16.00 and exaggerated in people with asthma).

Dangerous circadian patterns

Circadian patterns appear in several common disorders that frequently have a fatal outcome. It has been known for some time that the incidence of heart attacks, sudden death from heart disease, and strokes is significantly higher during the early morning hours than at other times of the day. A number of circadian mechanisms have been identified that contribute to this. The ability of the blood to break down clots that have formed (fibrinolytic activity) is lowest in the early morning. The tendency for blood platelets to stick together is greatest at that time. Adrenaline and cortisol levels and blood pressure are highest at that time, as is the activity of the enzyme renin

Perhaps most important of all, the tendency for arteries, such as the coronary arteries of the heart, to narrow is significantly greater in the morning than later in the day. This results in reduced blood flow to the parts supplied by the arteries and is partly or wholly due to the effect of adrenaline on the arteries. Studies of the efficiency of small doses of aspirin in preventing heart attacks have shown that this measure is most effective in the mornings. Aspirin acts by reducing platelet stickiness.

Light and the body clock

Body temperature varies slightly but predictably during a 24-hour cycle. There is a reduction in heat production and increased heat loss in the evening, while in the early morning heat production rises. Body temperature is at its lowest in the middle of the night and peaks around mid-afternoon.

Light input variations are undeniably important in bringing the behaviour patterns and other physiological functions into sync with the day and night rhythm, but scientists have wondered whether there is any evidence that such patterns are anything more than a simple response to the light variations? Do we, they ask, have an internal circadian rhythm clock? Careful trials suggest that we do.

People kept in isolation chambers, with no indication of time and no light clues to day or night continue to follow a regular sleep and wake cycle. This cycle does not, however, have a 24 hour periodicity but is nearer to 25 hours. Over the course of a month or so, people kept in such conditions get completely out of sync with day and night. If these subjects are exposed to artificial light–dark cycles of a periodicity of less or more than 24 hours and the periods of light exposure include phases of very bright light, of the intensity of morning daylight, their body temperature swings soon fall into sync with the new cycles. These may have a periodicity of anything from 21–28 hours. Exposure to bright light after the normal time of the lowest body temperature causes the cycle to lengthen while exposure to bright light before the usual time shortens it. These findings show that the natural period of the body clock is not caused by the day-night cycle but is simply synchronized by it. Interestingly, the word 'circadian' is derived from the Latin words *circa* meaning near to, or approximate, and *dies* meaning day.

Synchronization of a free-running periodic system by another of fixed but slightly different frequency occurs very easily, in engineering as well as in nature, and the linkage between the two systems need not be strong. A very small stimulus, repeatedly applied is usually all that is necessary.

Melatonin and the suprachiasmatic nuclei

The *pineal gland* is a small, oval, flattened body centrally placed in the brain immediately above the top of the brainstem. In some simple animals the pineal acts as a kind of primitive eye. Descartes taught that the pineal was the channel by which the mind interacted with the body, and for a long time the pineal was regarded by philosophers as the 'seat of the soul'. Until recently, physiologists knew little or nothing of the function of the pineal, and its interest in medicine was limited to the fact that it often acquired a few grains of calcium which showed up on X-rays and could be used to demonstrate shift of the midline of the brain to one side as a result of tumours. We now know that the pineal produces a hormone melatonin and that this is released during darkness.

From the retina of each eye about 1 million nerve fibres run back in the optic nerve to form the neurological visual 'pathways'. These major nerve tracts pass right back through the brain to end at the rear pole in the visual cortex – the part of the brain in which the processes of bringing vision into consciousness occur. Not all the fibres of the visual pathways, however, go to the visual cortex. A substantial number of them run to other regions of the brain. Among these are two tiny nerve cell nuclei, the suprachiasmatic nuclei, lying in an area near the midline of the lower surface of the brain (the hypothalamus). The suprachiasmatic nuclei (so called because they lie immediately above the chiasma, or partial crossing, of the optic nerves) seem to be the biological clocks for circadian rhythm. If they are destroyed by a nearby brain tumour, circadian rhythm is lost. Disease of the hypothalamus commonly causes disturbances of the normal pattern of sleep, often with somnolence during the day and insomnia at night. The suprachiasmatic nuclei contain receptors for melatonin and when melatonin reaches them the electrical activity of the suprachiasmatic nuclei is reduced.

Neurophysiologists are still arguing about the details of this mechanism, but the evidence adduced so far strongly suggests that the complementary action of the pineal in turning off the clock, and the retinal connections in turning it on, is the basis of synchronization with the day–night cycle. There is little doubt that the suprachiasmatic nuclei, acting though the hypothalamic connections, organize the timing of such events as sleeping and waking, body temperature variations, times of eating and so on.

biosphere the totality of those areas of the earth's surface and atmosphere in which living things are able to survive.

biotechnology the use of micro-organisms or biological processes for commercial, medical or social purposes. The earliest known examples of biotechnology are the fermentation of wines and the making of cheese.

biotin a water-soluble B vitamin concerned in the metabolism of fats and carbohydrates.

biparietal relating to both PARIETAL bones of the skull.

bipolar neuron a nerve cell with only two processes – one dendrite and one axon – arising from the cell body.

birth the act or process of being born. The expulsion of the baby from the uterus.

birth canal the exit route through which the baby is forced by the contraction of the womb. The canal consists of the widely stretched cervix of the uterus, the vagina, and the small and large external lips – the labia.

birth control a euphemism for contraception. Strictly speaking, the term also includes celibacy, sexual continence, sterilization, castration and abortion.

birthmarks benign tumours of skin blood vessels, including the temporary strawberry marks, portwine stains (capillary haemangiomas) and the conspicuous cavernous haemangiomas which are raised, lumpy and highly coloured and consists of a mass of medium-sized blood vessels and blood spaces.

bisexual 1 pertaining to BISEXUALITY. 2 a person who manifests bisexuality.

bisexuality the inclination for, or capability of, sexual intercourse with either men or women. Bisexual behaviour is relatively common; genuine emotional neutrality in the choice of sex objects is very rare.

bite a dental term describing the relationship of the teeth of the lower jaw (MANDIBLE) to those of the upper and how they come together (the occlusion).

bite-wing a dental X-ray film used within the mouth and having a paper projection that can be held between the teeth to secure it in place during the exposure.

black eye a collection of blood released into the tissues around the eye, just under the thin, and relatively transparent, eyelid skin. The medical term is rather more impressive – periorbital haematoma.

blackhead an accumulation of fatty sebaceous material in a sebaceous gland or hair follicle, with oxidation of the outer layer, causing a colour change from white to dark brown or black. Blackheads, or comedones, occur in the skin disorder ACNE.

blackout a common term for a temporary loss of vision or consciousness. This may be a harmless fainting attack or a brief period of visual loss caused by standing up suddenly. Both are due to transient shortage of blood to the brain (cerebral ischaemia).

bladder see URINARY BLADDER.

blast cell an immature or primitive cell from which mature, differentiated cells are derived. The term refers mainly to the progenitors of blood cells (haemopoietic cells). The presence of these in the circulating blood is a feature of acute leukaemias. A stem cell.

blastocoele the cavity formed in the centre of a mass of dividing cells in an embryo at the start of the BLASTULA stage.

blastocyst the state of the development of the embryo at about eight days after fertilization, when it consists of 50–100 cells and at which implantation in the wall of the womb occurs. At this stage it consists of a double-layered hollow sphere full of fluid. The outer layer, the trophoblast, forms the placenta, the inner layer forms the future fetus.

blastomeres the two cells produced when a zygote has completed its first division.

blastopore the opening into the ARCHENTERON in a GASTRULA.

blastula an early stage in the development of an embryo when it consists of a hollow ball. Also known as a blastosphere.

bleb a blister-like collection of fluid, within or under the epidermis of the skin, usually containing serum or blood.

bleeding time the time from the infliction of a very small wound, such as a prick, and the cessation of bleeding. Bleeding time is increased in PLATELET deficiency and after taking aspirin or other PROSTAGLANDIN inhibitors.

blind spot the projection into space of the optic nerve head (optic disc) on the RETINA. This consists solely of nerve fibres and has no receptor elements (rods or cones). The blind spot lies about 15° to the outer side of whatever point we are looking at because the optic disc lies about 15° to the inner side of the macula. The blind spot is not normally perceived, even in monocular vision, because it is remote from the point of visual fixation.

blister a fluid-filled swelling occurring within or just under the skin, usually as a result of heat injury or unaccustomed friction. The fluid is serum from the blood and is usually sterile.

blocking involuntary interruption of a train of thought by emotional upset or psychotic disorder.

blood a complex fluid vital to life and circulated by the pumping action of the heart. The average blood volume is 5 litres. It is a transport medium, especially for oxygen, which it carries in the red blood cells linked to the HAEMOGLOBIN with which they are filled. It also transports dissolved sugars, dissolved proteins such as ALBUMIN and GLOBULIN, protein constituents (AMINO ACIDS), fat-protein combinations (LIPOPROTEINS), emulsified fats (TRIGLYCERIDES), vitamins, minerals and hormones. Blood also carries waste products such as carbon dioxide, urea, lactic acid, and innumerable other substances. In addition to the countless red cells the blood carries enormous numbers of uncoloured cells most of which are concerned in the defence of the individual against infection and cancer. It also contains large numbers of small non-nucleated bodies called PLATELETS which are concerned with BLOOD CLOTTING (coagulation).

blood alcohol the amount of alcohol in the blood expressed as a quantity per given volume of blood. Blood alcohol levels are often expressed in milligrams of alcohol per 100 millilitres of blood.

blood–brain barrier the effective obstruction to the passage of certain substances from the blood to the brain cells and the cerebrospinal fluid. The basis of the blood–brain barrier is that the endothelial cells comprising the walls of the brain capillaries are more tightly joined together than elsewhere and their gaps are further occluded by ASTROCYTES. In other parts of the body the capillary endothelial cells have gaps between them through which quite large molecules can pass. Substances in the brain blood can reach the brain cells or the cerebrospinal fluid only by passing though the plasma membranes of the endothelial cells. The blood–brain barrier protects the brain against many dangers, but can interfere with attempts at treatment of brain conditions.

blood cells any of the formed elements in the BLOOD which constitute about half the blood volume.

blood cholesterol the level of CHOLESTEROL circulating in the blood in combination with other fats and proteins in the form of low and high density LIPOPROTEINS. Cholesterol levels vary with sex, age, diet and hereditary factors. In people who get most of their energy from carbohydrates cholesterol levels are low. The blood cholesterol varies considerably from time to time, so a single uncontrolled reading is of little significance.

blood clotting the cascaded sequence of changes which occur when blood comes in contact with damaged tissue and which culminates in the production of a solid seal in the damaged vessel. At least 13 factors are consecutively involved in a process that culminates in the conversion of the protein fibrinogen to the fibrin that forms the main constituent of the clot. This last stage is catalyzed by the enzyme thrombin. The smooth endothelial lining of the blood vessels normally prevents clotting within the circulation, but damage to this lining may allow thrombosis to occur.

blood count determination of the number of red and white blood cells per millilitre of blood. The white cell count usually includes a differential count, in which the percentages of the different kinds of white cells is estimated.

blood electrolytes simple inorganic compounds in solution in the blood that form charged particles called IONS. The concentration of ions is critical for normal body function and the movement of ions is fundamental to much of the basic functioning of all the cells of the body. Measurement of blood electrolytes is therefore an important investigation in many conditions.

blood gas analysis an invaluable investigation providing information of literally vital importance. Arterial oxygen saturation (PaO_2) and carbon dioxide levels ($PaCO_2$) indicate any inadequacy of oxygenation and of breathing.

blood glucose the levels of sugar in the circulating blood. Blood glucose is of critical importance in diabetes in which the ideal treatment is to keep the levels within the normal range of 3.5–5.2 mmol/l – an ideal seldom achieved.

blood groups see ABO BLOOD GROUPS, KELL BLOOD GROUP SYSTEM and RHESUS FACTOR.

blood pressure the pressure exerted on the artery walls and derived from the force of the contraction of the lower chambers of the heart (the VENTRICLES). Blood pressure changes constantly. Peak pressure is called the systolic pressure and the running pressure between beats is called the diastolic pressure. Blood pressure in measured in millimetres of mercury. A typical normal reading is 120/80.

blood sedimentation rate the rate at which the red cells settle when a column of blood is held vertically in a narrow tube. Also called the erythrocyte sedimentation rate (ESR). A rapid rate is a non-specific indicator of the presence of some inflammatory disease process.

blood serum the liquid part of the blood (blood plasma) with the protein which forms during clotting (fibrin) removed. Serum remains liquid.

blood sugar see BLOOD GLUCOSE.

blood transfusion the administration of blood, by instillation into a vein, to replace blood lost or to treat a failure of blood production. Before transfusion, the blood group of the recipient must be known and serum from the blood to be transfused is cross-matched with the recipient's blood cells to confirm compatibility. Sometimes the patient's own blood, collected at operation or obtained earlier, is used.

blood urea the levels of UREA in the blood. Normal kidney function keeps the blood urea levels low by excreting it in the urine. A high blood urea suggests kidney failure.

blood vessel any artery, arteriole, capillary, venule or vein.

blunt end an ending in DNA in which both strands of the molecule stop at the same base-pair so that no single strand protrudes.

blushing a transient reddening of the face, ears and neck, often spreading to the upper part of the chest, caused by a widening (dilatation) of small blood vessels in the skin so that more blood flows.

BMI *abbrev. for* BODY MASS INDEX.

BMR *abbrev. for* BASAL METABOLIC RATE.

body and soul a duality that has been accepted, largely without question, since the earliest times. The term soul is, however, indefinable in scientific terms but is taken to be an entity associated with the body but clearly distinguished from it. Various accounts of the properties of the soul have been asserted from time to time by theologians, but such assertions are not of a nature as can be verified.

body bag a plastic bag for transporting a human corpse.

body fuels glucose and, to a lesser extent, fatty acids and amino acids. These are oxidized to release chemical energy so that cells can carry out their many functions. When fuels are not provided from diet, the body first mobilizes its fat stores to provide fuel, and when these are exhausted, the muscles are broken down for fuel. Both fats and proteins are converted to glucose, when necessary.

body image the mental picture of the body provided by the association connections between the part of the brain concerned with body sensation (the sensory CORTEX in the postcentral GYRUS) and those parts concerned with the special senses. Body image is distorted in various conditions, especially anorexia nervosa.

body language the communication of information, usually of a personal nature, without the medium of speech, writing or other agreed codes. Body language involves a range of subtle or obvious physical attitudes, expressions, gestures and relative positions. It can, and often does, eloquently reflect current states of mind and attitudes towards others, whether positive or negative. Body language is often at variance with explicit verbal statement and in such cases is often the more reliable indicator.

body mass index (BMI) the weight in kilograms divided by a number obtained by taking the height in metres and multiplying it by itself. (kg/m^2). The BMI is a more satisfactory way of determining the risk of obesity than simple weight. The normal range of BMI is 19–25. Obesity is defined as a BMI of 27 or over. People with this figure show a significant excess of illness over those in the normal range.

body odour a socially unacceptable smell usually caused by the action of bacteria on the sweat produced by the sweat glands of the armpits and the groin areas (apocrine glands).

body packer a slang term for a person who smuggles narcotics in the intestines, usually contained in condoms. This is a dangerous trade with a high mortality from poisoning.

body piercing body decorations in the form of metal rings or studs applied to any part of the body especially the external ears, nose, lips, tongue, nipples, umbilicus, labia and clitoris. Body piercing is being adopted by an increasing number of women in the Western world, exposing them to a range of health risks including hepatitis and HIV infection, bleeding, shock, allergies, interference with surgery, burns from electrosurgical instrument, and interference with X ray and MRI investigations.

body sensations

Aristotle listed the senses as being limited to vision, hearing, taste, smell and touch. But even in his time it might have been possible to go further and to include the ability to perceive temperature, vibration, pressure, pin-prick and pain. This list of the sources and routes of information input to the brain is not exhaustive and the sensory receptors also include the transducers for head position in the inner ears, proprioception that informs us of the position of limbs, the stretch receptors in the muscles and tendons, receptors in the joints, and the pressure receptors in the neck arteries that inform the brain of the state of the blood pressure.

All the information passing to the brain, however varied, is coded in the form of repetitive nerve impulses and all these nerve impulses are the same. In this, sensory information differs radically from, for instance, the electrical information passing from an amplifier to a loudspeaker. Individual nerve impulses follow the 'all or none' law. They either occur fully or not at all. There is no question of amplitude modulation as in the case of audio signals in a microphone cable or a stream of digital information as in a computer. The only thing that can vary, when information is conveyed along a single nerve fibre, is the repetition frequency. The maximum repetition frequency of nerve impulses is not high – only a few hundred per second (Hz). Fibres in the acoustic nerve seldom fire at a frequency of more than 200 Hz. But in spite of this seemingly severe limitation, a considerable range of intensities (amplitude) can be represented to the brain by differences in nerve impulse frequency. Sound frequency (pitch) variations are conveyed by differences in the location of the fibres stimulated.

In general, information, in the nervous system, is conveyed in terms of which particular fibres are firing. Electronic information systems operate with single channels that are modulated in various ways to convey data. Biological information systems, on the other hand, use bundles of thousands or millions of parallel channels, each one representing to the brain a position or a particular modality of sensation. Each optic nerve contains about one million separate fibres, each one coming from a particular point on the retina and informing the brain that that point has, or has not, been stimulated by a photon of light. The point-to-point correspondence persists all the way from the retina to the visual cortex . The same kind of point-to-point correspondence occurs in all the senses. Medical illustrators have repeatedly drawn grotesquely distorted inverted human figures to represent the mapping on the sensory cortex of the brain of the skin sensation. Such figures are distorted because different weighting has to be given to the higher concentration of sensory representation in different parts of the body such as the tongue and the sex organs.

In electronic systems the coded information is converted back, at the receiving end, into a familiar form – sound, pictures, text on a screen. In the brain, no such conversion occurs. The coded information, itself, it perceived as sound, vision, touch, taste, smell, and so on. Thus it will be seen that the means by which information is conveyed to the brain also differs fundamentally from the thing it represents. The ultimate philosophical question is whether our perception of the 'outer world' bears any meaningful resemblance to what is really there.

Much of the information received by the brain gives rise to no conscious perception. This is fortunate, as we have quite enough to attend to without being constantly informed of such parameters as the level of oxygen and carbon dioxide in our blood or the degree of tension in the arteries supplying our intestines. Even the sensory modalities of which we are often fully aware do not, happily, intrude at all times on our consciousness. Touch and pressure sensation tend to get through to us only when a significant change occurs or when they become excessive. The stimulation of the nerve endings subserving pain may or may not even give rise to awareness of pain; much depends on the circumstances.

The nature of pain

Pain differs from the other kinds of sensation in that, for the psychologically normal, it is invariably unpleasant and usually has a major psychic component. Pain is a localized sensation caused by stimulation strong enough to damage tissue or to threaten damage to tissue. Unless chronic (long–lasting), it commonly serves as a warning of danger and prompts action tending to end it. The response to pain may be reflex, involuntary, and rapid, or conscious, deliberate and purposeful.

Persistent pain is usually associated with distress and anxiety and often with fear, and there may be physiological changes similar to those experienced during anger and aggression. The heart beats faster, the blood pressure and the rate of respiration rise, the pupils dilate and the skin sweats. There is an increased secretion of adrenaline from the adrenal glands and increased mobilization of glucose from the glycogen stores in the liver.

Our perception of the significance of pain is often more related to these secondary effects than to the intensity of the pain itself. If pain is separated from its mental component, as is possible by the use of drugs such as morphine, it may still be felt but may no longer be unpleasant. The distress caused by pain depends also, to a large extent, on our awareness of the cause and is modified by past experience. Even minor pain inflicted by an assailant may seem more severe than the same physical hurt resulting from an innocent cause such as an accident.

Origins of pain

The nerve ending subserving pain are called nociceptors. These do not appear to differ physically from one another, but different nociceptors seem to respond to different kinds of painful stimuli – mechanical, thermal or chemical. Tissue damage results in the release of various strongly stimulating substances such as prostaglandins and these are the principal stimulators of nociceptors, causing the nerve fibres to fire and conduct impulses to the brain. Drugs such as aspirin inhibit the enzymes that cause the release of prostaglandins from damaged cells and this is how they act as painkillers. Aspirin has no effect on pain caused by pinprick, in which nerve endings are directly stimulated without the intermediate stage of tissue damage.

Different nociceptors show different sensitivities. Some are stimulated by low-grade 'warning' events of insufficient force to cause actual pain. Others respond only to strong stimuli such as pricking, cutting or burning. The stronger the stimulus, the higher the frequency of the nerve impulse sequence sent to the brain. If a nerve fibre for pain impulses is stimulated at any point along its length, pain will be perceived in the area of the nerve ending. In the condition of post-herpetic pain following shingles, for instance, sensory nerves are stimulated near the spinal cord by an inflammatory reaction caused by herpes zoster viruses. This causes pain which is perceived as coming from the skin.

Although the nerves carrying pain impulses terminate in the brain, and give rise to neurological activity there, the pain is usually felt in the region in which the nerve endings are situated. If the conduction of pain nerve impulses is prevented, as, for instance, by injecting a local anaesthetic drug around the trunk of the sensory nerve or near the spinal cord, no pain will be felt, although the damaging events at the nerve ending are continuing unabated. Passage of nerve impulses may be blocked by other means such as the arrival of other nerve impulses caused by the stimulation of other sensory nerves. These second impulses may be stimulated by rubbing, scratching or stroking the skin, by electrical stimulation applied through the skin or by acupuncture.

The 'gate' theory

Computers operate by an elaboration of logical 'gates' through which a stream of electrical square-wave pulses (ones or zeros) passes or is blocked by a secondary controlling electrical signal of the same kind, or is compared with other similar pulses. Most physiologists now accept that the nervous system contains analogous arrangements of neurons operating as gates and that pain impulses travelling up the spinal chord pass through such gates and can be blocked, or allowed to pass, by controlling signals. It seems probable that all neuronal signals concerned with pain must pass through such gates. Nerve fibres carrying pain impulses may be large or small. Both affect the state of the gates. The small fibres tend to open the gates and large fibres to close them. Large fibre stimulation also sends messages to a higher level which in turn also acts to close the gate. Gates are also probably under the control of the brain.

Endorphins

Drugs like morphine act on specific receptor sites in the brain, which appear to have no other function. This seemingly remarkable coincidence led scientists to expect, and some to predict, that natural morphine-like substances must be produced by the body. In 1975 two such substances were isolated from the brain and called enkephalins (*enkephalon* is the Greek word for the brain). Later, several more of these active substances were found, all with the same opioid core of five amino acids. Because of their morphine-like chemical structure and properties and because they come from within the body, they have been named endorphins – a contraction of the phrase 'endogenous morphine-like substances'.

Endorphins are neurotransmitters with a wide range of functions. They control the perception of pain during highly stressful events, such as sudden severe injury; they help to regulate the action of the heart; they exert a controlling influence on hormones; and they have an action in reducing dangerous surgical shock from blood loss. They seem to be involved, in some way, in controlling mood, emotion and motivation. They act on the centres of the brain concerned with the heartbeat and the control of blood pressure and on the pituitary gland. One extraordinary research finding was that the levels of circulating endorphins were found to be higher after subjects took a tablet which they *believed* to contain a pain-killing drug, but which was in fact an inactive placebo. Long-distance runners are said to become addicted to their own endorphins.

body temperature control a thermostatic mechanism in the brainstem that monitors the blood temperature and responds to deviations from the normal 37°C either by increasing heat production by promoting shivering, or by increasing heat loss by promoting the widening of the blood vessels in the skin.

bodywork an informal term for any claimed therapy, such as massage, in which parts of the body are manipulated.

bolus a chewed-up quantity of food formed up into a ball by the action of the tongue and preferably in a state of mastication ready to be swallowed.

bonding the formation of a strong relationship, particularly that between a mother and her new-born child. Bonding is believed to be important for the future psychological well-being of the infant.

bonding, dental the application of strongly adhesive cosmetic or protective surface material to a tooth. Bonding material is often applied in a plastic state and smoothed off before hardening.

bone

Bones differ greatly from the other body tissues because the protein (collagen) framework of which they are made is impregnated with calcium phosphate hydroxyapatite crystals. In health, these salts account for about 60 per cent of the weight of the bone.

Demineralized bones retain their general shape but are as flexible as rubber. The collagen scaffolding (matrix) is laid down in well-organized strands to form struts and girders disposed in such a way as to withstand the normal stresses of standing and walking. If an engineer were to design a bone it would have to be made very much as evolution has accomplished it. The strength of bones depends as much on the collagen matrix as on the mineral salts. As we get older there is a progressive loss of both collagen and minerals.

The skeleton of the living body bears little resemblance to the dried bones of a skeleton in a museum. Living bone is in a dynamic state of flux with a constant interchange of its constituent materials such as calcium, phosphates, amino acids and fatty acids. There is also a constant consumption of glucose, oxygen and other materials. Calcium has many bodily functions in addition to the mineralization of bone, and there is a constant interchange of calcium between the bones and the blood. This is strictly regulated by hormones and by the action of the kidneys. Maintenance of correct blood calcium levels is vital to life and is more important than calcification of the bones. If, for any reason, there is a shortage of calcium it is always the bones that suffer. Inadequate calcium leads to softening of the bones – rickets in children and osteomalacia and osteoporosis in adults.

Osteoblasts and osteoclasts

Embedded in the collagen matrix of bones are many cells called osteoblasts. These synthesize collagen to maintain and repair bone structure. The osteoblasts are especially active during the period of body growth and after bone fractures. Throughout life, their activity is also greatly influenced by sex hormones, both male and female, which act on them to promote collagen formation and calcification. This is why women suffer more from osteoporosis than men. Women lose hormones after the menopause while men continue to secrete them. Another important stimulus to osteoblast activity and resultant bone bulk is physical loading of the bones – especially the weight-bearing involved in standing and walking. Astronauts living in conditions of zero gravity quickly lose bone mass, as do people lying in bed. Weight-bearing exercise in youth promotes strength and thickness in the bones, and such exercise throughout life minimizes the natural rate of decline in bone bulk. Men naturally have heavier bones then women and are less likely to reach a stage of dangerous rarefaction.

Bone growth, remodelling and fracture repair involves reabsorption as well as construction and this is the function of a different group of cells known as osteoclasts. These secrete digestive enzymes that break down collagen polymers to release the amino acids and the calcium salts. Osteoclasts are especially active after fractures, removing splinters of bone, cleaning up the break, preparing it for osteoblastic new bone formation, and even, over the years, gradually remodelling and realigning bones that have healed askew after fractures.

Bone growth

During childhood and adolescence all the bones of the body increase greatly in size. The amount of growth is genetically determined so long as adequate quantities of building materials – amino acids, fatty acids, vitamins and minerals – are provided. Body height increases mainly by growth of the long limb bones, and this occurs by virtue of special growth plates of cartilage, called the epiphyseal plates, which are situated near both ends of the bones. Throughout the whole growth period the cartilage is the site of a cooperative osteoclast and osteoblast activity. In each epiphysis, the edge of the cartilage nearer the centre of the bone is gradually converted to bone while, at the same time, new cartilage grows outward at the edge further from the centre. In this way the bone length progressively increases. Inadequate nutrition will stunt growth; excessive nutritional intake cannot increase height – only width. Once body growth is complete – about the age of 25 – the epiphyseal plates are converted wholly to bone and no further growth of long bones is possible.

Bone marrow

In addition to its skeletal function and its function as a calcium depot, bone also acts as the site of red and white blood cell production. The blood cells develop from stem cells in the red marrow of the flat bones – the ribs, breast-bone (sternum), shoulder-blades (scapulae), pelvic bones and skull – and in the bodies of the bones of the spine (the vertebrae). This process of blood cell production is continuous and is necessary to make up for the steady losses of cells in the circulation. The bone marrow therefore has a rich blood supply into which large quantities of red cells, white cells and platelets are poured. The marrow of the long bones does not normally produce blood cells and is filled mainly with yellow fat cells.

Control of bone growth

The activity of the cells in the epiphyses is under the control of the pituitary gland growth hormone somatotropin. So the amount of this hormone determines the extent of growth and the ultimate height of the individual. If, for any reason, the hormone is not produced during the growth period, the result is a dwarf; if excess hormone is produced during childhood or adolescence the result is a giant. Absence of somatotropin is rare and can be remedied artificially by injections of growth hormone. Excess growth hormone usually comes from a tumour of the hormone-secreting cells of the pituitary gland. Excess hormone after the epiphyses have fused produces the condition of acromegaly – a disorder in which bones not formed from epiphyseal plates – such as the jaw, skull, spine and hand and foot bones – continue to enlarge.

bone conduction the transmission of sound to the inner ear hearing mechanisms by way of bone rather than by the normal air, eardrum and middle ear route.

bone fractures bone breaks may be partial (greenstick), with intact skin (simple), with penetration of the skin (compound) or with involvement of blood vessels or nerves (complicated). They may be transverse, oblique or spiral or the bone may be shattered (comminuted).

bone marrow the substance contained within bone cavities. This is red in the flat bones and the vertebrae, and yellow from fat in adult long bones. The volume of the red marrow in young adults is about 15 l. The basic marrow stem cell differentiates into HAEMOGLOBIN-carrying red blood cells, the white blood cells of the immune system and the blood PLATELETS which are essential for BLOOD CLOTTING.

borborygmi bowel noises caused by the gurgling of gas through the almost liquid contents of the small bowel as they are passed along by the process of PERISTALSIS.

bot flies flies of the genera *Gasterophilus, Oestrus* or *Dermatobia* whose larvae can burrow into human skin, eyes or nasal openings. Parasitization by fly larvae is called myiasis.

Botox a brand name for a powder used to reconstitute a solution of botulinum A toxin complexed with haemagglutinin. This is used to treat a wide and increasing range of conditions.

bottle-feeding the popular alternative to breast-feeding for babies. Formula milk cannot be made identical to human milk and the bottle-fed baby is deprived of many valuable antibodies present in the mother's milk. The risks of contamination of the feed are also greater with bottle than with breast-feeding.

bowel the intestine. A tube, about 8 m long, which extends from the throat to the anus and consists of the OESOPHAGUS, STOMACH, DUODENUM, JEJUNUM, ILEUM, COLON, SIGMOID COLON, RECTUM and ANAL CANAL.

bowel movement DEFAECATION. The normal frequency of bowel movement varies between three times a day and three times a week.

Bowman's capsule the filtering unit in the kidney by which urine is formed from the blood. Sometimes called the malpighian corpuscle. (Sir William Bowman, 1816–92, English surgeon and ophthalmologist).

Bowman, William William Bowman was born in Nantwich in 1816. His medical training began with an apprenticeship to a Birmingham surgeon, after which he studied at King's College Hospital, London and in Europe. At the age of 23 he decided to devote himself exclusively to scientific research, in the course of which he was the first to describe the hollow capsule which surrounds each of the millions of glomeruli of the kidney. A year later, in 1840, he was appointed professor of physiology and general and morbid anatomy at King's. There he did not waste his time. In 1841 he became a Fellow of the Royal Society, in 1844 a Fellow of the Royal College of Surgeons, and in 1845 he published the first of five volumes of the major work *Physiological Anatomy and Physiology of Man* (1845–56).

Soon Bowman's interests took a radical turn. In 1849 he published *Lectures in the Parts Concerned in the Operations of the Eye*, took up an appointment as Surgeon to the Royal Ophthalmic Hospital, Moorfields and devoted the rest of his life to the study of the eye. Soon he became the undisputed leader, and the greatest exponent, of British ophthalmology. In 1854 he was awarded a Baronetcy. He died in 1892. To this day, Bowman is remembered, not by the urologists but by the ophthalmologists. For many years at the Ophthalmic Society of the United Kingdom (now the Royal College of Ophthalmologists), a distinguished Fellow has been invited to deliver the Bowman lecture.

brachial pertaining to the arm.

brachial artery the main artery supplying the arm with blood.

brachial plexus the complicated rearrangement of spinal nerves arising from the spinal cord in the neck and in the upper back, which occupies the armpit (axilla), and supplies the many muscles of the arm.

brachiation non-pedal locomotion effected by swinging by the arms from branch to branch.

brachy- *combining form denoting* short.

brachycephalic having a short, wide, almost spherical head.

brachydactyly abnormally short fingers or toes.

BRCA genes tumour suppressor genes, the mutation of which is responsible for an inherited predisposition to breast and ovarian cancer. In 1990 BRCA1 was found to be on the long arm of chromosome 17, and it was cloned in 1994. It is involved in about 5 per cent of all breast cancers. In these cases, and in many cases of invasive breast cancer, there is decreased expression of BRCA1. A considerable number of FRAMESHIFT or NONSENSE MUTATIONS have been found that confer high risk of cancer. BRCA2, located on chromosome 13, is less commonly a cause of inherited breast cancer.

brady- *combining form denoting* slow.

bradycardia a slow heart rate. In the healthy this often indicates a high degree of fitness, but bradycardia can be a sign of heart disease.

bradykinin a peptide that widens blood vessels (vasodilatation) and lowers blood pressure, increases capillary permeability and the secretion of saliva and mediates pain associated with inflammation. Bradykinin is inactivated by the angiotensin-converting enzyme.

braille a method of coding information using groups of six raised spots embossed on paper, to enable the blind to read through touch. (Louis Braille, 1809–52, French school teacher).

brain–computer analogy

The brain is composed of some 100,000,000,000 nerve cells (neurons) each consisting of a cell body, a long, often branched, output fibre called the axon and anything from one to several thousand short input fibres called dendrites. Connections between adjacent or widely separated neurons are numerous and complex and single neurons commonly receive several hundreds to several thousand inputs from other nerves. In general, the cell body and the dendrites receive incoming signals which are averaged (integrated) in the cell body, and the output signal leaves by way of the axon.

This almost incredible complexity of interconnection could hardly be more different from the inherent simplicity of the structure of the ordinary digital computer. Even so, the arrangement of stimulatory and inhibitory connections closely mimics the logical 'gates' that are the basis of digital computing. The biological system would allow all of the logical elements found in the central processing unit of the digital computer. AND, OR, NAND (AND with reversed output) and NOR (OR with reversed output) gates would be readily implemented by such an arrangement of neurons. But in view of the number of connections it seems that some kind of statistical gating system is normally involved. Nerve impulses, although electrical in nature, travel at very low speeds compared to the speed of propagation of electricity in computers and the timing of impulses in the brain is probably much less important than in digital computers where everything is organized by a clock (oscillator or pulse generator) running at several billions of cycles per second (Ghz). Nerve impulses follow each other at a repetition rate that varies. This is called frequency modulation (FM), a form of electrical signal much used in electronic communication but not as the basis of computing.

Tracing the brain connections

Another major difference between the brain and digital computers is that whereas computers are general purpose machines performing different functions in accordance with instructions provided in the software, the brain is largely 'hard wired' to carry out its functions. Major subdivisions of brain function – movement, sensation, speech, vision, hearing, and so on – are served by particular known parts of the brain and these are connected by nerve fibre bundles to the effector or sense organ concerned, as well as to other parts of the brain.

The tracing of these connections was a vital step in furthering the comparison between the brain and the computer and much of the early work was done by the Spanish anatomist Ramon Y Cajal. More recently, newer and better techniques have carried this work further. In the early 1950s it was found that a killed nerve cell, while degenerating, takes up a certain stain throughout its entire extent. In this way it is possible to show how particular cells in certain parts of the brain are connected to quite remote parts. The method has been used to work out in considerable detail the neuron map of the brain.

It is not enough, however, merely to demonstrate the structure of the brain. Neurons are functional units that may or may not be firing at any particular time. Much attention has been paid recently to this functional aspect. The brain fuel is glucose and this is consumed, in any particular area of the brain, at a rate that depends on the level of function in that part. Various methods have been devised to measure the rate of glucose consumption in different parts of the brain. Glucose molecules containing radioactive atoms can be made and the degree of radioactivity from local accumulation of radioactive substances can easily be measured. The most sensitive technique, to date, uses glucose labelled in this way with positron-emitting isotopes. These can be detected in the living brain by a positron emission tomography (PET) scanner outside the skull. If this is done in the course of any activity, such as hearing, seeing, smelling or performing mental arithmetic, the part of the brain especially involved can be determined.

brain structure and function

The only way in which we can derive any information about the outside world is by way of our sense organs – eyes, ears, nose, tongue and skin, muscle and tendon nerve endings – all of which send nerve impulses to the brain. And the only way in which we can respond to the outside world and change it is by brain action initiating the passage of nerve impulses to our muscles and glands. Thus, the brain is a responsive entity with an input and an output. Without input, it is questionable whether anything much would happen in the brain. Total sensory deprivation in a mature adult would probably lead to total immobility. We know that, in a child, sensory deprivation is seriously damaging and results in failure of brain development. It is clear that information input is essential if the brain is to be changed.

The brain is the central, and by far the most important, organ of the body and it is not fanciful to consider the rest of the body as essentially a support, protective, locomotory and effector system for the brain. The brain is the seat of consciousness, pleasure and emotion, and the information centre of the body. It is the storage site of everything we have learned since we were born, and contains a unique database that underlies our whole personality and capability. The brain also stores much information inherited from our ancestors that is manifested as instinct, patterns of response, and so on. The brain is a pleasure-seeking organ, intent upon its own gratification and directing us, most of the time, to act in such a way as to stimulate pleasure areas within it. For some people, this process seems an adequate explanation of the purpose of existence. But there are other aspects of brain operation about which we know little or nothing. Unlike many of the more easily understood brain functions, these higher aspects – consciousness, perception, thought, imagination – cannot be localized to any particular part of the brain.

General structure

The most obvious neurological difference between man and the lower animals is the great development of the two large, almost mirror-image masses known as the cerebral hemispheres. The two hemispheres are almost isolated from each other except for a massive multi-cable junction, called the corpus callosum, that connects them. The cerebral hemispheres are conspicuous for the complex infolding of the outer surface, the cortex . The effect of this infolding is to allow a very large area of cortex to be accommodated in a small space. The cortex is the most advanced part of the brain and it is here that the best-understood functions of the brain are located.

Running down to connect the middle of the under side of the cerebrum to the spinal cord is the brain stem. This is a thick stalk of nervous tissue containing the great longitudinal tracts of nerve fibre bundles running into and out of the spinal cord. The brain stem also contains the collections of cell bodies (nuclei) of most of the twelve pairs of nerves, the cranial nerves, which emerge directly from the brain. These nuclei are interconnected and are linked to other parts of the brain.

Lying under the undersurface of the cerebrum, at the back, is the cerebellum, a separate brain, distinguishable by its narrower surface corrugations. The cerebellum is largely concerned with unconscious, automatic functions such as balance and the control and coordination of voluntary movements. It receives numerous connections from the motor parts of the brain, from the balancing mechanisms in the inner ears and from the basal ganglia (see below).

For good reason, the brain is better protected than any other organ. It is cushioned in water, wrapped in three layers of membranes called the meninges and enclosed in a strong bony case, the skull. The brain has an exceptionally large requirement for fuel – a requirement that varies with brain activity – and can only function properly if provided with an unceasing supply of glucose, oxygen and other nutrients by way of the bloodstream. So the blood supply to the brain is profuse, consisting of four major arteries (the two carotids and the two vertebrals) which provide branches to all parts. Blockage of any of these branches, or of the main trunks, is always serious. Even more serious is bleeding from any of the cerebral arteries. Both blockage and bleeding cause brain tissue destruction manifested by the clinical condition known as stroke. This is a very common cause of death. Any interruption to the blood supply to the brain, even for short periods, is dangerous. Heart stoppage for only three or four minutes will cause permanent damage and death is inevitable if, at normal temperatures, the blood supply ceases for more than six to eight minutes.

Sensory input

The receptor organs of the four main modalities of information input – vision, hearing, smell and taste – connect directly to the brain by short nerve tracts. By way of these nerves, a mass of data is supplied to the conscious brain, and this is analyzed, correlated with existing stored data, stored and, if necessary, acted upon. At the same time, a mass of sensory information enters the brain from receptor nerve endings in the skin, muscles, tendons, joints and internal organs. This sensory input informs the brain about the state of the environment and about the relative position of the limbs. If the internal organs are disordered, information from them may be

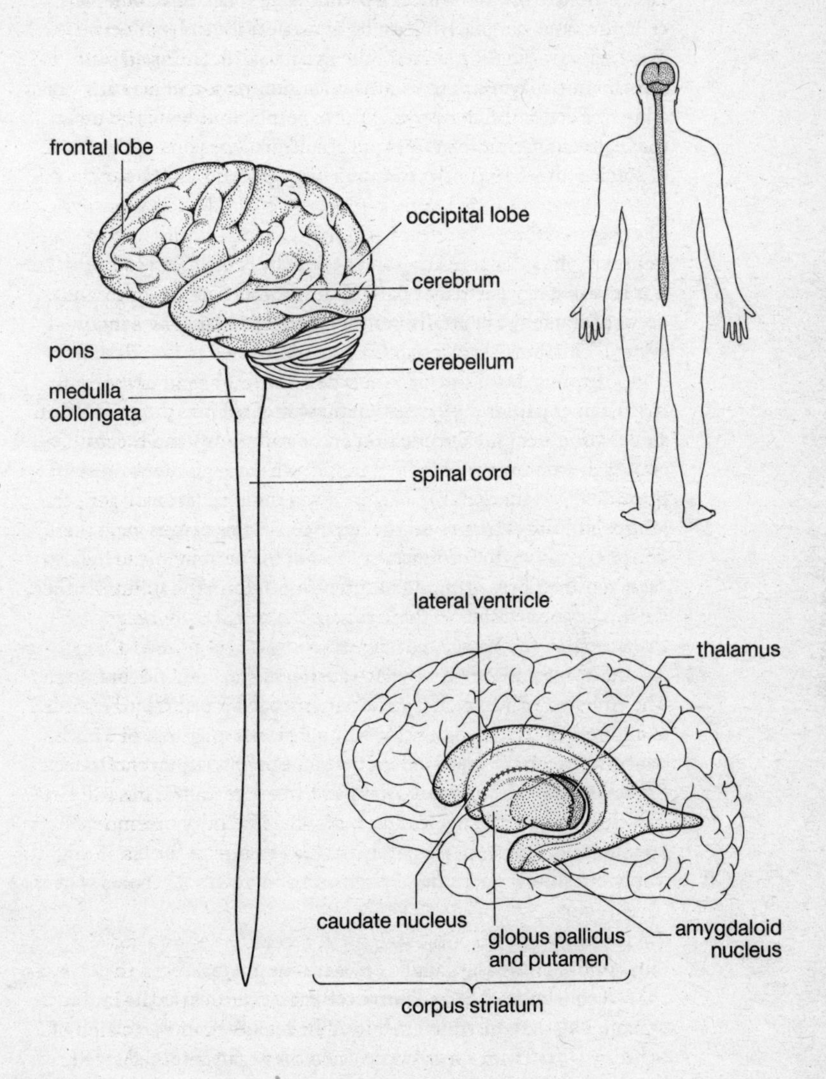

Fig. 2 **Brain structure and function**

supplied to the brain. Much of the incoming sensory information passes, first, to one or other of a pair of large collections of nerve cells, the basal ganglia, lying in the brain near its underside. The main sensory ganglion is called the thalamus. Incoming sensory information may result in unconscious automatic compensatory or adjusting action, often by way of the hypothalamus lying just under the thalamus. Other sensory input results in conscious awareness of some bodily function or state and prompts voluntary action.

The effector system

Voluntary movement of any part of the body is mediated by the brain. It is initiated in a particular part of the brain surface or cortex and is brought about by a centrally-placed part of the brain, the pyramidal system. This consists of a massive pair of inverted pyramids of nerve fibres running down through the whole vertical height of the brain. In the upper part of the brainstem these motor fibres cross from one side to the other ('the decussation of the pyramids'), and link to connecting neurons which then pass down through the brain stem into and down the spinal cord. The lower motor neurons. There, the long motor fibres trigger off the nerve cells whose axons form the peripheral nerves to the muscles. Most of the peripheral nerves also carry sensory fibres bringing in information from the skin and other parts. The connections of these run up the spinal cord to the thalamus and the sensory cortex.

Brain output is not limited to causing skeletal muscle contraction. The other output function is so important that it is commonly regarded as a system in its own right – the autonomic nervous system. This is concerned with the operation of the smooth (involuntary) muscle in the walls of arteries and hollow organs, with the control of the heart muscle and with causing glands to secrete. The autonomic nervous system is not, however, a separate nervous system as the name might imply, but simply a peripheral extension of the central nervous system.

The limbic system

Surrounding the basal ganglia on each side is a massive sickle-shaped collection of linked nerve cell masses known as the limbic system. This system, which includes the hypothalamus, is a kind of primitive brain with connections from the organs of sight, smell and hearing. The limbic system is concerned with the processing of information so that it can be stored in memory. It is also greatly concerned with the production of emotional states. Stimulation of various parts of the limbic system causes a variety of emotions – joy, fear, rage, sexual excitement and so on – and behaviour associated with emotion. If the hypothalamus is isolated from the limbic system the emotional components of the behaviour are absent.

The neuron

The nerve cell or neuron has much in common with other body cells. It has a nucleus and cytoplasm containing the same general cell organs (organelles) as other cells. It differs, however, in two main particulars – shape and excitability. Although the body of the nerve cell is of roughly the same general size as that of other cells, it possesses a fine extension in the form of a fibre called the axon, that may be many hundreds or thousands of times longer than any other type of body cell. This fibre is insulated by a fatty covering, the myelin sheath, and may be 60–100 cm long. Running into the cell body are numerous shorter branching fibres, called dendrites. The cell body seems to be concerned largely with the nutrition and maintenance of the axon and has little to do with the origination and propagation of the nerve impulse.

Excitability, the most important property of nerve tissue, depends on the presence of potassium within the axon and sodium outside it. Normally, the outside of the axon carries a strong positive charge and the inside is negative. This is because unlike charges attract one another, while like charges repel each other. When a nerve impulse reaches the cell at a particular point, positively charged sodium ions pour into the axon, making the inside locally positive. The outside, at this point immediately becomes negative and this zone of negativity moves along the axon in both directions. This is called depolarization and it is the basis of the nerve impulse.

The synapse

Nerve fibres do not directly contact one another but do so by way of special links, called synapses. The number of synapses in the brain almost defies belief and can only be estimated. So far as can be judged, the number is of the order of 100,000,000,000,000 (one hundred trillion). The main feature of the synapse is the actual gap between the two nerves. This gap is very narrow and lies between a slight widening, or bulb at the end of the axon, and the surface of another nerve cell or a muscle cell. The gap, of course, contains tissue fluid, containing plenty of sodium, but the nerve impulse cannot pass across it. When the impulse reaches the end of the axon it causes one of a range of particular chemical substances called neurotransmitters to be released by the axon. The substance quickly diffuses across the gap and its molecules fit into receptors on the other surface. This changes the local permeability of the membrane to sodium and sodium ions pass in causing the surrounding area to reverse its electrical charge (depolarize). As a result, a nerve impulse is propagated or a muscle fibre contracts.

Drugs that act on the brain often either act like neurotransmitters or block their action. Some synapses cause an impulse to pass to the next neuron, others are inhibitory and can prevent impulses passing in the second neuron. Integration of the excitatory and inhibitory effects will determine whether the second neuron will fire or not.

Much of what happens in the brain between the receipt of incoming stimuli and the resulting response is still unknown. This is especially true of the neurological activity that underlies phenomena such as perception, thought, emotion, memory and creativity. Perhaps significantly, the wiring and neurophysiology underlying these 'higher' activities is far less well understood than the brain areas and connections concerned with more primitive functions such as movement and sensation. So far as the higher functions are concerned it is often necessary to think of the brain as a 'black box'. Every part of the brain has a specific function and, although we know little of the neurological basis of consciousness there remains hardly any parts of the brain whose function is not at least partly understood. Possible conclusions from this are that the higher functions are, in their nature, essentially different from functions that can be localized and that they are products of the simultaneous functioning of many parts of the brain.

Some of the simpler connections between input to the nervous system and output from it are particularly well understood. When the number of synapses between input and output is small, the function involved is usually described as a reflex. The sudden withdrawal of the hand from contact with a red-hot iron is a reflex involving only three or four sets of synapses and occurs by way of the spinal cord without the brain being necessarily involved. The nervous system is interested in changes, especially sudden changes, and all the sense input systems are so designed that they send in strong signals in response to change but very weak signals when nothing much is happening. This is an efficient arrangement. We do not need to be constantly reminded that we are sitting in a chair, but the brain needs to know when we get up.

Cerebral cortex

A great deal of what goes on in the brain, especially in connection with sensation and movement, is understood. The location in the brain of the areas for movement and bodily sensation (motor and sensory functions), vision, hearing, speech, taste and smell has long been well known. These areas are located in specific zones in the outer layer of the brain – the cerebral cortex. The cortex is remarkable for its sheer size, which is concealed by the extraordinary degree of infolding. If it were fully spread out it would have about the same area as a full double sheet of a broadsheet newspaper. It contains some 50 billion nerve cell bodies of different types arranged in six layers.

Because most of the cortex consists of nerve cell bodies it is known as 'grey matter'. From these cell bodies, a mass of nerve fibres (axons) runs under the cortex to interconnect the various areas, to connect them to other collected masses of nerve cell bodies (nuclei), deep in the brain, and to connect to the spinal cord and hence to rest of the body by way of peripheral nerves. Nerve fibres, en masse, give a white appearance and are called 'white matter'.

Cortical functional areas

These are of basic importance. We know that mental experience of the various modalities of sensation, as well as volition concerned with movement, are associated with neurological activity in the appropriate areas. Adjacent to the main areas for the sensory functions are what are called 'association areas'. Perception of events in the world usually involve activity in the association areas as well as in the primary functional areas. The functions of the cortical areas is best illustrated by what happens when they are damaged.

Personality

The pre-frontal areas relate to personality and to some of the higher functions of mankind. The function of this part of the brain was revealed to the world in 1868 in a paper describing the case of Phineas Gage, a 'capable, God-fearing foreman' who, after a crowbar had been driven through the front of his brain by a gunpowder explosion, became 'irreverent, dissipated, irresponsible and vacillating.' Removal of the pre-frontal lobes has a calming and normalizing effect on people with severe psychiatric disorders but damages initiative and spontaneity and the inclination to make use of intelligence. Problem-solving capacity is affected by a defect of conceptualization. There is general loss of 'ego strength', of sensitivity and compassion. Pre-frontal lobotomy, as a treatment of psychiatric disorder, has long been abandoned. These areas are connected to the thalamus (see below) and the hypothalamus.

Movement

The primary motor areas lie near the centre of each side of the cortex. These areas are concerned with voluntary movement and are mapped out so that different parts serve different parts of the body. Damage to this area on one side causes paralysis of voluntary movement on the other side of the body. This is because the nerve fibres from this area cross to the other side before running down in the spinal cord. Irritative disturbances in this area cause major epileptic seizures. Adjoining areas are concerned with automatic functions such as the control of muscles in standing and the maintenance of posture, the relationship of head movement to eye movement, movements of the mouth, tongue, throat and larynx concerned in speech and swallowing. Damage here causes severe speech defects of a purely motor kind.

Skin, tendon and muscle sensation

Immediately behind the motor areas lie the sensory areas and the sensory association areas. The sensory areas receive massive bundles of nerve fibres carrying sensory information from every part of the body, especially the skin, the tendons and the muscles. Like the motor cortex, the sensory cortex is mapped out in terms of body areas. These fibres reach the cortex via the large sensory nucleus, the thalamus near the base of the brain, which is also connected to the hypothalamus. Destruction of parts of the sensory cortex causes loss of sensation in various parts of the opposite side of the body; irritative damage causes 'pins-and-needles' or a crawling sensation (formication) in corresponding parts of the skin. The sensory association areas are necessary for the correlation of sensation with other data, such as names and functions. Damage to these areas affects the ability to identify an object by feel alone.

Language

Below the motor and sensory areas are the areas concerned with the motor aspects of speech. But the whole function of language involves much more than speech and has a much wider representation in the cortex. Specific language areas are concerned with the relationship of speech to vision and hearing, with the perception of written language and with the perception of spoken language. Damage localized to these areas causes corresponding defects. There is, in consequence, a wide range of disorders of the language function, depending on the part of the cortex affected. Loss of production or comprehension of language is called aphasia and this may be expressive (motor), sensory or both. There may be word blindness, word deafness or the inability to communicate by writing (agraphia). Language is also concerned with cerebral dominance and is, in most people represented on the left side. In some left-handed people it is represented on the right side.

Hearing

Near the language areas are the areas for hearing. Destruction on both sides causes deafness, but loss of this area of the cortex on one side has little effect. The adjoining auditory association areas are concerned with sound associations. Damage in the association areas does not affect the perception of pure tones and sounds but seriously damages the ability to recognize sounds or to identify music. To a person affected in this way, all sounds, however diverse, seem to be alike. The creaking of a door is indistinguishable from the tinkle of a bell.

Vision

The visual area of the cortex is right at the back of the brain in the occipital region. If destroyed by disease or injury the person concerned will be completely blind although the eyes are entirely unaffected. If the visual association areas are destroyed the affected person will be blind but will deny it, and will rationalize the tendency to bump into things by saying 'It's too dark' or 'I've lost my glasses'. Damage to the visual association area can cause a variety of effects. Objects may seem distorted in size, form or colour; there may be hallucinations of light flashes, stars, geometric forms, or even of persons or animals. The latter may appear of normal size, or very tiny, or very large. There may be total inability to name, or to state the function of, objects seen, although such objects can at once be identified by feel, taste or smell. This is called visual agnosia.

brainstem the part of the brain consisting of the medulla oblongata and pons, which connects the main brain (cerebrum) to the spinal cord. The brainstem contains the 'vital centres' for respiration and heartbeat and the nuclei of most of the CRANIAL NERVES as well as massive motor and sensory nerve trunks passing to and from the cord.

brainwashing concentrated and sustained indoctrination designed to delete a person's fundamental beliefs and attitudes and replace them with new, imposed data. It is questionable whether this intention can ever be fully realized.

brain wave one of many periodic electric potentials generated in the brain and detectable by the electroencephalogram. Some of the brain waves are dominant and of recognizable frequency.

bran the fibrous outer coat of wheat grain normally removed in milling to make the flour more attractive to many palates. Bran is valuable in the treatment of constipation and other disorders of the large bowel.

Braxton-Hicks' contractions the common, irregular, painless and harmless contractions of the womb which occur throughout pregnancy. (John Braxton-Hicks, 1825–97, English gynaecologist).

breaking of the waters the release of amniotic fluid from the womb when the amniotic sac ruptures prior to the birth of a baby.

breast see MAMMARY GLAND.

breast augmentation a procedure in cosmetic plastic surgery designed to increase the size of the female breast usually by implanting a silicone oil-filled silicone rubber bag.

breastbone the STERNUM.

breast budding the first indication of puberty in the young girl.

breast enlargement this may occur in both sexes in the first week or so after birth, from maternal sex hormones acquired in the womb. In girls, the breasts enlarge at puberty under the influence of hormones. Breast enlargement is normal in pregnancy and especially during milk production (LACTATION). Adolescent boys may suffer breast enlargement (gynaecomastia).

breast-feeding the normal, and best, method of providing baby nutrition. The chemical constitution of breast milk is attuned to the digestive capacity of the baby and the nutritional balance is exactly what is required.

breast pump a device used to relieve engorged and painful breasts of excess milk, or to remove milk for use later.

breath-holding attacks a form of infantile blackmail imposed on indulgent parents by determined and manipulative young children. The attacks superficially resemble epileptic seizures, but, although the child may turn purple and even appear to lose consciousness, they are harmless.

breathing the automatic, and usually unconscious, process by which air is drawn into the LUNGS for the purpose of oxygenating the blood and disposing of carbon dioxide. Breathing involves a periodic increase in the volume of the chest occasioned by the raising and outward movement of the ribs and the flattening of the domed DIAPHRAGM. This is an active process involving muscle contraction and results in air being forced into the lungs by atmospheric pressure. Expiration is passive, the air escaping as a result of elastic recoil of the lungs and relaxation of the respiratory muscles.

breathlessness an automatic increase in the rate of respiration in response to a reduction in the levels of oxygen, and an increase in the levels of carbon dioxide, in the blood. Undue breathlessness, inappropriate to the degree of exertion, is an important sign of possible heart failure or of any other condition that leads to inadequate oxygenation of the blood. Breathlessness should be distinguished from hyperventilation which is usually either a response to acute anxiety or a voluntary activity. The medical term for breathlessness is dyspnoea.

breech delivery buttock-first birth.

bregma the point on the top of the skull at which the irregular junction lines of the bone (the coronal and sagittal sutures) meet. Also known as the sinciput.

Brenner, Sydney the South African scientist Sydney Brenner was born in Germiston, near Johannesburg, in 1927. He studied at the University of Witwatersrand, and then, in 1957, encouraged by Francis Crick, he joined the Molecular Biology Laboratory of the Medical Research Council, Cambridge, where he remained until 1996, when he moved to the USA. He was awarded the Nobel Prize for Physiology or Medicine in 2002.

Brenner worked with Crick to elucidate the human genetic code and by 1961 they had found the base triad sequence for each of the 20 amino acids. It was Brenner who suggested the term codon for the unit of three nucleotides coding for an amino acid. In 1960 Brenner discovered messenger RNA (mRNA), demonstrating the mechanism by which information from DNA is transferred to the ribosomes for the synthesis of proteins.

Brenner's work also encompassed techniques of genetic engineering to purify proteins, synthesize amino acids and clone genes. He has also studied the biology of tumours.

bridge a fixed support for false teeth which bridges across the gap between surviving natural teeth.

broad ligaments double folds of PERITONEUM hanging over the womb (uterus) and FALLOPIAN TUBES to form a partition in the pelvis.

Broca, Pierre Paul Pierre Paul Broca (1824–80) was a distinguished surgeon, anatomist and anthropologist, with degrees in surgery, pathology and anthropology and professorships concurrently at the Paris Faculty of Medicine and the Anthropological Institute. In 1861, in a lecture at the Paris Anthropological Society, Broca demonstrated the brain of a former patient. This man, M. Leborgne, had lost the ability to speak or to write although his intelligence and comprehension were unaffected. Prior to his death he had been able to communicate only by gestures, nods and facial expressions. Broca had conducted a postmortem examination and had found an area of damage the size of a golf ball in the left hemisphere of the brain. Broca's conviction that this was the cause of the language defect (aphasia) was disputed by many. But within a decade, other neurologists began to correlate brain damage in the same general area with various forms of aphasia.

The area of the brain affected is still called 'Broca's area' and the type of aphasia first described by him is called 'Broca's aphasia'.

The acceptance of Broca's findings prompted others to search for further correlations between recorded loss of function and subsequent observable brain damage. Later, in the course of brain surgery it became possible to discover the effects on the body of mild electrical stimulation of the cortex of the brain. By the end of the 19th century much of the function of the cortex of the brain was known.

Broca's area an area of the surface layer of the brain (cortex), on the left side near the front, concerned with speech.

broken neck fracture, often with dislocation, of any of the bony vertebrae of the neck. The importance is not so much the bony injury as the probable injury to the spinal cord lying within the bony canal. Such injury is likely to be serious and may be fatal. Survivors often suffer permanent paralysis below the level of the injury. Neurological damage can be increased by moving a person with a broken neck.

bromhidrosis profuse odorous sweat, especially from the feet, caused by the breakdown of short-chain FATTY ACIDS, especially isovaleric acid.

bronch- *combining form denoting* the air tubes of the lung (bronchi).

bronchial pertaining to any part of the branching system of breathing tubes (the bronchial tree).

bronchiole one of the many thin-walled, tubular branches of the bronchi, which extend the airway to the terminal air sacs (alveoli). Bronchi have cartilaginous rings, bronchioles do not.

broncho- *combining form denoting* a major air tube (BRONCHUS).

bronchoscopy direct visual inspection of the insides of the air tubes (bronchi), either through a hollow metal tube or by means of a fibre optic ENDOSCOPE.

bronchus a breathing tube. A branch of the windpipe (TRACHEA) or of another bronchus. The trachea divides into two main bronchi, one for each lung, and these, in turn, divide into further, smaller bronchi. See also BRONCHIOLES.

brown fat a kind of animal body fat more readily available for rapid conversion to heat than is normal yellow fat. It is believed that hibernating animals use their brown fat in the recovery from the winter state. Small human babies have deposits of brown fat around the spine.

Bruch's membrane an insulating membrane that separates the RETINA from the underlying CHOROID. In older people, Bruch's membrane is liable to develop splits or cracks through which blood vessels branching from the choroidal fine vessels may pass. This can lead to AGE-RELATED MACULAR DEGENERATION.

bruise the appearance caused by blood released into or under the skin, usually as a result of injury, but sometimes occurring spontaneously in case of bleeding disorders or disease of the blood vessels.

brutality a perennial characteristic of man, present from the earliest times and, although often latent in contemporary civilization, liable to reappear.

brush border the free surface of an intestine; epithelial cell that is covered by MICROVILLI. The name arises from the appearance on microscopy.

buccal relating to the cheek.

buccal smear a convenient way to obtain cells for chromosomal and other studies and to obtain a sample of DNA. The inside of the cheek is gently scraped with a spatula and the cells spread on a glass slide.

buck teeth undue protrusion of the central upper teeth. This can readily be put right by orthodontic treatment.

buffers chemical substances in the blood, such as combinations of a weak acid and a weak base, eg. lactic acid or bicarbonate, which act to limit changes in the composition, especially the acidity, by binding hydrogen ions. Buffers are important in maintaining constancy when acids are added or removed.

Buffon, Georges French nobleman (1707–88) who made the first major attempt at a systematized comparative anatomy. Buffon spent 53 years compiling a monumental 44-volume work *Natural History, General and*

Particular (1749–1804), in which he presented to an appreciative public a great deal of information about the relationships between animal and other species.

buffy coat the creamy layer of white blood cells that forms between the PLASMA and the column of packed red cells when anticoagulated blood is centrifuged.

buggery see SODOMY.

bulbo-urethral glands two small glands that open into the URETHRA just below the bladder in the male, and secrete a clear lubricant fluid that facilitates sexual intercourse. Also known as COWPER'S GLANDS.

bundle branch one of the two branches of conducting muscle tissue originating in the BUNDLE OF HIS and running down on each side of the septum between the ventricles near to its surfaces.

bundle of His the short bundle of specialized heart muscle fibres that conducts electrical impulses from the ATRIOVENTRICULAR NODE to the conducting bundle at the top of the interventricular septum. Also known as the atrioventricular bundle. (Wilhelm His, 1863–1934, German professor of medicine).

burp a belch.

bursa a small fibrous sac lined with a membrane which secretes a lubricating fluid (synovial membrane). Bursas are efficient protective and friction-reducing structures and occur around joints and in areas where tendons pass over bones.

bursa of Fabricius an organ in the cloacal-hind gut junction of birds. The justification for its entry is that early in the history of the elucidation of the cells of the immune system this organ was found to be important in the maturation of B lymphocytes (B for bursa). The equivalent in humans is probably the tonsils or lymphoid tissue in the intestine.

bursiform pouch-like.

Burt, Cyril Cyril Lodowic Burt (1833–1971) was one of Britain's most famous and respected psychologists. As a young man he met and was deeply influenced by Francis Galton whose ideas on the importance of heredity in determining individual ability he quickly assimilated and never abandoned. Burt's central interest was in intelligence and from the beginning of his career he was involved in the collection of data on intelligence and in formulating tests. In 1932 he was appointed professor of psychology at University College, London in succession to Charles Spearman, who had also taken a close interest in intelligence.

After his appointment Burt devoted himself mainly to writing, and produced a number of papers and books inspired by Galton's ideas on the central role of heredity. An increasing number of scientists, however were becoming critical of Burt's ideas. In response to these criticisms, Burt published a succession of papers, especially in 1956 and 1966, purporting to show, on the alleged evidence of comparative studies of a large number of twins reared together and apart, that intelligence was overwhelmingly innate and not the result of environmental factors. These papers had great influence on psychological and educational thought.

It was not until after his death in 1971 that it became virtually certain that much of the 'evidence' quoted in these papers had been fabricated and that his alleged collaborators, Miss Howard and Miss Conway, did not exist. Burt was the effective founder of educational psychology and a man of great achievement, who was knighted for his services to British psychology. But his almost religious adherence to the Galtonian tradition in the face of contrary evidence, and his inability to accept adverse criticism, indicate that he was not, at heart, a scientist.

buttock one of the twinned masses of powerful muscle at the base of the trunk, behind. Each buttock consists of three muscles – the gluteus maximus, gluteus medius and gluteus minimus. These muscles arise from the back of the bony pelvis and run into the upper end of the back of the thigh bone (femur) and act to straighten the flexed hip joint.

butyric pertaining to butter or derived from BUTYRIC ACID.

butyric acid a water-soluble saturated FATTY ACID occurring in animal milk fats. It has a strong rancid odour.

Cc

cachexia a state of severe muscle wasting and weakness occurring in the late stages of serious illnesses such as cancer. The usual condition of bodily decline in those dying after long debilitating illnesses. Cachexia is not due to malnutrition and research findings suggest that an important element in the causation may be selective depletion of the myosin heavy chain in myofibrillary proteins.

cadaver a corpse. The term may correctly be applied to any corpse, but tends to be confined to corpses used for anatomical dissection.

cadherins a family of calcium-dependent adhesion molecules found on the surface of cells. Cadherins are trans-membrane glycoproteins that mediate cell to cell junctions and contribute to the maintenance of cell shape. Cadherins differ in molecular weight at different locations, ranging from 120–135Kd.

caecum the large blind-ended pouch in the bowel at the beginning of the large intestine (COLON) in the lower right quadrant of the abdomen. The end of the ILEUM joins the caecum at the ileo-caecal valve. From the caecum protrudes the worm-like (vermiform) APPENDIX.

caeruloplasmin a copper-containing globulin formed in the liver.

Caesarian section an operation to deliver a baby through an incision in the abdomen, performed when natural delivery is impracticable or dangerous or urgency is necessary.

caffeine one of the most popular and widely used drugs of mild addiction. Caffeine is used, in the form of coffee, tea and Cola-flavoured drinks, by about half the population of the world. It elevates mood, controls drowsiness, decreases fatigue and increases capacity for work. Caffeine is incorporated in various drug formulations such as Cafergot and Migril for the treatment of MIGRAINE.

Cajal, Santiago Ramòn Y (1852–1934) see GOLGI, CAMILLO.

calcaneus the heel bone, or os calcis.

calcar a spur or spur-like projection from a bone or tendon.

calcareous chalky. Containing, or pertaining to, calcium or lime.

calcarine fissure a conspicuous groove in the visual area of the brain at the back (occipital cortex). Also called the calcarine sulcus.

calciferol vitamin D. A fat-soluble vitamin necessary for the absorption of calcium from the intestine. Deficiency causes RICKETS in infants and OSTEOMALACIA in adults.

calcification deposition of calcium salts, usually calcium hydroxyapatite crystals, in body tissues, especially when there has been prolonged inflammation or injury. Calcification is normal in bones and teeth.

calcitonin a hormone secreted by the THYROID gland, independently of the thyroid hormones, and concerned with the control of calcium levels in the blood. It acts on bone to interfere with the release of calcium. It is less important in calcium balance than the parathyroid glands.

calcium a mineral present in large quantity in the body, mainly in the form of calcium phosphate in the bones and the teeth. Electrically charged calcium atoms (ions) are present in the blood and body fluids and are essential for many physiological processes including cell membrane permeability, cell excitability, the initiation and transmission of electrical impulses, muscle contraction, cell shape and cell motility. Calcium is necessary for blood coagulation, the production of ATP, and enzyme actions. Calcium levels in the blood are kept with in narrow limits by feedback mechanisms.

calcium channels protein ducts in cell membranes that control the movement of calcium ions in and out of cells thereby affecting their function. Contraction of muscle cells is mediated by calcium ions.

calculus a stone of any kind formed abnormally in the body, mainly in the urinary system and the gall bladder. Calculi form in fluids in which high concentrations of chemical substances are dissolved. Their formation is encouraged by infection.

calculus (dental) see DENTAL CALCULUS.

caldesmons proteins that bind CALCIUM and CALMODULIN and are involved in cell adhesion, shape and motility.

calisthenics physical exercises to build up muscles and improve the efficiency of the heart and lungs.

calix 1 any cuplike structure.
2 one of the many divisions of the urine collecting structure of the pelvis of the kidney. The plural is calices.

callus 1 a collection of partly calcified tissue, formed in the blood clot around the site of a healing fracture. Callus is readily visible on X-ray and indicates that healing is under way.
2 a skin thickening (see CALLOSITY).

calmodulins small intracellular proteins that can bind to one to four calcium ions, producing complexes that can change the configuration of associated proteins such as enzymes. Calmodulins are second messengers. They react to changing calcium levels and are involved in such processes as cell proliferation, glucose metabolism and the release of neurotransmitters.

caloric pertaining to heat or calories.

calorie the amount of heat needed to raise 1 gram of water by 1°C. (Strictly, from 14.5°C to 15.5°C). For nutritional purposes the Calorie (or kilocalorie) is the amount of heat needed to raise 1000 grams of water by 1°C. The modern unit is the joule. One calorie is 4.2 joules.

calorific anything producing or able to produce heat, or pertaining to heat production.

calorimetry measurement of the energy value of foodstuffs or the energy expenditure of a person. Food is burnt in a special chamber called a bomb calorimeter and the heat rise measured. Human energy expenditure can be measured indirectly by assessing the amount of oxygen consumed.

calvarium the vault of the skull. The skull less the jaw and facial bones.

canaliculus a narrow channel in the body, such as a tear duct. The lacrimal canaliculi run from the inner corner of each eyelid to the LACRIMAL SAC.

canalization forming a channel. A blood vessel blocked by blood clot may, in time, recanalize.

canal of Schlemm a fine, circular, sometimes multi-channel passage, running round the periphery of the CORNEA (limbus) in the SCLERA. Aqueous humour from the eye drains out through the TRABECULAR MESHWORK into the canal of Schlemm. (Friedrich. S. Schlemm, 1795–1858, German anatomist).

cancellous of a spongy, porous, lattice-like structure. A term applied to the inner parts of bone.

cancer a disease of DNA. The term is used by the medical profession as a convenient and comprehensive label for all forms of malignant growths. There are two broad classes of cancers – those which arise from surface linings (carcinomas) and those which arise from solid tissues (sarcomas). Cancers spread by local invasion and by lymph and blood spread (metastases) and their degree of malignancy is a measure of the rapidity with which they spread.

canine tooth one of the two pairs of pointed teeth on either side of the two central pairs of incisor teeth. Canines are tearing teeth; incisors are cutting teeth.

cannabis a drug derived from the hemp plant. Marijuana is the dried leaves, flowers or stems of various species of the hemp grass *Cannabis*, especially *Cannabis sativa*, *Cannabis indica* and *Cannabis americana*. Cannabis resin contains the cannabinoid tetrahydrocannabinol which produces euphoria (the easy promotion of silly laughter or giggling) and an apparent heightening of all the senses, especially vision, with distortion of dimensions.

cannabism any toxic or other undesirable effect of excessive or habitual use of CANNABIS.

canthus the corner the eye where the upper and lower eyelids meet. In epicanthus, the upper lid margin curves over to conceal the canthus.

capacitiation changes that occur in the heads of spermatozoa when they are in the uterus or fallopian tube that lead to the release of enzymes that allow penetration of the egg. An oestrogen environment promotes capitation; progestins of the luteal phase of the menstrual cycle oppose it.

cap-binding protein a protein that binds to the ends of messenger RNA molecules and helps to speed up protein synthesis on RIBOSOMES.

capillary the smallest and most numerous of all the blood vessels. Capillaries form dense networks between the arteries and the veins, and it is only in the capillary beds that interchange of oxygen, carbon dioxide and nutrients can take place with the cells.

capitellum, capitulum any small rounded prominence or ending on a bone.

capsule any outer covering such as the tough, protective outer coat of solid organs including the kidneys, liver and spleen or the delicate outer membrane of the internal crystalline lens of the eye. Joints, too, have capsules which contribute to their stability and function.

caput 1 a head.
2 an abbreviation for CAPUT SUCCEDANIUM.

caput succedaneum a boggy (oedematous) swelling of a baby's scalp seen for a period after birth and caused by sustained scalp pressure against the edges of the widened (dilated) cervix. The caput disappears within hours or days.

carbohydrates

As the name implies, carbohydrates, are made from carbon and the constituents of water – hydrogen and oxygen. Carbohydrates are the main structural elements in plants, but in humans their main importance is as a source of energy – a fuel. Carbohydrates include sugars, starches and celluloses and are structurally classified into three groups – monosaccharides, disaccharides and polysaccharides. Starches and celluloses are polysaccharides. Carbohydrates do have minor, but important, structural functions in the body, especially the deoxyribose sugar molecules that link with phosphates to form the backbone of the double helix of DNA. They also link with protein to form substances, such as mucopolysaccharides, which have a structural function in cartilage and elsewhere.

The principle fuel of the body – glucose – is a carbohydrate, derived mainly from food. Glucose is a simple sugar, or monosaccharide, and can be derived from the breakdown, by digestion, of more complex carbohydrates such as the disaccharide table sugar (sucrose) or the polysaccharide starch. Polymerization of simple sugars to form polysaccharides is analogous to the linkage of amino acids to form polypeptides and proteins. Glucose is stored in the liver in the compact form of the polysaccharide glycogen. This consists of a long, branching chain of glucose molecules linked together. When glucose is needed for fuel, glycogen is broken down to release glucose molecules into the blood.

carbon the non-metallic element on which all organic chemistry is based and which is thus present in all organic matter. A carbon atom is capable of combining with up to four other atoms (tetravalent), including other carbon atoms; it is this property that allows so many compounds to be formed.

carbon cycle the important biological cycle in which carbon in carbon dioxide in the atmosphere is taken up by plants, incorporated, by photosynthesis, into carbohydrates which are eaten by animals, and the carbon then oxidized and finally returned to the atmosphere as carbon dioxide waste gas.

carbon dating a method of estimating the age of specimens of biological origin. Cosmic rays convert a small proportion of atmospheric nitrogen atoms into carbon-14 which is radioactive. These carbons form a small proportion of atmospheric carbon dioxide and are taken up by trees and plants during photosynthesis. Photosynthesis stops when trees are cut down and plants harvested for food. So human food, and thus human bone, contains some carbon-14, which has a half life of 5730 years and decays to stable carbon-12. The ratio of carbon-14 to carbon-12 in the specimen can easily be measured and, after calibration against known historic changes in atmospheric C-14, provides a fairly accurate estimate of the lapse of time since the food was eaten. Results are consistent up to about 40,000 years.

carbon dioxide a compound in which an atom of carbon is linked to two atoms of oxygen (CO_2). Carbon dioxide is a colourless, odourless gas and is one of the chief waste products of tissue metabolism.

carbonic anhydrases ENZYMEs that reversibly catalyse the breakdown of carbonic acid into carbon dioxide and water, thus allowing the easy transfer of carbon dioxide from the tissues to the atmosphere via the blood and the lungs. In the stomach a carbonic anhydrase frees hydrogen ions for the formation of hydrochloric acid; and in the pancreas they produce bicarbonate to neutralize duodenal acid from the stomach. See also CHLORIDE SHIFT.

carbon monoxide a simple, but poisonous, compound consisting of an atom of carbon linked to an atom of oxygen (CO). It is formed when carbon is oxidized in conditions of limited oxygen, as in the internal combustion engine.

carboxyl group the –COOH termination of an AMINO ACID to which the amino group of another amino acid attaches with the loss of a molecule of water when the two form a peptide bond.

carbopeptidase an enzyme present in pancreatic juice that splits off amino acids one by one from the carboxyl (C-terminal) end of a protein chain in the final stage of protein digestion. See also AMINOPEPTIDASES.

card-, cardio- *combining form denoting* heart.

cardia the opening of the lower end of the gullet (OESOPHAGUS) into the STOMACH.

cardiac 1 pertaining to the heart. 2 pertaining to the CARDIA.

cardiac achalasia failure of food to pass normally from the gullet into the stomach, without organic obstruction.

cardiac arrest complete cessation of the normal heart contractions so that the pumping of the blood stops. In about 30 per cent of cases the heart muscle is motionless (asystolic); in the remainder it is in a state of quivering, ineffectual fluttering (ventricular fibrillation). Asystolic cardiac arrest is treated by intravenous injection of VASOPRESSIN alone or followed by ADRENALINE; ventricular fibrillation is treated by electrical defibrillation.

cardiac bypass use of a temporary mechanical pump to maintain the blood circulation during a heart operation or to assist the heart during a period of recovery from heart disease or surgery.

cardiac massage see CARDIOPULMONARY RESUSCITATION.

cardiac output the volume of blood pumped by the heart in one minute. It is equal to the stroke volume – the output per beat – multiplied by the number of beats per minute.

cardiopulmonary pertaining to the heart and the lungs.

cardiopulmonary resuscitation combined heart compression (cardiac massage) and 'kiss of life' (mouth-to-mouth artificial respiration). Commonly abbreviated to CPR. New guidelines, issued by the International Liaison Committee on Resuscitation in August 2000, recommend that lay people should no longer check for a pulse before starting chest compressions. If no signs such as breathing, coughing, spontaneous motions, or movements in response to stimulation are present chest compressions should begin. If only one helper is present, chest compression alone, without mouth-to-mouth artificial respiration, is acceptable. Chest compressions generate enough force to clear most obstructions and therefore rescuers should now begin CPR immediately and let chest compressions clear the airway.

cardiorenal pertaining to the heart and the kidneys.

cardiorespiratory pertaining to the heart and the respiratory system.

cardiothoracic pertaining to the heart and the chest cavity and its contents.

cardiovascular relating to the heart and its connected closed circulatory system of blood vessels (arteries, arterioles, capillaries, venules and veins).

caries tooth decay. The term also refers to bone decay but is seldom used in this sense.

carina any keel-shaped ridge in the body, especially the ridge formed where the TRACHEA divides into the two bronchi.

carnal knowledge sexual intercourse.

carotene one of a group of orange pigments found in carrots and some other vegetables. Beta-carotene (provitamin A) is converted to vitamin A in the liver. This vitamin is needed for normal growth and development of bone and skin, for the development of the fetus and for the proper functioning of the RETINA.

carotenoids a large group of yellow or orange pigments occurring in plants some of which have antioxidant properties. Some of the carotenoids are carotenes.

carotid artery one of the paired arteries running up on either side of the front of the neck which, together with the two vertebral arteries, provide the whole blood supply to the head, including the brain.

carotid body a chemical receptor, situated at the first branch of each carotid artery, that monitors oxygen levels in the blood and regulates the rate of breathing accordingly.

carotid sinus a small widening of the wall of each CAROTID ARTERY, at the first branch (bifurcation), that contains pressure-sensitive nerve endings to monitor blood pressure and provide feedback data for its control.

carpal pertaining to the wrist or wrist bones (carpals).

carpal tunnel a restricted space on the front of the wrist, bounded by ligaments, through which pass the tendons which flex the fingers and wrist and one of the two sensory nerves to the hand, the median nerve.

carpometacarpal pertaining to the wrist bones (carpals) and the bones of the palms of the hands (metacarpals), especially to the joints betweem them.

carpopedal pertaining to the hands and feet.

carpus the wrist or the bones of the wrist.

carrier any molecule that, by attaching itself to a non-immunogenic molecule, can provide epitopes for helper T cells thus making the second molecule immunogenic.

Cartesian relating to the philosophy, methods or coordinates of (Des)cartes, who proposed the notion of a mind–body dualism ('ghost in the machine') which has haunted medical thought ever since, but which is now beginning to be rejected by many of those with enough interest to consider the matter. (René Descartes, 1596–1650, French mathematician and philosopher).

cartilage gristle. A dense form of connective tissue performing various functions in the body such as providing bearing surfaces in the joints, flexible linkages for the ribs, and a supportive tissue in which bone may be formed during growth.

cartilaginous of, pertaining to, or formed in, CARTILAGE.

cascade a physiological system in which the completion of one event has an outcome that initiates the next successive event. Blood coagulation, for instance, is a cascade involving more than a dozen successive events. In genetics a cascade system controls the order in which genes are expressed.

casein a protein derived from CASEINOGEN in milk by the action of renin in the stomach.

caseinogen the principal protein in milk, present in higher proportion in cow's milk than in human milk.

caspases a family of proteases, some of which are involved in APOPTOSIS.

castration the removal of the testicles (orchidectomy or orchiectomy), or, sometimes, of all the male external genitalia. The term is also occasionally used to refer to the removal of the ovaries in women. Castration can be valuable in the treatment of androgen dependent cancer of the prostate gland, even if widespread.

catabolism the breakdown of complex body molecules to simpler forms, often with the release of energy, as when muscle protein breaks down to amino acids or fats to glycerol and fatty acids. The opposite process is called anabolism and both processes are encompassed in METABOLISM. Compare ANABOLISM.

catalase an ENZYME found in the microbodies (peroxisomes) of cells and that promotes the reaction in which two molecules of hydrogen peroxide are converted to two molecules of water and one molecule of oxygen.

catalyst a chemical substance that promotes or accelerates a chemical reaction without itself being changed. Most biochemical catalysts are ENZYMES and almost all body chemistry depends on the catalytic action of thousands of different enzymes.

cataract opacification of the internal focusing lens of the eye (the crystalline lens) due to irreversible structural changes in the orderly arrangement of the fibres from which the lens is made as a result of aggregation of crystallin protein in the lens.

catecholamines the group of AMINES, which includes adrenaline, noradrenaline, dopamine and chemically related amines. These are derived from the amino acid tyrosine, and act as neurotransmitters or hormones.

categorical variables qualitative variables, variables that cannot meaningfully be expressed in numbers. eg. skin colour.

catharsis 1 purging of the bowels.
2 a psychoanalytic term meaning the release of anxiety and tension experienced when repressed matter, which has been 'poisoning' the mind, is brought into consciousness.

cathepsin any ENZYME that acts to split the interior PEPTIDE bonds of a protein, causing its decomposition.

cathepsis protein HYDROLYSIS by CATHEPSINS.

cation a positively charged ion, such as Na^+ (sodium) or K^+ (potassium), which is attracted towards a negative electrode (cathode). An atom which, in solution, takes a positive charge. Unlike charges attract; like charges repel.

caucasian a term commonly used to refer to a person who is neither black, brown, yellow or red. It is based on the anthropologically disreputable theory that humankind can be divided into five ethnic classes – Caucasians, Negroes, Mongols, Malaysians and Americans.

cauda equina the leash of spinal nerves hanging down in the spinal canal below the termination of the SPINAL CORD, at about the level of the first lumbar vertebra.

caudal pertaining to the tail end of the body. Denoting a tailward direction in anatomy. Although not externally visible, the human tail still exists, in a vestigial form, as the COCCYX.

caudate possessing a tail.

caudate nucleus a large collection of grey matter (nerve cell bodies) lying deep in the white matter of the lower part of the CEREBRUM on either side of the midline, and concerned with the control of movement. One of the basal ganglia of the brain.

caul a persistent AMNION membrane covering the baby's head at birth. Normally, the amnion ruptures before birth allowing the baby to pass through.

caval pertaining to the great veins, the VENA CAVAE.

cavernous sinus the large vein channel, lying immediately behind each eye socket (orbit) and on each side of the PITUITARY GLAND. Through it run a loop of the internal carotid artery, the nerve supplying the central part of the face with sensation, and several nerves supplying the muscles that move the eye. The sinuses receive veins from the face and orbits and communicate with each other and with other large veins surrounding the brain.

cavity, dental an area of tooth enamel destroyed by acids formed by the action of mouth bacteria on carbohydrate food particles accumulating around unbrushed teeth.

CCN proteins a family of six cellular signalling regulator proteins with a wide range of functions. They are concerned in cell proliferation and differentiation, wound healing and the production of bone, cartilage and blood vessels.

CD *abbrev. for* cluster of differentiation (or designation). This is a standard descriptive code applied to antigenic surface molecules on white cells, especially T LYMPHOCYTES, that are identified by a particular group of monoclonal antibodies. Well over 100 CD receptors have been described.

CD1 antigen a cell membrane complex found on THYMUS cells, on some peripheral B cells and on dendritic cells in the skin. They are coded for by five genes on chromosone 1 and are structurally similar to the MAJOR HISTOCOMPATIBILITY COMPLEX.

CD2 antigen a complex found on some bone marrow stem cells and on pre-thymocytes. The genes are on chromosome 1.

CD3 antigen a cell membrane complex of five polypeptide chains that is the signal transduction element of the T cell receptor on helper and cytotoxic T cells. Transduction occurs on interaction with antigen.

CD4 antigen a glycoprotein cell surface molecule on helper T cells (T LYMPHOCYTES) that recognizes major histocompatibility complex (MHC) class II molecules on antigen-presenting cells. Because of this, helper T cells are often referred to as CD4 cells.

CD5 antigen a membrane glycoprotein found on all human T cells, but on greater quantity on helper T cells. Found also on peritoneal and pleural B cells.

CD8 antigen a glycoprotein cell surface molecule on cytotoxic T cells (T LYMPHOCYTES) that recognizes major histocompatibility complex (MHC) class I molecules on target cells. Because of this, cytotoxic T cells are often referred to as CD8 cells.

CD14 antigen a lipopolysaccharide-binding protein found on the cell membranes of macrophages, monocytes, dendritic cells and other cells of the germinal follicles of the spleen. CD14 antigens are thought to be essential for the differentiation of the cells of the monocyte and myelocyte series. The genes are on chromosome 5.

celibacy the state of being unmarried and of avoiding sexual intercourse. Celibacy is mainly associated with the professional clergy of various religions. Roman Catholic priests, for instance, are forbidden marriage by canon law. Celibacy should be distinguished from chastity.

cells

Cells are immensely complex bodies engaged in constant physical and biochemical activity, but each one represents the simplest structural and functional unit into which an organism, such as the human being, can be divided. Cells vary greatly in size, from less than a thousandth of a millimetre across in the case of small micro-organisms, to several centimetres in the case of large avian and reptilian eggs. Human body cells lie in the range between about one hundredth of a millimetre to about a tenth of a millimetre. The largest human cell is the ovum. So almost all of them are of microscopic dimensions.

The outer cell wall – the cell membrane – is far more than simply a bag for the cell contents. It is a complex fat and protein structure that contains many specialized sites for the receipt of information from the external environment and from adjacent cells, and controllable protein channels (ports) through which ions and molecules can pass into and out of the cell.

cell cycle

When a dead tissue is cut into a thin slice and mounted on a microscope slide and stained, the part of the cell which appears central and densely stained is called the nucleus. The term comes from the Latin *nux*, meaning 'a nut'. The nucleus contains the chromatin (uncoiled chromosomes). The chromatin is, approximately, a two-metre-long delicate strand of DNA that contains the genes – unique sequence of bases that specify the proteins (mainly enzymes) that control our physical construction. The nucleus is enclosed in a nuclear envelope containing pores through which information from the DNA can be carried by messenger RNA. Surrounding the nucleus, within the cell, is the cytoplasm. This contains many important structures suspended in a fluid called the cytosol. These structures are called cell organelles (miniature organs) and include the mitochondria, the endoplasmic reticulum, the ribosomes, the golgi apparatus, the lysosomes and the filamentary cell skeleton (cytoskeleton).

cell cycle the sequence of events from the start of a cell division to the start of the division of a resulting daughter cell.

cell differentiation the process by which primitive dividing cells, all of which originally appear identical, alter and diversify to form different tissues and organs. See also HOMEOBOX GENES.

cell-free protein synthesis the artificial production of proteins by the addition of MESSENGER RNA to a medium containing enzymes, amino acids, ribosomes, transfer RNA and cofactors.

cell line the CLONE or clones of cells derived from a small piece of tissue grown in culture.

cell-mediated immunity action by the immune system involving T cells (T LYMPHOCYTES) and concerned with protection against viruses, fungi, tuberculosis and cancers and rejection of foreign grafted material. Cell-mediated immunity is not primarily effected by ANTIBODIES.

cellular pertaining to a CELL.

cellular respiration chemical processes within a cell that release energy required for cell meta-bolism and other activities. See KREBS CYCLE.

cellulite a lay term for the fatty deposits, with connective tissue strand dimpling, around the thighs and buttocks.

cellulose a complex polysaccharide forming the structural elements in plants and forming 'roughage' in many vegetable foodstuffs. Cellulose cannot be digested to simpler sugars and remains in the intestine.

cell organelles

The little organs of the cell are all essential for life and each kind serves a separate purpose. They were discovered and elucidated in various ways – by the use of the electron microscopy, by separation by powerful gravitational forces using the ultracentrifuge, by chemical analysis and by biochemical experiments to determine their function.

Mitochondria

These are tiny oval or rod-shaped bags each with a smooth outer membrane and a much-folded inner membrane. The cell cytoplasm contains hundreds of mitochondria, the number depending on the amount of energy required by the cell. Highly active cells, such as those in the liver, may contain as many as 1000. Mitochondria contain enzymes that accelerate the chemical processes by which energy is released from the combination of glucose and oxygen, with the production of carbon dioxide. This process is mediated by the chemical substance *adenosine triphosphate* (ATP). Mitochondria also contain a circular genome of DNA which has been completely sequenced. Mitochondrial DNA contains genes that are connected with muscle tissue and defects in these genes can cause certain forms of muscular dystrophy. The genome is unique in that it is inherited from the mother only. This fact has been of great interest to archaeologists and evolutionists.

Endoplasmic reticulum

This is the largest organelle and extends throughout most of the cell cytoplasm. It takes two forms – a series of flattened sacs covered with tiny granules and a network of smooth, non-granular, tubules of different diameter. These are called, respectively, the rough (granular) and the smooth (agranular) reticula. So far as can be judged from electron microscopic studies, the interiors of the two types are continuous with each other. The granules on the rough reticulum are much more important than their tiny size might suggest; they are the vital ribosomes (see below). The smooth reticulum has a number of functions, the most important being the synthesis of various fatty acids and phospholipids needed to form the cell membrane. In liver cells the smooth reticulum contains many enzymes that detoxify dangerous chemicals, converting them into safer conjugated and soluble forms that can be excreted from the body.

Ribosomes

As well as studding the endoplasmic reticulum, the ribosomes also exist as free bodies in the cell cytoplasm. Ribosomes are protein bodies containing nucleic acids. It is in the ribosomes that amino acids, free in solution in the cytosol, are selected in the right order, in accordance with the code carried by messenger RNA from the DNA, and are strung together to form new proteins, mostly enzymes. New proteins synthesized in the ribosomes attached to the endoplasmic reticulum, pass into the interior of the reticulum and are delivered to the Golgi apparatus; those formed by the free ribosomes are released into the cytosol ready to act immediately.

Golgi apparatus and vesicles

Several minutes after they are synthesized, proteins formed in the rough endoplasmic reticulum are transferred into another group of organelles known as the Golgi apparatus, named after the Italian histologist Camillo Golgi (1843–1926). The Golgi complexes are situated near the cell nucleus and consist of a series of flattened membranous sacs surrounded by a number of spherical bubbles or vesicles. These vesicles form initially on the surface of the rough reticulum in areas not coated with ribosomes. Proteins within the rough endoplasmic reticulum pass into these vesicles which then travel though the cytosol and fuse on to the surface of the Golgi complex, transferring their contents into the Golgi sacs. Secondary 'transfer' vesicles are now formed on the surface of the Golgi sacs and these are 'tagged' with a carbohydrate or phosphate group to indicate where they should go. Golgi vesicles have been called the 'traffic police' of the cell as they play a key role in directing the many proteins formed in the cell to their required destination.

Lysosomes

These are tiny oval or spherical bodies surrounded by a single membrane enclosing digestive enzymes and acids. A typical cell may contain several hundred lysosomes. Their function is to scavenge and remove cell debris including pieces of DNA and RNA and organelles that are damaged and no longer functioning. They do this by engulfing the material, breaking it down by means of the digestive enzymes – which work only in the highly acid environment of the inside of the lysosomes – and carrying it to the outer cell membrane where it is ejected from the cell. Some cells, such as the macrophages, are equipped to take in large particles of unwanted material such as bacteria and foreign protein. They do this by forming vesicles which then fuse with the lysosomes and transfer their contents into the lysosomes where they are digested and destroyed.

Cytoskeleton

Most cells contain a protein filamentous structure, known as the *cytoskeleton*, that maintains the cell shape and allows cell movement. The filaments are of different diameters and are solid or tubular. All cells contain fine filaments made of the contractile protein *actin*. The smallest filaments – the actin microfilaments – can be assembled and taken down rapidly, allowing the cell to change shape as required. The larger filaments are more permanent. Muscle cells are largely composed of the contractile protein myosin.

cell reproduction (division)

Mitosis, or normal cell division, results in two daughter cells genetically identical to the cell from which they are derived. In mitosis, each of the threads of chromatin – the DNA strands – is exactly duplicated and the sets of pairs of new threads separate to form the chromosomes of the daughter cells. This separation involves a temporary reorganization of the cell and much mechanical movement of chromosomes.

The first stage in cell division is the longitudinal replication of DNA. Following this division, which produces pairs of new chromatin threads from the original chromatin, the new paired threads coil up to form pairs of new chromosomes. At first, these are stuck together at the centromere to produce the characteristic X-shaped double or replicated chromosomes seen in the karyotype. At this stage, the replicated, X-shaped double chromosome is called a chromatid. These double chromosomes now form up in a rough circle around the equator of the cell, and a delicate temporary structure of fine strands or fibres of protein, radiating to the equator from each end of the cell, appears. This is called the spindle apparatus and its function is to pull apart the two chromosome rods that form the arms of the X of each double chromosome. As the strands of the spindle contract and apply traction, the two halves of the replicated chromosome are pulled apart, separating at the centromere, and are dragged to opposite ends of the cell.

In this way each end of the cell contains 46 single chromosomes. The chromosomes now uncoil and a nuclear membrane forms around the bundle of chromatin at each end of the cell. The spindle disappears. Now the centre of the cell begins to narrow until a dumb-bell shape is formed, and soon the cell divides into two daughter cells, each containing the full complement of 46 chromosomes in their extended form as chromatin. Barring accidents, the chromosomes in each of the daughter cells are identical in every way to the chromosomes in the mother cell. This remarkable process of duplication, with conservation of the genetic material, is going on in millions of cells in all our bodies, all the time. The spindle, or mitotic, apparatus exists only during mitosis.

A different form of cell division is necessary when the germ cells in the ovaries or testicles divide to form ova and sperm. This kind of division is called meiosis and, in addition to producing the half (haploid) number of chromosomes in the ova and sperm, it has another important feature – a remarkable process that ensures that when humans pass on to their offspring the characteristics inherited from their parents, their offspring (other than identical twins) will differ considerably from each other.

The first stage of meiosis is similar to that of mitosis. The chromatin strands, each representing a single chromosome, replicate longitudinally and the resulting pairs of daughter strands coil up and stick together to form X-shaped double chromosomes (chromatids). These now congregate in pairs with their original partners. The duplicated number one chromosome that came originally from the father aligns itself with the duplicated number one maternal chromosome, and so on for all 23 pairs. The partners in each pair of double chromosomes now twist intimately together and become closely aligned along their entire length. While in this relationship they exchange several short corresponding segments with each other. This occurs in a random manner so that the genetic material from the father and the mother becomes thoroughly mixed and new chromosomes are formed, each of which has a unique blend of the genes from both parents. This is called crossing-over.

The cell now divides, but in this division the X-shaped chromatids do not have their arms pulled apart as in mitosis. Instead, one of each of the pairs of the intact double chromosomes goes to each daughter cell. It is a matter of pure chance which of the pairs goes to which daughter cell. As a result, the randomness of the redistribution of genes already affected by crossing-over is further increased. Each daughter cell now has 23 chromatids. A second division occurs but this time the halves of the chromatids are pulled apart by a spindle, as in mitosis, and each daughter cell receives a single chromosome from each pair – a total of 23. A third division, producing a total of four sperm or ova from the original parent cell, does not alter the fact that each germ cell now has half the normal number of chromosomes – the haploid number.

Celsius scale a temperature scale in which the freezing point is 0°C and the boiling point 100°C. This corresponds to the old Centigrade scale and is a sensible alternative to the arbitrary Fahrenheit scale which had 32°F as freezing point and 212°F as boiling point. So the normal body temperature of 98.4°F has become 37°C. (Anders Celsius, 1701–44, Swedish astronomer).

cementoblast a cell that actively forms CEMENTUM.

cementocyte a cell found in the tiny spaces (lacunae) in the CEMENTUM and having processes radiating into the cementum CANALICULI.

cementum the layer of calcified substance covering a tooth below the gum line.

censor a Freudian idea for the supposed agency that distorts or symbolizes repressed unpleasant material in the unconscious so that it need not be directly recognized either in dreams or in waking awareness.

centigrade this is identical, in every respect, to the CELSIUS SCALE. Centigrade literally means '100 levels' and this is what the Celsius scale contains. The change to Celsius was purely for honorific reasons.

centiles see PERCENTILE.

centimorgan a unit of distance between two genes on a chromosome, representing a 1 per cent probability of recombination in a single meiotic event. Named after Thomas Hunt Morgan (1866–1945) the American Nobel-prizewinning geneticist often described as the father of modern genetics.

central canal the elongated tubular space within the spinal cord continuous with ventricles of the brain. The central canal is the continuation of the fourth ventricle and contains cerebrospinal fluid.

central dogma the proposition by Francis Crick (1916–2004) that, in genetics, the only possible progression was from DNA to RNA to protein. Embarrassingly, the discovery that retroviruses used RNA to make DNA demonstrated the riskiness of pronouncing dogmas in science.

central nervous system (CNS) the brain and its downward continuation, the spinal cord, which lies in the spinal canal within the spine (vertebral column). The central nervous system is entirely encased in bone and is contrasted with the peripheral nervous system, which consists of the 12 pairs of cranial nerves arising directly from the brain, the 31 pairs of spinal nerves running out of the spinal cord, and the AUTONOMIC NERVOUS SYSTEM.

central sulcus a conspicuous, almost vertical, groove on both sides of the brain that separates the frontal lobe from the parietal lobe and the motor area of the cortex from the sensory area. Also known as the fissure of Rolando. (Luigi Rolando, 1773–1831, Italian anatomist).

central venous pressure the pressure of blood in the right atrium. This is measured by an in-dwelling catheter carrying a pressure transducer. Central venous pressure readings provide valuable diagnostic information in a range of serious heart and lung conditions.

centrifuge a laboratory machine that subjects matter suspended in solution to powerful outward-tending forces by high-speed rotation. This allows particles of different mass to be separated into bands.

centrilobular situated at the centre of a LOBULE of an organ, such as the liver.

centriole a short, hollow, cylindrical ORGANELLE consisting of nine sets of microtubules and usually occurring in pairs set at right angles to each other. Centrioles are responsible for the production of the spindle apparatus that appears just before the separation of the chromosomes into two sets prior to cell division.

centromere the constriction in a chromosome at which the two identical halves (chromatids) of the newly longitudinally-divided chromosome are joined, and at which the chromosome attaches to the spindle fibre during division (mitosis). The centromere contains no genes.

centrosome a small mass of CYTOPLASM, lying near the nucleus of a cell and consisting of a pair of centrioles, which divides into two parts before cell division. These migrate to the poles of the cell and the spindle develops between them.

cephal-, caphalo- *combining form denoting* head.

cephalad situated towards the head.

cephalic relating to the head or in the direction of the head.

cephalometry measurement of the various diameters and other dimensions of the head.

cerebellum the smaller sub-brain lying below and behind the CEREBRUM. The cerebellum has long been thought to be concerned only with the coordination of information concerned with posture, balance and fine voluntary movement. Recent studies have shown, however, that the cerebellum functions to assist in many cognitive and perceptual processes. The cerebellum may also have a role to play in coordinating sensory input, and even in memory, attention and emotion.

cerebral 1 pertaining to the CEREBRUM or brain.

2 having the quality of intellectualism.

cerebral cortex the grey outer layer of the cerebral hemispheres, consisting of the layered masses of nerve cell bodies which perform the higher neurological functions.

cerebral hemispheres the two halves of the CEREBRUM joined by the CORPUS CALLOSUM.

cerebration thinking.

cerebrosides glucolipids present in nerve tissue, especially in myelin sheaths. Their metabolism requires HYDROLASES and it is inherited mutations of the genes for these enzymes that causes serious conditions such as Tay-Sachs disease.

cerebrospinal pertaining to both the brain and the spinal cord.

cerebrospinal fluid the watery fluid that bathes the brain and spinal cord and also circulates within the ventricles of the brain and the central canal of the cord.

cerebrovascular pertaining to the blood vessels supplying the brain.

cerebrum the largest, and most highly developed, part of the brain. It contains the neural structures for memory and personality, cerebration, volition, speech, vision, hearing, voluntary movement, all bodily sensation, smell, taste and other functions.

cerumen ear wax.

ceruminous glands modified eccrine sweat glands in the external ear canal that secrete ear wax.

cervical pertaining to a neck. This may be the neck of the body or the neck of an organ such as the womb. The noun, from the adjective 'cervical' is 'cervix'.

cervical rib a short, floating, rudimentary rib attached to the lowest neck vertebra on one or both sides. In about 10 per cent of cases the rib causes compression of arteries or nerves in the neck, leading to pain and tingling, or sometimes more serious effects, in the arm or hand.

cervix the neck of the womb (UTERUS).

CFCs see CHLOROFLUOROCARBONS.

chalones POLYPEPTIDEs or glycoproteins, released by actively-dividing cells that inhibits chromosomal reproduction (MITOSIS) in cells of the tissue in which they are formed, thus controlling a tendency to HYPERPLASIA.

change of life see MENOPAUSE.

chaos theory the mathematical conception that some phenomena that seem random may be of a deterministic order highly sensitive to initial conditions and perturbations. There is a growing appreciation that chaos may be a feature of many biological systems and that chaos theory may prove to have many applications in medicine.

chaperones chaperonins, intracellular proteins that assist in the correct folding of other proteins by means of hydrophobic surfaces that recognize and bind to exposed hydrophobic surfaces on misfolded proteins. Intracellular proteins with these abnormal configurations interact abnormally with other molecules, and many disease processes relate to misfolded proteins. Misfolding of proteins that are regulators of growth and differentiation, for instance, may be an important factor in the causation of cancers. Protein misfolding may be a factor in as many as half of all diseases. Increasing understanding of this phenomenon may lead to new and important therapies.

chastity abstention, usually by choice, from sexual activity, generally from a conviction that the state of virginity possesses or confers merit or constitutes a sacrifice to God. Chastity is voluntarily undertaken by monks and nuns.

chauvinism an unreasonable and offensive degree of expression of partisanship, patriotic sentiment or jingoism. The term derived from the name of a simple-minded French soldier Nicolas Chauvin who was loud in his expression of satisfaction for all things Napoleonic. Later the sense changed to denote undue partiality to any place or social group, and it is now sometimes narrowed to indicate a sense of male gender superiority.

cheekbone the zygomatic bone.

cheil- *combining form denoting* the lip.

cheilion the angle of the mouth.

cheiro- *combining form denoting* the hand.

chemical elements of the body

Over 98 per cent of the total body weight consists of the elements hydrogen, oxygen, nitrogen, carbon, calcium and phosphorus. Other elements, although present in very small quantities, are nevertheless essential. These include potassium (0.25 %), sulphur (0.2 %), sodium and chlorine (each 0.1 %) and magnesium (0.05 %). Other necessary elements are iron, copper, manganese, iodine, zinc and selenium. Water accounts for about 60 per cent of the body weight.

The 20 amino acids from which all body proteins are constructed are: alanine, isoleucine, leucine, methionine, phenylalanine, proline, tryptophan, valine, asparagine, cysteine, glutamine, glycine, serine, threonine, tyrosine, arginine, histidine, lysine, aspartic acid and glutamic acid.

DNA consists of a double helical backbone of a polymerized sugar-phosphate compound linked by pairs of bases. These bases are: adenine, thymine, guanine and cytosine. Adenine always links to thymine and guanine always links to cytosine. The order of these base pairs along the molecule is the genetic code. In RNA, thymine is replaced by a different base, uracil.

chemical messengers hormones.

chemo- *combining form denoting* chemistry or chemical. The term is also used informally as an abbreviation of CHEMOTHERAPY.

chemokines a family of structurally-related protein CYTOKINES that can induce activation and migration of specific types of white cell by chemotaxis. They have a fundamental role in inflammation and are concerned in the immune system protective responses to infecting organisms. Chemokines are also concerned in ANGIOGENESIS.

chemoreceptors nerve endings stimulated to produce nerve impulses by contact with chemical substances or as a result of changes in the local concentration of chemical substances. The receptors for smell and taste, those for oxygen in the carotid bodies and those for glucose in the pancreas are of this type.

chemotactic pertaining to CHEMOTAXIS.

chemotaxins agents that promote CHEMOTAXIS.

chemotaxis the movement of a cell or other living organism in a particular direction as a result of attraction by an increasing concentration of a chemical substance. Cells of the immune system find their prey by this means.

Cheyne-Stokes respiration periods of very shallow, almost imperceptible, breathing alternating with periods of deep breathing. This sequence often precedes death (John Cheyne, 1777–1836, Scottish physician; and William Stokes, 1804–78, Irish physician).

chiasma **1** the intersection and partial crossing of the optic nerves behind the eyes within the skull. The fibres on the outer halves of each optic nerve do not cross over; those on the inner halves of each nerve do. Also known as the optic chiasm.
2 the site at which a pair of homologous chromosomes exchange material during MEIOSIS.

child development

The first few years of life are critically important. The things that happen then to the developing infant and child colour and condition all the rest of its life. Early impressions go deep and have a life-long effect. Indeed, some authorities believe that early environmental influences play a major part in determining the whole personality. The consensus view, however, is that the outcome is the result of the interaction of environmental forces on the genetically determined organism. But there is no denying that major events, especially if traumatic, do have a life-long effect.

Growth and development are as much a matter of interior change as of a simple increase in the size of the body. The most important internal changes are in the brain and nervous system – changes which make it possible for the young person to register and store information, to acquire social skills and to mature emotionally. Above all, maturation of the nervous system allow increasing interaction with the environment and this means obtaining information. Information input is essential for brain development. Unless the brain receives normal input from all the sense organs, it will inevitably be retarded. Sensory deprivation is as damaging as deprivation of food. Well-fed babies deprived of human relationships and contacts soon fall behind their more fortunate contemporaries both in mind and in body.

A baby's head is about three quarters of the adult size and a quarter of its body length because most of the growth of the brain occurs before birth. An adult's head is only about one eighth of the body length. Brain growth and development continues at a rapid rate. By the end of the first year of life half of the life-time growth of the brain has been completed. Thereafter, most of the growth is in new nerve fibres and complicated connections, formed largely as a result of the effect of sensory input.

The causes of growth

In parallel with the development of the nervous system is the gradual maturation of the bodily organs. All of these are present at birth, but many are structurally and functionally immature. Their growth and development is brought about largely by the action of growth hormone, produced by the pituitary gland. This hormone has many specific effects. The body is made largely of protein, and growth hormone stimulated protein formation. It does this by increasing the supply of amino acids to the cells and by increasing the rate at which

DNA is consulted for details of the order in which these amino acids are put together to form proteins. Growth hormone increases the rate at which cells divide and thus reproduce. Finally, it acts on the growing zones in the long bones (the epiphyses) causing them progressively to extend. Growth hormone is given out in pulses, mainly during the night. Body growth is also greatly influenced by other hormones – thyroid hormone, insulin and the sex hormones. All these, acting together, bring about a steady increase in the bulk and dimensions of tissues and organs.

The newborn baby

As a result of genetic and intrauterine environmental influences all newborn babies are different. They differ in temperament and responsiveness and in the way they react to their environment. All possess a set of built-in reflexes that help them to survive. They are able to react to many stimuli, often in a self-protective way, and know, without instruction, how to feed from the breast or from a bottle equipped with a teat. Hearing is well established and different sounds produce different reactions. Sudden loud sounds are startling and are disliked. The sound of the female human voice tends to cause alert attention, while the lower tones of the male voice tend to be soothing. Babies are distressed by high-pitched crying sounds produced by others. Within a few weeks they are able to distinguish the sound of the mother's voice from those of other women.

The sense of smell is well developed at birth and strong prejudices in favour of pleasant smells and against unpleasant smells seem to be innate. Within a week or two of birth, babies can usually identify their mothers by smell. They also show definite taste preferences at birth, enjoying sweet flavours and rejecting sour or bitter substances. Vision is somewhat blurred at birth, but the ability to fix the eyes on an object of interest and to follow it is well established by two months. Babies have definite visual preferences, preferring to look at images with high contrast, bright colours and curved lines. The favourite object of gaze is the human face.

Early mental and social development

The first few weeks of life are spent mainly sleeping, crying and feeding. Crying increases to a maximum around 12 weeks because this is the only way the baby can respond to stimuli such as hunger, discomfort, pain, fear or over-stimulation. Later the baby develops other ways of responding and cries less. During the first year of life, the baby's perception of reality is limited to what can be seen, touched and sucked. Objects outside the field of vision no longer exist. But around 9–12 months infants begin to develop the idea that objects may have permanent existence, even when not seen. This idea is first related to the mother, who thus acquires great emotional importance. 'Peek-a-boo' games delight the child because of this. The mother's response to the baby's attempts at communication is also vital for normal development. If the mother is depressed or emotionally cold and cannot respond to the baby's expressions, future development and the ability to form close human attachments are likely to be seriously damaged.

By 15 months, the ability to recognize that a stranger is not the mother reaches a peak. At this stage the resulting anxiety may be severe. By two years the development of the nervous system allows the infant to recognize that there are many different people in the world and 'stranger anxiety' disappears. By this stage the child's perception of the mother may be so strong that even a photograph is recognized.

Early physical development

Many factors determine a baby's size and weight at birth, but the most important are the size, nutritional state and general health of the mother. Some weight is usually lost in the first few days after birth but this is made up in about 10 days and, thereafter weight is usually gained at a rate of about 30 g per day. If the baby is genetically destined to be large but is born small as a result of environmental effects, it will generally grow faster than average so as to catch up with its deficit. In most cases, such babies will have reached their full genetically-determined size by 18 months. Similarly, babies genetically coded for physical smallness, but who are large at birth as a result of such conditions as maternal diabetes, may seem to be failing to grow at an expected rate. In fact, they are reverting to type.

The normal average rate of growth is such that most babies are 50 per cent longer at the end of the first year and have trebled their birth weight. This rapid rate of growth soon declines, however, and by 2 years of age has reached the fairly constant rate characteristic of most of childhood. On average, infants and young children gain weight at a rate of 2–3 kg per year, and increase in height by 5–7.5 cm per year. In the early months of life almost half the calories in food are expended on growth. This proportion also declines rapidly, so that by about two years of age, only 3 per cent is devoted to growth.

Walking

Most babies can sit up at six months but the age at which they begin to walk varies widely within the range of normality and may be as early as about nine month or as late as 17 months. Different parts of the nervous system may mature at different rates and walking is such a complex accomplishment, calling for the effective action of so many different parts, that variability in the age of starting need occasion no surprise or concern.

Walking is a vital stage in development as it allows the child to explore and extend its environment and so greatly increase the possibilities of information capture. At the same time, the sense and experience of relative independence are important developmental factors. Conflicts commonly arise at this stage between parental concern for the safety of the child and the child's quest for knowledge of the limits of what it can control. Exploration and new experiences are important for development and a nice balance is necessary between restriction and safety. Inevitably there will be many vetoes and many stormy scenes.

Speech

By about two months a baby is capable of producing cooing sounds in response to sounds from the mother. By six months the baby can make spontaneous, repetitive babbling sounds and by one year the child is beginning to understand that there is some relationship between particular sounds and objects. Once it becomes apparent that uttering a particular sound can result in the acquisition of a desired object this idea is quickly reinforced. By 18 months some children have a vocabulary of as many as 50 words. A large vocabulary at this stage is encouraged by the use of simple normal language by the parents to refer to objects, rather than by making baby noises. The vocabulary of understood words is usually larger than the list of words that can be articulated. Around the second year there is usually a sudden and remarkable increase in the number of words that can be understood and spoken. Children with the good fortune to be brought up in a highly literate family now usually progress rapidly in power of expression and are soon prompted to enlarge their environment even more by learning to read.

chimera an organism that contains a mixture of genetically different cells derived from more than one ZYGOTE. A chimera may, for instance, occur as a result of fertilization by more than one spermatozoon; fusion of two zygotes; an ALLOGENEIC bone marrow graft; cell exchange between dizygotic twin fetuses; or combination of portions of embryos of different species. Compare MOSAICISM. The term derives from the name of a mythical monster with a lion's head, a goat's body, and a serpent's tail.

chimeric DNA recombinant DNA.

chirality the state of two molecules having identical structure except that they display 'handedness' (as in the right and left hand) and are mirror images of one another. Such pairs of molecules are also known as enantiomers or optical isomers. When dissolved in a fluid they rotate a plane-polarized beam in opposite directions.

chiro- see CHEIRO-.

chi squared test a series of statistical procedures used to test how closely the observed result of a trial or a statistical observation corresponds to an expectation, hypothesis or hoped-for result. If, for instance, the expected outcome of a trial is that half the participants will show a particular result, the expected frequency (E) is the total number of participants divided by 2. If Y is the number that show the result and N is the number that do not, then chi squared = $(Y-E)^2/E + (N-E)^2/E$.
A low chi squared value supports the expectation or hypothesis; a high value rejects it. Tables relating the value to statistical significance are available.

chlor, chloro- *combining form denoting* **1** that a hydrogen atom in a molecule has been replaced by a chlorine atom.
2 any chlorinated substance.
3 green.

chloride shift the movement of chloride (Cl^-) ions from the blood plasma into red blood cells to balance the outflow of bicarbonate ions from the cells. The bicarbonate is formed when carbon dioxide carried by the blood reacts with water to form carbonic acid. This process is greatly accelerated within the red cells by CARBONIC ANHYDRASE enzymes and the carbonic acid dissociates into hydrogen (H^+) ions and bicarbonate (HCO_3^-) ions. The red cell membrane is readily permeable to negative ions and most of the bicarbonate passes outward.

chlorofluorocarbons a range of compounds, composed of chlorine, fluorine, carbon and hydrogen, used as aerosol propellants and refrigerants. CFCs have become discredited because of their effect on the ozone layer.

choana a funnel-shaped opening, especially one of the internal openings of the nose into the PHARYNX.

choking partial or total obstruction of the main air passage (the LARYNX OR TRACHEA) by foreign body or external pressure. This induces a protective COUGH response which often clears the obstruction.

chol-, chole- *combining form denoting* BILE.

cholang- *combining form denoting* bile duct.

cholecalciferol colecalciferol, vitamin D_3, the natural form of the vitamin, formed in the skin by the action of the ultraviolet component of sunlight on 7-dehydrocholesterol. As a drug, the vitamin is usually formulated along with calcium for the treatment of calcium deficiency or OSTEOPOROSIS.

cholecyst- *combining form denoting* gall bladder.

cholecystectomy surgical removal of the gall bladder.

cholecystokinase an ENZYME that accelerates the breakdown of CHOLECYSTOKININ.

cholecystokinin a HORMONE released into the blood from the lining of the duodenum when fat and acid are present. It causes the gallbladder to contract and the sphincter of Oddi to relax, so sending bile into the duodenum to emulsify the fat, and stimulates the pancreas to secrete fat- and protein-splitting enzymes. The hormone also inhibits the motility of the stomach and the secretion of gastric acid.

choledocho- *combining form denoting* the common bile duct.

cholesterol an essential body ingredient found in all human cells, mainly as part of the structure of the cell membranes.
It is needed to form the essential steroid hormones, CORTISOL, corticosterone and ALDOSTERONE, the male and female sex hormones and the bile acids. It is synthesized in the liver and a large quantity of cholesterol passes down the bile duct into the intestine every day. Most of it is reabsorbed. A diet high in saturated fats encourages high blood cholesterol levels. Soluble dietary fibre and various drugs can bind intestinal cholesterol and prevent its reabsorption. Cholesterol is carried to the tissues in tiny cholesterol carriers called low density lipoproteins (LDLs). Oxidation of these allows cholesterol to be deposited in the walls of arteries causing dangerous narrowing (ATHEROSCLEROSIS).

choline one of the B vitamins necessary for the metabolism of fats and the protection of the liver against fatty deposition. The important NEUROTRANSMITTER acetylcholine is formed from it.

cholinergic 1 pertaining to nerves that release ACETYLCHOLINE at their endings, including the nerves to the voluntary muscles and all the PARASYMPATHETIC nerves. 2 having effects similar to those of acetylcholine.

cholinesterase an enzyme that rapidly breaks down acetylcholine to acetic acid and choline so that its action as a NEUROTRANSMITTER ceases.

Chomsky, Noam the linguist Noam Chomsky (1928–) believes that the ability to produce an endless variety of sentences cannot be explained by any theory involving post-natal input. Chomsky holds that such an ability necessitates inborn knowledge of the linguistic rules for sentence formation – a built-in system of formal grammar that lays down what is permitted and what is not. It does not seem possible to Chomsky that such rules can be acquired by environmental exposure.

Chomsky points out that sentences have 'surface' and 'deep' structures. Surface structures can vary widely – can, for instance, be expressed in the active or the passive voice – without any change in the essential meaning (the deep structure). Alternatively, a given sentence can mean two or more entirely different things – can have two or more deep structures. The sentence 'They are washing cloths' can mean 'These people are engaged in washing cloths' or 'These cloths are used for washing something or somebody'. The deep structure captures the functional relationships, such as that between the subject and the object, and specifies all the information needed to allow the correct interpretation of the sentence. Chomsky believes that the understanding of deep structures is an inherent capacity of the human brain and that the 'mental organ' that performs this function is manifested in many different languages.

chondral pertaining to CARTILAGE.

chondrification a change to CARTILAGE.

chondro- *combining form denoting* cartilage.

chondroblasts cells that differentiate from fibroblasts and mature into CHONDROCYTES.

chondrocostal pertaining to the COSTAL CARTILAGES and the ribs. Also known as costochondral.

chondrocytes cells that secrete the non-cellular matrix of CARTILAGE and become trapped in minute spaces within it.

chondrogenesis cartilage formation.

chondroitin a viscous compound of protein and carbohydrate (a GLYCOSAMINOGLYCAN or MUCOPOLYSACCHARIDE) found in crystalline lenses and corneas and in connective tissues. CHONDROITIN has been described as an intercellular glue.

chorio- *combining form denoting* CHORION or CHOROID.

chorion the outer of the two membranes that enclose the embryo. The inner is called the amnion.

chorionic gonadotrophin a hormone secreted by the placenta, throughout pregnancy, to maintain the CORPUS LUTEUM of the ovary. The corpus luteum produces the steroid hormones oestrogen and progesterone, especially the latter, and this prevents further ovulation and menstruation during the remainder of the pregnancy. At the end of pregnancy, the loss of the placenta allows for eventual resumption of ovulation. The hormone is available as a drug for the treatment of infertility, poor gonadal development and delayed puberty.

chorionic villi the finger-like projections from the CHORION into the wall of the womb at the site at which the PLACENTA is developing. Since both the chorionic villi and the embryo are derived from the same fertilized ovum a sample of the former provides material for genetic studies of the latter. Chorionic villus sampling has become an important method of early pre-natal screening for genetic defects.

choroid the densely pigmented layer of blood vessels lying just under the retina of the eye, contributing to its fuel and oxygen supply and optical efficiency.

choroid plexus pouch-like, blood-vessel-filled projections of the inner layer of the

MENINGES, the PIA MATER, into all four ventricles of the brain. Cerebrospinal fluid is continuously formed, mainly by secretion through the thin walls of the choroid plexuses.

Christmas factor one of the 20 or so factors necessary for the normal clotting of the blood. Christmas factor is Factor IX and its absence causes a form of HAEMOPHILIA sometimes called Christmas disease. (Named after Stephen Christmas, the patient in whom this deficiency was first found).

chrom- *combining form denoting* colour, pigment or stain.

chromatid one of the two duplicated copies of a chromosome produced by replication while still connected at the CENTROMERE before separation at the subsequent cell division. Each chromatid becomes a new chromosome.

chromatin DNA. The elongated, fine-stranded complex of roughly equal quantities of DNA and the protein histone, from which chromosomes are made by condensing into a coil. The individual chromosomes cannot be distinguished in a chromatin strand.

chromatocyte a pigmented cell.

chromatography a method of separating the components of a complex mixture, such as a gas, by passing it through selectively adsorbing media.

chromatolysis loss of the ability of a part of a cell to take up a stain from microscopic purposes. Nuclear chromatolysis implies dissolution of the nucleus.

chromatophore a pigment-containing cell.

chromogenesis the production of pigment.

chromogranins calcium-binding glycoproteins found in the secretory granules of all endocrine glands. They are cleaved to form biologically-active peptides.

chromophil readily capable of taking up biological stains or dyes. See also CHROMOPHOBE. Chromophil cells are those with granules that readily stain with dyes.

chromophobe resistant to biological stains. This property can be an important distinguishing characteristic in identifying some cells. Chromophobe cells are those with granules that do not take up dyes.

chromosomal sex the gender as determined by the nature of the sex chromosomes – female for two X chromosomes (XX) and male for one X and one Y (XY).

chromosomes

The physical basis of heredity is essentially the same in plants and animals, including humans. The cell is the basic building structure of almost all living things and almost every cell contains a central part – the nucleus – which consists largely of a mass of thread-like material called chromatin. The chromatin strands are remarkably long in relation to their thickness. Each one is a protein-coated molecule of DNA containing about 22,000 of the chemical groups we call genes strung along its length, in single file. Most of the time the chromatin remains bundled up like a loose ball of wool, but when the cell is about to divide it gets coiled up in a remarkably compact way to form a number of separate bodies called chromosomes. Loose chromatin takes stains poorly so it is almost impossible to see by normal light microscopy. But when it is coiled up to form chromosomes it stains readily. The term arose when these 'coloured bodies' were first distinguished under the microscope by means of specific stains. So the term 'chromosome' actually refers to a characteristic not present in life.

With the exception of sperm and eggs and their parent cells, each human cell contains 46 chromosomes, arranged as 23 pairs. If cells are photographed when they are dividing, pictures of the 46 chromosomes can be obtained and can be enlarged. The photographic print can then be cut up to separate them and they can be arranged in a standard, numbered order. This is called the karyotype. Each chromosome is a rod-like body, but when photographed for a karyotype in the process of division, all 46 of them are duplicated into two arms still stuck together to form a roughly X-shaped body. The slight constriction at the crossing of the arms of the chromosome is called the centromere. The position of the centromere, which often divides the double (replicated) chromosome into long and short arms, together with the size of the chromosome and its banding pattern, allows each one to be identified and numbered. Chromosomes are visible, as such, only during the stage of division of the cell, for it is only during division that the mass of chromatin is compacted, by coiling, to form solid-seeming bodies.

Autosomes and sex chromosomes

In women, the members of each pair of chromosomes appear identical, but in men this applies only to 22 of the pairs. The two chromosomes numbered 23 are called the sex chromosomes. In women they are also called the X chromosomes and look exactly the same. But in men, the sex chromosomes consist of one X (which looks just like a female X chromosome) and one Y, which is much smaller. It is this XY configuration that determines maleness, while the XX configuration determines femaleness. The chromosomes other than the sex chromosomes are called autosomes. So a characteristic caused by a gene on an autosome is called an autosomal characteristic and one caused by a gene on a sex chromosome is called a sex-linked characteristic. The term X-linked is also often used, because the Y chromosome has very few genes.

Reduction division

If sperm and eggs (ova) contained the normal number of 46 chromosomes, the penetration of an ovum by a sperm would result in a cell with 92 chromosomes. For this reason, the reproductive cells (ova and sperm) undergo, in the course of their maturation, a special series of divisions in which the net result is to reduce each pair of chromosomes to one. This is called reduction division or meiosis. So sperm and ova each contain only 23 chromosomes. This is called the haploid number of chromosomes. The division of normal body cells, in which the chromosome number remains 46 throughout, is called mitosis and the full complement of 46 chromosomes is called the diploid number. Meiosis results in the haploid number; fertilization restores the diploid number.

As soon as a sperm penetrates an egg, a dense membrane forms around the ovum to prevent the entry of any more. In this way the fertilized ovum has 46 chromosomes, one member of each pair of chromosomes being derived from the mother and the other member from the father. This applies to the sex chromosomes as well as to the autosomes. One X chromosome comes from the mother's ovum, the other sex chromosome (an X in the case of a girl, a Y in the case of a boy) comes from the father's sperm. During meiosis in the production of sperm, the separation of the sex chromosomes means that half of them will contain a Y chromosome and half an X chromosome. If a sperm with an X chromosome is the first to get through, the result will be a girl. If a sperm with a Y chromosome wins the race, the result will be a boy.

Homologous chromosomes

In normal body cells the chromosomes are present in 23 pairs. One member of each pair is a maternal chromosome and is matched with its fellow, which is a closely similar paternal chromosome. These pairs of chromosomes are called homologous chromosomes. The corresponding genes at any particular site (locus) in a pair of homologous chromosomes also form a pair. These paired genes are called alleles, and they may be identical, in which case the individual concerned is said to be homozygous for that gene, or they may be slightly different, in which case the individual is heterozygous for that gene.

Of the huge number of gene loci on our chromosomes we, all of us, are homozygous for some, heterozygous for others. In most cases, the distinction between homozygous and heterozygous alleles is a matter of indifference to us – they merely determine some characteristic such as our blood group. But many loci can have alleles that are far from unimportant and in these cases the difference between the homozygous and the heterozygous condition may make the difference between health and serious hereditary disease.

It is important to appreciate that a great many human characteristics are the result, not of a single gene, but of the effect of many different genes working in combination. This is called multifactorial inheritance and many of these characteristics are also determined to a variable degree by environmental influences.

chromosome analysis examination of stained CHROMOSOMES in a stage at which they are widely separated and easily visualized. The chromosomes are photographed and set out in matching pairs in an orderly arrangement known as a karyotype.

chronaxie the minimum time for which an adequate electric current must be applied to a nerve to produce a contraction of the associated muscle. An adequate current is one that is at least twice the threshold value.

chronic lasting for a long time. A chronic disorder may be mild or severe but will usually involve some long-term or permanent organic change in the body. From the Greek *chronos*, time.

chyl-, chylo- *combining form denoting* CHYLE.

chyle a milky alkaline fluid consisting of lymph and emulsified fat that is absorbed into fine ducts called lacteals in the lining of the intestine after a fatty meal. Chyle is carried by lymph vessels into the bloodstream.

chylomicrons microscopic globules, 80 to 1000 nanometres in diameter, of fat, phospholipids, cholesterol, fat-soluble vitamins and other materials. Chylomicrons are formed by the epithelium of the small intestine and are found in the blood during the ingestion of dietary fats, etc. The size of chylomicrons relates to the proportion of fats in the diet, being greatest after high-fat meals. Chylomicrons, and some of their contents are broken down in the liver and the constituents released.

chyme semifluid, partly digested food passed from the stomach into the small intestine for further digestion and absorption.

chymopoiesis conversion of food into CHYME.

chymotrypsin an ENZYME that breaks down (digests) protein to amino acids and simpler substances. It is secreted by the pancreas and released into the DUODENUM. The enzyme is also used to clean wounds and in an earlier form of cataract surgery to cut the suspensory ligament (zonules) of the cataractous lens.

CI *abbrev. for* CONFIDENCE INTERVAL.

cicatrix a scar. Scar tissue.

cicatrization formation of a scar.

cilia the microscopic hairlike processes extending from the surface of certain kinds of lining cells (ciliated epithelium) and capable of a rhythmical lashing motion. 2 eyelashes.

ciliary 1 pertaining to CILIA. 2 pertaining to the CILIARY BODY.

ciliary body the thickened ring of muscular and blood vascular tissue that forms the root of the IRIS and contains the focusing muscle of the eye. It is continuous with the CHOROID.

ciliary movement the rhythmical beating movement of CILIA on the surface of ciliated EPITHELIUM reminiscent of wind blowing across a field of ripe corn.

circadian exhibiting a 24 hour periodicity.

circadian rhythm a biological 24 hour cycle that applies to many physiological processes and variables and is synchronized to the day-night cycle occasioned by the rotation of the earth. See also BIORHYTHMS.

circulatory system

Lack of blood to any part of the body, with the failure to supply the vital materials it transports, is by far the most important cause of death and serious disease. Failure of blood perfusion, even for a very few minutes, through organs such as the brain or the heart is seriously damaging and often fatal. An adequate idea of the blood circulation, and of the pump – the heart – that maintains it, is fundamental to the understanding of the functioning of the human body.

Heart anatomy and function

The heart is a controlled pump, immediately responsive to every change in the body's needs for fuel. It maintains two separate but interconnected circulations and often does so, without flagging, for upwards of 80 years. It is constructed almost entirely of muscle, has four chambers, and is equipped with valves that ensure that the blood can move only in one particular direction. The two upper chambers are called atria and the lower chambers are called ventricles. The walls of the lower chambers are much thicker than those of the upper chambers because it is the ventricles that do the real work of pumping.

Functionally, there are two hearts, so separated in their purposes that doctors commonly refer to 'the right heart' or 'the left heart'. It is common for one side of the heart to fail while the other side remains functionally normal. The right side of the heart is less powerful than the left because it pumps blood only through the lungs. The left side pumps blood around all the rest of the body.

The heart muscle is of a kind unique in the body and has the property of spontaneous rhythmical contraction at a steady rate without any external stimulus. The muscle fibres of which the heart is constructed are joined together in a kind of branching network (a syncytium) so that contraction of one fibre sets off contraction in adjoining fibres. In addition, the heart muscle contains bundles of specialized muscle fibres that convey the electrical impulses associated with contraction in a systematic manner from the top of the organ to the bottom. As a result, the contraction starts at the top and proceeds downwards.

The heart works very hard and needs an excellent blood supply. Because the chambers have a 'water-tight' lining, blood must be supplied to the muscle from outside. This is the function of the coronary arteries. These are two small arteries that branch out from the main output artery of the body (the aorta) immediately after it leaves the heart. One coronary artery immediately divides into two large branches and the three vessels course over the surface of the heart like a crown – hence the name 'coronary'. From these arteries many smaller branches penetrate the muscle to supply it with blood. The coronary veins drain this blood into the right upper chamber.

Blood pumped from the heart always travels, immediately, in arteries. These are strong, elastic-walled vessels, readily able to withstand the pressure. Blood from the tissues is always returned to the heart by veins, which are less strong and have thin walls. Blood in veins is at a much lower pressure than blood in arteries.

From the powerful left ventricle blood is pumped into the aorta through a one-way valve, called the aortic valve. Just above the valve

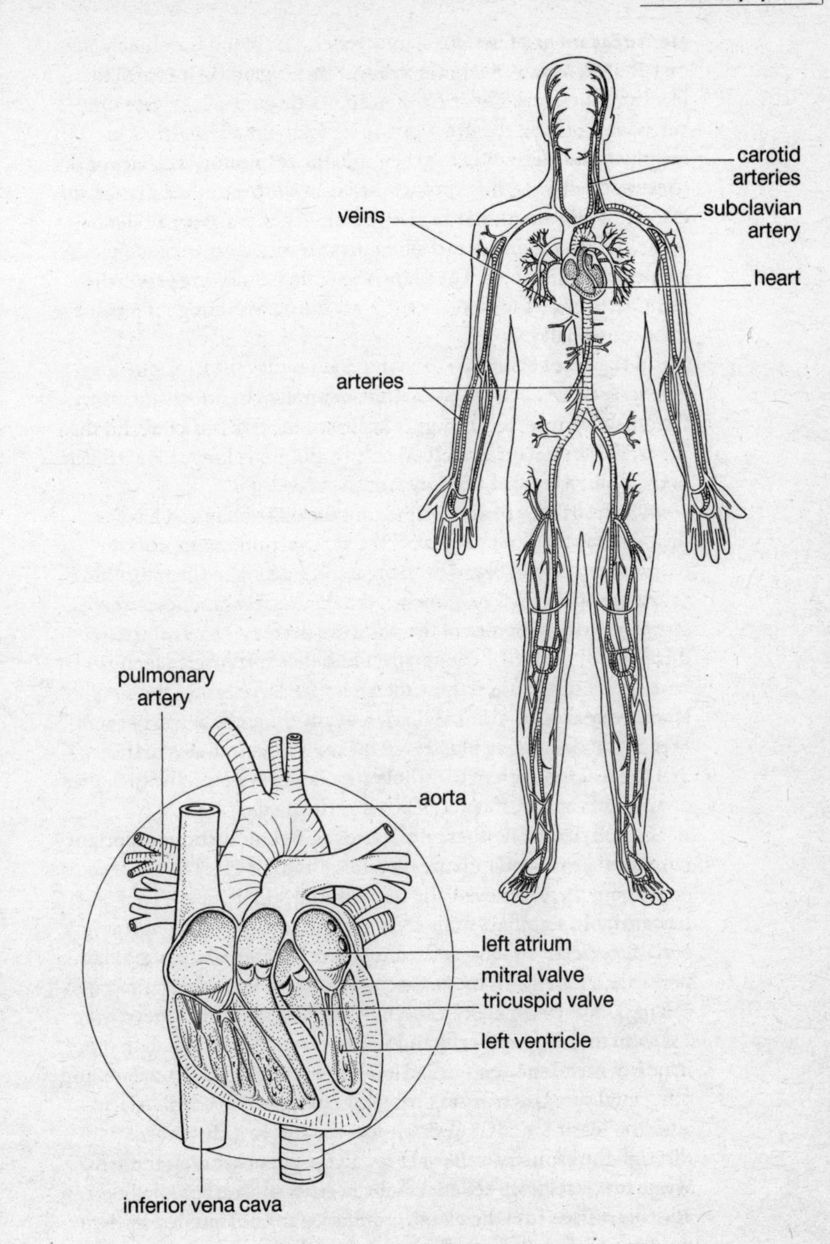

Fig. 3 **Circulatory system**

are the openings of the coronary arteries. The blood is immediately distributed, by way of large arteries, branching off the aorta, to the head and the arms. The aorta then arches through 180 degrees and runs down behind the heart, giving off many small branches to supply various tissues in the chest and the respiratory muscles of the chest wall. After passing through a hole in the back of the diaphragm into the abdomen, the aorta gives off branches to supply all the organs of the abdomen, including the intestines where blood picks up absorbed nutrients. The branches to the kidneys are especially prominent as all the blood in the body must pass though the kidneys many times a day.

In the lower abdomen the aorta divides into the two large iliac arteries, one for each leg. Each limb is supplied by one major artery – the brachial artery for the upper limb and the femoral artery for the lower. These arteries branch repeatedly to supply the many muscles of the limbs as well as the bones, tendons and skin.

At each branch of an artery the total cross-sectional area of the two branches exceeds that of the parent trunk. So there is a progressive drop in blood pressure and a slowing in the rate of flow as arteries form smaller branches. Small arterial branches are called arterioles and as the size of the branches decreases so does the thickness of the walls. The smallest, and most profuse, branches have very thin walls and these merge imperceptibly into the smallest blood-vessels of all – the capillaries. At any time, about five per cent of the total circulating blood is in the capillaries and it is in these that the real function of the whole circulatory system – the exchange of nutrients and waste products – is performed.

Capillaries are neither arteries nor veins but form an important intermediate class. Capillaries are ubiquitous. If every other structure in the body were removed, the shape would still be easily recognizable. Capillary walls consist of little more than single layers of flattened cells so loosely fixed together by their edges that small pores are left between them, through which water and certain cells can pass. Some capillaries, especially in the brain and kidneys, have a second inner layer of cells and these are less permeable than the majority, forming the so-called 'blood – brain barrier'. Nutrients and other vital substances from the blood – oxygen, glucose, fatty acids, amino acids, minerals and vitamins – are able to diffuse easily through the capillary walls or through the pores between the cells. Waste products from cell metabolism, especially carbon dioxide, can as easily diffuse into the blood. Capillaries are surrounded by tissue fluid in which the cells of the body are bathed and the initial interchange is between the blood and this fluid. At the same time, an interchange is going on between the tissue fluid and the cells. There is also a constant fluid interchange between the blood in the capillaries and the tissue fluid.

On the opposite side of each capillary bed from the arterioles lies the system of tiny veins (the venules) by which the blood is carried away. Venules join up to form larger veins and these, in turn run into the main veins returning the blood to the heart. Running parallel to each of the major arteries is a major vein in which the blood runs in the opposite direction.

The whole of the vein drainage system ends up in two massive veins, each called a vena cava. The upper (superior) vena cava drains the head, neck and arms and the lower (inferior) vena cava drains the lower part of the body. These direct the blood into the upper chamber on the right side of the heart – the right atrium. From there the blood passes down through a valve into the right ventricle which pumps it to the lungs by way of the lung (pulmonary) arteries. The lung circulation takes the same general form as that of the rest of the body except that the capillaries lie in intimate contact with the lung air sacs. This allows blood low in oxygen to be reoxygenated by oxygen diffusion, and allows carbon dioxide carried by the blood from the tissues to be passed out into the air sacs to be expired. Refreshed blood returns to the heart by four large pulmonary veins, but this time it is directed to the upper chamber on the left side – the left atrium. From here it passes down through a valve to the left ventricle to be pumped around the body.

Lymphatic system

In the capillary beds there is a net outflow of fluid from the capillaries into the tissue fluid of about 3 litres per day. Because of the higher pressure, more fluid passes out at the arteriolar side than returns at the venous side. Excess tissue fluid formed in this and other ways would soon cause water-logging (oedema) of the tissue were it not for the lymphatic drainage system by which it can be returned to the circulation. Lymph vessels have one-way valves, and external pressure on them by muscles maintain the flow back into veins at the root of the neck. These lymph vessels drain fluid through lymph nodes containing large collections of lymphocytes – cells of the immune system that can usually deal with infective organisms or other undesirable material. Any rise in the pressure in the veins, for any reason, will cause an increase in the tissue fluid. This may exceed the draining ability of the lymphatic system and oedema may result. Lymph vessel blockage by small parasitic worms called microfilaria causes elephantiasis.

circinate ring-shaped.

circumcision surgical removal of the male foreskin (prepuce) or of parts of the female external genitalia. Male circumcision is mostly done for ritual purposes, as in Jews and Muslims, but is occasionally necessary for medical reasons. Female circumcision is a culturally-determined but barbaric and heartless practice causing pain, mutilation and distress to millions of women. The procedure may be limited to removal of the clitoris or may also involve removal of all the external genitalia and stitching together of the raw surfaces so that they heal across and make sexual intercourse impossible (infibulation).

circumcorneal around the periphery of the CORNEA.

circumduction the movement of a limb that causes the hand or foot to describe a circle.

circumflex bent into the form of an arc or circle.

circumoral around the mouth.

circumvallate surrounded by a groove or by a raised ring as in the case of the circumvallate papillae on the tongue.

cis 1 in chemistry, being on the same side. Usually refers to two groups being on the same side of a ring or of a double bond.
2 in genetics, describing two sites on the same molecule of DNA. Compare TRANS.

cis- *prefix denoting* that two groups in a molecule are in the CIS CONFIGURATION. The prefix is usually italicized.

cis configuration having both of the dominant ALLELES, of two or more gene pairs, on one chromosome, and the recessive alleles on the other (HOMOLOGOUS) chromosome.

cisterna an enclosed space acting as a fluid reservoir. The cisterna magna is a space filled with cerebrospinal fluid lying between the CEREBELLUM and the MEDULLA OBLONGATA.

cistron a short length of DNA that codes for a protein subunit (a polypeptide), together with adjacent sequences that control its expression. It is the smallest unit that transmits genetic information.

citrate synthetase an enzyme that forms citrate from oxaloacetate and acetyl coenzyme A in the Krebs cycle.

citric acid cycle see KREBS CYCLE.

clairvoyance the claimed ability to perceive other than by the senses. Most if not all episodes of alleged clairvoyance are either illusory or fraudulent.

class switching the change, by a B cell, of the class of an antibody it produces (for example from IgM to IgG) without a change in its specificity.

-clast *combining form denoting* anything that breaks up or crushes.

claustrum any structure resembling a barrier.

clavicle the collar-bone, which runs from the upper and outer corner of the breastbone (sternum) to connect to a process on the outer side of the shoulder-blade (scapula).

clearance 1 the removal of a substance from the blood, usually by the kidneys.
2 the rate of such removal.

cleavage 1 the process of splitting, especially the repeated stages of cell division that produce a BLASTULA from an ovum that has been fertilized by a spermatozoon.
2 the breaking down of a complex molecule into smaller parts.
3 the vertical furrow between a woman's breasts visible when low-cut garments are worn. For inscrutable reasons this exerts an irresistible attraction on the gaze of most men.

climacteric the MENOPAUSE. The time in life after which reproduction is no longer possible. This is an exclusively female phenomenon; the male menopause is a journalistic fiction.

climax 1 a point of maximal intensity in a progression of events.
2 an ORGASM.

clinical psychology a discipline ancillary to medicine and surgery and devoted largely to the evaluation of psychological status and disorders by structured interviews and psychological tests and to helping people with psychological problems to adjust to the stresses and demands of life.

clinoid resembling a bed.

clitoris the female analogue of the penis. The principle erectile sexual organ in women and the main erogenic centre. It lies under the pubic bone at the front junction of the inner lips (labia minora) and immediately in front of the urethra, to which it is closely applied. The clitoris has a substantial nerve and blood supply. From the Greek *kleitoris*, little hill.

cloaca the combined urinary and faecal opening in the embryo before the two become separated. The term derives from the Latin *cloaca* a sewer.

clonal selection the production by an antigen of an expanding CLONE of lymphocytes from a single cell bearing a receptor complementary to the antigen.

clone 1 a perfect copy, or a population of perfect copies, of any organism. Cloning occurs when an organism reproduces non-sexually, so that the genetic content (genome) of each is identical.
2 a number of identical cells derived from a single cell by repetitive division.
3 a perfect copy, or any number of copies, of any DNA sequence, such as a gene, or any other nucleotide sequence.

clot a thick, coagulated, viscous mass, especially of blood elements. See also BLOOD CLOTTING.

clothing body coverings that serve a protective, heat-insulating, decorative and modesty-preserving function. The clothes a person wears are incorporated to a remarkable degree into other people's perception of that person. It is almost as if the clothes were part of the body and are to be identified with the person. Thus, a person of no great physical attractions may appear attractive if attractively dressed; a short person may appear taller by wearing shoes and a hat; a smart uniform confers authority; conformity to a particular fashion in dress confers acceptance by a group; and so on. This phenomenon has made preoccupation with dress one of the central concerns of a large section of humanity and has spawned a huge industry. It also tells us something about the relative importance of emotion and logic in the assessments most of us make of our fellow men and women.

clumping aggregation or clustering and adhering together.

co- *prefix signifying* with, together, jointly.

coagulase an enzyme secreted by the micro-organism *Staphylococcus aureus* that causes clotting in blood plasma by converting prothrombin to thrombin. This ability probably contributes to the tendency of the organism to form abscesses.

coagulation, blood see BLOOD CLOTTING.

coagulum a clot.

cobalamine vitamin B_{12}. The specific treatment for PERNICIOUS ANAEMIA.

coccygeal pertaining to the COCCYX.

coccyx the rudimentary tail bone, consisting of four small vertebrae fused together and joined to the curved SACRUM. From the resemblance of the bone to a cuckoo's beak.

cochlea the structure in the inner ear containing the coiled transducer, the organ of Corti, that converts sound energy into nerve impulse information. The cochlea resembles a snail shell.

cochlear nerve the branch of the 8th cranial nerve (ACOUSTIC NERVE) concerned with hearing.

coding strand the single strand of a separated double helix of DNA that has the same base sequence as the MESSENGER RNA (mRNA) formed from the complementary DNA strand. The coding strand is not used to form mRNA.

codominant alleles a pair of ALLELES both of which contribute to the phenotype, neither being dominant over the other.

codon a sequence of three consecutive nucleotides (a triplet) along a strand of DNA or messenger RNA that specifies a particular AMINO ACID or a stop signal during protein synthesis. The order of the codons along the DNA molecule determines the sequence of particular amino acids in the protein produced.

coeliac pertaining to the cavity of the abdomen.

coelom the body cavity of the developing embryo situated in the mesoderm.

coenzymes small organic molecules, acting as cofactors that must bind to an enzyme before it can function properly. Tightly-bound coenzymes are called prosthetic groups; loosely-bound coenzymes are more like cosubstrates. Most of the B vitamins are coenzymes.

cognition the mental processes by which knowledge is acquired. These include perception, reasoning and possibly intuition.

cognitive dissonance an inconsistency between beliefs or between belief and action. The perception of one's own cognitive dissonance results in changes in belief or behaviour and the more difficult this is the more powerfully the change is later rationalized.

cognitive functions those aspects of brain activity concerned with perception, sensation, reasoning, memory, imagination, ideation and the formation of new concepts. The precise mechanisms underlying these 'higher' functions remains obscure but it is clear that cognitive operations involve the cooperative activity of most, if not all of the many parts of the brain, even those which might seem to be unrelated. Each functional area has its nearby association area. For example, thinking that involves imaging will involve activity of the visual association area. Puzzling out a problem in semantics will involve the speech association areas.

cognitive psychology

Cognition is the mental act or process by which knowledge is acquired. The same word is used to mean the knowledge that results from such a process. Knowledge is acquired by perception, by the correlation of past remembered data with recent acquisitions in the act of creativity, by solving problems, and possibly by a process described as 'intuition'. All these elements are within the province of cognitive psychology which can be said to be the study of all human activities related to knowledge. More specifically, cognitive psychology is concerned with how knowledge – of whatever kind – is acquired by the individual, how it is stored, how correlated and how retrieved. Thus it is concerned with attention, with the processes of concept formation, with information processing, with memory and with the mental processes underlying speech (psycholinguistics).

To some extent, cognitive psychology is a reaction against behaviourism which it holds to be seriously incomplete as a theory of mental functioning. Unlike the behaviourist, the cognitive psychologist asks questions about mental events and hopes for explanations of them. The brain is conceived as an information processing system that operates on, and stores, the data acquired by the senses. Much of this activity is performed without conscious awareness.

Although cognitive psychology is very much concerned with the actual processes going on in the mind it does not investigate these by conscious introspection. What the cognitive psychologists do is to carry out experiments to measure and analyse human performance in carrying out mental tasks – experiments such as comparing the time taken for different mental activities such as pattern recognition, assessing how much data can be stored in short-term memory, measuring reaction times, assessing accuracy in mental tasks, noting characteristic errors made, and so on. On the basis of the analysis of these data, possible models of the underlying mental processes are constructed.

Although most cognitive psychologists are peripherally interested in neurophysiology, they do not pretend that these models represent the actual processes occurring in the brain. Nevertheless, as these models are refined, it is hoped that they will approach ever closer to reality.

cohort a group of persons all born on the same day. Cohort studies are valuable in medical and epidemiological research.

coitus copulation or the physical act of sex.

coitus interruptus withdrawal of the penis from the vagina prior to ejaculation. This is unreliable as a method of contraception.

cold agglutinin a substance found in blood serum that causes red blood cells to clump together (agglutination) if the blood is kept at low temperatures.

collagen an important protein structural element in the body. Collagen fibres are very strong and, formed into bundles which are often twisted together, make up much of the connective tissue of the body. Bones are made of collagen impregnated with inorganic calcium and phosphorus salts. Vitamin C is necessary for the cross-linking and full strength of the collagen molecule.

collagenases enzymes that can break down COLLAGEN and gelatin.

collagenous fibre white connective tissue fibres found in soft tissues and bone matrix.

collapse an abrupt failure of health, strength or psychological fortitude. The term is used more by the laity than by the medical profession.

collar-bone the CLAVICLE. This comparatively delicate bone is readily fractured either by direct violence or by a force applied indirectly, as in a fall on the outstretched arm.

collecting ducts the tubules in the kidney into each of which which several distal convoluted tubules empty their urine and deliver it to the pelvis of the kidney

colliquative 1 featuring a turning to liquid of solid tissue, often after death.
2 denoting excessive watery discharge.

colloid a substance in which particles are in suspension in a fluid medium. The particles are too small to settle by gravity or to be readily filtered. The colloid state lies between that of a solution and that of an emulsion.

colon the large intestine. It is called 'large' because of its diameter. Its main function is to conserve water by absorption from the bowel contents. It also promotes the growth of bacteria which synthesize vitamins.

colony stimulating factor one of a number of glycoprotein factors that allow and promote the reproduction and differentiation of blood cells and their precursor cells.

colorectal pertaining to the COLON and the RECTUM.

colostomy an artificial anus on the front wall of the abdomen, formed when the cut upper end of the colon is brought to the exterior. This is often necessary when the colon has to be cut through, as in the treatment of cancer. Evacuated bowel contents are collected in a waterproof bag. Colostomies are often temporary.

colostrum the yellowish, protein-rich, milk-like fluid secreted by the breasts for the first two or three days after the birth of a baby. Colostrum contains large fat globules and a usefully high content of antibodies.

colotomy a surgical cut (incision) into the COLON.

colour blindness an inaccurate term for a lack of perceptual sensitivity to certain colours. Absolute colour blindness is almost unknown. Most colour perception defects are for red or green or both. About 10 per cent of males have a colour perception defect, but this is rare in females.

colour vision the ability to perceive the world in colour. The light-sensitive rods of the retina are colour-blind, but the cones provide a maximal output signal to the fibres of the optic nerve for one of three spectral wavelengths, thus effectively discriminating between the three primary colours. The output from each of the three types of cones thus depends on the colour of the light falling upon then and, in particular, to the proximity of the colour to that of the spectral sensitivity of each type. The three types are mixed closely together. The process is the reciprocal of the way in which the sensation of different colours is achieved by a colour TV or computer monitor.

colpo- *combining form denoting* the vagina.

columnar epithelium a surface tissue of tall, column-like cells standing on a basement membrane. Some columnar epithelia are of smooth free surface, some have a covering of microvilli (brush border), some are ciliated (see CILIA) and some have mucus-secreting vacuoles. Columnar epithelium may be of a single layer (unilaminar), STRATIFIED, or pseudo-stratified. The latter is a histological illusion produced when the cell nuclei are at different levels.

colyones natural inhibitors of cell growth and proliferation. Colyones are coded for by colygenes.

coma a state of deep unconsciousness from which the affected person cannot be aroused even by strong stimulation. Coma can result from head injury, oxygen lack, interruption of the blood supply to the brain, poisoning and various disease states.

comatose in, or resembling, a state of coma.

comedo, comedone a BLACKHEAD. See also ACNE.

commissure 1 a line or point at which two things are joined.
2 a nerve fibre bundle passing from one side of the brain or spinal cord to the other to connect similar structures.

common bile duct the final duct carrying bile from the liver and the gall bladder, to the duodenum.

common hepatic duct the duct formed by the junction of the ducts from the lobes of the liver.

communication the transmission of information. Human communication is conducted through many different channels, some less obvious than others. These include speech, the written or printed word, body language, recorded and/or transmitted data, art, music, drama and film.

comparative anatomy the study of the similarities and differences between the body structure of different animals. Although external appearances may vary considerably, in many cases the similarities are much greater than the differences. This observation has been one of the principal reasons for the belief that we have evolved from common ancestors.

compensatory hypertrophy an increase in the size of an organ or volume of a tissue following loss or malfunction of the paired organ or loss of functioning tissue.

complement a collection of about 20 serum proteins involved in the immune system process by which the action of antibodies against the invading agent (the ANTIGEN) is completed. Complement combines with antigen-antibody complexes to bring about the breakdown of the antigen-bearing cell or molecule. Some of the serum proteins form enzyme-activated cascades to produce molecules involved in INFLAMMATION, PHAGOCYTOSIS and cell rupture.

complementary of a NUCLEOTIDE or nucleotide sequence the base, or bases, of which can link to one or more other bases to form a BASE PAIR or a sequence of base pairs.

complementary base pairs see BASE PAIRs.

complementary DNA a sample of DNA that has been produced from MESSENGER RNA after conversion into double-strand DNA.

complementary genes genes which, although not alleles, perform similar or opposing functions. Complementary genes may cooperate in the production of an effect or they may tend to oppose its production.

complementary pairs the pairs of bases that link together, like the rungs of a ladder, along the length of the DNA molecule. The whole process of DNA replication depends on the fact that, in DNA, adenine can only link to thymine and guanine can only link to cytosine

complete linkage the location of genes sufficiently closely to each other in a chromosome that they are always transmitted together to daughter cells. In the phenotype, the effects of such genes always occur together in the same individual.

compos mentis literally, of composed mind. Sane.

conation the mental processes characterized by aim, impulse, desire, will and striving. The functioning of the active part of the personality. Compare COGNITION and AFFECT.

conception 1 penetration of an OVUM by a SPERMATOZOON, with the initiation of a new individual and the state of pregnancy. The formation of a ZYGOTE.
2 the individual zygote or embryo so formed.

conceptus the product of CONCEPTION.

concha the visible, external part of the ear. Also known as the pinna or auricle.

conceptus the immediate result of conception. The earliest stages of the embryo.

concretion a solid mass of chalky or inorganic material formed in a cavity or tissue of the body. A CALCULUS.

conditional lethal mutation a mutation that will kill a cell under certain conditions but not under others.

conditioned reflex an automatic response to a stimulus which differs from that initially causing the response but which has become associated with it by repetition. The sight of food causes a dog to salivate. If a bell is rung every time the food appears, the bell alone will, in time, cause salivation.

conditioning an important element in human programming and behaviour. Conditioning is a form of learning in which a particular stimulus will eventually and reliably elicit a particular behavioral response.

condom a sheath of fine rubber, available in various sizes, that may be rolled onto the erect penis to serve as a contraceptive measure and as a protective against sexually-transmitted diseases.

conduction system a system of specialized heart muscle fibres that generates and distributes rhythmical contraction stimuli to all parts of the heart muscle. The system ensures an orderly sequence of contraction.

condyle a rounded prominence at the end of a long bone that gives attachment to tendons and articulates with the adjacent bone.

cones the tiny light-sensitive transducers of the RETINA that are present in greatest concentration in the central part, the macula. Cones are less sensitive than the more peripherally placed, colour-blind rods, but are capable of distinguishing three primary colours.

confidence interval (CI) a statistical term that quantifies uncertainty. In a clinical trial, the 95 per cent confidence interval (the interval usually employed) for any relevant variable is the range of values within which we can be 95 per cent sure that the true value lies for the entire population of people from which those patients participating in the trial are taken. The greater the number of patients on which the confidence interval is based the narrower it becomes.

confinement the period from the start of labour to the delivery of the afterbirth (placenta).

conflict the effect of the presence of two mutually incompatible wishes or emotions. Unacceptably unpleasant conflict leads to REPRESSION and this may be manifested as NEUROSIS.

conformational epitope adjacent amino acid strings at different points in a folded protein, making a complex site to which an antibody can bind. Conformational epitopes are less stable than LINEAR EPITOPES, a fact that explains why some allergies are outgrown in time. Allergies, such as peanut allergy, in which the antibody (IgE) binds to a linear epitope tend to persist.

conformational exposure a technique of identification of abnormal proteins in which CONFORMATIONAL EPITOPEs that become exposed when a protein is misfolded are identified.

congeal to clot or coagulate.

congener one of a group of chemical compounds with a common parent substance or derived from a common basic formula.

congenital present at birth and resulting from factors operating before birth. A congenital disorder need not be hereditary, although many are. Conditions acquired during fetal life are congenital as are those acquired during the process of birth.

conjugate coupled or joined in pairs or groups. Of covalently linked complexes of two or more molecules.

conjugate measurements measurements of distances between bony points, especially various measurements of the inlet and outlet of the female pelvis in connection with prospective childbirth.

conjugated protein a compound of a protein with a nonprotein.

conjugation 1 chemical combination or linkage of chemical groups to organic molecules, often to produce a water-soluble form and allow more ready excretion. 2 the exchange of genetic material between paired single-cell organisms, such as bacteria.

conjunctiva the transparent membrane attached around the CORNEA, covering the white of the eye and reflected back over the inner surfaces of the eyelids.

connectin a protein found on the surface of some cancer cells that binds to cytoskeleton ACTINS and glycoproteins on other cells.

connective tissue loose or dense collections of COLLAGEN fibres and many cells, in a liquid, gelatinous or solid medium. Connective tissue participates in the structure of organs or body tissue or binds them together. It includes cartilage, bone, tooth dentine and lymphoid tissue.

connexin-26 one of a number of gap junction connexin proteins. Gap junctions between cells allow the passage of ions and small molecules from one cell to another. The gene for connexin-26 is situated on chromosome 13.

consanguinity blood relationship. The term does not imply any particular degree of closeness and ranges from identical twin to remote cousin.

conscious awareness of one's existence, sensations, and environment. Capable of thought and perception.

consciousness full awareness of self and of one's environment. The conviction that it is possible to explain the sources of consciousness has spawned a small library of books purporting to do so.

consciousness, grades of a scale of degrees useful in clinical practice. Grade 0 = fully conscious; 1, responds to voice; 2, unconscious but reacts to minor applied pain; 3, unconscious but shows some reaction to a strong painful stimulus; 4, no response to any stimulus.

consensual 1 pertaining to the reflex response of an organ to the reflex action of another, usually paired, organ. For example, the constriction response of one pupil to light is accompanied by the constriction of the other. This is a consensual reflex. 2 involving common consent, as in the consent of both parties to an act of sexual intercourse.

conservative recombination recombination of broken segments of DNA without the synthesis of any new sequences.

conserved of genetic entities that remain unchanged between individuals of a species or between different species or over a period of time. The adjective is often qualified as in 'highly conserved'.

constipation unduly infrequent and difficult evacuation of the bowels. This disorder is often due to deliberate suppression of the desire to defaecate. It is almost unknown in people whose diet is largely vegetable with a high fibre content.

constitutive genes genes that express all the time. Most genes are switched on only when required.

constitutive mutation a mutation whose effect is to cause a normally regulated gene or group of genes to express themselves continuously.

constrict *v.* 1 to narrow or make smaller, to shrink or contract. 2 to squeeze or compress.

constriction 1 a narrowing. 2 the act or process of narrowing.

constrictor a muscle that contracts, narrows or compresses a part or organ of the body.

consummation the first act of sexual intercourse after marriage. Nowadays, this has mainly legal significance.

contact inhibition the control or cessation of cell growth and reproduction due to contact with adjacent cells. This important restraint is lost in cancer.

continence self-control or restraint, especially in relation to sexual activity. The antonym, incontinence, seems, nowadays, to be applied mainly to urination and defaecation.

contraception the prevention of CONCEPTION by avoiding fertile periods; by imposing a barrier between the sperms and the egg; by killing sperm; or by preventing the release of eggs from the ovaries. Intrauterine contraceptive devices (IUCDs) act by preventing implantation of fertilized ova but are also usually considered a form of contraception.

contractile capable of contracting or of causing contraction.

contractile ring a ring of actin filaments around the equator of a cell formed at the end of MITOSIS. Tightening of this ring lead to the separation of the two daughter cells.

contractility the ability to become shorter, as in a muscle cell or an anatomical muscle. Muscle contractility does not imply a reduction in the volume of the part, merely a change in shape with constant bulk.

contraction the primary function of muscle by which a change of shape brings the ends closer together. By contracting, muscles bring about movement of bones or other parts. The term comes from a Latin word meaning 'to draw together' so, strictly, a phrase such as 'isotonic contraction' is a contradiction in terms. Popular usage, however, will have it that a muscle can contract without shortening.

contractions a term usually applied to the periodic tightening and shortening of the muscle fibres in the womb (uterus) during labour which gradually bring about the expulsion of the baby. All muscles act by contraction.

contralateral referring to the opposite side. The term ipsilateral is used in referring to the same side.

contusion a bruise.

convergent evolution 1 the process in which phylogenetically distinct lineages acquire similar characteristics.
2 evolutionary changes in which descendants resemble each other more closely than their progenitors did.

convergent thinking analytical thinking that follows a set of rules, as in arithmetic, or in which the logical validity of the thought processes is checked and verified. Compare divergent or creative 'lateral' thinking, characterized by unorthodox mental processes but often productive of a number of different and sometimes valuable solutions.

convolutions the folded elevations, or gyri, of the brain into which the surface layer (cortex) is thrown so as to accommodate its great area.

copolymer a POLYMER consisting of repeated units of two or more subunits.

copro- *combining form denoting* faeces.

copulation a joining together in the act of COITUS. The physical element in sexual intercourse.

cor the heart.

coracoid a bony process on the outer side of the shoulder-blade (scapula) which projects forward under the outer end of the collar-bone (clavicle).

core octamer an elongated structure of four pairs of histone protein subunits forming a core around which DNA is wound. Histones are strongly basic proteins that show little variation in sequence from one species to another.

corium the layer of living skin under the mainly dead outer layer. The 'true' skin, containing nerve endings, sweat glands, and blood vessels.

cornea the outer, and principle, lens of the eye through which the coloured iris with its central hole (the pupil) can be seen. The cornea performs most of the focusing of the eye. Fine adjustment (ACCOMMODATION) is done by the internal crystalline lens.

corneal epithelium the thin, layered, outer 'skin' of the cornea.

corneal reflex 1 automatic blinking on light touch to the cornea. The reflex is sometimes used by anaesthetists and others as a test of the level of consciousness.
2 the position of the reflection of a small light on the cornea when it is directly regarded by the subject. If mid-pupillary in one eye and eccentric in the other, the subject has a squint.

corona any structure resembling a crown.

coronal relating to the crown of the head.

coronal plane a vertical anatomical plane that divides the standing body into front and rear halves. A plane lying in the direction of the side-to-side CORONAL SUTURE of the skull.

coronal section an imaginary, radiological or anatomically demonstrative cut made in the CORONAL PLANE.

coronal suture the irregular line of junction of the paired parietal bones of the skull with the frontal bone.

corona radiata 1 the radiating 'crown' of nerve fibre bundles running up from the INTERNAL CAPSULE of the brain to all parts of the CORTEX.
2 a layer of cells radiating outwards from the maturing OVUM and which persist for a time after ovulation.

coronary pertaining to a crown. The CORONARY ARTERIES arise from the main artery of the body immediately above the heart, and give off branches which spread like a crown, over the surface of the heart.

coronary arteries two important branches of the AORTA that supply the heart muscle with blood. The left coronary artery divides almost at once into two main trunks, so it is common for surgeons to refer to the three coronary arteries. Smaller branches of the coronary arteries spread over the surface of the heart and send twigs into the heart muscle. Obstruction of a coronary artery branch, by ATHEROSCLEROSIS and subsequent THROMBOSIS, is commoner than blockage of one of the main trunks. Such obstruction causes a heart attack by depriving a part of the heart muscle of its blood supply to cause local death of muscle tissue (myocardial infarction).

coronary thrombosis a heart attack caused by clotting of blood at the site of narrowing of a coronary artery, so that the heart muscle is locally deprived of blood and part of the muscle dies.

coronary veins the vessels that drain blood from the heart muscle, joining to form a vein that empties into the right ATRIUM.

corpora cavernosa the two parallel longitudinal columns of spongy tissue in the penis, capable of a remarkable increase in size under the influence of sexual interest or excitement when flooded with blood under pressure. Such swelling is an example of tumescence. See also ERECTION.

corpulence obesity. Being excessively fat.

corpus a body, usually in the sense of a bodily structure. The plural is corpora.

corpus albicans the white fibrous tissue body remaining in an ovary after the CORPUS LUTEUM has regressed.

corpus callosum the wide curved band of nerve fibres (white matter) that connects the two cerebral hemispheres.

corpuscle a general term for any small, discrete, microscopic structure such as a red blood cell, a sensory nerve ending, an OSTEOCYTE or a GLOMERULUS of a kidney.

corpus luteum a yellow mass of fatty material swelling out the empty GRAAFIAN FOLLICLE in the ovary after the egg (ovum) has been discharged. The cells of the corpus luteum secrete both oestrogens and progesterone and these hormones cause the lining of the womb to thicken and form a suitable bed for the fertilized ovum. If pregnancy does not occur the corpus luteum degenerates in less than two weeks.

corpus spongiosum the third erectile column of the penis, less effective than the CORPORA CAVERNOSA, and placed centrally and a little behind the others. The urine outlet tube (urethra) is surrounded by the corpus spongiosum.

corpus striatum a gray and white striped collection of nerve cell bodies and nerve fibres in the lower and outer part of each cerebral hemisphere. The corpus striatum is the largest subdivision of the BASAL GANGLIA and consists of the caudate and lentiform nuclei. It is concerned largely with control of movement.

corpuscle an old-fashioned term for a free-floating blood cell, such as a red or white blood cell. The term derives from the Latin *corpusculum*, the diminutive of the term for 'body'.

correlation the degree to which changes in variables reflect, or fail to reflect one another. Correlations are said to be positive when the variables change in the same direction and negative when they move in opposite directions. A common fault in statistics is to assume that correlations are significant when they are not, that is, to assume unjustifiably that changes in variables are causally related.

cortex the outer distinguishable zone of any solid organ. The cerebral cortex, for instance, is the outer layer of grey matter of the brain consisting of nerve cell bodies. The adrenal cortex is quite different in function from the inner part.

cortical pertaining to, or consisting of, a CORTEX.

corticoid an informal term for any steroid produced by the adrenal cortex. A corticosteroid.

corticospinal connecting the cerebral cortex and the spinal cord or pertaining to both. The corticospinal tracts are large bundles of nerve fibres largely concerned with voluntary movement.

corticosteroid hormones natural steroid hormones secreted by the cortex of the adrenal glands. These are cortisol, corticosterone, aldosterone and androsterone.

corticotrophin corticotropin, adrenocorticotrophic hormone (ACTH), a hormone produced by the PITUITARY GLAND which stimulates the adrenal cortex to secrete steroids in response to stress.

cortisol a hormone produced by the adrenal cortex. Also called hydrocortisone.

cosmesis consideration or concern for appearance.

cosmetic surgery a branch of plastic surgery devoted to the improvement or alteration of the human appearance. Cosmetic operations include those on the nose (rhinoplasty), the ears (otoplasty), the chin (mentoplasty) and the breasts (augmentation or reduction mammoplasty).

costa a rib.

costal cartilages the flexible cartilages by which the front ends of most of the ribs are connected to the breast bone (STERNUM).

costive suffering from, or causing, constipation.

costochondral pertaining to a rib and its cartilage.

cotyloid cavity see ACETABULUM.

covalent bond a chemical bond in which an electron is shared between the two atoms that are bonded.

Cowper's glands a pair of small mucus-secreting glands that open into the sphincter of the URETHRA immediately below the prostate gland in males. Also known as the bulbourethral glands. (William Cowper 1666–1709, English surgeon).

COX see CYCLO-OXYGENASE.

coxa the hip or hip joint.

cranial nerves the 12 pairs of nerves which spring directly from the brain and brain stem. They include the nerves for smell, sight, eye movement, facial movement and sensation, hearing, taste and head movement.

cranio- *combining form denoting* the bones that enclose the brain (the CRANIUM).

craniofacial pertaining to the cranium and the face.

cranium the skeleton of the head without the jaw bone.

C-reactive proteins a group of proteins that increase rapidly in amount in the blood during infections.

creatine a nitrogenous substance present in all muscle cells.

creatine kinases (CKs) enzymes that catalyze the bond between creatine and ATP to form creatine phosphate and ADP with the storage of energy in the phosphate bond. Creatine phosphate occurs mainly in muscle and contributes energy required for muscle contraction.

creatinine a breakdown product of the important nitrogenous metabolic substance CREATINE. Creatinine is a normal metabolic waste substance and is found in muscle and blood and excreted in the urine.

creationism the belief that the account of the creation of the world contained in the first chapter of the book of Genesis in the bible is literally true. The implication, often expressed, is that the scientific account, including the geological evidence, is false. Creationism denies Darwinian evolution, but a belief in, and knowledge of, evolution has become an essential component in the mental armamentarium of the biological scientist.

True creativity – the ability to produce something completely new – is, in a sense, antagonistic to conventional logical thought. This is especially so in science. Normal scientific thinking tends to be convergent – a synthetic process of putting together causally related phenomena in a conventional way. Creative thought is divergent and often appears irrational – a leap in the dark, a process of 'lateral thinking', of dreaming or fantasizing or engaging in free associations. Creativity is the realm of those whose minds are untrammelled by logic but who are, nevertheless capable of logic when logic is appropriate.

Creativity is a mysterious activity about which few generalizations are possible. The processes adopted by creative artists and scientists vary greatly. People of comparable status appear to create with greatly varying degrees of difficulty. Mozart composed masterpieces at top speed, with no hesitation and few amendments; Brahms would weep and groan in the agony of his composition. Mozart could see every note in the score in his mind before picking up his pen; other composers can do nothing without an instrument.

Experience shows that important problems are often solved, not at the height of concentrated thought, but in the period of relaxation afterwards. Solutions are often apparent on waking from a night's sleep. Clearly, much creativity is a function of the unconscious mind which, once supplied with the necessary data and given time to work on it, may come up with a surprising answer. Einstein, noted for the fundamental novelty of some of his most important ideas, pithily remarked that the really creative scientists are those with access to their dreams. To produce something completely new, it is often necessary to forget conventional wisdom.

The nature of the creative process

It is said that the difference between a craftsman and an artist is that the craftsman knows exactly what the outcome of his effort will be before he starts, while an artist must wait until the work is complete before discovering what has been achieved. The latter process is probably most typical of the creative artist, but there are those who know in advance exactly what they want to achieve and appear merely to be concerned with the means of doing so. Most novelists, on being asked about the creative process, agree that they have no idea, in advance, how a book is going to evolve. They need an idea to get them started and then, if all goes well, the characters take over. Many agree that, unless this happens, the book never comes to life.

Some artists find that the production of a work of art involves recognizable successive stages such as preparation, incubation, inspiration and elaboration. Others experience all these stages repeatedly in the same act of creation, or describe other stages. Some begin in a state of confusion with many fragmentary ideas competing for attention in the mind. For these people, the nature of the new creation very gradually appears. As ideas are put together or rejected, some kind of definitive entity solidifies from the mist.

Freudian and Jungian psychologists have had plenty to say about the creative process, but their theories have been no more enlightening than those of other persuasions. Freud, initially, saw creativity as the working out of unconscious desires (wish fulfilment). Later, as his ideas changed, he came to see the creative act as a process of defence by the ego against indictments by the superego. Jung, in his life-long preoccupation with symbols, saw creativity as an unconscious symbol-making process.

It would be unreasonable to expect to be able to understand the creative process at a psychological level. The most that one can say is that nothing can come out that has not previously gone in, but that the possibilities of synthesis, by the interaction of new with stored data, are infinite. The components, before synthesis, may be familiar; once incorporated into a new creation they may no longer be identifiable, so that it may seem that something completely new has been made.

Can creativity be measured?

One possible measure of creativity is to assess what is sometimes called ideational fluency – the number of different ideas a person can generate in a particular context. One might, for instance, ask a test subject to suggest as many uses as possible of an empty Champagne bottle. Other tests might involve interpretation of Rorschach ink blots or writing an account of the story behind various posed photographs. A major difficulty in such testing is the highly subjective nature of the examiner's response to the subject's answers. Some answers might seem highly original to one examiner and banal to another.

Creativity and intelligence

Psychologists who believe that intelligence is a complex of a large number of separate definable abilities, will usually include among these a group of abilities that are essentially creative in nature. The idea that creativity is a component of intelligence is not, however, universally accepted. This may merely reflect the difficulty in adequately defining intelligence, but many people with an intuitive idea of what intelligence is, insist that intelligence and creativity are quite distinct qualities.

Studies made to try to correlate intelligence and creativity have produced conflicting results. Some studies seem to show a close correlation, others to show that the two qualities are almost independent of each other. This suggests that the originators of different trials may have had different ideas of the nature of creativity. The consensus of opinion, however, is that the two correlate well at low and average levels, but that when exceptionally gifted people are studied intelligence and creativity are disparate talents that may even be mutually incompatible.

We know that many adults of great originality were unremarkable as children. We also know that the highest achievers in science – the people who make the major advances – do not, in general, have exceptionally high IQs. They tend to be people of high average ability whose interest and imagination is caught by a particular subject, which they have thoroughly grasped, and who then concentrate very hard on it.

Creativity and the right side of the brain

In the great majority of people the left hemisphere of the brain contains the nerve centres for speech, language and language-related functions. It also controls movement of the right side of the body, including right hand activities. Because speech, language and writing are so central to our higher activities, the left hemisphere is called the dominant hemisphere. In a small proportion of people the right hemisphere is dominant in this way.

In most people, the right side of the brain is concerned with a wide range of non-verbal activities – things like spacial relationships, patterns, styles, design, data synthesis, metaphors, new combinations of ideas, and so on. The right brain is intuitive rather than logical, holistic rather than specific, and relational rather than factual. In short, all the important functions concerned with general artistic creativity are centred in the right side of the brain. This draws an interesting comparison between verbal creativity and the other forms of creative art. Verbal creativity must, of course, substantially involve the left side. Most would agree that there is a fundamental difference between the literary and the plastic arts.

Since the relative development of the different parts of the brain determines our abilities, we can infer that highly creative people have outstanding development in certain parts of the right side of their brains, relative to other parts. It is not then surprising that, in groups of such people, there may sometimes be a divergence between creativity and intelligence.

High levels of development in certain parts of the brain do not happen by chance. We know that innate abilities, or the structural basis for the development of such abilities, are often inherited. In such cases there will also inevitably be powerful early environmental influences operating to encourage the constant use and development of these faculties – a process associated with the development of the part of the brain concerned.

An infinite capacity for taking pains

Children showing outstanding early ability – child prodigies – are of special interest to creativity researchers. They are also of great interest to the general public in their power to evoke wonder and amazement. Most child prodigies seem to have been very one-sided geniuses and very ordinary in other respects. A clear distinction should be made between prodigies who are otherwise normal and those – the so-called 'idiot savants' who are retarded in other respects.

The general view that child prodigies achieve their remarkable success without great labour is almost certainly wrong. Studies of many musical composers have shown that most of them worked intensively for at least ten years before producing anything of merit. Even Mozart was drilled ruthlessly in composition by his father before, at the age of 12, he showed the first signs of his supreme mastery of the art. There is probably more truth than we imagine in Thomas Edison's witty remark: 'Genius is 1 per cent inspiration and 99 per cent perspiration'.

Research has shown that ordinary people can achieve feats of memory comparable to those of child prodigies if they work at them hard enough. After about 1000 hours of practice, normal people can learn to memorize lists of as many as 80 items. Without practice the limit is 8 or 10. It seems probable that in many child prodigies, unusual achievement is the result of an exceptional quality of mind that allows single-minded concentration on the recording and organization of experience so that great achievement is possible without help or even against opposition. When the mathematician Pascal was a child, his father, anxious he should study the classics, deprived him of the mathematical textbooks in which he was showing interest. So young Pascal secretly worked out and wrote his own textbook of geometry.

cremaster a thin layer of muscle looping over the SPERMATIC CORD and continuous with the internal oblique muscle of the abdominal wall. Its action is to draw up the testicle.

cremation disposal of bodies by burning. Nowadays, the great majority of people dying in Britain are cremated. In most historic traditions, cremation was considered more honourable than burial.

crenation the shrivelling of red blood cells that occurs when they are placed in a solution of greater concentration than that of the cell contents (a hypertonic solution) and water passes out of them.

creole a fully-formed language that has developed from a pidgin to become a primary vehicle of communication.

cribriform perforated like a sieve. The cribriform plate of the ethmoid bone allows the tiny nerve fibres of the nerve of smell (olfactory nerve) to pass though from the cranial cavity into the upper part of the nose.

Crick, Francis Francis Harry Compton Crick (1916–2004) received his basic scientific education as a student of physics at University College, London. During World War II he worked for the Admiralty on the development of magnetic mines. His interests later turned to biology, and in 1947 he moved to Cambridge where for two years he worked at Strangeways Research Laboratory. He then moved to the Medical Research Unit at the Cavendish Laboratory, where he worked under Max Perutz and Lawrence Bragg using X-ray diffraction methods to study the structure of proteins. James Watson joined him there in 1951 and an unofficial collaboration arose between the two men on the structure of DNA.

After the publication of the *Nature* paper in 1953, and the award of a PhD, Crick spent the next twenty years in Cambridge working on the problem of how DNA carries the genetic information and how this is translated into proteins. Crick was the first to suggest the role of transfer RNA in the assembly of amino acids into proteins. The basis of the genetic code, as triplets of bases, was finally established by Crick, in 1961, and he went on to determine the code for some of the 20 amino acids occurring in proteins. In 1962, Crick, James Watson and Maurice Wilkins were awarded the Nobel Prize for physiology or medicine.

Crick's book, *Of Molecules and Men*, was published in 1966. In 1977 he moved from Cambridge to take up a position in the Salk Institute in San Diego, California, where he embarked on a third scientific career in research into brain function. In 1981 he published a book, *Life Itself*, in which he speculated on the possibility that life may have originated outside the solar system.

cricoid 1 ring-shaped.
2 a ring-shaped cartilage in the voice-box (larynx).

crista a crest or ridge.

Cro-Magnon man one of a group of prehistoric, but anatomically modern, humans that lived in what is now the Dordogne and in other parts of France and Italy between about 30,000 and 10,000 years ago. The name derives from that of the cave in which the remains were first found.

cross-matching a test of the compatibility of blood intended to be transfused. Serum from the donor's blood is mixed with red cells from the recipient's blood. If the bloods are incompatible, the red cells will clump together (agglutination). See also BLOOD TRANSFUSION.

crossing-over the exchange of short lengths of CHROMATIDS between homologous pairs of chromosomes during one of the stages of division (meiosis) that occurs when the eggs (ova) and sperm are being formed. Crossing-over is one of the ways in which a random redistribution of genes occurs and ensures that the combinations of genes in each sperm or egg differs from the combinations in the cells of the parents.

cross-sectional plane a horizontal plane that divides the standing body into upper and lower parts. Also known as a transverse plane.

crown the visible part of a tooth. The part covered by enamel.

cruciate cross-shaped.

crus any leg-like structure.

crying the uttering of inarticulate sobbing or wailing sounds, associated with the secretion of tears and often with facial contortion, that expresses the emotion, usually of grief or sadness but sometimes of joy. Crying in babies and infants is prompted by minor distressful stimuli and has value in exercising the respiratory muscles, but may, if excessive, cause severe parental stress.

cryo- *combining form denoting* cold.

cryobiology the study of the effects of low temperatures on cells, tissues and organisms, including methods of using cold so as virtually to halt the processes of ageing and deterioration in living structures without causing serious damage.

cryonics freezing and storing the human body soon after death to preserve it indefinitely, in the hope that future scientific advances will allow correction of the process that caused the death, so that life can be restored.

cryoglobulins GLOBULINS that precipitate from solution and become visible on cooling.

cryoprecipitates substances isolated or purified from a solution by lowering the temperature or by freezing and then thawing. Cryoglobulin is demonstrated, and the antihaemophilic factor, Factor VIII, is obtained, in this way from blood plasma.

crypt any small recess, pit or cavity in the body.

crypto- *combining form denoting* hidden.

cryptorchidism, cryptorchism undescended testicle. The testicles develop in the abdomen and a testicle that fails to descend before puberty remains permanently sterile. Such a testicle is also liable to develop cancer.

crypts of Lieberkühn plain secreting glands of microscopic dimension lying in their millions between the villi of the MUCOUS MEMBRANE of the JEJUNUM and ILEUM that secrete mucus and the digestive enzymes of intestinal juice. The enzymes originate in Paneth cells that line the base of the crypts. (Johann Nathaniel Lieberkühn, 1711–56, German anatomist and microscopist).

crystalline lens the internal, fine-focusing, lens of the eye, which lies immediately behind the iris diaphragm and is suspended by a delicate ligament from the CILIARY BODY. In youth the lens is elastic and changes shape easily. Elasticity, and range of focusing power, fall off almost linearly with age.

crystal violet one of the many dyes used as a tissue and micro-organism stain for microscopic examination.

CSF *abbrev. for* cerebrospinal fluid.

CTLA4-Ig a fusion protein specific for the B7 surface receptor on T cells that has been shown to be capable of persuading T cells to recognize severely mismatched allogeneic transplanted organs as 'self' even in the absence of immunosuppressive drugs.

cuboid one of the bones of the foot. It lies on the outer side immediately in front of the large heel bone, the CALCANEUS, and behind the fourth and fifth metatarsal bones.

cubitus the elbow, especially the soft tissues of the elbow in front of the joint.

cubitus valgus an elbow deformity in which the forearm is tilted outward to an abnormal degree when the arms are by the sides. Some degree of such tilt, known as the 'carrying angle', is normal in women.

cultural competence possession of the knowledge and skills required to manage cross-cultural relationships effectively.

cuneiform 1 wedge-shaped.
2 one of the three wedge-shaped bones in the foot.

cunnus the VULVA.

cusp a projecting point.

cuspid a tooth with only one point on the crown. A canine tooth.

cutaneous pertaining to the skin.

cuticle 1 the epidermis or outer layer of the skin.
2 the narrow strip of thickened epidermis at the base of a fingernail or toenail.
3 the sheath of a hair follicle.

cutis the skin as a whole. The CORIUM.

Cuvier, Georges French Baron (1769–1832) who made a major step forward in the understanding of the principles of comparative anatomy, on a scientific basis. Cuvier, inspired by BUFFON, became interested in comparisons between the anatomies of different animals. He was a true scientist, declining to form theories until he had sufficient observable fact on which to base them – a sound principle, previously and subsequently often neglected. Cuvier's 9-volume *The Animal Kingdom Arranged in accordance with Structure* (1817–30) was widely influential, and he is regarded as the father of comparative anatomy.

cyanocobalamin vitamin B_{12}. This vitamin is necessary for the normal metabolism of carbohydrates, fats and proteins, for blood cell formation and for nerve function. It is used in the treatment of PERNICIOUS ANAEMIA and SPRUE.

cyanosis blueness of the skin from insufficient oxygen in the blood. Fully oxygenated blood is bright red and imparts a healthy pinkness to the skin. Blood low in oxygen is dark reddish-blue and, through the skin, looks a dusky blue.

Cyanosis may be due to lung disease, heart failure or disorders, especially congenital heart disease, in which, blood is shunted away from the lungs. 'Blue babies' have cyanosis.

cybernetics the study of the control and communication systems common to machines and animals, including the human being. The study of the analogies between complex feedback control systems and human physiology has been fruitful to both disciplines.

cyclic AMP a modified form of adenosine monophosphate in which a PHOSPHODIESTER BOND links the 5'- and 3'-carbons of the sugar within the molecule. Cyclic amp is a chemical messenger within the cell which, when external hormones reach the cell membrane, conveys information to the interior to initiate an appropriate response. It is sometimes called a 'second messenger'. It plays a key role in controlling biological processes. It activates protein kinases and controls GLYCOGEN synthesis and breakdown.

cyclins regulatory subunits of the kinases involved in the eukaryotic cell cycle. Cyclins are proteins whose concentration in the cells increases and decreases in phase with the cell cycle. Passage through the cycle is controlled by cyclin-dependent kynase complexes which are inactive unless associated with a cyclin.

cyclo- *combining form denoting* circular, cyclical, or the CILIARY BODY of the eye.

cyclo-oxygenase prostaglandin synthase, the enzyme that converts arachidonic acid to prostaglandins which are commonly mediators of pain.

cyclopia a congenital deformity featuring fusion of the eye sockets and the eyes, so that there appears to be only a single median eye.

cyesis pregnancy.

cysteine an AMINO ACID present in most body proteins.

cystic duct a narrow tube connecting the gallbladder to the common bile duct.

cysto- *combining form denoting* a bladder, sac or cyst.

cystoid cyst-like.

cytochemistry 1 the chemistry of cells.
2 an analysis of the chemical composition of cell components by staining properties and other means.

cytochrome an iron-containing protein electron carrier capable of easily being alternately oxidized and reduced.

cytochrome P450 a family of enzymes responsible for the detoxification and elimination of foreign substances including many drugs by hydroxylation and increasing their solubility.

cytokines a general term for a range of proteins of low molecular weight that exert a stimulating or inhibiting influence on the proliferation, differentiation and function of cells of the immune system by binding to specific receptors on the surfaces of these cells. Cytokines include INTERLEUKINS and INTERFERONS.

cytokinesis the movement of the CYTOPLASM during cell division.

cytology 1 the study of cells.
2 an abbreviation of the phrase 'exfoliative cytology' the examination of isolated cells, obtained from cervical smears, sputum or elsewhere, to determine whether or not they are cancerous.

cytomegalic characterized by enlarged cells.

cytoplasm the part of a cell outside the nucleus and inside the cell membrane.

cytoplasmic inheritance the genetic effects of DNA situated in MITOCHONDRIA.

cytoplasmic streaming the movement of cytoplasm in currents from one part of a cell to another as an internal transport system and in amoeboid movement of the entire cell.

cytosine a pyrimidine base, one of those forming the genetic code of DNA and RNA.

cytoskeleton a complex network of ACTIN filaments within the nucleated cell. Unlike the bony skeleton in vertebrates, this skeleton has contractile properties and can alter the shape, size and even movement, of the cell. The cytoskeleton is also concerned with the adhesion of adjacent cells.

cytosol the cell contents situated between the cell membrane and the nucleus, less the endoplasmic reticulum, the mitochondria and the other structured organelles.

cytotoxicity the property of being able to cause damage to, or death of, cells.

cytotoxic T cells T lymphocytes (usually CD8 cells) that kill target cells when they identify foreign MHC molecules on their cell membranes.

Dd

dacryo- *combining form denoting* tears or the lacrimal system.

dactyl a finger or toe. A digit.

dalton a unit of molecular mass roughly equal to the mass of a hydrogen atom. See also KILODALTON.

D-antigens the rhesus (Rh) ANTIGEN present in the red blood cells of 85 per cent of people, who are said to be rhesus positive. This antigen is inherited as an AUTOSOMAL dominant. See also RHESUS FACTOR.

dark adaptation the gradual acquisition of the ability to see in dim light that normally occurs in conditions of poor illumination. Dark adaptation becomes defective (night blindness) in vitamin A deficiency because this vitamin is necessary for the production of retinal VISUAL PURPLE.

dartos muscle a thin layer of muscle lying immediately under the skin of the SCROTUM. The dartos tightens in the cold causing the skin to wrinkle and the testicles to rise.

Darwin, Charles Charles Robert Darwin (1809–82) was the son of a successful English physician and grandson of the poet-physician Erasmus Darwin and of the porcelain manufacturer Josiah Wedgwood. He studied medicine but, shocked at the prospect of surgery without anaesthesia, decided he could not practice. He was then destined for the Church but his interests as an amateur naturalist soon displaced all theological fervour. In 1831 he was offered a place as ship's naturalist on HMS *Beagle*, which was about to set out on a voyage of scientific exploration. The voyage lasted for five years and was a torment of seasickness and ill-health for Darwin. In spite of this he was able to apply his keen naturalist's observation to the multitude of species he encountered and he returned to England with a mass of data. On the basis of his observations he later formulated a theory of evolution that was to make his name revered in the annals of science.

Darwin was a man of great character and unlimited generosity. When Alfred Russel Wallace sent him a paper closely embodying his own ideas, he made no attempt to publish quickly, but magnanimously circulated Wallace's paper to other interested scientists and collaborated with Wallace in the first presentation of the theory to the Linnaean Society. The controversy aroused by the publication of Darwin's *Origin of Species* – which was regarded by some as an attack on religion – lasted for many years and was deeply distressing to the author. Even so, in 1871 he published *The Descent of Man* in which he applied evolutionary principles to *Homo sapiens*. By the time of his death his ideas had gained almost universal acceptance.

deafness partial or complete loss of hearing. Deafness may be conductive or sensorineural. Conductive deafness results from disorders of the external ear, eardrum, middle ear and acoustical link to the inner ear; sensorineural (nerve deafness) results form disorders in the inner ear – the cochlea or acoustic nerve.

deaminase an ENZYME that brings about the breakdown of amino compounds.

deamination removal of the amino group from a molecule. When an NH_2 group is replaced by an oxygen atom a ketone is formed and the process is described as oxidative deamination. If the amino group is terminal, the process should, strictly, be called deamidation.

death

Death is the cessation of all biological functions and occurs because cells are deprived of the substances needed for continuing operation or are unable to use them because of poisoning or degeneration. Different functions cease at different times and some persist for hours after the heart beat and respiratory movements have stopped. Even after the heart has stopped beating, the organ remains alive and can be transplanted into another person to sustain life for years.

Many cells of the body continue to survive for a time after somatic death but, because all cells need oxygen and fuel and these are provided by a functioning circulation and respiratory system, the failure of the heart and lungs to maintain the supply is soon followed by cell death. Bacteria carry enzymes that break down tissue molecules to simpler substances and it is these that are mainly responsible for the final biological consequence of death – the return of the body to its chemical elements. Nowadays, it is common for us not to wait for bacterial action but to accelerate the process by burning.

The fear of death

There are few certainties in life but of one thing we can be confidently sure – that we will die. Young people do not really believe this, which is, perhaps, just as well, and are able to live as if they were immortal. This is a healthy and proper attitude. The elderly, having many more intimations of mortality, usually take a more realistic view, and recognize that they are not going to be spared the fate that has already overtaken so many of their contemporaries.

A study of attitudes to death has shown that 90 per cent of University students hardly ever consider the matter in relation to themselves, while in the case of older people 70 per cent often do. Fear of death is, however, usually much more acute in a young person when the imminent prospect must be faced, or in an older person suddenly stuck by a heart attack or facing extreme danger. For such, and for the young, the approach of death is the ultimate crisis, beside which everything else is insignificant. But for people who have lived out their life span, death is natural and normal and is usually accepted as such. Such people die without struggle or resentment and, if their affairs are in order, usually with an easy and accepting mind.

The experience of approaching death after illness or severe injury, even in the young, is usually well ordered by a beneficent nature. The effect on the brain, of the factors that are causing the death, brings its own panacea so that distress, horror and anxiety appear all to be abolished. This has been the experience of doctors and other observers of the dying through the ages. Some have even reported it of their own experience. William Hunter (1718–83), brother of the great anatomist John Hunter, and himself an experienced surgeon, said on his deathbed: 'If I had strength enough to hold a pen, I would write how easy and pleasant a thing it is to die.'

The wise French essayist Michele de Montaigne (1533–92) also put the matter well: 'It is not without reason we are taught to take notice of our sleep for the resemblance it hath with death. How easily we pass from waking to sleeping; with how little interest we lose the knowledge of light and of ourselves. For, touching the instant or moment of the passage, it is not to be feared that it should bring any travail or displeasure with it, forasmuch as we can have neither sense nor feeling...'

death rate the ratio of the number of deaths to the total of the population concerned.

death wish Freud's 'thanatos', which, like so many of his concepts, was derived from classical mythology. This idea, conceived late in his career, proposed that responses such as denial and rejection of pleasure or the repeated seeking of extreme danger indicated a general wish or instinct for death.

decalcification loss of calcium and other mineral salts from the normally mineralized tissues, bone and teeth. This occurs in OSTEOMALACIA and in OSTEOPOSOSIS.

decarboxylases enzymes that promote the freeing of CO_2 from –COOH. These enzymes are involved in the synthesis of amine regulators and neurotransmitters such as serotonin and dopamine.

decibel a logarithmic unit of comparison between a standard power level and an observed level. The decibel is not a unit of sound intensity but of power. It is, however, widely used to compare a noise level with a very low standard reference level near the limit of audibility, and to compare electrical power levels. A tenth of a bel.

decidua the thick lining of the womb (endometrium) during pregnancy with its associated membranes that are cast off with the PLACENTA after the birth of the baby.

deciduous shed or falling at a particular time or stage of growth. Sometimes applied to the primary teeth.

decomposition separation into chemical constituents or simpler compounds often as a result of bacterial enzymatic action.

decubitus the reclining position.

decussation a crossing so as to form an X, especially of tracts of nerve fibres.

deduction a method of logical reasoning by which new information can be inferred from a consideration of established data. In deduction conclusions follow necessarily from premises and cannot be false if the premises are true. Compare INDUCTION.

defaecation, defecation voluntary or involuntary emptying of the RECTUM so as to relieve oneself of accumulated faeces. Stretching of the wall of the rectum causes a conscious desire to defaecate, but if this is prevented by voluntary decision the rectal wall relaxes and the desire fades until the next movement of faeces from the colon.

Deliberate inhibition is a common cause of constipation.

defence mechanisms methods of coping with anxiety caused by conflict between desires and socially approved behaviour. Common defence mechanisms include:

Rationalization
This is one of the commonest and most widely used defence mechanism. The method is to find a plausible reason for an action or omission so that the true emotional reason is effectively concealed. One might insist that one simply cannot find time to consult a doctor when, in fact, one is terrified of discovering that one has cancer.

Denial
Pretending that the problem does not exist, typified by an attitude of cheerfulness in the face of adversity or danger. Denial of profound emotion is one of the ways of coping with grief.

Repression
Repression is not quite the same as denial, since in repression, something that causes us unpleasant or painful feelings is simply forgotten. This is an active protective process to spare us continual humiliation, regret, pain or discomfiture.

Projection
Here, a painful sense of personal defect is projected onto another person, or group of people. People may be astonished to be accused, by friends, of faults they are sure they do not possess but of which they are thoroughly familiar in the accuser. Dishonest people may, for instance, attribute dishonesty to others.

Substitution
Emotion can be so strong that it must have an outlet. When the logical outlet – against the cause – is impossible or unwise, the emotion may be directed against something, or someone, else. Anger against fate may be vented on an innocent person. Anger against the boss may be directed against the spouse.

Splitting
Some people cope with anxiety by splitting the world into the good and the bad, identifying strongly with, and hoping for the support of, the good group, and blaming the bad group for everything. This may cause racism.

Dissociation or Conversion
Strong emotion, especially fear, is disposed of by its conversion into a physical symptom such as loss of sensation in, or paralysis of, a limb.

Sublimation
Emotional needs, such as the sexual, that cannot for some reason be gratified in the most direct and obvious way, may be satisfied by devoting oneself to some other purpose, such as various forms of voluntary social service.

deep to referring to any part of the body which is nearer the centre than the part referred to. Thus the muscles of the abdominal wall are deep to the skin, and the peritoneum is deep to the muscles.

defibrillation the restoration of the normal beat rhythm in a heart which is in a state of rapid, ineffectual twitching – one kind of CARDIAC ARREST. A strong pulse of electric energy (about 300 joules) is passed across the heart from two metal electrodes pressed to the chest.

defibrillator an electrical device for applying sudden high-energy shocks to the heart in the attempt to convert VENTRICULAR FIBRILLATION into normal heart rhythm. See also DEFIBRILLATION.

deformity the state of being misshapen or distorted in body.

degeneracy in the genetic code this is a reference to the redundancy of codons arising from the fact that four bases, taken three at a time, offer 64 possibilities, while it is necessary to code for only 20 amino acids and three stop signals. The effect is that in many cases a change in the third base of a codon will not change the amino acid selected.

degeneration structural regression of body tissue or organs, from disease, ageing or misuse, which leads to functional impairment, usually progressive.

deglutition swallowing.

degradation the breaking down of compounds into simple molecules, such as when a protein is degraded by enzymatic action to polypeptides or amino acids.

dehydration a reduction in the normal water content of the body. This is usually due to excessive fluid loss by sweating, vomiting or diarrhoea which is not balanced by an appropriate increase in intake.

dehydroepiandrosterone quantitatively the principal male sex hormone (ANDROGEN), of the adrenal cortex. Output declines with age, a decline thought by some to be causally related to ageing and to the development of various diseases.

dehydrogenases a large number of enzymes that activate oxidation-reduction reactions by the removal of a pair of hydrogen atoms from a molecule.

déjà vu the sudden mistaken conviction that a current experience has happened before. There is a compelling sense of familiarity and often a persuasion, almost always immediately disappointed, that one knows what is round the next corner.

deletion in genetics, the removal of a segment of DNA with joining up of the cut ends, as in the loss of a segment of a chromosome. Deletion of a single BASE PAIR is one of the kinds of point mutation. Deletion of a base pair triplet (codon) will result in a protein with a missing amino acid.

deliquescent having the property of taking up water from the atmosphere in sufficient quantity to dissolve itself.

delirium a mental disturbance from disorder of brain function caused by high fever, head injury, drug intoxication, drug overdosage or drug withdrawal. There is confusion, disorientation, restlessness, trembling, fearfulness, DELUSION and disorder of sensation (HALLUCINATION). Occasionally there is maniacal excitement.

delivery the process of being delivered of a child in childbirth.

delta cells cells in the pancreas or intestine that secrete SOMATOSTATINS.

delta wave a low-frequency brain wave, recordable on the electroencephalogram, that originates in the frontal part of the brain during deep sleep in normal adults.

deltoid triangular. Shaped like the triangular Greek letter 'D'.

deltoid muscle the large, triangular 'shoulder-pad' muscle which raises the arm sideways.

deltoid ligament the strong triangular ligament, on the inner side of the ankle, which helps to bind the foot to the leg. A torn deltoid ligament may leave an unstable ankle that 'goes over' easily.

delusion a fixed belief, unassailable by reason, in something manifestly absurd or untrue. Psychotic delusions include delusions of persecution, of grandeur, of disease, of abnormality of body shape, of unworthiness, of unreality and of being malignly influenced by others.

dementia a syndrome of failing memory and progressive loss of intellectual power due to continuing degenerative disease of the brain. About half are believed to be due to Alzheimer's disease and about one third to small repeated strokes.

demise death.

denaturation 1 alteration in the folding pattern of a protein by heat or chemical reaction from its physiological conformation to an inactive shape. Various non-covalent bonds are disrupted.
2 in the case of DNA or RNA the conversion from a double-stranded structure to a single stranded structure, usually by heating. This is an essential stage in the POLYMERASE CHAIN REACTION.

dendrite one of the usually numerous branches of a nerve cell that carry impulses toward the cell body, having received signals from the axons of other neurons. Dendrites allow the most complex interconnection between nerve cells, as in the brain, so that elaborate control arrangements over the passage of nerve impulses are made possible. Recent research suggests that sections of some dendrites can function independently.

dendritic cell see LANGERHANS CELL.

dendron see DENDRITE.

dent- *combining form denoting* tooth or DENTAL.

dental pertaining to the teeth or to dentistry.

dentine the hard, calcified tissue that makes up the greater thickness of the tooth. It is denser and harder than bone, but softer than the outer enamel coating and contains tubules of cells which connect the inner pulp of the tooth to the surface.

dentinoma a benign tumour of tooth DENTINE.

dentition pertaining to the teeth. The primary dentition consists of 20 teeth, the secondary, or permanent, dentition, usually 32.

dentulous possessing teeth. The opposite is edentulous.

deoxy- *prefix denoting* removal of an oxygen atom from a molecule.

deoxygenation removal of oxygen.

deoxyribonuclease an enzyme that cuts DNA strands by breaking PHOSPHODIESTER BONDS.

deoxyribonucleic acid see DNA.

deoxyribose a sugar, part of the 'backbone' of the DNA double helix, deoxyribonucleic acid. In RNA, the equivalent sugar is ribose.

depigmentation loss of normal pigmentation.

depolarizing capable of bringing about depolarization.

depolarization the immediate cause of the formation of a nerve impulse. Nerve fibres normally carry a positive charge of some 70 millivolts on the outside of the fibre, which is balanced by an equal negative charge on the inside. When movement of potassium ions causes a local reversal of this polarization, the fibre is said to be depolarized. A zone of depolarization then passes along the fibre. This is the nerve impulse.

depression sadness or unhappiness, usually persistent. This may be a normal reaction to unpleasant events or environment or may be the result of a genuine depressive illness.

deprivation failure to obtain or to be provided with a sufficiency of the material, intellectual or spiritual requirements for normal development and happiness.

deprivation syndrome a state of developmental retardation, both physical and emotional, and sometimes intellectual, resulting from early parental rejection. The effect is lifelong and may involve grave psychosocial disadvantage.

derepressed the state of a gene that is turned on. Also described as induced. See also REPRESSED.

derma- *combining form denoting* skin.

dermatitis inflammation of the skin from any cause. Dermatitis is not a specific disease, but any one of a large range of inflammatory disorders featuring redness, blister formation, swelling, weeping, crusting and itching.

dermis the true skin (cutis vera) or corium. The dermis lies under the EPIDERMIS.

Descartes, René the French philosopher Rene Descartes (1596–1650) who was responsible for a revolution in philosophic thought that promoted acceptance of the importance of science. His clear and easily accessible writings helped to turn thought away from imaginative, but unsupported, assertion into the direction of observation and rationalism. Descartes was determined to believe only those things about which he could be entirely certain. To this end he used various hypotheses to try to test his beliefs. One of these was the hypothesis that an evil genius existed whose whole purpose was to deceive him. This method seemed powerful but immediately raised the difficulty that the evil genius might be deceiving him into believing that he, the philosopher, existed, when, in fact, he did not. Descartes' answer to this was the logically dubious axiom '*Cogito ergo sum* I think, therefore I am (I exist)'.

Descartes held that human beings are composed of two kinds of substances – mind and body. The mind was dimensionless and indivisible but conscious, capable of volition, understanding, imagination, perception and will. The body was an entity extended in space and infinitely divisible. The mind, being dimensionless was capable, in theory, of surviving the death of the body. These two parts, he believed, were capable of interacting causally and the mind was able to make the body move by moving a small part of the brain. This it did by way of the pineal gland. Such movements could also cause sensation and emotion.

Some of Descartes' ideas on the mind–body problem have had to be set aside in the light of more recent physiological knowledge, but the general notion of the mind and the body as distinct, but causally-related, entities was almost unquestioned for three centuries. It is now being seriously challenged by those who recognize that the mind is an epiphenomenon of brain activity, the two being so intimately inter-related as to be more usefully considered to be two aspects of one entity.

descending tracts bundles of motor nerve fibres that carry nerve impulses from the brain down through the through spinal cord.

designer baby a baby derived from an embryo selected for a particular purpose or from one whose genome has been modified for a particular purpose.

desmosomes protein plaques on cell membranes connected to intermediate filaments, that form contact areas linking adjacent cells, especially epithelial cells, and between cells and the extracellular matrix.

detoxification 1 the alteration of a substance in the body to a non-poisonous form, either as a spontaneous biochemical reaction or as a result of medical treatment with an antidote.

2 the process of treating a person for an addiction to a drug such as alcohol or heroin. This usage is largely metaphorical; in practice the process involves prohibition rather than removal of a toxic substance.

detrusor 1 any entity that pushes down.

2 the muscle of the bladder.

detumescence a return to normal, from a swollen state, of an organ or part, especially the penis.

deuteranomaly partial DEUTERANOPIA.

deuteranopia a form of colour blindness (colour perception defect) causing a tendency to confuse blues and greens, and greens and reds and with a reduced sensitivity to green.

Devonian period the period in geological time from 408 million years ago to 360 million years ago.

de Vries, Hugo Marie in 1900, Hugo Marie de Vries (1848–1935), professor of botany at the University of Amsterdam, had effectively worked out laws of inheritance by research on plants. Before publishing, he reviewed the literature to see if anything had been done on the subject. It was with mixed feelings that he discovered a paper by Gregor Mendel, and realized that the laws had been discovered a quarter of a century before. De Vries was not the only scientist to be thus disappointed. Two others, Erich von Tschermak and Carl Correns, unknown to him or to each other, had independently worked out the laws and then had found Mendel's paper. In the best tradition of science, all three immediately acknowledged Mendel's priority and referred to their own work only in confirmation.

But de Vries was able to go further. In 1901 he proposed a new theory – the mutation theory. Aware that new varieties of plants and animals that bred true occasionally occurred, he suggested that such alterations in hereditary characteristics were the result of actual permanent changes – mutations – that had taken place in the hereditary material of the parent. This theory has since been universally accepted and remains the explanation for the one remaining gap in Darwin's theory – the origin of the spontaneous changes in species.

dextrocardia the congenital anomaly in which the apex of the heart points to the right instead of the left. Dextrocardia is often associated with a similar mirror-image reversal of the abdominal organs.

diacylglycerol (DAG) an intracellular second messenger formed from phosphoinositides on receipt of stimuli from specific surface receptors.

diakinesis one of the stages in the process of division of eggs and sperm which ensures that the number of CHROMOSOMES is halved (meiosis). In diakinesis the chromosomes shorten and thicken and the spindle fibres form, ready for the separation of the chromosomes.

dialysis separation of substances in solution by using membranes through which only molecules below a particular size can pass. Dialysis is the basis of artificial kidney machines.

diapedesis the passing of blood cells through the intact CAPILLARY wall into the tissue spaces. Diapedesis is a feature of INFLAMMATION.

diaphoresis heavy perspiration, especially when medically induced.

diaphragm 1 the dome-shaped muscular and tendinous partition that separates the cavity of the chest from the cavity of the abdomen. When the muscle contracts the dome flattens, thereby increasing the volume of the chest. 2 any partitioning structure, such as the iris diaphragm of the eye.

diaphysis the shaft of a long bone. Distinguish from EPIPHYSIS, the growth zone at the ends of a long bone.

diarrhoea the result of unduly rapid transit of the bowel contents so that there is insufficient time for reabsorption of water to firm up the faeces. In consequence, the stools are loose and liquid and are passed more frequently than normal. The commonest causes are irritation from a bowel infection and psychological factors, as in the IRRITABLE BOWEL SYNDROME.

diarthrosis a freely movable joint with a SYNOVIAL MEMBRANE.

diastase an ENZYME capable of breaking down starch. An amylase.

diastole the period in the heart cycle when the main pumping chambers (the ventricles) are relaxed and filling with blood from the upper chambers (the atria).

diastolic pertaining to DIASTOLE. The diastolic blood pressure is the pressure during diastole and is the lower of the two figures measured. The peak pressure is called the SYSTOLIC pressure.

diathesis an inherited predisposition to a disease or condition.

dicephaly two-headed. A gross monstrous congenital anomaly.

dichromatism partial colour blindness in which only two of the primary colours can be perceived.

dicrotic having two waves, especially of a pulse.

diencephalon the central, lower part of the brain that contains the BASAL GANGLIA, THE THALAMUS, the HYPOTHALAMUS, the PITUITARY gland.

diet

With the exception of the original single fertilized cell from which we developed, every molecule in the body has been acquired from outside, mostly by way of the mouth. In a quite literal sense our bodies are made of what we eat. It is, however, important to bear in mind what happens to the food we eat before it is assimilated into our bodies. A particular amino acid is exactly the same as any other sample of the same amino acid regardless of differences of source. Vitamin C derived from a fresh orange is identical in every way to vitamin C derived from a cheap multivitamin pill, and the effect of the two molecules on the body is identical. Since nearly all the food we eat is broken down, in the process of digestion, to simple dietary elements of this kind – sugars, fatty acids, amino acids, minerals and vitamins – the source of these elements is, physiologically, a matter of indifference. There is, however, as we shall see, another sense in which the quality of our diet is very important.

Food

Food is the source of three groups of substances required by the body – fuel to be burnt (oxidized) to provide energy for all the physiological functions; constructional materials for growth, maintenance and repair; and atoms and molecules needed for many biochemical cell functions.

The standard body fuels are glucose and fatty acids. These are derived from the carbohydrates in the diet – the sugars and starches – and from the fats (triglycerides). Although glucose is the principal fuel, comparatively little glucose is stored – and that mainly in the liver, as a polymer called glycogen, that can break down easily to release free glucose molecules. The body's long-term energy stores are in the form of fats. The glycerine (glycerol) and fatty acids absorbed after a meal are carried to the fat cells, mainly under the skin, where they are resynthesized to triglycerides consisting of three fatty acids linked to a 'backbone' molecule of glycerol. Between meals, some of the fat in these stores is broken down to glycerol and fatty acids and these are released into the blood for use as fuel in the cells. Amino acids from the muscles are used as fuel only when the other fuel sources are almost exhausted.

The main body-building 'brick' is the amino acid. The 20 different amino acids are derived from the protein in the diet by digestion by the protein-splitting (proteolytic) enzymes of the stomach, pancreas and small intestine. Some of the necessary amino acids can be synthesized by the body. Those that cannot are called essential amino acids because they must be provided in the diet. Amino acids are small, fairly simple molecules that link together readily to form polypeptides. These in turn, link up to form proteins. The order in which the amino acids are linked, and their identity, determine the type of protein and these data are prescribed in the DNA in terms of the order of the sequence of bases. Amino acids are present in all cells, both free and linked to form proteins. Other structural elements derived from the food include the minerals calcium and phosphorus which contribute rigidity to the protein scaffolding of the bones.

Necessary biochemicals include the vitamins, the essential amino acids, the minerals calcium, phosphorus, magnesium, sodium potassium, fluorine, and some metallic elements such as iron, copper, zinc, selenium and manganese. Apart from some of the vitamins which are synthesized in the body, we have no other source of any of these substances and they must be provided in the diet.

Vitamins and their function

Vitamins are organic substances required in very small amounts for specific chemical reactions in cells. These reactions cannot use more than the quantity needed and excess has no useful effect and can sometimes be dangerous. A normal, well-balanced diet will provide all the vitamins needed. Some vitamins are stored in large quantity in the body; some are hardly stored at all. Some, such as vitamin K are even manufactured in the body. These facts have a bearing on how readily deficiency can occur.

Vitamins have a wide range of functions. Vitamin A promotes normal production of a substance called retinol-binding protein and is needed for growth, maintenance of surface tissues (epithelia), normal vision and reproduction. All the members of the large B group are co-enzymes – chemicals necessary to allow particular enzymes to work so that cell function can proceed normally. Vitamin C is necessary for the proper synthesis of collagen and thus for healthy body structure. Vitamin D is a hormone that regulates calcium deposition in bone. Vitamin E stabilizes cell membranes. Vitamin K is necessary for the normal production of the blood clotting factors (Factors II, VII, IX and X) by the liver.

Free radicals are highly active chemical groups capable of damaging body tissue. They are produced by many agencies including disease, radiation and smoking. Vitamins C and E, and some other dietary elements such as flavonoids, act as antioxidants, 'mopping up' free radicals. Their role in a wide range of conditions associated with free radical damage is the subject of much research.

Effects of starvation

Starvation is the result of insufficient food or of failure to absorb or utilize food. The chief effect is weight loss and wasting of the body. First the fat stores are consumed, then the muscles, including the heart muscle, and the other organs, especially the liver. There are also the effects of vitamin deficiencies, especially those of vitamin A – dryness of the eyes (xerophthalmia) and blindness from corneal melting. Unrelieved starvation leads to uncontrollable diarrhoea from changes in the intestines, widespread infections from immune deficiency resulting from shortage of antibody protein, and death from heart failure as a result of loss of heart muscle. It usually takes many weeks to die of starvation. The commonest cause of starvation in the Western world is anorexia nervosa.

Diet and disease

The coronary arteries supply the heart muscle with blood. If they are narrowed by plaques of cholesterol-containing material (atheroma), the supply of blood to the heart muscle may be insufficient to allow normal function under increased demand. This is called ischaemia and it causes angina pectoris and severe limitation of activity. If a blood clot forms on top of a plaque and obstruct the artery completely (coronary thrombosis), or if the narrowed artery goes into spasm, the result is a heart attack with grave or even fatal results.

Scientific trials have repeatedly shown that men with a high proportion of saturated fats in their diets are more likely than average to suffer heart disease from narrowing of the coronary arteries.

Protein intake appears to have no effect on coronary artery disease but it is difficult to avoid saturated fats when consuming a high protein diet. Surveys have also shown that those with a high intake of fruit and vegetables, starch (potatoes and flour) and other complex carbohydrates are less likely than average to develop coronary heart disease. The type of the tissue fat deposits reflects the type of the dietary fat intake.

Blood cholesterol varies more, from person to person, than any other blood constituent and appears to be the most important of the known risk factors for coronary artery disease. In more than twenty studies in different countries, total blood cholesterol was found to be related directly to the development of coronary heart disease. The higher the blood cholesterol, the more likely the development of coronary problems. The association held for both sexes, and in every population studied the risk was higher for people with higher levels of blood cholesterol.

High total blood cholesterol implies high levels of the cholesterol carriers in the blood. These are called low density lipoproteins (LDLs), and the risk is associated with the level of LDLs in the blood, rather than with the absolute amounts of cholesterol. A high intake of polyunsaturated fats, relative to saturated fats, lowers the levels of low density lipoproteins. The level of saturated fats (animal and dairy product fats that are usually solid at low room temperatures) should be no more than 10 per cent of the total calorie intake.

Energy release from body fuels

In every cell in the body, a remarkable cyclical biochemical process is going on continuously to free the energy needed for all the numerous cell activities, such as formation of proteins, muscle contraction and movement of materials within and between cells. This is called the Krebs cycle after Sir Hans Adolf Krebs (1900–81) who first described it in 1937. Krebs was awarded the Nobel Prize for Physiology or Medicine in 1953.

The Kerbs cycle, or citric acid cycle, is complicated. In very simplified form it involves a sequence of enzymes found in the mitochondria which act on the three fuel molecules – glucose, fatty acids and amino acids to produce carbon dioxide and to build up a substance of central importance called adenosine triphosphate (ATP) from a simpler form adenosine diphosphate (ADP). ATP, which has three phosphorus atoms in the molecule, and ADP, which has two, are present in every living cell from bacteria to humans. In the synthesis of a molecule of ATP from a molecule of ADP, a large amount of energy, derived from the oxidation of the food fuels, must be added. This energy, stored in the ATP, is soon released where it is needed in the cell, when ATP breaks down to ADP. With each turn of the Krebs cycle an ADP molecule has a phosphorus and some oxygen atoms added (oxidative phosphorylation), plus energy, to change it back to a molecule of ATP.

dietary fibre a group of complex carbohydrates that includes plant cellulose, lignin, pectins and gums. These polysaccharides resist digestion and thus cannot be absorbed, but remain in the intestine until excreted, providing a useful sense of fullness or satiety. Fibre is of value in the management of OBESITY. It bulks out the stool and is useful in the treatment of constipation and diverticulitis. Dietary fibre reduces the risk of colorectal cancer possibly by removing carcinogens. Some soluble fibres bind bile cholesterol and prevent it from being reabsorbed. This can lower blood cholesterol. High fibre foods include vegetables and fruits, bran, beans, peas and nuts.

dietetic pertaining to diet.

dietetics the science of the principles of nutrition and their application in the pursuit of health. Dietetics includes the scientific selection of meals for people with digestive, metabolic and malnutritional disorders.

dietician a person trained in DIETETICS.

differential blood count an assessment of the percentage numbers of the various types of white cells present in the blood – the neutrophil polymorphs (40–75%), lymphocytes (20–45%), monocytes (2–10%), eosinophils (up to 6%) and basophils (up to 1%). Changes in the normal percentages are usually significant.

differential splicing see ALTERNATIVE SPLICING.

differentiation the process by which stem cells acquire the special characteristics of the tissues into which they are developing.

diffusion the natural tendency for dissolved substances to move from areas of high concentration in a solution to areas of low concentration. This is a consequence of the random motion of atoms, molecules or ions.

digastric 1 of a muscle having two bellies connected by a thinner tendinous part. 2 a muscle that acts to open the mouth by moving the jaw bone (mandible) down.

digestion the conversion of food into a form suitable for absorption and use by the body. This involves both mechanical reduction to a finer consistency and chemical breakdown to simpler substances.

digestive system

The digestive system is best regarded as a fuel and materials processing factory that, in a topological and functional sense, lies outside the body. The input is a wide range of edible material and the product is a comparatively small list of absorbable materials – mainly glucose, amino acids, fatty acids, vitamins and minerals – that are needed by the body. The amount of the residue of unabsorbable material varies with the composition of the diet, but is also small. Most of the content of the faeces does not come from the diet unless this is unusually high in unabsorbable cellulose. Foodstuffs are chemically complex and considerable physical and biochemical processing is needed before they are reduced to the state in which they can be absorbed and assimilated.

It may seem strange to regard the main part of the digestive system – the intestinal tract – as being outside the body, but in a practical sense this is true. The inner surface of the gut, from the mouth to the anus, is continuous with the surface of the skin. Unabsorbable objects and materials can pass through the intestinal tract without ever entering the body proper and this is of practical importance in surgery and in bacteriology.

Mechanical processing

The first stage in the processing of food is its reduction to a suitable chunk size and consistency for swallowing. This is the function of the teeth, the muscles of mastication, the tongue and the salivary glands. Teeth are of different shape and have different functions. The central eight sharp-edged teeth, the incisors, are cutters and are used to sever food by biting. On either side of the incisors are the pointed canines and pre-molars. These are used, in conjunction with the fingers, to tear food. Food divided in these ways is then shifted by the tongue to lie between the molar teeth, which have irregular upper surfaces of much greater area than the other teeth and are used as crushers and grinders.

By the time food reaches the grinding stage it must be mixed with a lubricating and compacting fluid if it is to retain its form as a bolus. This fluid is provided by the three pairs of salivary glands – the parotid glands in the cheeks, the sub-lingual glands immediately under the tongue in the floor of the mouth, and the submandibular glands deeply set within the curve of the jaw bone (mandible). The presence of food in the mouth (or even, sometimes, the contemplation of food) prompts these glands to produce an alkaline fluid which runs through short ducts into the mouth. Saliva also contains an enzyme, amylase, that quickly promotes the breakdown of the carbohydrate polysaccharide starch to simpler (and sweeter) sugars. A piece of chewed bread or potato held in the mouth soon becomes sweet, as the disaccharide maltose is released by the enzyme action.

Swallowing

Swallowing (deglutition) is partly voluntary and partly involuntary and, ideally, should not occur until food has been thoroughly broken down in the mouth and mixed with saliva to form a kind of rough paste. Once the bolus is in suitable form it is shifted to the centre of the mouth by the tongue. A complex sequence of muscular actions follows which is controlled by nerve centres in the brainstem. First the bolus is pushed to the back of the mouth by the middle part of the tongue. Simultaneously, the rear part of the tongue flattens and the soft palate presses firmly upwards to seal off the mouth from the cavity of the nose. Once the bolus enters the upper part of the throat (pharynx) swallowing becomes largely involuntary, although it can be inhibited by violent action.

The muscles surrounding the pharynx relax and at the same time the entrance to the voice box (larynx) is closed off by a leaf-shaped cartilage (the epiglottis) lying over it. Breathing is temporarily stopped and the vocal cords are pressed tightly together. Next, the muscle ring at the upper end of the gullet (oesophagus) relaxes and

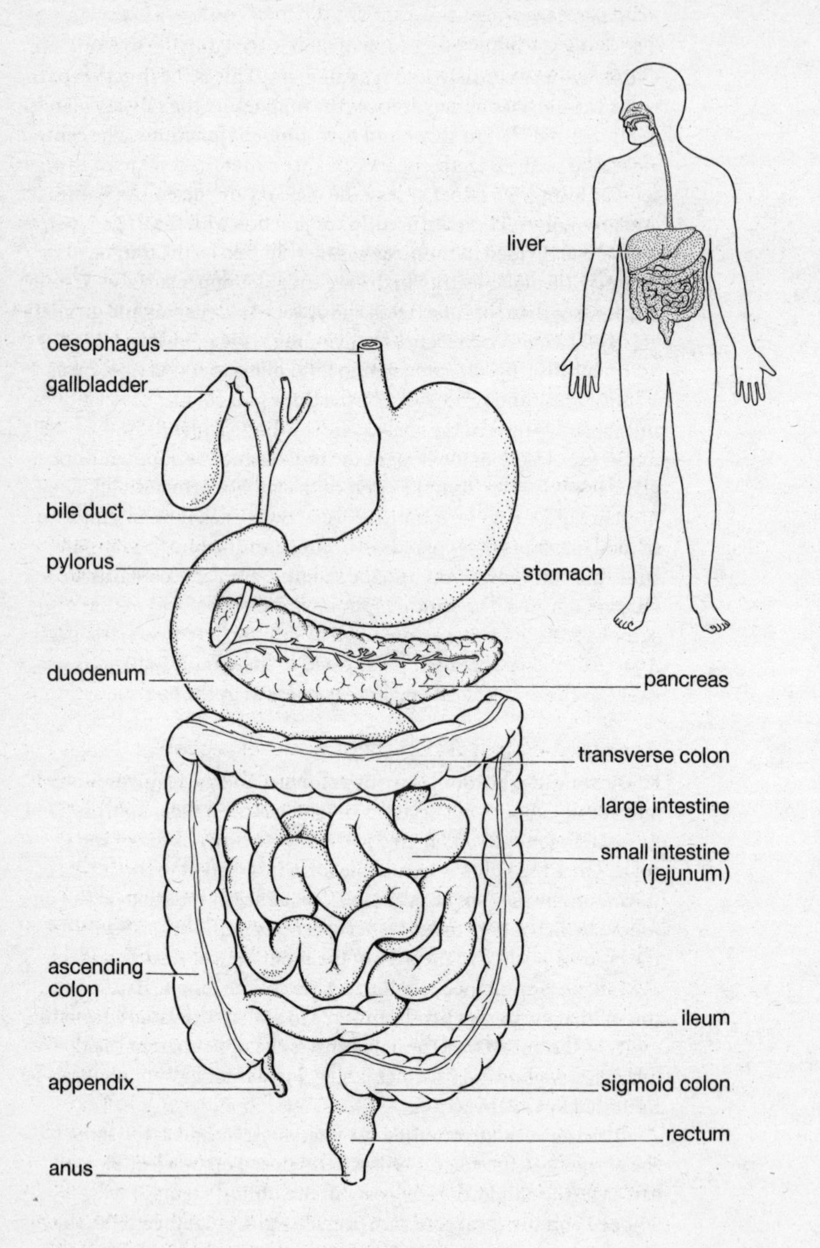

Fig. 4 **Digestive system**

food passes down. As this happens the vocal cords relax and breathing is resumed. The bolus is now carried rapidly down the oesophagus by the action of peristalsis. Gravity is unnecessary; it is possible to swallow efficiently while standing on one's head.

Stomach

This is more than simply a bag-like reservoir for food. The stomach also has important digestive functions. It is the widest part of the digestive tract and lies immediately under the diaphragm towards the left side. Its outlet to the duodenum is narrow and in this region the three muscular coats of the stomach wall thicken to form a strong muscular ring called the pylorus that controls emptying. The stomach lining is thick and thrown into deep folds and contains many glands. These secrete an important protective mucus, that prevents the stomach from digesting itself, a digestive enzyme, pepsin, that can break down protein and a strong acid, hydrochloric acid.

The stomach is highly active and is seldom still, constantly churning the contents to ensure thorough mixing. About two litres of acid are secreted each day. Hydrochloric acid has various functions. It kills germs in the food and prevents food putrefaction in the intestine, it activates the enzyme pepsin from its precursor, pepsinogen, and it acts on the cell walls of food to release the cell contents so that the protein can be acted on by the pepsin.

Small intestine

The adjective 'small' refers to the diameter of this part of the gut, not to its length, which is some 6 metres. The first short part is called the duodenum because it is about 12 finger-breadths long. The other successive parts are the jejunum where most digested material is absorbed and the ileum, where considerable water is re-absorbed. The bile duct from the liver and gall bladder and the duct from the large gland, the pancreas, enter the duodenum at the same point. The lining of the small intestine also contains glands that secrete digestive enzymes.

When food leaves the stomach, the pepsin has already acted on the protein, breaking it down to polypeptides. In the duodenum the polypeptides meet enzymes from the pancreas, mainly trypsin, that complete the breakdown to amino acids. The pancreatic juice is highly alkaline and this helps to neutralize the strong acid leaving the stomach. Pancreatic juice also contains enzymes that break down complex carbohydrates and fats. Not all carbohydrates can be split by human digestive enzymes. Polysaccharides like cellulose and various fibres such as pectin are not digested and pass on through. Soluble fibre, however, serves a useful purpose in binding cholesterol in the bile and in reducing the incidence of various diseases of the large intestine. The other carbohydrates in the diet are easily broken down to monosaccharides by the pancreatic enzymes. Pancreatic amylase splits starch to dextrin and then to maltose. Pancreatic maltase then splits maltose into molecules of the monosaccharide glucose. The disaccharide lactose is split to the monosaccharides glucose and galactose, and the disaccharide cane sugar (sucrose) is split to glucose and fructose. These monosaccharides are easily absorbed.

The most important function of bile is to emulsify fats so that they can more readily be acted on by the fat-splitting enzymes. Fats (triglycerides) are broken down to glycerol and fatty acids. The latter react with the alkaline intestinal juice to form soaps and these, too, have an emulsifying action on further fats.

The lining of the small intestine is covered with millions of fine finger-like processes called villi which greatly increase the total surface area. Each villus contains fine blood vessels and lymph vessels and the wall is so thin that small molecules can easily pass through into the bloodstream or, in the case of fats, into the lymph vessels. This movement is called absorption and after a meal a large quantity of simple sugars, amino acids and fatty acids, together with minerals and vitamins (also small molecules) pass into the blood. Most of the blood returning from the intestine passes directly to the liver where further processing occurs and many of the absorbed materials are temporarily stored.

Colon

The colon or large intestine is about 1.5 metres long but is much wider than the small intestine. Its lining is smooth and free from villi. It secretes no enzymes and produces only mucus. The main function of the colon is to absorb water and concentrate and firm up the faeces. Towards its lower end is a wider segment, called the rectum. When colon contents move into the rectum there is a conscious desire to defaecate. The contents of the lower colon consist largely of bacteria, but also include cells cast off from the intestinal lining and some cellulose residue from the diet. The colour is derived from the bile. Colonic bacteria synthesize small quantities of certain vitamins which are absorbed. If diet is adequate this is an unimportant source.

Segmentation and peristalsis

The wall of the intestine contains three layers of muscle running in different directions. The orderly contraction of these muscles is controlled by a network of nerves also running in the wall. The commonest kind of muscle action in the bowel produces an effect known as segmentation, the purpose of which is to mix the bowel contents with enzymes and to promote absorption. In this process, short segments of circular wall muscle of the intestine contract and squeeze the contents. At the same time, the segments between these contracting zones relax. The relaxed zones now contract and the contracted zone relax. Segmentation ensures thorough mixing but has no effect on the net movement of the bowel contents.

After digestion and absorption are largely complete the bowel contents must be moved on. The mechanism that effects this is called peristalsis. Because the intestine is so long an effective process for moving the contents is needed if blockage is to be prevented. In peristalsis, a number of separate short segments of bowel wall contract while, at the same time, the segments lying on the downward (distal) side of the contracting segments relax. The same process then occurs repeatedly a little further along the intestine. The effect of this is to squeeze the bowel contents in the direction of the anus. If peristalsis fails, as in certain severe intestinal diseases, obstruction of the bowel occurs.

Intestinal waste disposal

Food residue, mainly cellulose, forms only a small proportion of the faeces. Only people on a very high fibre diet have much food residue from cellulose (for which we have no digestive enzymes). The faeces consist of about two-thirds water and one-third solid matter. The latter consists largely of bacteria that have multiplied in the colon, some cellular debris scraped from the lining of the bowel (epithelium), bile pigments and small amounts of salts.

Defaecation, the process of emptying the lower bowel, is initiated by the stretching of the wall of the rectum by the mass movement of faeces into it from the colon. Stretching causes a conscious desire to defaecate, but this is readily inhibited by voluntary decision and is not renewed until the next movement of faeces from the colon. Undue inhibition leads to constipation and hardening of the stools by progressive water absorption. Voluntary defaecation involves relaxing the anal sphincters, taking a deep breath, suppressing the breathing and contracting the abdominal muscles so that the pressure is applied to the outside of the rectum. The anal sphincter is then relaxed to allow passage of the bowel contents.

dilatation widening. This may be a normal process or may imply stretching beyond normal dimensions, either as part of a disease process or as a deliberate surgical act.

dioptre a measure of lens power. The reciprocal of the focal length in metres. A lens of 1 m focal length has a power of 1 dioptre. A lens of 50 cm focal length has a power of 2 dioptres. One of 33.3 cm focal length has a power of 3 dioptres. Dioptres are used by opticians in preference to focal lengths because they can immediately be added together or subtracted during vision testing.

dioxins contaminant byproducts of the manufacture of chlorinated phenols used in herbicides, wood preservatives, dyes and other products. Dioxins are highly toxic and can cause cancer and fetal malformations. See also AGENT ORANGE.

diphosphate a chemical compound containing two phosphate groups.

dipl-, diplo- *combining form denoting* double.

diploblastic formed from two of the germ layers of the embryo.

diploe the spongy layer of bone between the hard outer and inner layers of the vault of the skull (cranium).

diploid having an identical (homologous) pair of chromosomes for each characteristic except sex. This is the normal state of most body cells. Eggs and sperm, however, have only a single set of half the number of chromosomes, and are said to be haploid. Red blood cells have no chromosomes.

diplopia double vision. The perception of two images of a single object. This occurs in squint (strabismus) when both eyes are not aligned on the object of interest. Diplopia with one eye is rare but possible.

dipsogenic causing thirst.

dipsomania an uncontrollable, often episodic, craving for, and abuse of, alcohol. Between episodes the affected person may avoid alcohol altogether.

-dipsia *combining form denoting* thirst or drinking.

disaccharide one of the class of common sugars, including milk sugar (lactose) and cane sugar (sucrose), that can be broken down by hydrolysis, under the action of enzymes, to yield two monosaccharides.

dis- *prefix denoting* not, un-, away, apart, scattered, reverse.

disarticulation a separation at a joint.

discharge an abnormal outflow of body fluid, most commonly of pus mixed with normal secretions, or of normal secretions in abnormal amount. Discharge may occur from any body orifice or from a wound.

disc, intervertebral see INTERVERTEBRAL DISC.

disclosing agents stains that reveal PLAQUE on the teeth to encourage regular toothbrushing and flossing.

discontinuous gene a gene in which the coding information is split up between two or more strand lengths called EXONS, which are separated from each other by non-coding sections called INTRONS. Many genes are discontinuous.

discontinuous replication the synthesis of DNA in short lengths that later join up to form a continuous sequence.

discordance difference of phenotype in two individuals of identical, or near-identical, genotype.

disease 1 any abnormal condition of the body or part of it, arising from any cause. 2 a specific disorder that features a recognizable complex of physical signs, symptoms and effects. All diseases can be attributed to causes, known or unknown, that include heredity, environment, infection, new growth (neoplasia) or diet.

disjunction the separation movement of members of pairs of chromosomes to opposite poles of a cell in the process of cell division.

dislocation separation, especially the disarticulation of the bearing surfaces of a joint with damage to the capsule and to the ligaments that hold the joint together.

disorientation bewilderment or confusion about the current state of the real world and of the affected person's relationship to it. Awareness of time, place and person are usually lost in that order.

displacement activity one of the psychological DEFENCE MECHANISMS. Frustrated emotions aroused by a person, idea or object are transferred to another person, idea or object.

Thus an aggrieved employee might take out his resentment on a punch-bag.

distal situated at a point beyond, or away from, any reference point such as the centre of the body. Thus, the hand is distal to the elbow. Compare PROXIMAL.

distoclusion an improper relationship of the teeth of the upper jaw to those of the lower, in which those of the lower teeth are placed further forward than (more DISTAL to) the corresponding upper teeth.

distention an expansion or swelling from an increase in internal pressure.

distraction a pulling apart.

disulphide bond a covalent linkage between sulphhydryl groups on two cysteine residues in proteins or between these groups on the same protein.

diuresis an unusually or abnormally large output of urine.

diurnal pertaining to a day. Occurring daily or in a day.

divarication 1 a divergence at a wide angle. 2 the point at which divergence occurs.

divergence 1 the act or state of moving off in different directions from a point.
2 the departure from each other of two processes, modes of action or courses of evolution.
3 in genetics, the degree, usually expressed as a percentage, to which two related DNA lengths differ in nucleotide sequences, or two similar proteins differ in amino acid sequence.

divergent transcription transcription started by two PROMOTERs acting in opposite directions so that transcription proceeds simultaneously in both directions from a point.

dizygotic derived from two separately fertilized eggs (ova). The term is used especially to refer to non-identical twins. Distinguished from monozygotic twins, who are derived from a single fertilized ovum and are identical.

DNA *abbrev. for* deoxyribonucleic acid. The very long molecule that winds up to form a CHROMOSOME and that contains the complete code for the automatic construction of the body. The molecule has a double helix skeleton of alternating sugars (deoxyribose) and phosphates. Between the two helices, lying like rungs in a ladder, are a succession of linked pairs of the four bases adenine, thymine, guanine and cytosine. The molecules of adenine and guanine are larger than thymine and cytosine and so, to keep the rungs of equal length, adenine links only with thymine and guanine only with cytosine. This arrangement allows automatic replication of the molecule. The sequence of bases along the molecule, taken in groups of three (codons), is the genetic code. Each CODON specifies a particular amino acid to be selected, and the sequence of these, in the polypeptides formed, determines the nature of the protein (usually an ENZYME) synthesized. Polypeptide formation occurs indirectly by way of MESSENGER RNA and TRANSFER RNA. Periodicity of DNA is defined as the number of base pairs per turn of the double helix. See Fig. 5 on page 136.

DNA cloning a recombinant DNA procedure in which a gene, or other fragment of DNA or of complementary DNA is incorporated in a cloning vector and then inserted into suitable host cells for culturing and reproduction.

DNA fingerprinting the recording of a pattern of bands on transparent film, corresponding to the unique sequence of regions in the DNA (core sequences) of an individual. DNA fragments, obtained from a DNA sample by cutting it with restriction enzymes, are separated on a sheet of gel by ELECTROPHORESIS. The fragments are then denatured into single strands and the gel is blotted onto a membrane of nylon or nitrocellulose which fixes the fragments in place. Radioactive probes, complementary to the core sequences, are then added. These bind to any fragment containing the core sequence. The membrane is laid on a sheet of photographic film and a pattern of bands is produced by the action of the radiation. The arrangement of the banding pattern is unique to each unrelated person but parents and their offspring have common features. Patterns from different individuals, or from different samples from the same individual can be compared. The method can be used as a means of positive identification or of paternity testing. Only a tiny sample of blood, semen or of any body tissue is needed.

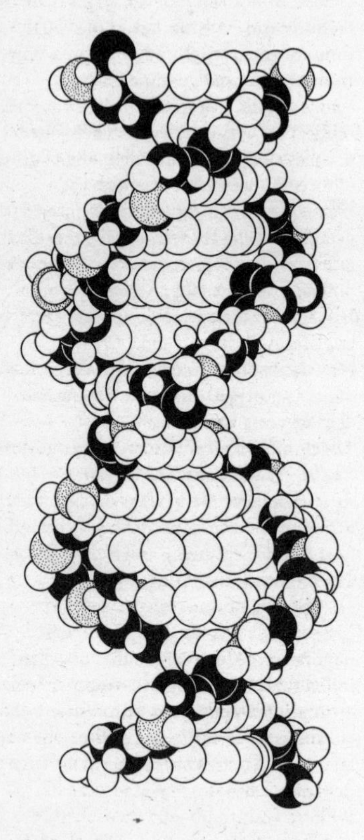

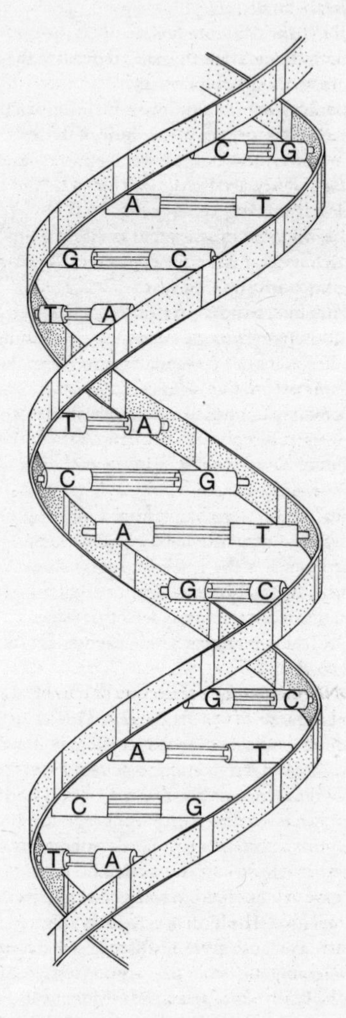
C
G
A
T
G
C
T
A
T
A
C
G
A
T
G
C
G
C
A
T
C
G
T
A
A-adenine
G-guanine
T-thymine
C-cytosine
sugar
phosphate

Fig. 5 **DNA**

DNA helicases enzymes that bring about the unwinding of DNA strands prior to replication.

DNA hybridization the use of radioactive known segments of DNA to determine the presence of complementary strands in a sample so as to determine identity, or to assess the degree of similarity between two individuals.

DNA library a collection of cloned DNA fragments that can be used as probes.

DNA ligases enzymes that reconstitutes the double strand in a DNA molecule that has a discontinuity in one strand. DNA ligases join up the broken strand by catalyzing a PHOSPHODIESTER BOND at this point. From the Latin *ligare*, to bind.

DNA methylation the addition of a methyl group to the cytosine ring that occurs in DNA when it is replicated in a dividing cell. The process forms methyl cytosine and is catalysed by enzymes called DNA methyl-transferases. In humans, the change occurs only to those cytosines that preceded guanines in the DNA sequence. Methyl cytosine has a strong tendency to deaminate to form thymidine, and if this mutation is not repaired it remains. It is the commonest kind of genetic variation (polymorphism) in human populations.

DNA microarray a collection of tens of thousands of DNA single-strand molecular probes capable of detecting specific genes or measuring gene expression in a sample of tissue. A gene is expressed when it is transcribed into messenger RNA (mRNA) and forms a protein. Currently, DNA microarrays are mainly used as research tools but they have great potential as diagnostic devices and as a reliable means of predicting a patients' susceptibility to various diseases.

DNA polymerases enzymes that bring about the synthesis of a daughter strand of DNA on the basis of a complementary DNA template. They are involved in DNA replication and repair, and act by adding deoxynucleotide triphosphates to the 3'-OH group of the new DNA strand. These enzymes not only synthesize new DNA but proof-read the new strand and remove incorrect nucleotides and replace them with the correct ones.

DNA polymerase inhibitor a drug that acts against viruses by interfering with the action of the enzymes viruses use to build up their own DNA. Examples of this class of drugs are acyclovir (Soothelip, Zovirax), ganciclovir (Cymevene, Virgan), valganciclovir (Valtrex) and foscarnet (Foscavir).

DNA probe a DNA sequence labelled with a radioactive element used to identify the position of a segment with the complementary sequence by binding to it. DNA probes can also be used to identify the presence of complementary sequences in a mix of fragments.

DNA replicase an enzyme specific to DNA replication.

DNA replication the formation of new and, hopefully, identical copies of complete genomes. DNA replication occurs every time a cell divides to form two daughter cells. The copying of DNA is a largely automatic process in which the base adenine always links with thymine and the base guanine always links with cytosine. The effect of this is that if the double helix is split longitudinally so that the base pairs that form the rungs of the ladder are separated, the single strands so formed will attract to themselves appropriate bases from the immediate environment so that perfect replicas, running in the opposite direction, are formed. Each single strand acts as a template for the second strand. A second sugar-phosphate backbone, attached to the other side of each of the newly attached bases, is simultaneously acquired.

Under the influence of enzymes, DNA unwinds and the two strands separate over short lengths to form numerous replication forks, each of which is called a replicon. The separated strands are temporarily sealed with protein to prevent re-attachment. A short RNA sequence called a primer is formed for each strand at the fork. These primers provide a free 3'-OH end on which the new complementary sequence can be formed along the strand. The LEADING STRAND is synthesized continuously in the 5' to 3' direction, working towards the fork direction with removal of the RNA primers as the parental duplex is unwound. The LAGGING STRAND is synthesized discontinuously in the opposite direction as short fragments called Okazaki fragments. Lagging strand synthesis requires extension of the primer, then removal of the primers and gap filling. At least 20 different enzymes and factors, including DNA helicases, DNA polymerases, RNA primases, DNA TOPOISOMERASES and DNA ligases are involved in the complex process of DNA replication.

DNAse or Dnase an enzyme that breaks down (hydrolyzes) bonds in DNA, releasing component nucleotides.

DNA sequencing the determination of the sequence of base pairs in a length of DNA.

DNR *abbrev. for* do not resuscitate. An instruction to refrain from energetic measures to restore the heart beat and the breathing in those people with terminal, irreversible illness in which death is expected, who suffer cardiac arrest. This has been the unwritten rule of many doctors and was enacted in American legislation in 1988.

doctor-assisted suicide the cooperation of a medical practitioner in bringing about the voluntary death of a patient suffering from painful or distressing terminal illness. Polls of doctors have shown that at least half agree that doctor-assisted suicide should be legalized for carefully-selected cases.

Dobzhansky, Theodosius the American scientist Theodosius Dobzhansky (1900–75) was born in Nemirov in the Ukraine, studied zoology in the University at Kiev, graduating in 1921. He taught zoology there for a time and then taught genetics in Leningrad University. In 1927 he emigrated to the USA to take up a Fellowship at Columbia University, New York. He worked at the California Institute of Technology from 1929–40, in which year he was appointed professor of zoology at Columbia University, New York. From 1962–71 he worked at the Rockefeller Institute then he moved to the University of California at Davis.

Dobzhansky had long been interested in the problem of the process of speciation during evolution. His principal contribution to human biology was to show how genetics and natural selection work together to solve this problem. He showed that in any large population there was a huge amount of genetic variation including many potentially lethal recessive genes. This genetic variability, however, conferred the advantage of allowing natural selection to work effectively in condition of varying environments. Indeed, without this diversity Darwinian evolution could never occur. Dobzhansky's book *Genetics and the Origin of Species*, published in 1937, demonstrated the clear link between the work of Darwin and Mendelian genetics. Dobzhansky also demonstrated that, in the course of evolution there were periods when speciation is incomplete.

Other highly influential books by Dobzhansky include the philosophical work *The Biological Basis of Human Freedom* (1956) and the anthropological study *Mankind Evolving* (1962).

dolichocephalic having a skull longer than it is broad.

domain **1** of a protein, a discrete length of the amino acid sequence with a distinct tertiary structure and that is known to be associated with a specific function.
2 of a chromosome, a region in which supercoiling occurs independently of other domains; or a region that includes a gene of raised sensitivity to degradation by DNASE I.

dominance the power of a gene to exert its influence whether the other member of the gene pair is identical or dissimilar. GENES occur in pairs at corresponding positions (loci) on each of the paired CHROMOSOMES. A gene that has its effect only if paired with an ALLELE of the same kind is said to be RECESSIVE. The effect of a dominant gene paired with a recessive gene will be the same as if both genes had been identical to the dominant gene, but every cell in the affected person's body, including those producing sperm and eggs, contains the recessive gene. Such a person is said to be HETEROZYGOUS for that gene. When the sperm and eggs are produced, only one of the pair of chromosomes is included, so there is a 50/50 chance that this will be the one with the recessive gene. Should a sperm with the recessive gene fertilize an egg which also has the recessive gene, the recessive characteristic will be expressed because there is no other genetic material for the characteristic.

dominant see DOMINANCE.

dominant hemisphere the left half of the brain in almost all right-handed people and 85 per cent of left-handed people. This is the hemisphere concerned with language and logical thought and containing the motor areas for voluntary use of the right side of the body. In 15 per cent of left-handed people, the right hemisphere is dominant and subserves speech.

donor a person, or cadaver, from whom blood, tissue or an organ is taken for transfusion or transplantation into another.

dopamine a monoamine NEUROTRANSMITTER and hormone with an adrenaline-like action. Dopamine is the principal neurotransmitter in the extrapyramidal system. It is formed in the brain from the amino acid tyrosine via dopa. Dopamine is the precursor of noradrenaline. It is also concerned with mood, memory and food intake. Excess is associated with psychiatric disorders. Dopamine is converted into at least 30 other substances some of which are hallucinogenic.

dopamine receptors nerve cell membrane binding sites that are activated by dopamine molecules. Some are excitatory and some inhibitory, but both classes contain many subtypes. Some of these receptors mediate behavioural effects.

dope a slang expression for any narcotic or addictive drug.

dorsal relating to the back of the body or towards the back. Compare VENTRAL.

dorsal root ganglia collections of the bodies of sensory spinal nerve cells lying outside but alongside the spinal cord, one for each spinal segment. The sensory spinal nerves contain the long dendrites of these cells; the axons pass into rear part of the spinal cord. Also known as posterior root ganglia.

dorsiflexion a bending backwards of any part.

double helix the commonest three-dimensional structure for cellular DNA. Two strands are wound round each other in a helical manner and are attached to each other by hydrogen bonds between complementary bases.

double jointed an informal term referring to a person with an unusual range of movement at one or more joints.

double vision see DIPLOPIA.

doula a person, such as a woman who has experienced normal childbirth who provides emotional, physical, and informational support to a woman during and after labour.

Down's syndrome a major genetic disorder caused by the presence of an extra chromosome 21 (trisomy 21). People with Down's syndrome have oval, down-sloping eyelid openings, a large, protruding tongue and small ears. There is always some degree of learning difficulty, but this need not be severe and many people with Down's syndrome are able to engage in simple employment.

downstream in genetics, at a stage in the sequence of processes in the expression of a gene that is nearer the final protein product than any stage in the direction of the DNA. The term is also used to mean in the direction of the 3'-end of a chain of bases in DNA (see PHOSPHODIESTER BOND). In both cases the opposite sense is called 'upstream'.

Dravidian language family the language group predominant in southern India, northern Sri Lanka, Pakistan, Afghanistan and some parts of the Far East. It includes such languages as Tamil, Malayalam, Kannada, Telugu, Kurukh and Brahui and is spoken by over 100 million people.

dream analysis a process purporting to derive information about the state of the unconscious mind by an interpretation of the symbolic significance of the dream content.

Humans have always been deeply interested in dreams and their meaning, and many extraordinary things have been believed about dreams. In some cultures it is believed that the soul leaves the body during dreams and that it is dangerous to wake a dreaming person. Many believe that dreams foretell the future. A dream of a wife's adultery is taken, by some, to justify repudiating her. Other peoples believe that if a man dreams that a girl encourages his sexual advances, the dreamer is entitled to enjoy her sexual favours. We need not flatter ourselves that we are above such childishness. In Britain and America, dream books – full of transparent nonsense – still enjoy a ready sale and are consulted by many people on all kinds of important matters. Such books are popular all over the world.

Historically, the most prevalent view about dreams was that they predict the future. Written records of dream interpretation date back at least 2000 years and are found in the archives of the ancient Egyptians, the Babylonians, the ancient Indians, the Sumerians, the ancient Greeks and Romans, and others. Dream interpretation appears to have been an everyday affair for the Greeks and the Romans. Many enterprising soothsayers and priests made a profitable business of it. The Greeks also incorporated a kind of dream therapy into their medical armamentarium. Certain dreams were held to be curative and many sufferers sacrificed in Aesculapian temples in the hope of enjoying such advantages.

The bible, especially the Old Testament, is rich in dream interpretation and prophecy. The story of Joseph and his multi-coloured dream coat, in the 37th chapter of Genesis, is a case in point. Joseph merely recounted his dreams; it was his brothers and father who interpreted them. There are strong implications in this story that dreams of obvious symbolic content were taken to be both prophetic and to reflect the wishes of the dreamer.

In spite of the prevalence of belief in magic, there were some who were able to examine the subject more dispassionately. Aristotle (384–322 BC), with remarkably modern insight, was able to point out the connection between dreams and previously experienced external events and to recognize the way in which normal sensory data could be distorted by emotional factors. Likewise, Marcus Tullius Cicero (106–43 BC), in his book *De diviniatione*, mounted a scathing attack on the dream superstitions of the age. Six hundred year later, the prophet Mohammed (c. 570–632), the founder of the Muslim religion, distressed at the extent to which the lives of the people were being influenced by dream divinations, expressly forbade the practice.

These few voices of common sense did not, however, have much effect on the mass of public superstition and it was not until well into the 19th century that thinking people began to recognize that dreams were more likely to be a reflection of what was going on in the mind than what might later happen outside the body.

The content of dreams

Each of us dreams every night, but only a small proportion of our dreams are remembered. Most are immediately forgotten on waking. It is, however, possible to increase the quantity of remembered material by making notes or using a recorder as soon as one wakes.

Information on the content of dreams can be obtained only from reports of dreams and so is not particularly reliable. But many different reports have so much in common that some of the general characteristics of dreams can be determined. Dreams commonly feature a familiar location or ordinary everyday surroundings, rarely a bizarre or exotic setting. Visual imagery is by far the most common, but most dreams have some auditory features. Most seem to reflect, or relate to, events, emotions and thoughts that have been experienced in the recent past – often within a day or two. The content can often be recognized as a mixture of past experience, personal interests, wishes, inclinations and stimuli currently operating on the body. A full bladder, for instance, will frequently make its tensions known in a dream, and full vasa deferens may lead to erotic dreams involving orgasm. The emotional content of dreams is usually unpleasant, the commonest emotions experienced being anxiety or fear. Anger is fairly common.

Freud on dreams

To Freud, dreams were the 'royal road' to the unconscious mind, revealing to the initiated all the buried secrets of the inner life. It seemed to him that thinking during sleep involved less repression than conscious thought so that the unconscious preoccupation with sex and hostility had freer rein. Freud suggested that every dream has a manifest content – the remembered details – and a latent content – the repressed infantile, sexual and aggressive wishes of the dreamer. It was, however, necessary to disguise the real nature of the inner life because it was so shocking that it would, otherwise, wake the dreamer, so the details had to be distorted or represented symbolically.

Sometimes a number of latent elements were represented by a single manifest element – a process Freud called condensation. Sometimes emotions felt towards one person or object were transferred, in the dream, to another person or object. This was called displacement. Freud also claimed that people unwittingly altered the accounts or recollection of their dreams so as to make more sense of them. This complex of processes, by which the latent content is converted into the manifest content, Freud called the dreamwork. Few people outside psychoanalytic circles now give much credence to Freud's ideas on the symbolic interpretation of dreams.

REM sleep and dreams

In 1953 two American physiologists, E. Aserinsky and N. Kleitman, were studying the behaviour of the eye during sleep by a method of continuous recording of the electrical changes caused by eye movements (the electro-oculogram). They were struck by the fact that their recordings showed periods during which the eyes of their subjects were moving rapidly together. Further investigation by them and by other scientists confirmed that this finding was universal and was associated with other changes. It is now well known that about an hour after we fall asleep, our voluntary muscles, which have been normally tense, suddenly relax, our eyes move rapidly from side to side under our lids, our breathing deepens, our heart-rate becomes irregular, and the electrical pattern of our brain waves (the electro-encephalogram or EEG) comes to resemble that of an alert awake person. In 95 per cent of men, the penis becomes erect.

The phenomenon is called rapid eye movement (REM) sleep, and each period of REM sleep lasts for 5–15 minutes. 80–90 per cent of people wakened from REM sleep report that they have been having vivid dreams. The dreaming appears to be continuous during REM sleep, and reports of long dreams follow long periods of uninterrupted REM sleep. These episodes of REM sleep are followed by more quiescent periods of deeper sleep during which the EEG pattern becomes much slower and with higher amplitude waves and the muscles become somewhat more tense. People wakened during this phase of sleep seldom admit to having real dreams but only to having been in a state of calm thoughtfulness. These two phases alternate on a 30–90 minute cycle, REM sleep constituting, on average, about 20 per cent of the total sleeping time. The time spent in REM sleep increases towards the end of an undisturbed night.

REM sleep is important to us. When attempts are made to deprive a person of REM sleep by waking him as soon as the REM stage starts, a remarkable phenomenon is observed. On each successive night in which this is done there is an increase in the number of times the subject tries to enter REM sleep. To begin with, the subject may have to be wakened 7–8 times. By the fifth night he may have to be wakened 30 times. If the subject is then left undisturbed he spends a far higher than normal proportion of the time, during the next few consecutive nights, in REM sleep. Deprivation of REM sleep has striking effects during the waking hours, and causes irritability, anxiety, loss of concentration, suspiciousness, apathy, even a tendency to hallucinations.

In babies and infants REM sleep persists for half the sleep cycle – more than twice as long as in adults. This probably implies that infants dream much more than adults do, but the point cannot be established for certain. By the age of two, REM sleep periods total about three hours a day. Some scientists believe that REM sleep in babies is necessary for the normal growth and functional connection of nerves. Old people, too, sometimes show an increase in the proportion of sleeping time spent in REM sleep. REM sleep is not limited to humans. Many of the higher animals, including all mammals, exhibit it and this together with other evidence suggests that at least the more highly developed of these have a rich dream life.

The undeniable need for REM sleep does not necessarily imply that we need to dream. So far, research has not been able to throw light on whether the dreaming is a central part of the phenomenon or merely a kind of by-product. This interesting question remains unanswered.

Modern theories of dreams

Ideas on the function of dreaming are still entirely speculative. One suggestion is that dreaming is a way of helping to maintain brain function during sleep, when the amount of external stimulation is much reduced. We know that the brain needs constant stimulation to continue to function normally. Some neurophysiologists reject that dreams have any psychological meaning or importance. In their view, dreams are simply the result of random stimulation of the regions of the brain (areas of the cerebral cortex) which are concerned with conscious experience. When this happens, an attempt has to be made to make sense of the resulting impressions and these, they say, are our dreams.

It was suggested, by Francis Crick, molecular biologist turned neurophysiologist, and his colleague Graeme Mitchison of Cambridge University, that dreams are the result of the brain's actions in trying to erase false associations by disconnecting erroneous neural links between brain cells. This process, they say, serves the essential function of allowing correct memory processing. The bizarre content of dreams is the record of material that is to be purged from the memory.

The American neurophysiologist Jonathan Winson believes that, on the basis of experimental evidence in lower animals, previously stored information vital to the animal's survival is accessed during 3 sleep and integrated with past experience so as to modify future behaviour in a direction that increases the chances of survival. He believes that our dreams may serve a similar function. He suggests that this mechanism may have been inherited from less evolved species and that the largely visual content of our dreams reflects the absence of speech in lower animals. Winson cites other evidence, derived from studies of the reported REM dreams of subjects with marital problems. The contents of these dreams, he points out, is strongly related to the way in which the subject is coping with his or her real-life crisis.

drop attack a tendency to fall suddenly, without warning, and without loss of consciousness. Drop attacks may be due to a temporary shortage of blood to the brain and should be investigated.

dropsy an old-fashioned term for a collection of fluid in the tissues (OEDEMA).

drowning death from suffocation as a result of exclusion of air from the lungs by fluid, usually water. This may result from fluid produced within the lungs themselves (pulmonary oedema).

drug 1 any substance used as medication or for the diagnosis of disease.
2 a popular term for any narcotic or addictive substance.

drug abuse the use of any drug, for recreational or pleasure purposes, which is currently disapproved of by the majority of the members of a society. 'Hard' drugs are those liable to cause major emotional and physical dependency and an alteration in the social functioning of the user. See also COCAINE, DRUG DEPENDENCE, ECSTASY, HEROIN and MARIJUANA.

drug dependence a syndrome featuring persistent usage of the drug, difficulty in stopping and withdrawal symptoms. Drug dependent people will go to great lengths to maintain access to the drug, often resorting to crime. Drug dependence is not limited to dependence on illegal drugs.

Dryopithecus an extinct primate genus whose remains have been found in many areas including Europe, East Africa, India, Pakistan and China. The genus flourished in the Miocene epoch some 7–25 million years ago.

duct a tube or passage, especially one leading from a gland, through which a fluid or semisolid substance is conveyed.

ductus arteriosus a short shunting artery lying between the main artery to the lungs (pulmonary artery) and the main artery to the body (aorta). During fetal life blood need not pass through the lungs and this vessel acts as a bypass. It normally closes soon after birth. If it fails to do so (patent ductus arteriosus) the blood is insufficiently oxygenated and there may be interference with growth and development and additional strain on the heart.

duodenum the C-shaped first part of the small intestine into which the stomach empties. The ducts from the GALL BLADDER and PANCREAS enter the duodenum. The duodenum is said to be 12 finger-breadths long – hence the name.

duplex a double stranded length of nucleic acid. The term is commonly used to refer to DNA in its double helical form, as distinct from a single strand of DNA.

duplex DNA normal, double-stranded DNA.

dura mater a tough fibrous membrane, the outer of the three layers of the MENINGES that cover the brain and the spinal cord. The dura mater lies over the ARACHNOID and the PIA MATER.

dwarfism abnormal shortness of stature. This may be of genetic origin or it may result from glandular defects such as pituitary growth hormone deficiency, primary thyroid deficiency (cretinism), precocious puberty or adrenal gland insufficiency. It also results from various metabolic disorders.

dynein one of a family of MOTOR PROTEINS that move along microfilaments and microtubules. Dynein moves along microtubules by successively making and breaking new bonds. The movement of cilia and flagella is occasioned by dynein.

dys- *prefix denoting* disordered, defective, difficult, bad or abnormal.

dysfunction any disorder or abnormality of operation or performance especially of any part of the body.

dysgenesis any abnormality of development.

dyslexia abnormal difficulty in reading or in comprehension of what is read, in a person of normal intelligence and emotional stability and who has had normal educational and cultural opportunities. The condition is familial and heritable and there is evidence of differences in areas of the brain that are also affected in acquired alexia. The basic difficulty appears to be in processing the sounds of speech and in the awareness that words can be broken down into smaller units of sound. Many dyslexic children respond well to remedial help especially if provided early.

dysmelia deficiency or abnormality of the development of the limbs.

dysmorphophobia a conviction that other people are aware of a physical defect in the sufferer which is in fact non existent. Dysmorphophobia may be a presenting symptom of schizophrenia or a neurotic disorder. Dysmorphophobia by proxy has been described in a woman who, in three consecutive pregnancies, became convinced that the unborn child would suffer some physical defect. As a result she sought termination on all three occasions.

dysphagia difficulty in swallowing.

dysphasia impairment of speech or of the production or comprehension of spoken or written language.

dysphoria a state of unhappiness, anxiety and restlessness. The opposite of euphoria.

dysplasia an abnormal alteration in a tissue due to abnormality in the function of the component cells, but excluding cancer. There may be absence of growth, abnormal increase in growth or abnormalities in cell structure. Dysplasia in an epithelium commonly progresses to cancer.

dyspnoea difficult, laboured or obstructed breathing.

dyspraxia a disturbance of voluntary movement.

dystocia abnormal labour from failure of the expulsive power of the womb, from obstruction to the birth passage or from abnormalities in the size, shape or presentation of the baby.

dystrophy a vague term applied to conditions in which tissues fail to grow normally, or to maintain their normal, healthy, functioning state. See CORNEAL DYSTROPHY and MUSCULAR DYSTROPHY.

dystrophin a large, rod-shaped structural protein situated in the sub-sarcolemmal region of the muscle fibre membrane. A mutation of the dystrophin gene that eliminates dystrophin production causes a form of muscular dystrophy.

dysuria pain on passing urine. This is most commonly due to a urinary infection and is usually associated with undue frequency in the desire to urinate.

Ee

eardrum the tympanic membrane that separates the inner end of the external auditory canal (the meatus) from the middle ear. The outer side of the drum is covered with thin skin and to the inner side is attached the malleus, first of the three tiny bones, the auditory ossicles.

ear lobe the soft, pendulous fleshy tissue at the lowest point of the external ear (pinna).

early antigens antigens on the surface of cells that are coded for by viral genes and appear very soon after cell invasion and before synthesis of viral nucleic acid starts.

ear wax the secretion of the ceruminous glands in the skin of the outer ear canal. Wax is a deterrent to small insects and traps dust. Over-production can lead to blockage of the canal and deafness which is easily corrected by wax removal.

ecbolic causing the pregnant womb to contract and expel its contents. Oxytocic.

eccentric located away from, or deviating from, the centre or from the usual position.

eccrine exocrine, secreting externally, especially on to the surface of the skin, as in the case of a sweat gland.

ecdysis the shedding of the outer layer of the skin. The term is sometimes used as a synonym for EXFOLIATION in humans.

echinate prickly or covered with spines.

echino- *combining form denoting* spiny or prickly.

echinocyte a red blood cell with a crenated or spiky surface.

ecosystem 1 the interaction between a biological community and its non-living environment. 2 a largely self-sufficient area of the world requiring input from outside of itself of little more than water and solar energy.

ecstasy a popular name for the drug 3,4-methylene dioxymethamphetamine (MDMA), a hallucinogenic amphetamine with effects that are a combination of those of LSD and amphetamine (amfetamine). Ecstasy is widely used to promote an appropriate state of mind at 'rave' all-night dance session, but the combination of strenuous physical exercise and the direct toxic effect of the drug has led to a number of deaths in young people. Such deaths result from an uncontrolled rise in body temperature (hyperthermia), kidney failure, muscle breakdown (rhabdomyolysis) and sometimes liver failure. Urgent measures to reduce body core temperature can save life. The drug can also precipitate a persistent paranoid PSYCHOSIS. Claims that ecstasy can damage the dopamine system of the brain and cause Parkinson's disease have been discredited.

ectasia permanent widening, distension or ballooning of any tubular organ or part. 'Broken veins' are small ectatic skin blood vessels.

ecto- *prefix denoting* outside or external.

ectoderm the outermost of the three primary germ layers of an embryo, the others being the MESODERM and the ENDODERM. The ectoderm develops into the skin, the nervous system, and the sense organs.

ectogenesis having origin and undergoing early growth outside the body, as in the case of IN VITRO FERTILIZATION.

ectomorph one of the arbitrarily classified body types (SOMATOTYPES). Ectomorphic people are lean and moderately muscled.

ectoparasite any organism living on the outside of another organism and depending on it for nutrition. Lice, ticks and mites may be human ectoparasites.

ectopia malposition of an organ or structure.

ectopic situated in a place remote from the usual location. An ectopic pregnancy is one occurring outside the womb, often in the FALLOPIAN TUBE.

ectopic expression the expression of a gene in a tissue in which it is not normally expressed. This occurs only in artificial situations such as in transgenic animals.

ectopic heart beat a contraction of the heart VENTRICLES occurring prematurely so as to disturb the regular rhythm. This usually results in a compensatory pause, experienced by the subject as a 'palpitation'.

ectopic hormone secretion hormone production at body sites not normally capable of hormone synthesis. In most cases ectopic hormone production occurs in tumours, many of which produce hormones.

ectopic pregnancy a dangerous complication of pregnancy in which the fertilized egg (ovum) becomes implanted in an abnormal site, such as the FALLOPIAN TUBE or in the pelvis or abdomen, instead of in the womb lining. The great danger is severe, and sometimes life-threatening, bleeding (haemorrhage). Treatment is by urgent operation to remove the growing embryo. This is now mainly done by laparoscopic surgery.

eczema the effect of a number of different causes and a feature of many different kinds of skin inflammation (dermatitis). It features itching, scaly red patches and small fluid-filled blisters which burst, releasing serum, so that the skin becomes moist, 'weeping' and crusty.

edentulous toothless.

education the provision of instruction and information, ideally in such a manner as to inculcate in the recipient the desire to continue the process, spontaneously, throughout life. Literally, a 'drawing out', education may be regarded as probably the most important non-physical activity in which a human being can engage. As it is a programming for life it should be of the highest quality available. Propositions about the kind of education from which a particular person may best benefit tend to reflect previous educational advantage or disadvantage rather than innate capacity or incapacity. Cultural influences also have a large effect in determining acceptance of, or resistance to, a particular aspect of education. For these reasons it can be a serious mistake to make premature judgements about the educability of an individual. There are plenty of examples of highly successful 'late developers'. Since early influences are, in general, the most powerful and persistent, the quality of infant and primary school education, especially in its social and moral aspects, should be of paramount concern. Ideally, teachers of such groups should be more highly qualified and more rigorously selected than those engaged in secondary or tertiary education. Older children have better defences against defective programming.

efferent 1 directed away from a central organ or part.
2 nerve impulses travelling away from the central nervous system to a peripheral effector.

egestion defaecation.

egg the OVUM or female reproductive cell (GAMETE). The egg contains half the chromosomes required by the new individual, and the other half are supplied by the sperm at the moment of fertilization. The egg is a very large cell, about one tenth of a millimetre in diameter, and much larger than a sperm. This is because it contains nutritive material (yolk) to supply the embryo in its earliest stages before it can establish a supply from the mother via the placenta. If more than one egg is produced and fertilized, a multiple pregnancy results, but the offspring are not identical since half the chromosomes in each come from different sperm, with different genetic material. If a fertilized ovum divides, and each of the two halves forms a new individual, these will be identical twins, with identical chromosomes.

ego 1 the Latin word for 'I'.
2 a person's consciousness of self.
3 in Freudian terms, a kind of rational internal person largely at the mercy of the 'id' (German for 'it') with its wicked and mainly sexual drives, but sometimes saved from disaster by the virtuous 'super-ego'.

egoism the philosophic theory that the motive for all conduct is the promotion of one's own interest rather than the interest of others. This is a seductive hypothesis, easily supported by argument and hard to attack. Although widely rejected it continues to challenge other views.

egomania a pathological degree of preoccupation with self. An abnormal degree of self-esteem.

eicosanoids a general term for a range of modified polyunsaturated fatty acids, products of the metabolism of arachidonic acid. They are hormones and include prostaglandins, prostacyclins, thromboxanes and leukotrienes.

eicosapentaenoic acid a polyunsaturated fatty acid found in fish oils which is able to reduce the tendency to blood clotting.

eidetic strikingly vivid, detailed and accurate, allowing an extraordinarily lifelike imaging, or sometimes rehearing, of past experience.

ejaculation the forceful emission of seminal fluid, caused by muscular contraction, at the time of the male ORGASM. Premature ejaculation is a common disorder.

ejaculatory duct the continuation of the vas deferens beyond the point at which the seminal vesicle duct enters. The ejaculatory duct enters the urethra in the prostate gland.

elastins extracellular scleroproteins capable of two-way stretch and rapid recoil. The amount of stretch is limited by the tethering effect of interwoven inelastic collagen fibrils. Elastins are present in quantity in the lungs, large arteries, the ligaments of the spine and in the skin.

electric charge the state of any particle or body on which there is an imbalance between electrons and protons. An excess of electrons causes a negative charge; a deficiency of electrons causes a positive charge. An understanding of electric charge, ELECTRIC FORCE and ELECTRIC POTENTIAL DIFFERENCE is fundamental to the understanding of much of physiology.

electric force the force existing between bodies bearing electric charges that are not equal. Bodies with unlike charges experience mutual attraction; those with like charges repel mutually.

electric potential difference a difference between two points in the local concentration of electrons in each. Potential difference tends to cause an electric current to flow.

electro- *combining form denoting* electricity or electric.

electrocardiogram (ECG) the tracing on paper, representing the electrical events associated with the heartbeats, produced by the ELECTROCARDIOGRAPH.

electrocardiograph an instrument consisting of a series of electrical cables (leads) with a lead-switching device, a high-gain, low-noise, balanced differential amplifier and a moving coil rotary transducer that converts the amplified signal into a varying trace on a calibrated strip of moving paper. The leads are connected to low-resistance contacts on the chest of the subject and the input to the device is derived from the minute electrical currents that flow towards and away from them as a result of the heart's contractions and relaxations.

electrocardiography recording of the rapidly varying electric currents which can be detected as varying voltage differences between different points on the surface of the body, as a result of heart muscle contraction. The electrocardiograph (ECG) tracings show patterns highly indicative of a wide variety of heart disorders. Modern ECG machines usually carry out an automatic analysis of the waveform and suggest a diagnosis.

electrocution death from the passage of an electric current. Judicial electrocution employs a current of several amperes passing between the head and one leg by means of a high voltage and moistened electrodes to give low contact resistance.

electroencephalogram the multichannel tracing on paper of the output of the electroencephalograph, representing the electrical activity of the brain as occurring between pairs of electrodes in contact with the scalp and representing the algebraic sum of an immense amount of underlying electrical activity.

electroencephalography the process of making a multiple tracing, by voltmeter-operated pens, of the electrical activity of the brain. The multiple readings are of the constantly varying voltage differences occurring between pairs of points on the scalp of the subject. The electroencephalograph (EEG) is affected by sleep, hyperventilation, drugs, concussion, brain injury, brain tumours, bleeding within the brain (cerebral haemorrhage), brain inflammation (encephalitis), epilepsy and various psychiatric conditions. It also assists in the determination of brain death.

electrogenic pump an active transport system that uses energy to separate electric charges and produce a potential difference.

electrolysis the decomposition of a solution by the passage of an electric current to separate charged particles (ions). Water can be separated into the gases hydrogen and oxygen by this means. The term is used for electrical destruction of unwanted hair follicles, but the main destructive effect is one of heating.

electrolyte any substance which, when dissolved in water, separates into pairs of particles (ions) of opposite charge. For example, sodium chloride (common salt) when dissolved in water forms positive ions of sodium and negative ions of chloride. The electrolytes include salts, acids, alkalis and metal oxides.

electrolyte balance the critical balance between the concentration in the cells, and that in the tissue fluid surrounding the cells, of the various inorganic IONS. The electrolytes mainly in the cells are potassium, magnesium, sulphate and phosphate. Those in the surrounding fluid are mainly sodium, chloride and bicarbonate. This balance is essential to life and is maintained by the active pumping action of the cell membranes.

electrons negatively-charged subatomic particles existing in the space surrounding the nucleus of an atom.

electron carrier a compound, such as a coenzyme, capable of taking up electrons from a molecule and transferring them to another, thereby undergoing reversible reduction and oxidation. Most electron carriers are PROSTHETIC GROUPS.

electron microscopy a method of producing a greatly enlarged image of very small objects by using a beam of accelerated electrons instead of light. Modern instruments enable objects smaller than 1nm (one millionth of a millimetre) to be seen. This is almost down to atomic level. Focusing is done by means of magnetic fields obtained from charged plates or current-carrying coils. These fields act as lenses. Electron microscopes are essential tools in medical research and diagnosis.

electrophoresis separation of charged particles in a solution (ions) by the application of an electric current. This can be done in a thin layer of solution on paper or in a gel. Ions of low weight move more quickly than those of high weight, so separation occurs and can be demonstrated by staining. The method is widely used in medicine to identify and measure the proteins present in the blood including the ANTIBODIES (IMMUNOGLOBULINS). It is extensively used in genetic work such as DNA fingerprinting. Electrophoresis is remarkably sensitive. Pieces of DNA, for instance, that differ in length from each other by only one base pair can be separated into discrete bands by this method.

electrovalent bond the chemical bond between an ANION and a CATION. Compounds formed by electrovalent bonds ionize when dissolved in water, with dissociation of the cations and anions.

emaciation the state of extreme thinness from absence of body fat and muscle wasting usually resulting from malnutrition, widespread cancer or other debilitating disease.

embalming the preservation of dead bodies by draining out most of the blood and replacing it with disinfecting fluids such as formalin so as to retard the bacterial activity that leads to putrefaction.

embolus any material carried in the bloodstream to a point where it causes obstruction to the blood flow. Emboli are commonly blood clots but may consist of crystals of CHOLESTEROL from plaques of atheroma in larger arteries, clumps of infected material, air or nitrogen, bone marrow, fat or tumour cells.

embryo an organism in its earliest stages of development, especially before it has reached a stage at which it can be distinguished from other species. The human embryo is so called up to the eighth week after fertilization. After that it is called a fetus.

embryogenesis the stage from the fertilization of the ovum to the formation of the three primary germ layers of the early embryo.

embryology

Embryology is the branch of science concerned with the process of physical development of the body, from the time of fertilization of the egg (ovum) to the time of birth.

In the first three days after fertilization and the fusion of the sperm and egg nuclei, the egg cell splits into two four times, within its membrane, to form a 32-cell mass called a morula. At this stage none of the cells have become specialized and it is impossible to tell which of them will develop into the fetus and which will become the placenta. These cells are called 'totipotential'. Soon, however, the first signs of differentiation appear as the cell mass organizes itself into a hollow, fluid-filled ball called a blastocyst. Inside this, at one side, is a small collection of cells that will become the fetus. Alongside this is the yolk sac derived from the egg. This provides nourishment for the growing embryo until the placental circulation is established and nutrition is provided by the mother.

The cells on the outer side of the blastocyst develop tiny projections called microvilli and these become the placenta. These outer cells have a powerful tendency to burrow into things and the 6-day embryo is now at the stage at which it is capable of implanting itself into the lining of the womb. If it does not reach the womb it will still try to burrow into any nearby tissue and this is how pregnancies sometimes occur in the fallopian tubes or even in the abdominal cavity (ectopic pregnancies).

Embryos at these early stages are remarkably tough. Those at the blastocyst stage can be chopped in two to produce twins. In fact, each cell of an eight-cell embryo is capable of developing into a complete individual if the cells are separated from each other at this stage. Such an embryo can be transferred to another person or animal and does not cause immunological rejection problems. In theory, it could even be transferred to the body of a man but its burrowing tendency would almost certainly cause dangerous bleeding.

About two weeks after fertilization, the embryo is fully implanted and is buried in the womb lining. The outer layer of the blastocyst forms into a membranous sac enclosing the growing embryo and the fluid around it. The small processes on the outside (the villi) branch and acquire tiny blood vessels. Now known as the trophoblast, this layer produces first the chorionic villi which then develops into the placenta. The cells of the chorionic villi are genetically identical to all the others in the embryo and can be sampled, to provide genetic information, without harming the embryo proper.

The next stage is the separation of the inner cell mass into three distinct layers. Each of these will eventually develop into a specific group of adult tissues. The outer layer is the ectoderm, the middle is the mesoderm, and the inner is the endoderm. Definite head and tail ends appear and a longitudinal groove forms along the line of the back into which cells from the ectoderm sink to form a sunken tube in the mesoderm. This tube will form the spine, the central nervous system and the eyes. The ectoderm becomes the skin, the hair, the nails and the breasts. From the middle of the three layers of the embryo, the mesoderm, develops the skeleton, the muscles, the heart and blood vessels, the kidneys and the sex organs. The inner layer, the endoderm, becomes the digestive system, the liver and the glands that produce the digestive enzymes.

In the process of differentiation, cells gradually lose their universal potentiality and collect together to form distinct and specific tissues with particular structures and functions. Some acquire the ability to contract and become muscle cells; some develop a secretory function and become glands; others become able to respond to stimuli by producing electrical impulses and become nerve cells. This process of development of different cell types and their incorporation into tissues and organs is called morphogenesis.

While this rapid differentiation is going on, the embryo becomes suspended in the surrounding fluid by a stalk of cells that connects it to the centre of the region of chorionic villi forming the placenta. This stalk encloses the remainder of the yolk sac and eventually becomes the umbilical cord. Soon blood vessels develop within this stalk and before long these connect the developing circulatory system of the embryo to the placenta which is intimately related to the blood vessels in the wall of the womb and thus to the circulation of the mother.

The size of the fetus

By the fifth week of pregnancy the sac surrounding the embryo is about 8 mm in diameter and this grows at a rate of about 1 mm a day. At 10 weeks the body of the embryo itself is about 2.5 cm long, measured from the crown of the head to the rump. By now it is a recognisable human male or female. The face is formed but the eyelids are fused together. The brain is in a very primitive state, with a smooth outline and little development of the outer layer – the cerebral cortex – that subserves thought, awareness and consciousness. By three months, the fetus is about 5 cm long (crown to rump) and by four months it is about 10 cm long. In the sixth month, the fetus is up to 20 cm long (head to toe) and weighs up to 800 g. A fetus born at this stage is unlikely to survive, but with increasing maturity the probability of survival increases. Nowadays, because of advances in fetal care and incubation, most fetuses over 2,000 g survive outside the womb.

Throughout pregnancy the fetus floats freely in the growing quantity of fluid that surrounds it, anchored by the umbilical cord. The fluid is called amniotic fluid and at full term its volume is about one litre. It cushions the fetus against physical shock and against the pressures of the contracting womb, it maintains a constant temperature and allows some fetal movement. Amniotic fluid is constantly swallowed by the fetus and excreted as urine, so it contains material from which much information about the health of the fetus can be obtained by sampling through a needle (amniocentesis). As the fetus and the volume of amniotic fluid increases, the womb (uterus) must grow to accommodate them. Eventually, the top of the uterus rises almost to the top of the abdominal cavity, displacing the bowels. At this stage, the pregnant mother suffers considerable discomfort.

Fetal circulation

The lungs of the fetus do not expand until after birth so an alternative source of oxygen is required. All the fetal oxygen and nutritional requirements are provided by way of the placenta, an organ in which the mother's blood comes into close proximity to, while not actually mixing with, that of the fetus. The fetal heart is not required to pump all its blood to its own lungs (which are largely bypassed during pregnancy). Instead, both sides of the fetal heart cooperate to pump its blood to all parts of its body and along the umbilical cord to the placenta. Blood returning to the fetus from the placenta is rich in oxygen and nutrients and this blood returns to the heart partly by way of the liver. After birth, the lung bypass channel, and the vessels to the umbilical cord, close.

The placenta at term is a thick, disc-shaped object about 15–20 cm in diameter firmly attached to the inside of the uterus. The mother's blood accesses the placenta from the uterine side and the umbilical cord comes off from the free surface. Oxygen, carbon dioxide, sugars, amino acids, fats, vitamins, minerals, as well as many drugs, pass freely across the placental barrier.

embryopathy any developmental or biochemical disorder of an EMBRYO.

embryo research the use of early human embryos for studies into the early detection and possible correction of genetic defects and the relief of human infertility. Cloning experiments, the alteration of the genetic pattern and attempts at hybridization are prohibited. Early embryos of two weeks gestational age have no organs or nervous system and are incapable of any perception or consciousness.

embryonic pertaining to, or to the state of being, an EMBRYO.

emesis vomiting.

emetic any substance that causes vomiting.

emission a discharge of something. A nocturnal emission is an involuntary EJACULATION of semen, usually with orgasm, during sleep.

emmetropia the state of the normal eye, with relaxed ACCOMMODATION, in which light rays from a distance (parallel rays) focus accurately on the retina giving perfect vision.

emotion any state of arousal in response to external events or memories of such events that affect, or threaten to affect, personal advantage. Emotion is never purely mental but is always associated with bodily changes such as the secretion of ADRENALINE and cortisol and their effects. The limbic system and the hypothalamus of the brain are the mediators of emotional expression and feeling. The external expression of emotional content is known as 'affect'. Repressed emotions are associated with psychosomatic disease. The most important, in this context, are anger, a sense of dependency, and fear.

empathy the state said to exist between two people when one is able to experience the same emotion as the other as a result of identical responses to an event and the adoption of an identical outlook.

en- *prefix denoting* in or into.

enamel the hard outer covering of the crown of a tooth.

enatiomers molecules that are identical to each other except that they are mirror images of each other. See also CHIRALITY and STEREOISOMERISM.

encapsulated enclosed in a protective, or strengthening, membrane or coating.

enceinte pregnant.

encephalins see ENDORPHINS.

encephalitis inflammation of the brain, most commonly from infection, usually by viruses.

encephalo- *combining form denoting* the brain.

encephalomyelitis inflammation of the brain and spinal cord (myelitis).

encephalon the brain.

encephalopathy any degenerative or other non-inflammatory disorder affecting the brain in a widespread manner.

endemic occurring continuously in a particular population. Literally, 'among the people'. See also EPIDEMIC and PANDEMIC.

endermic within or through the skin.

end labelling the linking of a radioactively-labelled sequence to the end of a DNA strand.

endo- *prefix denoting* within or inside.

endocardium the heart lining that also covers the heart valves.

endocrine system

The endocrine system is second only in importance to the central nervous system as a communication and control system for the body. The term 'endocrine' refers to glands that secrete their products, the hormones, directly into the bloodstream. The hormones, or 'chemical messengers' are thus carried to all parts of the body where they may influence the action of millions of different cells – the target cells for the particular hormone. Hormones are powerful substances that act in very low concentration. They are taken up by receptors on the cell membranes – proteins to which they bind specifically thereby starting a sequence of events in the cell that alter its function.

The endocrine system automatically controls and integrates many important body functions. These include body growth, the rate of build up and breakdown (metabolism) of tissues, the action of the heart, the tension of the blood vessels and hence the blood pressure, body temperature, the water content of the body, digestion and absorption of food, the development of the secondary sexual characteristics, reproduction and the body's response to stress. Although the elements of the system – the endocrine glands – are scattered throughout the body, the action of them all is tightly controlled by a part of the brain called the hypothalamus working through the pituitary gland. These two structures are the coordinating elements for all the other endocrine glands the thyroid, parathyroid and adrenal glands, the islet cells of the pancreas and the sex glands. The endocrine system does not work in isolation but is intimately linked to the central nervous system, and each profoundly influences the other.

Hormones

There are three classes of hormones. Many of them are proteins or peptides, and they vary greatly in size, ranging from small peptides containing only three amino acids to small proteins containing hundreds of amino acids. In most cases they are formed initially as large peptides, or prohormones, from which the active hormones are split off by enzymes. The second class, the amine hormones, include the adrenaline and the hormones from the thyroid gland. Adrenaline is called a catecholamine. The third class, the steroid hormones, are formed from cholesterol and have chemical structures made up of four interconnected rings of carbon atoms. They are produced by the adrenal glands, the sex glands and the afterbirth (placenta) in pregnancy.

Hypothalamus

A conspicuous feature of the centre of the underside of the brain is the pea-sized pituitary gland, hanging down from the brain by a narrow stalk of blood vessels and nerves, and accommodated in a bony hollow, the sella turcica, or 'turkish saddle', on the base of the skull, just behind the back wall of the nose. Immediately above the stalk of the pituitary is the region of the brain called the hypothalamus. This contains several groups of aggregated nerve cells (brain nuclei) most of which are connected directly by nerve fibres to the pituitary, but some of which are connected to other parts of the brain.

The hypothalamus, although very much a part of the nervous system, is also an endocrine gland. It produces two hormones, oxytocin and vasopressin that act remotely, and other hormones that act only on the pituitary gland to control the output of its own hormones. The hypothalamus also acts on the pituitary by way of the nerve fibres running in the pituitary stalk. The hypothalamus is itself part of a feedback hormone control system. Since blood passes to every part of the body, including the hypothalamus, hormones in the blood affect any cells carrying hormone receptors. The cells of the hypothalamus do carry such receptors and the effect of blood hormones on these is to exert a controlling negative feedback influence, damping down oversecretion from any of the endocrine glands.

The hypothalamus is especially important as it is the point at which the neural and hormonal systems of the body interact. Many parts of the brain connect by nerve fibres to the hypothalamus, which constantly receives information relating to such matters as the current state of bodily and mental stress, perception of danger and the need for physical activity. Such electrical brain action causes changes in the hypothalamus which are coordinated with other hormonal information and result in messages being sent to the pituitary gland. This, in turn, sends hormones to any or all of the other endocrine organs to prompt them into activity. Depending on the current general body situation, the result may vary from an outpouring of adrenaline and cortisol in a situation of high stress, to a calming reduction in the production of thyroid or sex hormones.

Whether emotion causes adrenaline and steroid hormone production or the hormones causes the emotion is a question not yet resolved. The two are normally inseparable. Some scientists believe that all emotions are the necessary concomitants of hypothalamic and endocrine action prompted by mentally significant information. Drugs which block the action of certain hormones can largely eliminate the emotional response without in any way altering the intellectual awareness of the stimulating information.

Pituitary gland

The pituitary gland has been described as the 'leader of the endocrine orchestra'. This rather fanciful metaphor might be extended. If the pituitary is the leader, the hypothalamus is the conductor. The pituitary is divided into two lobes and most of the hormones are secreted by the front lobe. The rear lobe is really an outgrowth of the hypothalamus and consists mainly of nerve tissue. The rear lobe, and the hypothalamus, secrete the two hormones vasopressin and oxytocin. All the other pituitary gland hormones are secreted by the front lobe. Some of these hormones act directly to produce an effect, but most of them act to stimulate the other endocrine organs into producing their own hormones.

The pituitary hormones from the front lobe are: growth hormone, which directly controls growth up to early adult life; thyroid-stimulating hormone which controls the output of thyroid hormones from the thyroid gland; adrenocorticotrophic hormone (ACTH), which controls the output of cortisol from the adrenal glands; follicle-stimulating hormone which controls the production of eggs (ova) from the ovary; luteinizing hormone, which is necessary to maintain pregnancy after fertilization; prolactin, which promotes milk production at the end of pregnancy; melanocyte-stimulating hormone which stimulates the growth of pigment cells in the skin.

Hormones from the rear lobe are: vasopressin, also called the antidiuretic hormone, which increases the reabsorption of water in the kidneys and controls water loss; oxytocin, which releases milk from the breast and causes the womb muscle to contract at the end of pregnancy.

The action of pituitary hormones can most easily be understood by considering the effects of pituitary under- or over-action.

Pituitary disorders

If, for any reason, the pituitary fails early in life, there is severe retardation of body growth with dwarfism, failure of sexual maturity, sterility, atrophy of the adrenal gland with failure of production of adrenal hormones and atrophy of the thyroid gland resulting in slowing of all body processes. Pituitary failure later in life causes marked loss of weight, severe weakness, and the effects of undersecretion of the thyroid and adrenal glands.

Pituitary growth hormone affects not only the growth of the long bones, but also the growth of many tissues and organs of the body. Overproduction of growth hormone, as a result of a tumour of the cells in the pituitary that produce the hormone, has different effects, depending on when this occurs. Excess growth hormone before puberty causes gigantism. If the excess occurs after the epiphyses have fused, the condition of acromegaly results.

Pituitary tumours can also result in oversecretion of the hormone prolactin. This results in milk production from the breasts of either sex. Milk secretion is normally associated with underfunction of the sex glands and a prolactin-producing tumour results not only in absence of menstruation and relative sterility in women, but also in loss of libido or impotence in men. Because of the close relationship of the pituitary gland to the crossing of the optic nerves (the optic chiasma) a tumour of this kind often causes blindness unless detected early and treated surgically.

Loss of the hormone vasopressin from the rear lobe of the gland results in the condition of diabetes insipidus in which large volumes

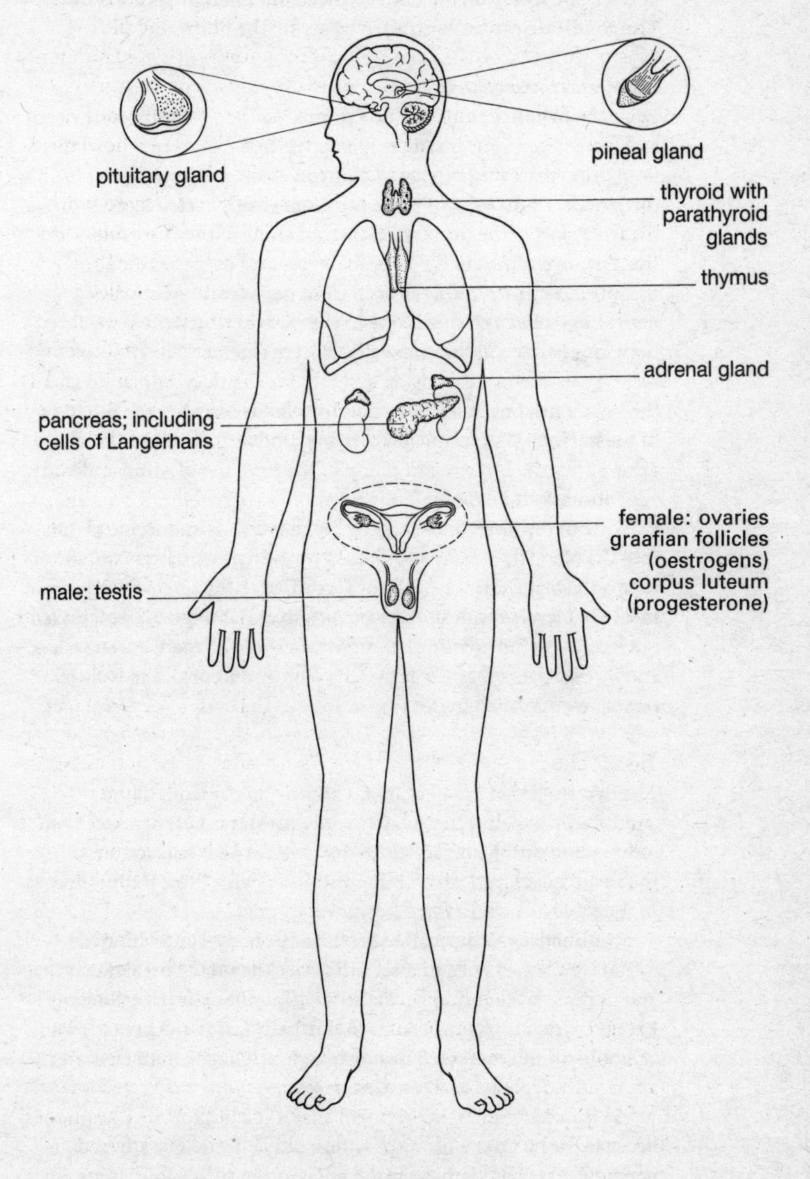

Fig. 6 **Endocrine system**

of water are lost from the body in the urine and there is great thirst. The condition can be controlled by giving the hormone.

Other endocrine organs
Once the hypothalamus has determined the body's immediate needs and has relayed appropriate information to the pituitary gland there is an immediate outpouring of control hormones from the latter into the bloodstream. Most of these hormones are secreted by cells in the front part of the pituitary gland but some of them are released from nerve endings in the hypothalamus and in the rear lobe of the pituitary. Hormones released from nerve endings are called neurohormones and this is one of the ways in which the central nervous system and the endocrine system interact. The nervous system can also act directly on several of the endocrine glands, and the hormones produced by endocrine glands can alter the function of the nervous system and affect many kinds of behaviour. This exemplifies the inter-relationship of the function of mind and body.

Pituitary and hypothalamic hormones control the endocrine glands, but the hormones secreted by the various endocrine glands into the bloodstream are monitored by the hypothalamus which then adjusts its hormone output according to the levels found. High measured levels result in reduced output and low levels in increased output. This kind of automatic control is called negative feedback and it results in a very stable system. The output of all the endocrine organs is controlled in this way so long as the body's needs are met.

Thyroid gland
The thyroid lies in the neck, like a bow tie on the front of the windpipe, just below the 'Adam's apple' (the larynx). It produces two iodine-containing hormones, thyroxine (T4) with four iodine atoms in the molecule, and tri-iodothyronine (T3) with three iodine atoms in the molecule, and a third hormone, calcitonin (see below). T4 and T3 act directly on almost all the cells in the body, controlling the rate at which they burn up fuel and hence the rate of breakdown and buildup of chemical substances within the cells (metabolism). Excess thyroid hormone causes abnormally fast tissue breakdown (anabolism) and increased heat production. Glucose and fat stores are rapidly depleted and muscles waste.

In the condition of hyperthyroidism, the symptoms are numerous because the hormone has such widespread effects. The affected person is hyperactive, has a rapid pulse, often with palpitations, is jumpy, anxious and emotionally labile, dislikes warm weather and has sweaty hands. The appetite is often good but weight is lost rapidly. There is often protrusion of the eyes (exophthalmos) causing a characteristic staring appearance and the upper eyelids lag behind on looking down. Hyperthyroidism affects women far more often than men and is usually caused by disease of the thyroid gland, but may sometimes result from excess production of the thyroid-stimulating hormone of the pituitary.

Inadequate output of thyroid hormone in adults is called hypothyroidism. This may be caused by thyroid gland disease, by various drugs and by overdosage of thyroid hormone (usually taken in an attempt to lose weight). The excess hormone causes the pituitary to shut down its production of thyroid stimulating hormone. Hypothyroidism causes physical and mental slowing, weight gain, undue sensitivity to cold, a hoarse voice, dry, flaky skin, loss of hair and of the outer parts of the eyebrows and puffiness of the tissues (myxoedema). There may be deafness, depression and even psychosis. Lack of thyroid hormone from birth causes cretinism with irreversible brain damage unless the condition is treated very early. In older children inadequate thyroid hormone leads to failure of growth and development.

The third hormone, calcitonin, although secreted by cells in the thyroid gland, has nothing to do with metabolism. It acts on bone, to limit the release of calcium into the blood. The cells in the thyroid gland which produce calcitonin also monitor blood calcium levels continuously and secrete calcitonin accordingly. If the level falls they produce less calcitonin, so the bones release more calcium; if the blood calcium level rises these cells produce more calcitonin. In this way the level of blood calcium, the constancy of which is critically important for nerve and muscle function, is kept within narrow limits. Strict maintenance of blood calcium is so important that another hormone mechanism is also involved. This is the function of the parathyroid glands.

Parathyroid glands

These are four inconspicuous, bean-shaped structures, each about half a centimetre long, partly-buried in the substance of the thyroid gland. Their secretion, the hormone parathormone, also regulates calcium levels in the blood. If the level of calcium in the blood drops, the parathyroids secrete more parathormone which acts on the bones to increase the rate of release of calcium so that the level of blood calcium rises again. The hormone also acts on the tubules of the kidneys to reduce calcium loss and on the lining of the intestine to increase calcium absorption. Abnormally high levels of parathormone, as may sometimes occur from a tumour of one of the parathyroid glands, leads to excessive loss of calcium from bones, with serious softening and distortion. Insufficient parathormone results in a dangerous drop in blood calcium with abnormal muscle excitability leading to uncontrollable spasms of contraction. This is called tetany.

Adrenal glands

The adrenals sit like triangular caps one on top of each kidney. Each gland has two parts with different functions. The inner part, or medulla produces adrenaline, and the outer part or cortex produces three kinds of steroid hormones – cortisol, aldosterone and male sex hormones (androgens). Cortisol prompts cells to increase protein synthesis – mainly enzymes and structural proteins – by acting on messenger RNA and it stimulates the conversion of amino acids to the fuel glucose. These actions help the body to react to stress. Aldosterone controls water balance by affecting the rate of reabsorption of water in the kidneys . The male sex hormones (androgens), which are produced by women as well as by men, strongly stimulate protein synthesis in many parts of the body (anabolic effect) and are important in determining body growth and development. This effect is greater in men than in women because of the additional powerful androgens produced by the testicles.

Adrenaline is produced when unusual efforts are required of the body, especially in emergency situations. It increases the heart rate, raises the blood pressure, speeds the rate and depth of breathing, widens the arteries to the muscles and narrows those to the intestines, mobilizes glucose for fuel and increases alertness and excitement. It has been called the hormone of 'fright, fight or flight'.

Pancreas

The pancreas has two functions; it produces digestive enzymes (exocrine function) and it synthesizes the hormones insulin and glucagon which, respectively, lower and raise the amounts of sugar (glucose) in the blood (endocrine function). Insulin promotes the movement of the vital fuel glucose through cell membranes into cells. If deficient, glucose accumulates in the blood. Muscle protein is converted to glucose and fats are burnt in excess producing toxic acidic byproducts. There is severe wasting of muscles and the urine is loaded with glucose which the body is trying to dispose of. This condition, type I diabetes (diabetes mellitus), can be controlled by injections of insulin.

Glucagon acts in a manner opposite to that of insulin. It causes liver glycogen to break down to glucose, so increasing the amount of sugar in the bloodstream. Glucagon also mobilizes fatty acids for energy purposes.

Sex glands
Sexual differentiation at puberty is initiated, in both females and males, by production, by the pituitary, of hormones called gonadotrophins. These cause the ovaries and the testicles to increase production of their own hormones, respectively oestrogen and testosterone. In girls, oestrogen causes breast development, growth of pubic hair, widening of the pelvis, deposition of fat under the skin in certain areas to produce female contours, and the onset of ovulation and menstruation. In boys testosterone causes the penis, testicles and scrotum to enlarge, the beard and pubic hair to appear, the prostate gland and seminal vesicles to mature, and the larynx to increase in size so that the voice deepens.

In both sexes there is a considerable anabolic effect at puberty with a spurt in body growth and weight gain. Body weight may almost double during this period. The increase in boys is mainly due to increase in muscle bulk. In girls, weight gain is also caused by the large increase in fat deposition.

endocytosis a method by which a large molecule, such as a protein, can enter a cell. The plasma membrane of the cell is invaginated by the molecule so that an internal vesicle is formed containing the molecule. This then breaks off from the surface membrane and moves into the interior of the cell. Such vesicles are coated with a protein called clathrin.

endoderm the innermost of the three primary germ layers of an EMBRYO. The endoderm develops into the INTESTINAL TRACT and its associated structures and glands, the respiratory and urinary tracts and most of the endocrine glands. See also MESODERM and ECTODERM.

endogenous occurring without an obvious cause external to the body, and believed to result from an internal cause. A person depressed by a succession of calamities is said to have exogenous depression. In the absence of external causes, depression is said to be endogenous.

endogenous opioids the morphine-like peptides, ENDORPHIN, enkephalin and dynorphin, produced by the body in response to stressful events.

endolymph the fluid within the LABYRINTH of the inner ear.

endometrium the inner mucous membrane lining of the womb (uterus) that undergoes changes in structure and thickness at different stages of the menstrual cycle, and much of which is shed at menstruation.

endomorph one of the arbitrary body types (Sheldon's SOMATOTYPES). Endomorphic people are short and stout with prominent abdomens.

endoneurium the CONNECTIVE TISSUE surrounding individual nerve fibres.

endonucleases enzymes that can split DNA or RNA at any point along the molecule by cutting PHOSPHODIESTER BONDS.

endoplasmic reticulum, ER a structure of highly convoluted, interconnected, membranes within cells. The rough ER has RIBOSOMES attached for protein synthesis; the smooth ER has no ribosomes and carries out lipid synthesis. The endoplasmic reticulum is also the site of many other cell metabolic reactions.

endorphins a number of morphine-like peptide substances naturally produced in the body and for which morphine receptors exist in the brain. Many of these active substances have been found, all with the same opoid core of five amino acids. They are neurotransmitters and have a wide range of functions. They help to regulate heart action, general hormone function, the mechanisms of shock from blood loss and the perception of pain, and are probably involved in controlling mood, emotion and motivation. They are thought to be produced under various circumstances in which acute relief of pain or mental distress is required. At least some of the endorphins are produced by the PITUITARY gland as part of the precursor of the ACTH molecule. Endorphins are fragments cleaved from the beta-lipotropin component of proopiomelanocortin (POMC). The term derives from the phrase 'endogenous morphine-like substances'.

endoscopy direct visual examination of any part of the interior of the body by means of an optical viewing instrument (endoscope) introduced through a natural orifice or through a small surgical incision.

endothelins a range of peptide substances, containing 21 amino acids, and believed to be the most powerful known constrictors of arteries. Very tiny amounts will raise the blood pressure. One of them, ENDOTHELIN-1, derived from the ENDOTHELIUM of blood vessels, also causes constriction of BRONCHI, inhibits the release of RENIN from the kidneys and increases the strength of the heart beat. Endothelins are formed continuously in the larger arteries. Increased output can be caused by oxidized low-density lipoproteins. Endothelin receptors are ubiquitous

endothelin-1 a peptide of 21 amino acids produced by the endothelial cells of arteries that has a powerful arterial tightening (vasoconstrictor) function. It also induces MITOSIS in cells. It has a short half-life of 1–2 minutes and its production is inhibited by endothelial-derived relaxing factor and stimulated by ANGIOTENSIN II, vasopressin, and thrombin.

endothelium the single layer of flattened cells that lines the blood vessels, the heart and some of the cavities of the body. Endothelium is no longer regarded as simply a non-stick lining. It is now known to be a physiologically and biochemically dynamic structure that exerts a regulatory control on the tone of blood vessels and on the blood flow, mainly by the release of the hormone NITRIC OXIDE. It regulates the growth and repair of blood vessels; modulated the contraction and relaxation of the heart; protects blood cells from damage; helps to control the inflammatory process; prevents blood clotting within vessels; and offers a selectively permeable barrier to the passage through it of molecules.

endothelium-derived relaxing factor a substance such as nitric oxide produced by the lining cells of arteries that acts on the circularly placed smooth muscle in the walls of arteries to relax it and so promote widening (vasodilatation). Its action lasts for only a few seconds.

endothermic of a chemical reaction that absorbs heat.

endotracheal within the windpipe (TRACHEA).

endotracheal tube a curved plastic tube, some 20–25 cm long, inserted through the mouth into the upper part of the windpipe (trachea), by way of the voice-box (larynx) so as to maintain a reliable AIRWAY in an anaesthetized or otherwise unconscious person. Most endotracheal tubes are cuffed with a small balloon that can be inflated to form a seal and prevent inhalation of secretions or vomit.

endplate the expanded termination of a motor nerve AXON that forms a SYNAPSE with a muscle fibre. See also MOTOR ENDPLATE.

endplate potential the depolarization of the MOTOR ENDPLATE by acetylcholine that triggers the action potential and hence the contraction of the muscle fibre.

energy the capacity of a body to do work. Energy occurs in several forms – potential as in a compressed spring or a mass in a high position, kinetic as in motion, chemical as in petroleum and nuclear as in the binding forces of the atomic nucleus. Its effect, when manifested, is to bring about a change of some kind. The unit of energy is the joule. The calorie is still used in nutrition studies. It is the heat required to raise 1 gram of water by 1°C. Calorific value of food is measured in kilocalories (1000 calories). 1 calorie = 4.2 joules. Short-term energy storage in humans is in ATP, long-term storage is in fat.

enhancer a regulatory DNA sequence located remotely from the gene it influences. When particular proteins bind to the enhancer, the rate of transcription of the gene concerned is altered.

enkephalins see ENDORPHINS.

enophthalmos a sinking backwards of the eyeball into its bony socket so that the lid margins cannot easily be held the normal distance apart and there is an obvious narrowing of the lid aperture. Enophthalmos is the opposite of EXOPHTHALMOS. Enophthalmos results from loss of fat within the eye socket.

ensiform shaped like a broad sword blade or leaf.

enter-, entero- *combining form denoting* the intestines.

enteral within the gut (gastrointestinal tract).

enteric pertaining to the small intestine.

enteric nervous system a collection of neurons in the intestine that can function independently of the central nervous system and has been described as the 'brain of the gut'. This system is responsible for intestinal motility including PERISTALSIS, the secretory function of the intestine, the control of blood flow in the intestinal wall and the regulation of intestinal immune and inflammatory reactions.

enterogastrones hormones released by the intestinal tract that inhibit stomach activity.

enterokinase see ENTEROPEPTIDASE.

enterohepatic circulation the absorption of bile from the intestine and its return to the liver to be resecreted into the bowel.

enteron the intestine.

enteropathy any disease of the intestinal tract.

enteropeptidase a protein-digesting enzyme in the epithelial lining of the duodenum that converts pancreatic trypsinogen into active trypsin. Also known as enterokinase.

enuresis bedwetting. The involuntary passage of urine, especially during sleep.

environment and the human being

The interaction of human beings with our physical surroundings geographic, climatic and biological have been major factors in determining our culture and evolution. But the physical milieu is only part of the environment which continues to exert its effects on us. The environment also includes economic, social, educational, political and historical elements. All these influences have been operating on humans from the time we evolved into the genus *Homo*, and they continue to do so. For centuries, we have been altering our environment to our own supposed advantage. While many of these changes are indeed advantageous, increasingly our impact on the environment has been to the detriment of both.

Contemporary concern with ecology tends to cause us to limit our definition of the environment to our natural habitat. This is an insufficient concept, and it is best to think of our environment as the source of all incoming information. In this context, the word 'information', too, must be broadly defined. Information can take many forms and is not to be restricted to verbal data. We derive information every moment of the day by the various routes of sensory input – eyes, ears, nose, mouth, skin. Much of this information is necessary for our survival. All of it is potentially capable of changing us and all of it constitutes the totality of the effect of our environment upon us.

The child and the environment

It is impossible to exaggerate the importance of this constant stream of ambient information on the young child. Numerous studies and a great deal of clinical experience have shown that information deprivation, either as a result of environmental deficiency or of a defect in a sense organ, is always damaging to the long-term interests of the child. An eye that is unable to see, for whatever reason, during the first year of life remains virtually blind (amblyopia) even if the defect is later corrected. Visual experience is necessary for the development of vision. Early deafness leads to a permanent hearing defect, even if the cause is corrected, and to a serious speech defect. Early deprivation of human or other contact is damaging to normal emotional development. Maternal cuddling may be one of the most important things that happen to us. Babies react automatically to achieve this vital contact with something soft and warm – which need not necessarily be the natural mother, or even, in the early stages, a human being.

The effects of early malnutrition

Severe and prolonged restriction of food intake in early childhood is a common major environmental influence affecting millions of children in economically backward countries. Medically, the condition is known as protein-energy malnutrition. In simple terms, these children do not get enough food. The worst case is known as nutritional marasmus in which the infant, usually under one year of age, fails to grow, has a body weight less than 60 per cent of normal, is wizened and shrunken and has a low chance of survival. The outlook in children with prolonged protein-energy malnutrition is poor. Most of them suffer permanent loss of stature, some have nutritional dwarfism. The physical growth of the brain is retarded and intelligence is likely to be impaired, especially if the subsequent environment is poor.

The emotional and educational environment

Every stage in childhood development requires an appropriate informational environment for normal internal maturation to occur. The absence of such an environment will inevitably lead to mental or emotional stunting. Happily, even at an early age, children are capable of searching for, and finding, stimulation. The young child is avid for new experience – new data – and soaks up information. To a large extent, the earlier this information is acquired, the more influential it is. Data acquired early in life tends to stay with us, and affect us, permanently. This is so even if, in the light of later experience, we see the early data to be irrational. It is as if the inherent plasticity of the young organism allows a deeper and more permanent 'impression' to be made.

The differences, in the outcome, in any particular context, of stimulus-rich and stimulus-poor environments are usually self-evident. In the limited context of verbal facility, for instance, the child born into and brought up in a highly articulate, educated and highly communicative family has a life-long verbal advantage over a child born into a family of low educational attainment and interest. The child in the educationally-deprived family may, however, have compensatory advantages from a different kind of environmental richness.

The environment in adolescence

Young people grow faster and mature earlier than they did 100 years ago. These changes are common to children in all countries with a reasonable standard of living. The average final adult height has been increasing at the rate of about 1 cm per decade for the past century. The age at which sexual maturation occurs has been becoming progressively lower. Since at least 1830, the average age at which European girls start to menstruate (the *menarche*) has been decreasing by about 4 months every decade.

The causes of these changes are mainly environmental, especially nutritional. Some genetic factors are also probably involved, especially the progressive increase in out-breeding that has occurred during the period. It seems probable that the maximum achievable height is genetically determined, but that the achievement of that maximum is a matter of environmental influences. There is good reason to believe that the former retardation of the menarche to the age of 17, 18 or even later was the result mainly of nutritional deficiency.

The healthy adolescent is an organism in a highly active inter-relationship with its environment. During this period an immense amount of information is ambient. By now, the individual has become more selective and is able to remain unaffected by many of the more obvious environmental influences. It is, however, often difficult to recognize when we are being influenced, and many of the psychological effects of the environment are subtle but pervasive.

Socio-economic factors

The effect of these on the individual can hardly be exaggerated. Social and economic levels tend to correlate fairly closely with educational and nutritional standards and to responsiveness to health advice. It is notable, for instance, that cigarette smoking and the wide spectrum of resultant diseases are now predominantly features of the lower socio-economic groups.

Socio-economic factors are important determinants of physical size and the menarche. This is partly due to nutritional differences and partly to differences in the standards of health. Persistent psychological stress and emotional disturbances are also determinants of physical development. Intelligence, too, while partly determined by genetics, is substantially affected by socio-economic factors. The beneficial effects on intelligence of good opportunity and improved environment have been strikingly shown in some American studies. Average IQ scores were higher in black people living in New York and Illinois than in white people living in the relatively economically disadvantaged states of Mississippi and Georgia. The average scores of the black people living in the north was consistently related to the length of their residence there. Happily, in many parts of the Western World, these socio-economic differences have become much less marked than formerly and in a few countries they have almost disappeared.

Effects of the damaged environment

None of us can remain unaffected by ecological changes especially when, as is usual, such changes are for the worse. Rampant industrialization and the exploitation of natural resources have been allowed to accelerate for two centuries with little or no regard to their damaging effects on our physical environment. These effects – atmospheric pollution, loss of natural amenity, deforestation, plant disease, loss of plant species, soil erosion, and so on, have become commonplace. But because the human organism is inherently adaptable, self-regulatory, and capable of repairing much of the damage caused to it by such environmental influences, we are often less conscious of the extent of the damage done to ourselves.

Humans rely too heavily on their own powers of internal adaptation (homeostasis) and tacitly assume that these powers will see them through the effects of any bodily insult, however severe and prolonged. In this we are mistaken. Modern medicine now recognizes the extent to which the body is damaged when the physical environment is damaged, and now devotes an entire medical speciality – environmental medicine – to the study.

A wide spectrum of disease and permanent, often irremediable, body damage can be attributed to the effects of the damaged environment. Conditions almost wholly caused by it include respiratory diseases such as emphysema and chronic bronchitis; the range of the dust-inhalation diseases causing permanent structural damage to the lungs (the pneumoconioses); brain damage from lead poisoning; and a wide range of cancers caused by such diverse environmental influences as cigarette smoke, alcohol, ionizing radiation, asbestos dust inhalation, fungal toxins such as the *Aflatoxin* that grows on moist peanuts and grains, and literally thousands of industrial chemicals. Ozone from the effects of sunlight on nitrogen oxides produced by car exhausts is a growing menace to health and in an increasing number of areas, urban and rural, it is becoming impossible to avoid such effects.

Large areas of the world are being poisoned and are being made poisonous to humans. Already the health of millions of people in the industrialized world has been damaged. We must urgently face up to the danger and critically inspect our values.

enzymes

Enzymes are organic catalysts – molecules that can enormously accelerate chemical reactions without themselves being changed. The energy in fuels can be released instantly at very high temperatures, by burning them. This is not acceptable in the body so fuels have, somehow, to be oxidized at low temperatures. This can only be done by enzyme action. Thousands of similar chemical reactions are going on all the time in our bodies – reactions connected with the breakdown of food to a form that can be absorbed and utilized (digestion); with the repair and replenishment of body tissues; with the transport of vital materials such as oxygen and fuels throughout the body; with the synthesis of new organic molecules; with their storage and release; with body growth; with the control of gene action to generate new enzymes; with the repair of damaged DNA – the list is endless.

Enzymes bind to the molecules whose action they control, causing the formation of weak bonds that impose strains on these molecules and allow chemical breaks or linkages that could not otherwise occur. Enzymes also position the reactant substances in relation to each other in the correct orientation so as to promote linkages.

enzyme inhibitor a molecule that can block the active site on a substrate or can compete with substrate. In either case the normal function of the enzyme is interfered with.

enzymology the branch of biochemistry concerned with the structure and function of enzymes and coenzymes.

eosin a red dye commonly used to stain bacteria or tissue slices, for microscopic examination.

eosinophil a kind of white blood cell (leukocyte) containing granules of toxic proteins that readily stain with EOSIN. As in the cases of other classes of white cell (e.g. basophil, neutrophil) this is an adjective that has become a noun.

eosinophilia an increase in the number of EOSINOPHIL cells in the blood or in the lungs (pulmonary eosinophilia). Eosinophilia is characteristic of worm infestation, reactions to certain drugs and allergic conditions.

epi- *prefix denoting* upon, over, outside, next to, near or beside.

epicanthus a condition in which the margin of the upper eyelid curves round and downwards on the inner side so as to conceal the inner corner of the eye. Epicanthus is common in babies and may cause an illusory appearance of in-turning of the eye.

epicardium the inner layer of the bag surrounding the heart (the PERICARDIUM).

epicritic pertaining to fine degrees of discrimination, as in delicate touch sensation.

epidemic the occurrence of a large number of cases of a particular disease in a given population within a period of a few weeks. Epidemics occur when a population contains many susceptible people. This is why epidemics often occur at intervals of several years.

epidemiology the study of the occurrence, in populations, of the whole range of conditions that affect health. It includes the study of the attack rate of the various diseases (incidence) and the number of people suffering from each condition at any one time (prevalence). Industrial and environmental health problems are also an important aspect of epidemiology.

epidermis the structurally simple outermost layer of the skin, containing no nerves, blood vessels, or hair follicles, and acting as a rapidly replaceable surface. The deepest layer of the epidermis is the basal cell layer. Above this is the 'prickle cell' layer. The epidermis is 'stratified', the layers of cells becoming flatter towards the surface. The outermost cells of the epidermis are dead and are continuously shed.

epidermoid having the characteristics of, or pertaining to, the EPIDERMIS.

epididymis the long, elaborately coiled tube that lies behind the testicle and connects it to the VAS DEFERENS. Spermatozoa mature during their long journey through this tube and the vas deferens. They also receive stabilizing glycoproteins from the epididymal fluid.

epidural external to the DURA MATER, the outer of the three MENINGES covering the brain and spinal cord.

epigastrium the central upper region of the abdomen. Epigastric pain is a feature of duodenitis or duodenal ulcer.

epiglottis a leaf-like cartilaginous structure, down behind the back of the tongue, which acts as a kind of lid to cover the entrance to the voice-box (larynx) and prevent food or liquid entering it during the act of swallowing.

epigenetic relating to inheritable changes in the pattern of gene expression caused by factors other than changes in the nucleotide sequence of genes. It is now recognized that cancer is caused both by genetic and epigenetic changes and that these two factors are intimately interrelated in the development of tumours.

epilation removal of the hair by the roots by any means, such as plucking, waxing, chemical destruction or electrolysis.

epineurium the connective tissue sheath of a nerve.

epiphenomenalism the belief that mental events are solely a consequence of physical events, specifically neural activity, and never the causes of them. Once considered heretical, the view is now widely held by scientists.

epiphysis the growing sector at the end of a long bone. During the period of growth, the epiphysis is separated by a plate of CARTILAGE from the shaft of the bone. The edge of this plate nearest the shaft becomes progressively converted into bone, while the other edge develops new cartilage. In this way, the bone lengthens.

epiploon the GREATER OMENTUM.

episiotomy a deliberate cut made during childbirth in the margin of the vaginal opening so as to enlarge it, facilitate delivery of the baby and prevent tearing of the tissues backwards towards the anus.

epistaxis nose bleed.

epithelialization the process of covering of a raw surface with EPITHELIUM. The final stage in healing.

epithelium the non-stick coating cell layer for all surfaces of the body except the insides of blood and lymph vessels. Epithelium may be single-layered, or 'stratified' and in several layers, with the cells becoming flatter and more scaly towards the surface, as in the skin. It may be covered with fine wafting hair-like structures (cilia), as in the respiratory tract, and it may contain mucus-secreting 'goblet' cells. See also ENDOTHELIUM and EPIDERMIS.

epitope an immunologically active discrete site on an ANTIGEN to which an ANTIBODY or a B or T cell receptor actually binds. See also LINEAR EPITOPE and CONFORMATIONAL EPITOPE.

erectile capable of becoming erect. Pertaining to tissue containing many blood vessels or blood spaces that can fill with blood and become rigid. The corpus spongiosum and corpora cavernosa of the penis are the most conspicuous examples of erectile tissue.

erection a temporary state of enlargement and rigidity of the penis, due to engorgement with blood under pressure (tumescence), which, given adequate lubrication, makes insertion into the vagina possible.

erector pili (arrector pili) muscle a tiny muscle capable of changing the angle of a hair follicle so as to cause hair to 'stand on end'. Each follicle has one of these muscles which function under extremes of emotion.

ergocalciferol vitamin D_2. This is produced in the body by the action of ultraviolet light on ergosterol. Vitamin D is necessary for the normal mineralization of bone. Deficiency leads to rickets in growing children and bone softening (osteomalacia) in adults.

ergonomics the scientific study of humans in relation to their working environment and the application of science to improve working conditions. The increasing application of complex technology has resulted in increasing human discomfort, difficulties and dangers. Ergonomics seeks to solve such problems.

erogenous zones parts of the body which, when stimulated by touch in a suitable context, are especially liable to arouse sexual interest or desire.

eroticism 1 the elements in thought, pictorial imagery or literature which tend to arouse sexual excitement or desire.

2 actual sexual arousal.

3 a greater than average disposition for sex and all its manifestations.

4 sexual interest or excitement prompted by contemplation, or stimulation, of areas of the body not normally associated with sexuality. The terms anal and oral eroticism are used in two senses – in reference to adult physical sexual activity and in a theoretical Freudian psychoanalytic sense.

eructation belching.

eryth-, erythro- *combining form denoting* red or red blood cell (erythrocyte).

erythro- *combining form denoting* red.

erythroblast a primitive, nucleated red blood cell. A stage in the development of the normal non-nucleated red cell (ERYTHROCYTE) found in the circulating blood.

erythrocyte a red blood cell. 'Erythro' means 'red', and 'cyte' means 'cell'. Erythrocytes are flattened discs, slightly hollowed on each side (biconcave) and about 7 thousandths of a millimetre in diameter. They contain HAEMOGLOBIN and their main function is to transport OXYGEN from the lungs to the tissues.

erythropoiesis red cell production.

erythropoietins hormones, produced mainly in the kidneys, that stimulates red blood cell production (erythropoiesis) in the bone marrow. The amount of hormone produced is based on monitoring of the blood flowing through the kidneys for oxygen concentration.

erythropsia red vision. A visual disorder in which all objects appear red.

eschar an area of dead, separated tissue (SLOUGH) produced by skin damage by a caustic substance or a burn.

esophoria a latent tendency for the eyes to turn involuntarily inwards. When a person with esophoria has one eye covered, that eye will turn inwards. As soon as the cover is removed, the eye straightens and single binocular vision is restored. Compare EXOPHORIA.

esotropia convergent squint, or STRABISMUS. Only one eye looks directly at the object of regard, the other being turned inwards. Esotropia in children calls for urgent treatment to avoid amblyopia. Compare EXOTROPIA.

ESR *abbrev. for* erythrocyte sedimentation rate. See BLOOD SEDIMENTATION RATE.

essential 1 absolutely required.
2 of unknown cause.
3 (of amino acids) necessary for normal growth but not synthesized in the body.
4 extracted from, or related to an extract from, a plant (as in essential oil). Of the essence of.

essential amino acids those AMINO ACIDS that cannot be synthesized in the body and must be included in the diet. They are arginine, histidine, isoleucine, leucine, lysine, methionine, phenylalanine, threonine, tryptophan and valine.

essential fatty acids fatty acids that cannot be synthesized in the body and that must be present in the diet for health. They are LINOLEIC acid, linolenic acid and arachidonic acid, and have at least two conjugated double bonds. These fatty acids are precursors of eicosanoids and other regulatory substances.

essential nutrient any food substance required for health or growth that is not synthesized in the body and must be present in the diet.

ethanoic acid acetic acid.

ethanol the chemical name for ethyl alcohol, the main constituent of alcoholic drinks.

ethmoid bone the delicate, T-shaped, spongy bone forming the roof and upper sides of the nose and the inner walls of the eye sockets (orbits). The upper part of the ethmoid is perforated (cribriform plate) to allow passage of the fibres of the olfactory nerve, and has a central downward extension forming the upper part of the partition (septum) of the nose. The bone contains numerous air cells (ethmoidal sinuses).

ethnography the subdivision of anthropology concerned with the direct study of cultural groups such as tribes, communities, social classes or any other definable class of people. Ethnography involves fieldwork in which observations are made and recorded, and then a written account and analysis of the data.

ethnology the subdivision of anthropology concerned with the origins, spread and inter-relationships of the races of human beings.

ethology the study of the behaviour of animals in their natural habitats.

ethyl alcohol see ALCOHOL.

etiquette a set of rules for the conduct of human beings in particular social and professional contexts. Conformity to the current etiquette is mandatory if one is to be generally accepted; non-conformists, unless exceptionally valuable in other respects, usually suffer disadvantage or rejection. Some items or codes of etiquette have a logical and utilitarian basis and help to regulate social conduct. Some, such as a slavish adherence to precedence on social occasions, confer exclusiveness on a group. But many are arbitrary and meaningless, with no greater justification than the desire of one human being to establish some kind of ascendancy over another.

eu- *prefix denoting* good, well, normal.

eugenics the study or practice of trying to improve the human race by encouraging the breeding of those with desired characteristics (positive eugenics) or by discouraging the breeding of those whose characteristics are deemed undesirable (negative eugenics). The concept implies that there exists some person or institution capable of making such decisions. It also implies possible grave interference with human rights. For these reasons, the principles, which have long been successfully applied to domestic animals, have never been adopted for humans except by despots such as Adolf Hitler.

eugeria the state of a high quality of life in old age. Eugeria should be the normal condition of old people, but is often precluded by physical illness or injury, psychological deficit or disturbance, and respect for defective cultural stereotypes. The term is derived from the Greek *eus-* good, and *geras* old age.

euglobulins globular proteins that are insoluble in water but soluble in saline solutions. Most of the body's globulins are euglobulins.

eukaryote any organism whose cell or cells contains a well defined nucleus with a nuclear membrane in which the genetic material is carried in the chromosomes. Only bacteria and blue-green algae are not eukaryotes. The word is also spelled eucaryote. See also PROKARYOTE.

eunuch a man castrated before puberty. This results in the loss of the male sex hormones and the failure of development of the secondary sexual characteristics – the beard, the deeper voice from enlarged larynx, the adult-sized penis, and the characteristic male body shape.

eupepsia good digestion.

euphenics improvement of the whole person (phenotype) through planned control and manipulation by means other than EUGENICS.

euphoria a strong feeling of well-being or happiness. The term is sometimes used to mean an abnormally exaggerated feeling of elation.

euploid 1 having a number of chromosomes that is an exact multiple of the HAPLOID number.
2 having the normal number of chromosomes.

euryon one of the two most widely separated points in the transverse diameter of the skull.

Eustachian tube a short passage leading backwards from the back of the nose, just above the soft palate, on either side, to the cavity of the middle ear. This allows air to pass to or from the middle ear cavity to balance the pressure on either side of the eardrum. (Bartolomeo Eustachi, Italian anatomist, circa 1520–74).

euthanasia mercy killing.

euthyroid having normal thyroid gland function. The term is used when thyroid function is restored to normal after either thyrotoxicosis or hypothyroidism.

evagination a turning inside out.

Eve hypothesis the suggestion that all mankind is descended from an African woman who lived 200,000 years ago. Support for the hypothesis comes from a study of mitochondrial DNA which is inherited only from the mother. Knowledge of the average rate of the occurrence of mutations allows backward extrapolation in time and an estimate of the date of the universal maternal ancestor. The hypothesis has been critizised.

eversion a turning outwards.

evoked responses signals on the ELECTROENCEPHALOGRAM prompted by visual or auditory sensory input and detected by special electronic averaging techniques. Evoked responses can provide direct, objective evidence that conduction is occurring along the nerve tracts connecting the sense organ with the brain.

evolution the theory that all living organisms have developed in complexity, from a simple life form. Evolution occurs by the natural selection of those who, by the fortune of spontaneous random changes (mutations), happen to be best suited to their contemporary environment, to survive and reproduce. It does not occur by the passing on to offspring of characteristics acquired during the lifetime of an individual.

The idea that humans evolved from a non-human species was widely rejected until end of the 19th century. But as the evidence of comparative anatomy and of the fossil record accumulated, this conclusion gradually became irresistible. It soon became apparent to most scientists that both man and the apes had developed separately from a common ancestor. Few now believe that there is a 'missing link' between the apes and man. The palaeoanthropological evidence of man's evolution is open to various interpretations and many 'facts' deduced by the experts have been challenged by others. There is, however, a large consensus of agreement about the broad lines of the evolution of humans.

evolutionary biology a paradigm for education and study based on the recognition that a failure to take due account of evolution when viewing human anatomy, physiology, psychology and pathology will result in a truncated, less valid and less fruitful understanding. It should, for instance, be recognized that the biological norm for infants and children is to be exposed to many infections – a condition that may be necessary for the 'normal' development of the immune system; and that since throughout almost the whole of human evolution life expectancy has been no more than about 30 years, systems have evolved which are largely indifferent to conditions that do not arise until well after that age. Affluence may have much more widespread malign effects than merely obesity and that these may include greatly increased susceptibility to various cancers.

excision cutting off and removing completely.

excision enzyme an enzyme that removes short damaged segments from the deoxyribonucleic acid (DNA) molecule of a cell.

excision repair correction of DNA damage by the removal of the damaged length of a single strand by one of a number of enzymes capable of recognizing damaged nucleotides, and resynthesizing the correct nucleotide sequence complementary to the normal strand. The repair is catalyzed by DNA polymerase I and DNA ligase.

excitability the property of any cell or tissue capable of producing an electric signal.

excitable membrane a membrane capable of producing an action potential.

excitatory amino acids aspartic acid and glutamic acid – amino acids that act on voltage-gated ion channels in the plasma membranes of cells of the nervous system to cause them to fire.

excitatory synapse a synapse which, on activation, increases the likelihood of an action potential in the post-synaptic neuron or increases the frequency of firing of the post-synaptic neuron.

excrescence any projection of abnormal tissue from a surface, such as a wart, heart valve vegetations or a nasal polyp.

excretion removal from the body of the waste products of metabolism.

exercise

It need hardly be mentioned that exercise is desirable, but there is more to the advantages of exercise than meets the eye in the shape of bulkier muscles. A study of the effects of exercise is a study in the unity and the dynamic inter-relationships of all the major systems of the body – the skeleton, the muscles, the heart and circulation, the respiratory system, the brain and nervous system, even the endocrine system. The body is, above all, a responsive entity and the response to exercise is much more widespread and detailed than is generally supposed.

Humans evolved as regularly exercising animals for whom physical and mental fitness were often a matter of life or death. We should not, therefore, be surprised that exercise is important in promoting health and that lack of exercise is damaging. During adolescence and early adult life the quantity and type of exercise taken has a permanent, life-long bearing on the physical state and quality of operation of most of the body systems and it is difficult, later, to compensate for lack of exercise during this period. But whatever the early experience, exercise remains important throughout the whole of life. Physiologists no longer subscribe to the defeatist idea that little physical improvement is possible after middle age.

Exercise and the heart

Exercise causes a large increase in the pumped output of the heart. Heart output averages about 5 litres per minute during rest, and this increases by several times during exertion. A reasonably fit person may, during exercise, have an output of 15–20 litres per minute, and a trained athlete can achieve an output of as much as 35 litres per minute. Such an increase implies that a major improvement in heart performance and efficiency can be achieved simply by exertion. The great increase in heart output is necessary to provide the markedly increased blood flow through the muscles without which exercise would be impossible. With training, the main improvement is in the volume of blood ejected with each beat (the stroke volume).

Such improvement in heart efficiency implies actual structural changes and these are, in fact, considerable. The heart muscle fibres become enlarged by an increase in the number of sarcomeres, the number of mitochondria in each muscle fibre increases, the branches of the coronary arteries enlarge and the ratio of capillaries to muscle fibres becomes greater by the budding out of new vessels. As a result of these changes, the resting heart rate becomes lower, the blood pressure drops, there is an increase in the total blood volume and the recovery to normal after exertion is more rapid. All these effects lessen the risks of coronary artery heart disease.

Exercise and the respiratory system

Similar effects to the changes in the heart muscle are observed in the muscles of respiration. Useful exercise is necessarily associated with breathlessness, which is simply more forceful action of the muscles of respiration – those between the ribs (the intercostal muscles) and the diaphragm. In strenuous exercise the accessory muscles of respiration – the muscles of the neck that are able to pull up the whole rib cage – are also brought into action. Lung expansion is greater because, with exercise, the ability to increase the maximum volume of the chest improves. The effect of these changes – together with the changes in the heart – is an increase in the amount of oxygen that can be supplied in a given time, to the tissues.

Exercise and fuel supply

The energy required by the muscles during exercise calls for the mobilization of large quantities of fuel in the form of glucose and fatty acids. The body's glucose stores are small and the fuel is compacted into long-chain glucose polymers called glycogen. This is stored in the muscles and in the liver and can readily release single molecules of glucose. Muscle glycogen is the first to be mobilized and for the first 10 minutes or so of exertion this is the main source of fuel. All the muscle glycogen is not, however, consumed at this stage. After about 10 minutes the muscles begin to use a higher proportion of the glucose and fatty acids brought in the bloodstream by the increased blood flow associated with exercise. To prevent the blood glucose from dropping dangerously (hypoglycaemia) glucose is released from glycogen in the liver. But liver stores are limited and, as exercise continues, more fatty acid is consumed. When liver glycogen threatens to be exhausted, the liver makes up the deficit in glucose by converting amino acids, glycerine (glycerol) and lactate into glucose. The glycerol comes from the breakdown of fats (triglycerides) and this also releases fatty acids which are used as fuel. Prolonged exercise causes a moderate drop in the levels of glucose in the blood and a drop in the secretion of insulin. Cortisol and growth hormone are secreted in increased amounts during exercise. Fatigue occurs when the glycogen stores in the muscles become depleted.

Exercise and elderly people

In people in their seventies and eighties, 10 months of moderate but regular exercise will usually increase oxygen consumption per unit heartbeat by 30 per cent. Increased oxygen consumption is an indication of an increase in cell metabolic processes but also reflects an improvement in heart function, especially in the output per beat. The vital capacity of the lungs – the maximum amount of air that can be exhaled after a maximal inspiration – can be improved by 20 per cent and the lung ventilation at maximum exercise can be increased by 36 per cent. The aerobic capacity of most elderly men who exercise regularly is about the same as that of the average non-exercising 40 year old.

Bone bulk and exercise

Twenty-five per cent or so of elderly people, mostly women, suffer hip and other fractures on minor trauma as a result of osteoporosis. Both men and women suffer progressive loss of bone mass with age but this is accelerated by inactivity and can be minimized by regular exercise. The compressive forces produced by walking or running and the effect of muscle tension on tendons provide a stimulus to osteoblast activity to maintain the bulk of bone. Athletes have significantly increased bulk in the bones involved in their particular activity. People lying in bed or astronauts in conditions of zero gravity can lose 1 per cent of bone volume per week.

Exercise and morale

Exercise induces a feeling of well-being. That this fact has been little used in the treatment of depression reflects the unfortunate effect of depression in inhibiting the will to exertion. But if people with mild to moderate depression can be persuaded to undertake exercise, the results are excellent. Depression can be cured by exercise alone, even in depressives who have failed to respond to drug treatment. Several factors account for the effect of exercise on the mind, but body and mind are so intimately inter-related that it is hardly surprising that an increase in general bodily efficiency and ease of function should be associated with a more healthy function of the mind. Factors which one would naturally expect to improve the general equanimity include an improved oxygen supply to the brain leading to an increase in the capacity for action; an awareness of growing physical powers; the sense of physical well-being; pride in performance; and relief of embarrassment over physical deficiencies.

Endorphins and enkephalins are natural morphine-like substances (opioids) each consisting of a sequence of five amino acids. They are produced by the body during, or even in anticipation of, painful or severely stressful experiences. It is these substances that prevent us from feeling pain at the time of severe injury and they have all the properties of the opium alkaloids. Taxing physical exercise is a sufficient stimulus for the production of natural opioids and this has given rise to the idea that exercise euphoria, or even addiction, may be mediated in this way. Common experience suggests that activities such as regular long-distance running is addictive. Many long-distance runners experience a 'high'. Whether this is a major factor in the mood-elevating effect of exercise is uncertain.

exfoliation shedding of cells from a surface, such as the skin. In exfoliative dermatitis, much of the surface of the skin peels off or is shed.

exoccipital of the bony masses on the occipital bone on either side of the FORAMEN MAGNUM.

exocrine glands simple glands, such as sweat, sebaceous, or mucous glands which secrete on to a surface. Endocrine glands, by contrast, secrete into the bloodstream.

exocytosis the movement of peptides or proteins out of a cell into the extracellular fluid, in tiny membranous vesicles that pass through the plasma membrane after fusing with it. Most molecules leave cells by exocytosis.

exoenzyme an enzyme that operates outside the cell in which it was formed.

exogamy breeding between organisms that are not closely related. Outbreeding, as distinct from inbreeding.

exogenous having an external origin or cause.

exon the segment of deoxyribonucleic acid (DNA) in a gene that codes for some part of the messenger ribonucleic acid (RNA). Any segment that is represented in the RNA product. Segments that do not code for RNA are called introns.

exonucleases enzymes that break down DNA and RNA molecules, starting at the ends of the molecular chains. Exonucleases split off nucleotides one at a time from the end of a chain and may be specific for either end.

exopeptidase an enzyme that acts to split off AMINO ACIDS or sometimes dipeptides from the ends of a protein molecule.

exophoria a latent tendency for the eyes to turn outwards (diverge), but which is usually controlled by the binocular fusion power. If one eye is covered in exophoria it will turn outwards under the cover. Compare EXOTROPIA, ESOPHORIA.

exophthalmos protrusion of an eyeball. This forces the eyelids apart and causes a staring appearance. Exophthalmos is caused by an increase in the bulk of the contents of the bony eye socket (ORBIT) from any cause.

exoplasmic face a surface, such as that of a cell membrane, that faces away from the cytoplasm. The exoplasmic face of a cell faces outwards; that of a mitchondrion faces in towards its lumen.

exotropia divergence of the lines of vision of the two eyes. Divergent squint. In exotropia one eye points at the object of regard, the other is directed outwards. In exotropia acquired in adult life there is usually double vision (diplopia). In childhood, the false image may be suppressed and long-term visual acuity may be affected in one eye. See also ESOTROPIA.

expected date of delivery an estimate of the date on which a baby will be born. This is an arbitrary calculation based on a statistical average gestation period of 266 days, counted as 280 days from the date of the first day of the last menstrual period.

expiration 1 breathing out.
2 death.

expressivity the degree of severity shown by an AUTOSOMAL dominant trait in any particular affected individual. The main feature of expressivity is its variability.

extension straightening at a joint. The opposite of flexion.

extensor muscle, extensor a muscle that, on contraction, straightens a joint. A flexor bends a joint.

external auditory meatus the skin-lined external passage of the ear, leading from the outside to the eardrum (tympanic membrane).

external carotid artery a major artery that springs from the common carotid artery in the neck and supplies blood to the front of the neck, face and scalp and the side of the head and the ear. The artery also supplies the DURA MATER.

external genitalia the penis and scrotum in the male; the labia majora and minora, clitoris and mons pubis in the female.

extero- *prefix denoting* outside.

exteroceptor any sense organ receiving external stimuli.

extra- *prefix denoting* outside, beyond, additional.

extra-articular outside a joint.

extracardiac outside the heart.

extracellular in the space surrounding a cell or collection of cells. The extracellular tissue fluid is the medium of transfer of nutrient materials and oxygen, and of waste products, between the cells and the blood.

extracellular matrix non-living material secreted by cells that fills spaces between the cells in a tissue, protecting them and helping to hold them together. The extracellular matrix may be semifluid or rigidly solid and hard as in bone. It is composed of polysaccharides and fibrous and adhesive proteins and includes collagens, elastin, reticulin, glycoproteins, proteoglycans, fibronectin, laminins and osteopontin.

extracorporeal outside the body.

extracranial located or occurring outside the cranium.

extrapyramidal system the part of the central nervous system, supplementary to the voluntary motor (pyramidal) system, and concerned with gross movement, posture and the coordination of large muscle groups. The pyramidal system contains the main motor control pathways and includes the basal ganglia, the reticular formation and descending nerve tracts outside the pyramidal system.

extrasensory perception the claimed ability to obtain information without the use of the normal channels of communication. ESP is said to include telepathy, clairvoyance, precognition and retrocognition. There is no respectable scientific evidence for extrasensory perception, but a mass of anecdotal 'proof'.

extrauterine located or occurring outside the womb (uterus), especially of a pregnancy. See also ECTOPIC PREGNANCY.

extravascular outside a blood vessel or the circulatory system.

extravert a person whose concerns are directed outward rather than inward, who is positive, active, optimistic, gregarious, impulsive, fond of excitement, often aggressive, and sometimes unreliable. The concept was invented by Carl Jung who also described the opposite personality type, the introvert. Also known as extrovert.

extrinsic extraneous to a body or system. Originating from outside.

extroversion turning inside out.

eye tooth see CANINE TOOTH.

Ff

F *abbrev. for*, or symbol for, the element fluorine.

fab the monovalent antigen-binding fragment of an antibody molecule obtained when the protein is digested with the enzyme papain. It consists of an intact light chain and the immediately-associated terminal and domains of a heavy chain.

facet a small flat surface on a bone or tooth or other hard body. A facet may be natural, as on the arches of the vertebrae, or the result of wear.

facial pertaining to the face.

facial artery one of the main branches of the external carotid artery. The facial artery crosses the lower edge of the jaw bone about half way back and divides into numerous branches to supply the skin and muscles of the face and the lining of the mouth, throat and tonsil.

facial bones these are the upper jaw (maxilla), the cheek bone (zygoma), the nasal bone in the upper part of the nose, the lacrimal bone at the inner corner of the eye, the hard palate (palatine bone) and parts of the deeper ETHMOID and SPHENOID bones. The lower jaw bone is called the mandible.

facial index the ratio of the length of the face to its width multiplied by 100.

facial nerve one of the 7th of the 12 pairs of CRANIAL NERVES. Each facial nerve supplies the muscles of the face on its own side. Loss of function in a facial nerve causes partial or total paralysis of one side of the face. This is called BELL'S PALSY.

facies the facial expression or appearance characteristic of a particular medical condition or state. Typical facies are found in many conditions and may assist a doctor in diagnosis.

facilitated transport movement of an ion or molecule across a cell membrane, at a rate greater than that of simple diffusion, by means of a transporter protein.

facilitation the general DEPOLARIZATION and passage of an impulse that occurs in a nerve cell when excitatory inputs at SYNAPSES exceed inhibitory inputs.

factor 1 any kind of biological material that causes a particular effect.
2 an effector whose function is known but which has not yet been chemically identified.
3 one of the components in the blood coagulation cascade.

factor VIII a protein (globulin) necessary for the proper clotting of the blood. The absence of factor VIII causes HAEMOPHILIA but it can be isolated from donated blood and given to haemophiliacs to control their bleeding tendency.

factor IX one of the many factors necessary for blood clotting. Absence of factor IX occurs as a result of an X-LINKED gene mutation and causes Christmas disease, a form of HAEMOPHILIA almost identical to factor VIII deficiency haemophilia.

factor XII the factor that initiates the sequence of reactions that ends in blood clotting (coagulation). Also known as the Hageman factor.

factor XIII an enzyme in the blood that catalyses cross-linking between molecules of fibrin so as to strengthen the forming blood clot.

facultative capable of adapting in response to changing environments.

FAD see FLAVINE ADENINE DINUCLEOTIDE.

faecalith a faece that has become so hard as to resemble a stone.

faeces the natural effluent from the intestinal tract. Faeces consist mainly of dead bacteria, but also contain cells cast off from the lining of the intestine, mucus secretion from the cells of the intestinal wall, bile from the liver, which colours the faeces, and a small amount of food residue, mostly cellulose.

Fahrenheit scale the temperature scale formerly used in medicine but now replaced by the CELSIUS SCALE. In the Fahrenheit scale, the melting point of ice is 32°F, and the boiling point of water is 212°F. Normal body temperature is about 98°F. To convert Fahrenheit to Celsius, subtract 32 and multiply by 0.555 or 5/9.

fainting temporary loss of consciousness due to brain deprivation of an adequate supply of blood and thus oxygen and glucose. This results from a reduction in blood pressure, either because the heart is pumping too slowly or less efficiently or because the arteries of the body have widened. If a fainting person is allowed to lie and the legs raised, recovery will be rapid. It is dangerous to keep a fainting person upright.

falciform curved or sickle-shaped.

falciform ligament a large sickle-shaped fold of PERITONEUM extending from the front of the top of the liver to the UMBILICUS.

Fallopian tube the open-ended tube along which eggs (ova) travel from the ovaries to the womb (uterus) and in which fertilization must occur if pregnancy is to result. The open end of each Fallopian tube has finger-like processes that sweep over the surface of the ovary at the time of ovulation, wafting the egg into the tube. Also known as a uterine tube. (Gabriele Fallopio, Italian anatomist, 1523–63).

false rib any of the five pairs of lower ribs that are attached to cartilages rather than directly to the breastbone (sternum). See also FLOATING RIB.

falx a sickle-shaped structure. The falx cerebri is the curved, vertical partition of DURA MATER that occupies the longitudinal fissure between the two cerebral hemispheres.

familial 1 occurring in some families but not in others, as a result of genetic transmission. The term is usually applied to diseases. 2 occurring more often in a particular family than would happen by chance.

family balancing the wish to choose the sex of an unborn child on the basis of how many children of each sex already exist in a family. Measures to elect a gender are currently considered at least unethical and, in some countries, are illegal.

fart see FLATUS.

fascia tendon-like fibrous connective tissue arranged in sheets or layers under the skin, between the muscles and around the organs, the blood vessel and the nerves. Fascial sheaths form compartments throughout the body. Some fascia is dense and tough, some delicate. Much of it contains fat cells. The 'superficial fascia' just under the skin is one of the main fat stores of the body.

fasciculus a bundle of fibres, especially nerve fibres, running together with common origins and functions.

fast fibre a voluntary muscle fibre capable of contracting more quickly than a slow fibre and with myosin of high ASTPase activity.

fasting refraining from taking food. So long as water is taken, reasonably well-nourished people can safely fast for several days. Once the fat stores are depleted, however, the voluntary muscles are consumed and soon become severely wasted. Many people find moderate regular fasting a useful aid to health.

fatigue physical or mental tiredness. Physical fatigue is due to accumulation in the muscles of the breakdown products of fuel consumption and energy production (metabolism). Mental fatigue is usually the result of boredom, frustration, anxiety, over-long concentration on a single task or dislike of a particular activity. Fatigue commonly involves both kinds.

fat mobilization increased breakdown of fats (triglycerides) with the release of fatty acids and glycerol into the blood.

fats (lipids)

Body fats are oily substances found, in variable quantity, in bag-like cells under the skin, within the abdomen, and elsewhere. Ideally, fats should not form important gross structural elements in the body, but they are important as energy stores. Fats are known as triglycerides because they consist of three fatty acids linked to a backbone of glycerol (glycerine). These fatty acids may be the same or different and may be saturated or unsaturated. A saturated fatty acid has only single bonds between carbon atoms; an unsaturated fatty acid has double bonds between some of the carbon atoms. The degree of saturation determines the melting point so that many polyunsaturated fats are liquid at room temperature while saturated fats are solid at low temperatures. Unsaturated fats, such as those that are found in vegetable and fish oils, are believed to be less harmful to health.

Fats, especially cholesterol, are important constituents of cell membranes, which consist of two layers of fat molecules penetrated by scattered proteins that act as ports. These membranes delineate a cell from its surroundings and divide a cell into functional compartments. Other important steroid fats include a number of hormones, such as those produced by the adrenal glands and the sex glands.

fat soluble able to be dissolved in fats or fat solvents. Four of the vitamins, A, D, E and K, are fat soluble.

fate map a diagram of an embryo showing the regions that will eventually form particular tissues in the adult.

fatty acids a large group of monobasic acids found in animals and plants. They are hydrocarbon chains and are saturated or unsaturated aliphatic compounds with an even number of carbon atoms. Chain lengths range up to nearly 30 carbon atoms. Fatty acids with more than about 8 carbon atoms in the chain occur most commonly as constituents of glycerides, phospholipids and sterols. The most abundant fatty acids are palmitic, stearic and oleic acids. Glucose and fatty acids are the two main fuel substances of the body. FATS are glyceride esters of fatty acids.

fauces the narrowed space at the back of the mouth and PHARYNX, under the soft palate and between the soft palatine arches (the pillars of the fauces) on either side from which the throat opens out. The term appears to be a plural of the singular word faux but is, in fact, a singular entity. From Latin, *fauces*, throat.

Fc a crystallizable, non-antigen-binding fragment of an antibody molecule obtained by brief digestion with the enzyme papain. It consists of the C-terminal part of both heavy chains, the part which binds to FC RECEPTORS.

Fc receptors cell surface receptors of a specific chemical 'shape' to bind to the Fc heavy chain terminals of the appropriate classes of immunoglobulin molecules (antibodies). IgE molecules, for instance, bind to the Fc receptors on MAST CELLS and blood eosinophil cells. IgG molecules bind to Fc receptors on phagocyte cell membranes.

fear the response to a real or imagined perception of danger. An abnormal degree of fear, or a fear inappropriate to its cause is called a phobia. Fear is accompanied by physical symptoms such as rapid heart action, muscle tension especially in the abdomen, dryness of the throat and sweating. These symptoms are mainly caused by ADRENALINE. It is believed that separate nuclei in the amygdala mediate different aspects of fear-conditioned behaviour.

febrile pertaining to, or featuring, a fever. From Latin *febris*, fever.

fecundation fertilization. Impregnation.

feedback a feature of biological and other control systems in which some of the information from the output is returned to the input to exert either a potentiating effect (positive feedback) or a dampening and regularizing effect (negative feedback). Too much positive feedback produces a runaway effect often with oscillation.

fellatio oral stimulation of the penis. From the Latin *fellatus*, to suck.

felo de se the act of committing SUICIDE.

female symbol ♀ the universally-used symbol for the female.

feminization 1 the effects of the sex hormones, oestradiol (estradiol) and oestrone (estrone), from the ovary, bringing about the normal female secondary sexual characteristics. 2 the development of female secondary sexual characteristics in the male, as when a tumour of an adrenal gland causes the production of mainly female sex hormones (oestrogens). In testicular feminization, the male sex hormone does not bind properly to the tissue receptors. In the absence of male sex hormones before puberty, as in men castrated early, feminization occurs.

femoral pertaining to the thigh or the thigh bone (FEMUR). From Latin *femor*, thigh.

femoral artery the main artery of the leg in the area between the groin and the back of the knee, where it divides into two and passes down to supply the lower leg. The femoral artery supplies all the muscles and other structures of the thigh with blood.

femoral nerve one of the main nerves of the leg. It branches widely to run into the group of large muscles on the front of the thigh and to carry back sensation from the skin on the front and inner aspects.

femoral ring the abdominal opening of the femoral canal.

femoral vein the large vein draining the whole of the leg that lies alongside the FEMORAL ARTERY.

femur the thigh bone. The upper end of the femur forms a ball and socket joint with the side of the pelvis. The lower end widens to provide the upper bearing surface of the knee joint.

fenestra an opening or window between two chambers or body spaces, or an opening made in a plaster cast or dressing to allow examination or drainage.

fenestrated having windows or window-like openings.

ferritin the principal IRON-binding protein of the body. Ferritin acts as an iron store in the liver and other tissues. Each ferritin molecule can hold up to 4500 iron atoms and the amount of iron in ferritin molecules accurately reflects the total iron stores of the body. Ferritin also protects against the toxic effects of excess iron.

fertility the power to reproduce or the possession of such power. See also INFERTILITY.

fertilization the union of the spermatozoon with the egg (ovum) so that the full complement of chromosomes is made up and the process of cell division, to form a new individual, started.

fetal pertaining to a FETUS. (See note there about spelling).

fetal haemoglobin the form of HAEMOGLOBIN normally present in the red blood cells of the fetus. The haemoglobin alpha chains are identical to those of the adult, but the gamma chains are slightly different.

fetal position a position, resembling that of the fetus in the womb, sometimes adopted by a child or adult in a state of distress or withdrawal. The position is one in which the body is drawn into itself. The head is bent forward, the spine is curved, the arms are crossed over the chest and the hips and knees are fully bent (flexed).

fetation the state of pregnancy or the development of a fetus.

fetishism sexual interest aroused by an object, such as an article of clothing, or by a part of the body not normally considered sexually significant. Fetishism is essentially a male disorder and, if severe, the affected person will prefer contact with the object to contact with the owner and will often use the object to assist in masturbation.

fetor oris bad breath. Halitosis.

fetus the developing individual from about the eighth or tenth week of life in the womb until the time of birth. The fetus has all the recognizable external characteristics of a human being. At 10 weeks, the fetus measures about 2.5 cm from the crown of the head to the rump. The face is formed but the eyelids are fused together. The brain is in a primitive state, incapable of any meaningful form of consciousness. By 3 months, the fetus is about 5 cm long (crown to rump) and by 4 months it is about 10 cm long. In the 6th month, the fetus is up to 20 cm long and weighs up to 800 g. Survival outside the womb at this stage is unlikely. Most fetuses over 2 000 grams survive if properly managed in an incubator. From the Latin *fetus*, an offspring. The common spelling 'foetus' is incorrect and is used only by journalists who should know better.

fibre, dietary see DIETARY FIBRE.

fibril a fine, slender fibre.

fibrin an insoluble protein that forms as a fibrous network when the blood protein fibrinogen interacts with THROMBIN. Fibrin is the basis of a blood clot and the end product of a complex cascade of reactions set in motion by injury to a blood vessel.

fibrinase Factor XIII, an enzyme in the blood coagulation cascade that catalyzes the formation of side links between fibrin molecules so as to create a mesh of polymerized FIBRIN that stabilizes the blood clot. Fibrinase is also known as the fibrin-stabilizing factor. See also Factor VIII, IX, XII.

fibrinogen a protein in the blood that is converted to FIBRIN by the action of THROMBIN in the presence of ionized calcium, thereby bringing about coagulation of blood.

fibrinoid 1 resembling fibrin.
2 a homogeneous, refractile, non-cellular material resembling FIBRIN that is found in the walls of blood vessels and elsewhere in certain disease processes. Fibrinoid is also found as a layer between the PLACENTA and the womb.

fibrinolysin an enzyme that can break down FIBRIN.

fibrinous 1 rich in FIBRIN.
2 having the nature of fibrin.

fibroblast a cell that generates the protein COLLAGEN, a major component of connective tissue and the main structural material of the body. Fibroblasts are important in wound healing. They can readily be cultured artificially.

fibroblast growth factors a family of peptide regulators produced by virtually all cells. Their range of actions is wide. Only some are stimulators of the collagen-producing cells, the fibroblasts. Other effects include strong stimulation of vascular endothelial cell proliferation. Some are produced in relatively large quantities by brain tumours. Hope for people paralysed by spinal cord injuries has been raised by the discovery that a measure of rejoining in severed spinal cord tracts in rats can be achieved using a fibrin 'glue' containing a fibroblast growth factor.

fibrocartilage a tough form of CARTILAGE containing many thick bundles of COLLAGEN fibres.

fibronectins a family of glycoproteins occurring on cell surfaces, in most basement membranes and in blood and other body fluids. They are binding molecules that link to specific receptors on and within cells. They are involved in platelet aggregation and in the determination of cell shape and the formation of tissues. Fibronectin inhibitor drugs have been developed.

fibroplasia the formation and spread of fibrous tissue, as occurs in wound healing.

fibrosis scarring and thickening of any tissue or organ by the replacement of the original structure by simple collagenous FIBROUS TISSUE. This usually follows injury or inflammation. Fibrosis is the body's main healing process and the scar tissue formed is usually strong.

fibrous protein a class of insoluble proteins in the form of collagen fibrils that constitute the main structural elements of the body, especially as bone matrix, tendons, ligaments and other connective tissue. See also FIBROUS TISSUE.

fibrous tissue a simple, strong structural or repair tissue consisting of twisted strands of COLLAGEN and laid down by cells known as fibroblasts. These are among the commonest cells in the body and occur everywhere. Tissue damaged beyond recovery by disease processes is replaced by fibrous tissue (scar tissue).

fibrovascular pertaining to fibrous tissue with a good blood supply.

fibula the slender bone on the outer side of the main bone of the lower leg (tibia). The fibula is fixed to the tibia by ligaments and helps to form the ankle joint, below, but plays little part in weight-bearing.

fight or flight response the general activation of the sympathetic nervous system in response to stress.

filiform thread-like, as in the case of the filiform papillae of the tongue or filiform warts.

filtrate the portion of material placed in a filter that passes through.

filtration the movement of fluid virtually free of large molecules such as proteins, across capillary membranes under the influence of hydrostatic pressure.

fimbriated fringed with finger-like processes, as at the open end of the FALLOPIAN TUBE.

fingerprint 1 the unique pattern printed by the ridges of epidermis on the pulpy surfaces of the ends of the fingers and thumbs.
2 of a protein, the pattern of fragments exposed by electrophoresis after splitting with a proteolytic enzyme such as trypsin.
3 of DNA, a pattern of varying-length (polymorphic) restriction fragments that differs from one individual to another and that can be used as a means of unique identification.
4 of a protein, the pattern of fragments produced on a plane surface when a protein is digested by a protein-splitting enzyme. See also DNA FINGERPRINTING.

first aid measures taken by those at the scene of an accident, or those present when a medical emergency occurs, to minimize the risk to the victim before the arrival of a medically qualified person. The essentials are to ensure free breathing (secure the airway), to prevent unnecessary loss of blood, to avoid unnecessary displacement of blood from the heart and brain (treat shock), to splint fractures and to reduce the risk of infection.

first messenger a HORMONE or NEUROTRANSMITTER operating outside a cell and interacting with a cell membrane receptor. The second messenger operates within the cell.

Fischer, Edmond the American biochemist Edmond M. Fischer was born in Shanghai in 1920 and was educated in Switzerland. He worked as a laboratory assistant, did research in Switzerland and in 1950 moved to the USA where he became a research Fellow at the Rockefeller Foundation. In 1953 he became an assistant professor of biochemistry at the University of Washington and later was appointed full professor, a position he held until 1990.

Fischer's most important contribution to human biology arose from work he did with Edwin Krebs in 1955 and 1956. Together, these men discovered the biochemical mechanism by which the enzyme that frees glucose molecules from the polymer glycogen is activated. This occurs when the enzyme, glucogen phosphorylase, is donated a phosphate group from ATP in a reaction catalyzed by another enzyme which the two workers named protein kinase. Activation is stopped when the phosphate group is removed from glucogen phosphorylase by another enzyme they called protein phosphatase. The two researchers showed that the addition and removal of the phosphate group altered the shape of the glucogen phosphorylase rendering it active or inactive.

This advance was seminal, in that it demonstrated the vital role of enzymes in many physiological processes and directed attention to the way many enzymes functioned. The advance led to the award of the 1992 Nobel Prize to Krebs and Fischer.

fission splitting into parts. 1 the asexual reproductive process by which a single-celled organism or a single cell in a multicellular organism splits into two daughter cells.

2 an atomic event in which the nucleus of an atom splits into fragments, with the loss of a small quantity of matter and the evolution of radiational energy of at least 100 million electron volts (mV).

fissure a deep groove or furrow that divides an organ, such as the brain, into lobes.

fistula an abnormal communication between any part of the interior of the body and the surface of the skin, or between two internal organs. Fistulas may be present at birth (congenital) or may arise as a result of disease processes such as abscesses or cancer.

fit a sudden acute attack of any disorder, especially an epileptic seizure.

fitness the ability to undertake sustained physical exertion without undue breathlessness. Fitness is associated with a sense of physical and mental well being. The achievement of fitness is possible only by making regular demands on the body to perform physical tasks. As fitness improves the bulk and strength of the voluntary muscles and the force and pumping efficiency of the heart muscle increase. The respiratory muscles perform more effectively. The subject is able to perform more work within the limits of the rate at which oxygen is supplied by the lungs and circulation (aerobic exercise). Recovery from fatigue is more rapid, a higher degree of muscle tension can be attained, the muscles are able to utilize glucose and fatty acids in the presence of less insulin, and the liver is better able to maintain the supply of glucose to the blood, and hence to the muscles, during strenuous exercise. The energy-producing elements in the muscle cells (the mitochondria) increase in size and number.

fixation **1** the accurate alignment of one or both eyes on a small object.
2 a psychoanalytic term meaning an excessively close attachment to an object or person, of a kind appropriate to an earlier, immature, stage of development.

flaccid flabby, limp or soft. Lacking firmness vigour or energy.

flagellation **1** the arrangement of whip-like structures (flagella) on an organism.
2 whipping, either for purposes of masochistic sexual arousal, or as a ritual means of atoning for sin.

flagellum a whip-like structure protruding from the surface of a cell that provides locomotion. The plural is flagella.

flatline an informal term for the state of a person whose medical monitoring equipment shows a flat line rather than the normal peaks and troughs. Such a person has recently died or is very near to death.

flatulence a sense of fullness and discomfort in the upper abdomen associated with a desire to belch, and usually caused by air swallowing. See also AEROPHAGY.

flatus gas discharged by way of the anus. The gas is a mixture of odourless nitrogen, carbon dioxide, hydrogen and methane, and a varying quantity of hydrogen sulphide, which is said to smell like rotten eggs. Hydrogen and methane are both inflammable, but the risk to non-smokers is small. The average person farts about 20 times a day.

flavin one of a range of water soluble yellow pigments that includes the vitamin RIBOFLAVIN. Flavins occur in the tissues as coenzymes of FLAVOPROTEINS.

flavine adenine dinucleotide a coenzyme that transfers hydrogen from one substrate to another. It is derived from the B vitamin riboflavine.

flavoproteins one of the subclasses in the class of proteins known as the chromoproteins because they are combined with a coloured group. In this case the coloured group is RIBOFLAVINE (riboflavin). The flavoproteins act as enzymes concerned with tissue respiration.

Fleming, Alexander in 1928 Alexander Fleming (1881–55), a Scottish bacteriologist working in London, noticed that one of a number of bacterial culture plates that had accumulated in the laboratory sink had grown a colony of mould. This was a common event on old, exposed plates but Fleming also observed that around the mould there was a zone almost completely free from bacterial growth. Intrigued, Fleming carried out many careful experiments using the mould on the plate and reported his findings in a paper published in 1929 in the *British Journal of Experimental Pathology*. This very full paper describes how he subcultured the mould and found that the broth in which he grew it at room temperature had acquired the property of preventing the reproduction of many common bacteria. For convenience he gave the name 'penicillin' to the filtered broth in which the mould – a *penicillium* species – had been grown. He proved it non-toxic by injecting it into a mouse and confirmed that it was not irritant when applied to the human conjunctiva every hour for a day. He showed that it did not interfere with the action of white blood cells.

Fleming thought that the new substance might be useful as a surface dressing for septic wounds but did not anticipate its use as an antibiotic for internal use. In 1940, referring to its possible use as a local antiseptic, he said that the trouble of making it did not seem worth while. The subsequent development of the drug had to wait until 1938 when Ernest Chain (1906–79) and Howard Florey (1898–1968), after reading Fleming's paper, succeeded in purifying enough penicillin to conduct human trials and demonstrate the amazing properties of this new drug. Penicillin was found to be present in several moulds of the *Penicillium* and *Aspergillus* genera. It was soon shown to be almost completely non-toxic to the body, and was the first fully successful and safe antibiotic for acute internal bacterial infections in humans. In 1945 Fleming, Chain and Florey shared the Nobel Prize for Physiology or Medicine.

Flemming, Walther in 1882, Walther Flemming (1843–1905), professor of anatomy at Kiel University in Germany published a book called *Cell Substance, Nucleus, and Cell Division*. Flemming had been using stains to show up the details of normally transparent cells and had demonstrated thread-like bodies, in cell nuclei, that stained deeply. These he called CHROMOSOMES, (literally, 'coloured bodies') because of the ease with which they were stained. No sooner had Mendel's work been publicized by de Vries than biologists began to realize that the behaviour of the chromosomes, as described by Flemming, could account for the transmission of hereditary characteristics. Flemming also described the process of MITOSIS in detail.

By 1902 it had been recognized that chromosomes were present in pairs in normal body cells, but that in the single cells constituting sperm and eggs only one member of each pair was present. During fertilization, the fusion of sperm and egg restored the number of chromosomes to normal, half of them coming from the father and half from the mother. This observation was immediately seen to support Mendel's ideas of the movement of 'hereditary units'.

Since there were thousands of inheritable characteristics and only 23 pairs of chromosomes in man, this would mean that each chromosome must carry the genetic material for a large number of features. So it was postulated that there were many separate units on each chromosome, each one controlling an individual characteristic. These units were later called GENES.

flexion 1 the act of bending of a joint or other part or the state of being bent.
2 pertaining to a bent part as in flexion deformity.

flexion reflex a sudden automatic withdrawal movement occurring in response to a painful stimulus and effected by the contraction of the flexor muscles of all the joints on the same side.

flexor a muscle that bends (flexes) a joint. A muscle that straightens (extends) a joint is called an extensor.

flexure a bend, curve, angle or fold. The hepatic flexure in the large intestine (colon) is the angle near the liver between the vertical ascending colon and the roughly horizontal transverse colon.

floaters semitransparent, shadowy bodies seen in the field of vision, usually remote from the point of observation, and moving rapidly with eye movement. For centuries, floaters have been called 'muscae volitantes' because of their resemblance to flitting flies.

Most floaters are shadows of developmental remnants in the jelly-like VITREOUS HUMOUR of the eye and are harmless. Sudden onset of very conspicuous dark floaters, especially if accompanied by flashes of light (phosphenes) suggest an incipient retinal detachment.

floating rib one of the two lowest pairs of short ribs that have no attachment at the front to the breastbone or the COSTAL CARTILAGES but end among the abdominal muscles.

flocculent having a fluffy, woolly or cloudy appearance, especially of a precipitate or suspension in a fluid.

flocculus a small projecting lobe of the CEREBELLUM lying on each side of its front surface on either side of the brainstem.

flora 1 the entire plant life of a region. 2 in medicine, the term is used to refer to the entire bacterial life of a region of the body, as in 'intestinal flora', 'oral flora', 'skin flora' or 'normal flora' (commensals). Although often free-moving, micro-organisms were not classified under fauna. This convenient usage originated at a time when all living things were either flora or fauna. It no longer complies with current biological classification; the bacteria and the cyanobacteria now have a kingdom of their own (Monera).

florid flushed, of ruddy complexion, rosy.

flow autoregulation the intrinsic capacity of arterioles to respond to changes in arterial pressure by a change in bore so as to maintain constancy of blood flow.

fluid-mosaic model the concept of the cell membrane as a two-layer lipid structure with embedded penetrating protein molecules capable of moving about. This model has for some time been unquestioned.

focal localized and circumscribed.

foetus the widely-used but incorrect spelling of FETUS.

folates see FOLIC ACID.

folic acid a vitamin of the B group originally derived from spinach leaves, hence the name (Latin *folium*, a leaf). The vitamin is necessary for the synthesis of DNA and red blood cells. Deficiency causes megaloblastic anaemia. Folic acid is plentiful in leafy vegetables and in liver but is also produced by bacteria in the bowel and then absorbed into the circulation. Deficiency may occur after antibiotic treatment. Folic acid taken immediately before pregnancy and during the first few weeks can virtually eliminate the risk of embryonic neural tube defects and resulting spina bifida or anencephaly in the baby. It will also reduce the risk of cleft palate.

follicle a sac-like depression or cavity, glandular or cystic in nature. Follicles may secrete new tissues as in the case of hair follicles, which synthesize and extrude hairs, and the Graafian follicles of the ovaries, which contain the eggs (ova) prior to ovulation.

follicle-stimulating hormone (FSH) a pituitary gland hormone that stimulates the Graafian follicles in the ovaries to produce eggs (ova), and the lining cells of the tubules in the testicles to produce sperm (spermatozoa). FSH may be used to treat infertility due to failure of ovulation (anovulatory infertility) or to low sperm counts. The secretion rate of follicle-stimulating hormone is controlled by feed-back of a polypeptide substance called inhibin, which is produced by the ovaries and the testicles.

follicular pertaining to, or resembling, a follicle.

fomites anything that has been in contact with a person suffering from an infectious disease, and which may transmit the infection to others. Fomites include sheets, towels, dressings, clothes, face flannels, crockery and cutlery, books and papers.

fontanelles the gaps between the bones of the vault of the growing skull of the baby and young infant that can be felt, as soft depressions on the top of the head, by gentle pressure with the fingers. The front (anterior) fontanelle lies at the junction of the two forehead (frontal) bones and the two side (parietal) bones. The rear (posterior) fontanelle lies between the two parietal bones and the single rear occipital bone. The fontanelles are covered by scalp and skin and allow moulding of the skull during birth. Both have usually closed by about 14 months. There are two other very small fontanelles on either side of the head.

food additives substances, numbered in thousands, added to food for purposes of preservation, appearance, flavour, texture or nutritional value. Without additives, much food would soon be spoiled and wasted. Common additives include vitamins, minerals and trace elements in bread, cereals, milk, margarine, table salt, fruit drinks and baby foods. Flavouring and colourings include sugar, salt, mustard, pepper, monosodium glutamate and tartrazine. Preservatives include salt, sugar, sodium nitrite, sodium benzoate, and the anti-oxidants BHT (butylated hydroxytoluene) and BHA (butylated hydroxyanisole).

footprinting a technique for finding the protein-binding regions on DNA or RNA. These binding sites are important as protein binding can control the expression of nearby genes.

foramen a natural hole in a bone for the passage of a nerve, artery or a vein or other anatomical structure.

foramen magnum the large, almost circular hole in the centre of the base of the skull through which the spinal cord, the continuation of the medulla oblongata, passes.

foramen ovale a valve-like opening in the inner wall (septum) between the right and left upper chambers (atria) of the heart of the fetus. Before birth, about three quarters of the blood returning from the body to the right side of the heart is shunted through the foramen ovale to the left side. After birth, the pressure on the left side rises and the foramen valve closes. Soon it fuses shut.

forced expiratory volume (FEV) a measurement made with an instrument that measures expired air flow (a recording SPIROMETER). Forced expiratory volume is the maximum volume of air that can be breathed out in one second. The FEV is then compared with the maximum amount that can be breathed out in a single breath, however prolonged (the VITAL CAPACITY). The ratio FEV/VC should exceed 70 per cent in healthy individuals but is much reduced in people with any condition that narrows the air tubes (obstructive airway disease), such as ASTHMA or chronic BRONCHITIS.

forced vital capacity the amount of air that can be expelled from the lungs by breathing out for as long as possible after a full inspiration. See also FORCED EXPIRATORY VOLUME.

foreskin the prepuce or hood of thin skin that covers the bulb (glans) of the penis. Surgical removal of the foreskin is called CIRCUMCISION and this is widely practised, usually for ritual or cultural reasons. The medical indications for circumcision are few.

forgetfulness the usual natural consequence of inattention or of overburdening a well-stocked mind with trivia. Failure to retrieve a memory becomes more likely the larger the number of similar memories that depend on the same cues. Abnormal (pathological) forgetfulness is the main feature of dementia and of amnesia.

fornix an arch, especially one of the pair of arch-like bands of white fibres lying under the corpus callosum of the brain, or the space between the upper walls of the vagina and the neck of the womb.

fossa a furrow or depression, especially in bone.

fourth ventricle the centrally placed, rearmost of the four fluid-filled spaces in the brain. This tent-shaped cavity for cerebrospinal fluid lies immediately behind the PONS of the brainstem and immediately in front of the CEREBELLUM. It communicates with the aqueduct of the midbrain, in front, and with the central canal of the spinal cord, below.

fovea any shallow cup-like depression. The fovea centralis of the RETINA is the central area of the macula lutea, of highest resolution and free of visible blood vessels. The highest concentration of cones is in the fovea.

foveal vision perception of objects whose images fall on the FOVEA centralis – the most discriminating and colour-sensitive part of the RETINA. Also known as photopic vision.

fracture a break, usually of a bone. This occurs when excessive force is applied to a healthy bone or when lesser force is applied to a bone generally weakened by a disease such as osteoporosis, or locally weakened by a tumour or cyst. Such a fracture is called a pathological fracture.

frame-shift mutation a genetic mutation caused by the addition or deletion of a number of NUCLEOTIDES other than three. Because the genetic code consists of codons (groups of three nucleotides), such a change shifts the reading frame for translation so all adjacent codons are changed and a completely new set is read into the messenger ribonucleic acid (mRNA).

Franklin, Rosalind the English biophysicist Rosalind Franklin (1920–58) was educated at Cambridge. From 1947–50 she worked in a State chemical laboratory in Paris. In 1951 she was appointed to a London University research post at King's College. There, her director John Randall asked her to work on the structure of DNA. There can be no question that the X-ray diffraction work on crystalline DNA she performed at King's College in the early 1950s was essential in providing Francis Crick and James D Watson with evidence they needed to determine the chemical structure of DNA.

Franklin was not pleased to learn that the same goal was being pursued in the adjoining laboratory of Maurice Wilkins. Nor was she impressed by the early attempts made in Cambridge by Crick and Watson to solve the problem. But James Watson saw one of her X-ray photographs of hydrated DNA at a seminar she gave in 1952 and at once recognized that the molecule was a helix. Franklin also preceded Crick and Watson in recognizing that the phosphate groups in the nucleotide chains were outside the helix, not within it. In March 1953 Franklin was working on a draft paper describing a double helix model for the DNA molecule when she heard the news from Cambridge. The same year Franklin moved to Birkbeck College where she worked until her early death from cancer.

There were features of the model proposed by Watson and published in the paper in *Nature* on 25 April, 1953 that Franklin had not appreciated. She had not realized that the chains ran in opposite directions and she had not appreciated the vital fact of base pairing. It was the latter recognition by Watson and Crick that made their model so extraordinarily important and that led to the unparalleled expansion of knowledge that sprang from their discovery. Had Franklin lived to 1962, however, it seems more than likely that she would have shared the Nobel Prize with Watson, Crick and Wilkins.

fraternal twins non-identical twins produced by the simultaneous fertilization of two different eggs by different sperm (dizygotic twins). In spite of the etymology, fraternal twins need not be male or even of the same sex.

freckle a small brownish or yellowish skin blemish due to local aggregation of cells containing melanin, the normal skin pigment. Melanin-containing cells enlarge under the influence of sunlight, especially in the fair-skinned and this is usually permanent.

freedom of will a faculty we all believe we possess but which, on philosophic analysis, appears to be an illusion.

free radicals highly chemically active atoms or group of atoms capable of free existence, under special conditions, for very short periods, each having at least one unpaired electron in the outer shell. Oxygen free radicals can be very damaging to DNA and proteins and to the fat in cell membranes where a free radical chain reaction can be set up.

frenulum any small fold of mucous membrane.

frenum a membranous fold or sheet, such as the fold of mucous membrane under the tongue, that restrains movement or provides support.

frequency coding a means by which the central nervous system, limited by the all-or-none properties of nerve impulse conduction, is able to convey information about varying intensity of signals. It does this by employing frequency modulation (FM). The frequency of impulses varies with the strength of the stimulus.

frequency distribution a table or histogram showing the number of times each value of a particular variable occurs in a sample.

Freudian theory a set of propositions about human personality and behaviour derived from observations of patients engaged in free association at a time in social history when the expression of sexuality was normally repressed. Such expression, often in symbolic form, convinced Freud that sex was at the basis of most psychopathology. He asserted that the uncovering of repressed unpleasant early experiences would disperse the psychopathology which he claimed they had caused. He proposed arbitrary divisions of the mind into superego, ego and id. He asserted that infants pass though three stages – oral (birth to 18 months), anal (2–5 years) and phallic (5 years onward), and that the personality could be fixed at any of these stages with serious consequences, curable only by psychoanalysis. He proposed the Oedipus complex and the castration complex. Freud's ideas and discoveries continue to have wide influence, but are not now generally believed to have any scientific basis. There is little convincing evidence that the application of his theories in psychoanalysis has any specific value in the treatment of psychological disturbance.

Freud, Sigmund Sigmund Freud was born in Freiberg, Moravia (now Czechoslovakia) in 1856 but spent most of his life in Vienna. He was a child prodigy who read Shakespeare and Goethe at the age of eight and was versed in the Greek, Latin, French and German classics. In spite of a taste for philosophy, he studied medicine and qualified in 1881. After some excellent early work on aphasia, infant cerebral palsy, neurological connections in the nervous system and the discovery of the medical uses of cocaine, he worked for some time in Paris with the French neurologist Jean-Martin Charcot (1825–93). Fascinated by Charcot's work on hysteria, Freud turned his attention to the psychological basis of mental disorder, and at first used hypnotism to elicit painful memories that seemed to be at the root of many psychological problems.

This work led him to the idea of repressed memories and impulses and to a profound consideration of the nature of the unconscious mind. Later he abandoned hypnotism and adopted the method of free association. Freud's interest in the significance of dreams as a source of insight into the unconscious mind led to the publication in 1900 of *The Interpretation of Dreams*. The book was almost ignored. In 1903 he published *Three Essays on the Theory of Sexuality*. At first this caused a storm of abuse, but gradually his ideas gained acceptance and a group of young men collected around him, meeting weekly to study and apply his ideas. This group developed into the International Psycho-Analytical Association. Among these men were Adler and Jung, both of whom eventually found his ideas on infant sexuality unacceptable and broke away to form their own schools of psychology. Freud continued to work and write for almost another forty years, constantly revising and amending his theories. During this period he was elected a Fellow of the Royal Society and was awarded the Goethe Prize for Literature.

On the German invasion of Austria in 1838 the risk to Freud of Nazi anti-semitic persecution became critical and, with great difficulty, he was smuggled out of the country and brought to London where he died in 1939 from cancer of the upper jaw from which he had suffered stoically for over 16 years.

frigidity an informal term meaning loss, in a woman, of sexual desire or of the ability to be sexually aroused, or to achieve an orgasm. The term, the usage of which is almost confined to disappointed men, is now considered pejorative and sexist. The condition may be a reflection of lack of affection, or the expression of it, by the partner, or may be due to recent childbirth, pain on intercourse, fatigue, depression, fear of pregnancy or psychological trauma following rape. Some drugs, especially those given for high blood pressure, for depression and for insomnia, reduce libido.

frontal pertaining to the forehead or to the FRONTAL BONE.

frontal bone the skull bone that forms the forehead and the roofs of the eye sockets (orbits). The frontal bone contains the frontal sinuses.

frontal lobe the large, foremost part of the brain. This part constitutes about one-third of the brain and is much more fully developed than in any other primate. Damage to the frontal lobe causes general disturbance of thinking, impairment of initiative and spontaneity, loss of strength of personality, and, if the rear part of the lobe is affected, paralysis.

frontal nerve a sensory nerve that emerges from the eye socket, as a branch of the OPHTHALMIC NERVE, and curls up over the upper edge of the bone to supply the skin of the upper eyelid, the forehead, and the front of the scalp.

frontal sinus one of a pair of mucous membrane-lined air spaces of variable size lying behind and above the level of the eyebrows.

fructose one of the simplest forms of sugar (a monosaccharide) and derived from fruit, sugar cane, honey and sugar beet. Fructose, linked to another monosaccharide, glucose, form the disaccharide sucrose which is the common domestic sugar. Fructose is sweeter than glucose, but has the same energy value. It is readily absorbed. Fructose provides direct energy for spermatozoa and is found in seminal fluid.

frustration the emotion resulting when aims or intentions are blocked. Frustration is inevitable but some people aim for goals inherently beyond their capacity and suffer a much higher level of frustration than others. Others aim for mutually incompatible, or equally attractive but mutually exclusive, goals. Frustration breeds anger which may lead to aggression and this is often displaced and directed against an inappropriate target. Displaced anger is a common cause of marital discord.

FSH *abbrev. for* FOLLICLE-STIMULATING HORMONE.

functional site the binding site on an ALLOSTEARIC PROTEIN which, when activated, allows the protein to carry out its biochemical function. Also known as active site.

fundamentalism a religious movement that insists on the literal truth of everything in the bible, and denies the claims of science when these contradict biblical assertions. In particular, Christian fundamentalists believe in the miracles, the virgin birth, the resurrection and the divinity of Christ.

fundus the inner surface of an organ furthest from the opening. The fundus of the eye is the inner area covered with the RETINA. The fundus of the uterus is the inner surface of the dome of the womb.

fungiform papilla one of the broad, flat, slightly raised, nipple-shaped protuberances scattered over the top and sides of the tongue.

funiculus any cord-like structure.

funnel chest a condition in which the breastbone is hollowed backwards, especially at its lower end. This greatly reduces the front-to-back dimension of the chest and the heart is displaced to the left and may be compressed. There is restricted chest expansion and reduced VITAL CAPACITY.

funny bone a popular term for the part of the back of the elbow containing a groove in which the ulnar nerve runs near the surface and is easily accessible to trauma. The severe pain experienced if the nerve is struck will evoke a conspicuous physical response that may seem funny to an unsympathetic observer. The victim is unlikely to be amused.

furuncle a boil.

furunculosis widespread or repeated boils.

fusiform spindle-shaped. Tapering to a point at each end.

Gg

G1 the period in the cell cycle between the last MITOSIS and the start of DNA replication.

G2 the period between the end of DNA replication and the start of the next MITOSIS.

GABA gamma-aminobutyric acid. GABA is a NEUROTRANSMITTER substance derived from glutamic acid that performs important dampening (inhibitory) functions in the brain.

gait the particular way in which a person walks. From the Middle English gate, a way or passage.

galact-, galacto- *combining form denoting* milk or milky.

galactose a monosaccharide sugar that is a constituent of LACTOSE, the main sugar of milk. Also known as cerebrose.

Galen, Claudius the Greek physician Claudius Galen (c.130–201) was born in Pergamum in Asia Minor and studied the works of Hippocrates and Aristotle at Corinth and Alexandria. He was surgeon to the gladiators at Pergamum and was personal physician to Marcus Aurelius and other Roman emperors. Galen was a highly intelligent and successful man most of whose ideas were the result of imagination and respect for authority rather than observation and rational deduction. Almost everything he taught was wrong.

Galen's chief dogma was that the body, like all nature, was composed of four elements – earth, air, fire and water, and that everything that happened in the body, from illness to the manifestations of the personality, were governed by the four humours – blood, phlegm, black bile and yellow bile. These he elaborated into a vast and nonsensical system that was to dominate medical thinking for 1500 years. An important consequence of the entirely imaginary humoral theory was the idea that blood-letting was a valuable therapeutic measure for almost all diseases. This was one of Galen's central dogmas and was responsible for the deaths of countless millions right up until the 19th century.

Galen taught that blood originated in the liver and was consumed in the other organs; and that it passes through tiny pores in the wall between the two sides of the heart and was mixed with air on the left side. He taught that the pulsation of the arteries served the same purpose as breathing, and seemed to be confused as to whether or not arteries contained air – as the name implies. He denied that the heart was a muscle and insisted, contrary to the visible evidence, that it lay in the exact centre of the body. He believed that the brain generated a 'vital spirit' that passed through hollow nerves to the muscles, which it then activated. This idea was still universally credited as late as the end of the 18th century. He taught that phlegm originated in the brain rather than in the lining of the nose. He stated that cataracts were caused by a 'humour' from the brain that solidified behind the lens and insisted, against all evidence, that the lens occupied the centre of the eyeball.

Galen repeatedly claimed respect, even reverence, for observed and demonstrable fact, but, in fact, his ideas were always dominated by his preconceived beliefs. His writings were so powerful, so explanatory, so plausible and, above all, so authoritative, that they remained the standard texts until the 16th century. His views were accepted by the Church and scientists who pointed out his errors were liable to be burned at the stake.

gall the old term for BILE, but still preserved in the word GALL BLADDER and GALLSTONES.

gall bladder the small, fig-shaped bag, lying on the under side of the liver, into which bile secreted by the liver passes to be stored and concentrated. When fatty food enters the beginning of the small intestine (the DUODENUM), the gall bladder empties into it, by way of the common bile duct.

Galton, Francis Francis Galton (1822–1911) was a child prodigy who, at the age of four could write to a friend: '... I can read any English book. I can say all the Latin substantives... I can cast up any sum in addition... I read French a little and I know the clock.' He studied medicine and graduated from Cambridge in 1844 then spent some years exploring Africa and writing books on his experiences. He then turned his attention to meteorology and in 1863 wrote a book on the subject that provided the basis for the modern methods of weather mapping. He coined the term 'anticyclone'. After reading the *Origin of Species* he became fascinated with anthropology and heredity and devoted the rest of his life to these subjects.

He was the first to use statistical methods in biology and demonstrated that the distribution of intelligence in man formed the now well-known bell-shaped curve – that was used so much by Gauss as to become known as the Gaussian distribution curve. Galton worked out the system of fingerprint classification and identification now used by forensic scientists. He carried out the first serious studies of the effects of environmental influences on identical twins. He held posts in the Royal Society, the Anthropological Institute, the Royal Institution and the British Association for the Advancement of Science. He was knighted in 1909.

Galton's enthusiasms tended to carry him away in crankish directions and he pursued his ideas on eugenics – which were often misunderstood – with religious fervour. He bequeathed the sum of £45,000 to be devoted to the study of the subject at London University.

Galvani, Luigi Luigi Galvani (1737–98) was a lecturer in medicine at the University of Bologna who, in 1775, became professor of anatomy. In 1771 he discovered, by accident, that the muscles in frog's legs twitched strongly whenever they were brought into contact with the free ends of two joined dissimilar metals. Galvani had already demonstrated that the muscles would twitch when connected to an electrical generating machine or to a charged Leyden jar, and was convinced that electricity was involved in muscle contraction. But this new observation convinced him that the electricity came from the muscle rather than from the metals. On the basis of this he formulated a theory of animal electricity to which he clung tenaciously for years.

Ironically, Galvani was wrong about the source of electricity in this case – Volta demonstrated in 1794 that connected dissimilar metals produced an electric current – and died a disappointed man. Later research, however, showed that there was much more in his idea of animal electricity than anyone at the time could possibly have foreseen.

gam-, gamo- *combining form denoting* joined, especially sexually united.

gamete a cell, such as a sperm or ovum, possessing half the normal number of chromosomes (haploid) and capable of fusing with another gamete in the process of fertilization, so that the full (diploid) number of chromosomes is made up. From the Greek *gamos*, marriage.

gamete intrafallopian transfer (GIFT) direct placement of up to three harvested and incubated human eggs, together with a quantity of spermatozoa, into a fallopian tube via a fine catheter. GIFT, which implies *in vivo* fertilization, is a method of achieving pregnancy that in some cases offers a better alternative to *in vitro* fertilization (IVF).

gametocyte a cell from which a GAMETE is developed by division. In the ovary, a gametocyte is an oocyte; in the testicle it is a spermatocyte.

gametogenesis the production of gametes. See GAMETE.

gamma the third letter of the Greek alphabet. Often used in biology to denote a particular class.

gamma-aminobutyric acid see GABA.

gamma globulin a group of soluble proteins, present in the blood, most of which are IMMUNOGLOBULINS (antibodies), and which show the greatest mobility towards the cathode during ELECTROPHORESIS. Gamma globulin provides the body's main antibody defence against infection. For this reason it is produced commercially from human plasma and used for passive protection against many infections.

gamma-glutamyl transferase an enzyme widely distributed in body tissues and released into the blood when tissue, especially liver tissue, is damaged. Increased levels occur in liver damage from any cause. Measurements can be used as an index of alcohol abuse. Increased blood levels may sometimes occur without liver cell damage.

ganglion any large, discrete collection of nerve cell bodies, from which bundles of nerve fibres emerge.

ganglioside a compound of fat and carbohydrate (glycolipid) that is an important component in cell membranes.

gangrene death of tissue, usually as a result of loss of an adequate blood supply. Gangrene is most commonly caused by disease of arteries that cause narrowing and obstruction, but may result from any other cause of arterial obstruction.

gap junction a protein-lined conduit between adjacent cells through which ions and small molecules can pass.

Gasserian ganglion a large collection of nerve cells forming the sensory root of the 5th cranial (trigeminal) nerve. Also known as the semilunar ganglion. (Johan Ludwig Gasser, 1723–65, Austrian anatomist).

gastr-, gastro- *combining form denoting* the stomach, belly or abdomen.

gastric pertaining to the stomach. The word derives from the Greek *gaster*, the belly.

gastric juice the watery mixture of hydrochloric acid, pepsin and mucin secreted by the glands in the lining of the stomach. Gastric juice has a powerful digestive action on protein and is also protective against many infective organisms.

gastrin a peptide hormone secreted by the stomach on the stimulus of the sight, smell or contemplation of food. The hormone is released into the blood. The entry of protein into the stomach stimulates even more gastrin production and this hormone returns to the stomach to stimulate the production of acid and pepsin (GASTRIC JUICE) from the cells of the stomach lining (gastric mucosa).

gastrocnemius the main muscle forming the bulge of the calf. The gastrocnemius arises by two heads from the back of the lower end of the thigh bone (femur) and is inserted, with the SOLEUS muscle, by way of the ACHILLES TENDON into the back of the heel bone (CALCANEUS). Its action is to extend the ankle joint in walking and standing on tiptoe.

gastrocolic pertaining to the stomach and the colon. A gastrocolic fistula is an abnormal connection between the inside of the stomach and the inside of the colon.

gastroduodenal pertaining to the stomach and the DUODENUM.

gastroepiploic artery one of a pair of arteries that runs round the greater curvature of the stomach.

gastroileal reflex the increased peristaltic activity of the ileum prompted by emptying of the stomach.

gastrointestinal pertaining to the stomach and intestines.

gastro-oesophageal pertaining to the stomach and the OESOPHAGUS.

gastrosplenic ligament a fold of PERITONEUM running from the stomach to the spleen. Also known as the gastrosplenic OMENTUM.

gastrula the stage in the development of an EMBRYO following the BLASTULA stage when the ECTODERM, ENDODERM and primitive GUT have developed.

Gaussian distribution normal distribution. The distribution of characteristics found in large populations subject to many causes of variability. The graph of the Gaussian distribution of any characteristic (such as body height) is a symmetrical bell shape, centred on the mean. (Johann Karl Friedrich Gauss, 1777–1855, German mathematician).

G banding a method of identifying the different members of a haploid set of metaphase chromosomes by means of staining that reveals their individual pattern of stripes.

gel a largely liquid colloid, retained in a semisolid state by molecular chains, usually cross-linked.

-gen *combining form denoting* something that produces or creates, as in PATHOGEN.

gender a classification of organisms based on their sex. From the Latin *genus*, a kind.

gender identity the inherent sense that one belongs to a particular sex. In almost all cases that sex corresponds to the anatomical sex, but for a small minority, the gender identity is for the opposite anatomical sex.

gender role all behaviour that conveys to others, consciously or otherwise, a person's GENDER IDENTITY as male or female.

gene the physical unit of heredity, represented as a continuous sequence of bases, arranged in a code, in groups of three (codons), along the length of a DNA molecule (nucleic acid). The gene is the transcription code for a sequence of AMINO ACIDS linked to form a single POLYPEPTIDE chain and includes lengths on either side of the coding region known as the leader and the trailer, and non-coding sequences (INTRONS) that intervene between the coding segments. The latter are called EXONS. Exons tend to be conserved throughout a long evolutionary period; introns may vary considerably in length. The length of a gene is largely determined by the introns. The function of genes can be altered by changes (MUTATIONS) in the base sequences and their operation is regulated by adjacent, or even remote, parts of the DNA molecule. All genes are present in all nucleated cells, but only genes relevant to the particular cell are 'switched on' as required.

gene action the operation of genes in determining the whole body constitution (phenotype) of an individual.

gene cloning the identification and isolation of a gene and the production of a number of identical copies of it.

gene control the whole range of factors and processes that regulate GENE EXPRESSION.

gene expression the process in which information coded in the length of DNA designated as a gene becomes manifest in a new physical form, especially as a protein, usually an enzyme.

gene family a group of genes whose exons are related, having been derived from an ancestral gene.

gene frequency the number of occurrences of a particular gene in a population.

gene library a collection of cloned deoxyribonucleic acid (DNA) fragments together with information about the gene function. See also GENETIC MAPPING.

gene silencing the inhibition of transcription of a gene. DNA METHYLATION is an important cause of gene silencing and when this affects tumour-suppressor genes, it may be a cause of cancer.

genera the plural of GENUS.

general anaesthesia a state of unconsciousness and immobility, brought about by drugs, so as to allow surgical operations or other physical procedures to be performed without pain or awareness.

general anaesthetics drugs used to induce and maintain GENERAL ANAESTHESIA.

generation 1 a single stage of reproductive descent in the history of an organism. 2 the average or normal time between the birth of parents and the birth of their offspring. In humans, this may be taken to be 25 years.

generative pertaining to reproduction.

gene redundancy the presence of many copies of the same gene within a cell.

gene regulation the operation of the mechanisms by which genes are switched on or off, so that cells are specialized so as to perform different kinds of functions, and are made to respond to environmental and signalling changes. See OPERON.

genesis origins, beginnings or the process of being formed.

gene suppression the situation in which a normal PHENOTYPE develops in an individual or cell with a mutant gene due to a second mutation either in the same gene or in a different gene.

genetic code

The great sequence of genes strung out along the DNA of our chromosomes may be considered as a program for the construction of the human body. But this is an inadequate description because the program is essentially self-reading. DNA is a program for the construction of proteins – a code for the identity and sequence of the constituents of proteins – the amino acids. Because this program is divided up into lengths called genes, it is known as the genetic code.

Cracking the genetic code was not only a triumph of scientific deduction but also a revolution in the understanding of genetics and the foundation of a new science – molecular biology. The code is concerned with one thing only – the order of the sequence of the amino acids necessary to produce proteins. The human genome – the entire gene collection – contains the codes for about 20,000 different proteins. A genome works like a complex, parallel-processing computer, of which different parts are operating at different times, regulating each other's activity in generating the appropriate proteins – mostly enzymes – to cause cells to function and, during body growth and repair, to take the appropriate form. This behaviour of the genome is the basis of cell differentiation and is the ultimate explanation of how our bodies are put together.

The secret of the genetic code lies in the sequence of bases in the DNA molecule. It is necessary to have a code that can specify 20 different amino acids by means of four chemical bases. The order of these acids is represented by the order in which the code for each amino acid occurs along the DNA molecule. A code consisting of all combinations of two out of the four bases would not be enough as this gives only 16 possible combinations and 20 are needed. But a permutation of three bases out of the four gives 64 combinations, so a triplet of bases is sufficient for the purpose, and it is this that is used. Sequences of four or more bases would work equally well, but as three are sufficient, nature has settled on this most economical solution. So each amino acid of the 20 has its own designated triplet of bases called a codon. This allows some redundancy and a few spare codons that are used to designate start and stop points in the comma-less sequence.

genetic counselling the process of trying to determine, for the purpose of advising prospective parents, the probability that a future child will suffer from a particular genetic disorder known to occur in the family. Genetic counsellors must have a detailed knowledge of the principles of genetics and of the nature of the conditions caused by gene defect. They must be skilled in constructing family pedigrees and in interpreting them. Counselling is best done in specialized centres where the necessary expertise can be obtained.

genetic dogma the idea, proposed by Francis Crick and James Watson, that genetic information is invariably transferred in the direction DNA to RNA to protein. The discovery of reverse transcription organisms, such as HIV, in which information passes from RNA to DNA, knocked this theory on the head.

genetic engineering the deliberate alteration, for practical purposes, of the GENOME of a cell so as to change its hereditable characteristics. This is done mainly by recombinant DNA techniques using gene copies obtained by the POLYMERASE CHAIN REACTION. Enzymes (restriction enzymes) are used to cut the nucleic acid molecule at determinable positions and short lengths of DNA from another organism are inserted. The second cell will now contain genes for the property or characteristic borrowed from the first cell.

genetic fingerprinting see DNA FINGERPRINTING.

genetic inheritance

Each human being is the unique outcome of an immensely complex inheritance acted upon by a no less complex environment. In terms of its effect, half of our inheritance can be said to come from our parents, a quarter from our grandparents, an eighth from our great-grandparents, and so on. The gene total possessed by each one of us, however, differs radically from that of our parents and even more from that of earlier forebears, for it has been progressively altered, both by mixing and by substitution, with each generation. Most of these substitutions are harmless or beneficial but many can be damaging.

Today we understand, as never before, the part played by genetics in determining our bodily characteristics and producing bodily disorders. So it is important for us to understand the way in which these characteristics, beneficial and damaging, are passed on to us from our parents. Most of what we consider normal characteristics are complex features involving the interplay of many genetic factors and many environmental influences. Because of their complexity, these are difficult to study. To see genetics operating directly and obviously it is necessary to consider those conditions caused by abnormal genes. Any change in the base sequence of a gene is called a mutation.

Single factor inheritance

Many disorders are due to a mutation in a single gene – an error in the genetic code sequence of bases in the DNA. Such disorders are inherited in a simple manner and it is possible to predict with accuracy the risk of such a disorder recurring in a family in which it is known already to have occurred. If the gene concerned is situated on any but the two sex chromosomes, the effect it produces (the trait) is called an autosomal trait. If the gene is on one of the sex chromosomes, the effect is called a sex-linked trait. In either case, the trait may be dominant or recessive.

Dominant genes

An autosomal dominant trait is one in which the condition will appear even if the gene with the mutation is present on only one of the autosomal chromosome pair. The condition occurs even if the corresponding gene on the other chromosome is normal. This combination of a defective with a normal gene is known as the heterozygous state, and this is the usual state, because most of these single gene conditions are rare and the chances of both parents carrying the gene are small. A few autosomal dominant conditions are common and some people suffering from them will have the gene at both loci. These people are said to be homozygous. The effect of the double dose of the gene is, however, the same as if only one affected gene is present.

Diseases caused by dominant genes include Marfan's syndrome, polycystic kidney disease, Huntington's chorea, familial high cholesterol levels, achondroplasia (a form of dwarfism), brittle bone disease and hereditary spherocytosis (a form of anaemia involving abnormal red blood cells).

Most people with an autosomal dominant disorder will have an affected parent. If not, the condition is probably due to a new mutation that has altered one of the gene pairs in a sperm or ovum cell of one of the parents. This is a fairly common way for autosomal dominant conditions to arise. If a person with a heterozygous autosomal dominant condition mates with a normal person, there is a fifty-fifty chance with each new child that the condition will be inherited. This is because only one gene is needed to produce the disorder and, on average, half the sperm, or ova, of the affected person, will contain the gene. If a parent is homozygous for an autosomal dominant gene, all the offspring will, of course, suffer from the condition.

The extent to which autosomal dominant genes produce their effect can vary quite considerably. The resulting variation in the severity of the condition concerned is described as its expressivity. Similarly, the effect of some dominant genes can be modified by other genes and by environmental effects. The result is also a variability in the effect of the gene. This variability is called penetrance.

Recessive genes

Autosomal recessive traits appear only if the mutant gene concerned is present at the corresponding loci on both chromosomes – the homozygous state. A single recessive mutant gene coupled with a normal gene – the heterozygous state – produces no effect, or, occasionally, a very minor effect. Most of the disorders caused by gene mutations involve defective production of an enzyme needed for the normal functioning of one of the numerous biochemical processes of the body. Almost all human biochemistry functions under the influence of enzymes – proteins that act as catalysts to speed up chemical processes without themselves being changed. The great majority of genes specify enzymes and a defective gene will result in a protein with reduced or even absent enzyme action.

In heterozygotes, one of the pair of genes specifies the normal enzyme, so body cells produce about half the normal amount. Regulatory mechanisms, however, ensure that the biochemical processes operate adequately and it is rare for any discernible effect to occur. If both genes are defective, however, none of the normal enzyme is produced and the effects are often very serious, producing conditions such as cystic fibrosis, albinism, sickle cell anaemia, beta-thalassaemia, phenylketonuria, Wilson's disease (a disorder in which copper accumulates in the body causing brain damage), various types of congenital blindness, deaf-mutism, and many other metabolic disorders due to defective or absent enzymes.

Recessive conditions are rare. Most of the offspring of people with autosomal recessive traits do not inherit the condition, because such traits can only be inherited if the affected individual mates with a person who either has the condition or who is heterozygous for it. If two people with an autosomal recessive condition mate, all the children will have the condition because all four genes concerned are affected and any combination will involve two affected genes. If a person homozygous for a recessive condition mates with a heterozygous person, there will be a fifty-fifty chance, with each child, that the condition will appear. If two heterozygous people mate there will be a fifty-fifty chance, both for the sperm and the egg, that it will carry the gene. So, on average, a quarter of the children will inherit two normal genes and will be normal, a half will inherit one defective gene and will be normal, and a quarter will inherit two defective genes and will be affected.

Because recessive conditions are rare, the parents of children with such conditions are often related and have inherited the same defective gene from a common ancestor. First cousins have a one in eight chance of being heterozygous for the same recessive gene.

So if there is a family history of a recessive condition, marrying a cousin greatly increases the risk of offspring developing the disease.

Sex-linked disorders

These are conditions caused by genes situated on one of the sex chromosomes, the X or the Y chromosome. Genes on the X chromosome produce X-linked characteristics or disorders, those on the Y chromosome produce Y-linked features. Since only males have Y chromosomes, Y-linked features occur only in males and all the sons of a man with a Y-linked disorder would inherit it. The Y chromosome is, however, very small and there are no positively proved single gene Y-linked disorders in humans. Since all known sex-linked conditions arise from genes on the X chromosome, the terms sex-linked and X-linked, in practice, mean the same thing.

X-linked conditions may be dominant or recessive. Only a few dominant X-linked conditions have been found. One is a form of rickets that resists treatment with vitamin D. An affected male will have the gene on his solitary X chromosome and will therefore pass on the condition to all his daughters but to none of his sons (to whom he passes on only Y chromosomes). An affected female will have the gene on one of her two X chromosomes. Offspring, of either sex, will therefore have a fifty-fifty chance of acquiring the trait.

X-linked recessive inheritance is especially interesting. Because women have two X chromosomes they may carry either one mutant gene (the heterozygous state) or two (homozygous). If they are homozygous, they will suffer from the condition, but this is very rare as it implies that the father had the condition and the mother had at least one mutant gene for the condition. So X-linked recessive conditions are almost always heterozygous in women, who remain unaffected but are carriers of the gene. Men, on the other hand, have only one X chromosome and show the full effects of any X-linked mutant gene – the Y chromosome, being different, has little effect. So, X-linked recessive conditions predominantly affect males and are transmitted by healthy females, who are carriers of the gene, to their sons. An affected man cannot transmit the disease to his sons, because he passes on only Y chromosomes to sons. His X chromosomes, carrying the gene, are passed on only to daughters.

If a woman carrying an X-linked recessive gene for a disease on one of her X chromosomes mates with a normal man, each child will have a fifty-fifty chance of acquiring the mutant gene. Thus, on average, half her sons will have the disease and half her daughters will be carriers of the gene. X-linked recessive diseases include haemophilia, Duchenne muscular dystrophy, colour blindness, testicular feminization syndrome, and Fabry's disease (a degenerative brain disorder).

Multifactorial inheritance

Unfortunately, we cannot apply these simple rules to the majority of human characteristics such as intelligence, height, weight, family resemblance, and so on. Such complex traits are all, to some extent, genetically determined, but they are the result of many different genes acting together and are also influenced by environmental factors. For these reasons they do not show simple Mendelian inheritance. A predisposition, or liability, to many common body disorders is also inherited in a similar way as a so-called 'polygene' trait. These conditions are not directly caused by the genes, but people who inherit them show a higher than average tendency to develop the conditions, and they often run in the family. Disorders occurring as a result of multifactorial causes include asthma, schizophrenia, cleft palate, high blood pressure, coronary artery disease, and duodenal and stomach ulcers.

Whole chromosome disorders

Some of the most serious genetic disorders occur as a result of gross abnormalities affecting the number or structure of whole chromosomes. Such disorders are much commoner than is generally supposed. About half of all fetuses that suffer spontaneous abortion (miscarriage) show abnormal chromosomes, and about one live-born baby in 200 has a chromosome abnormality. Many chromosome abnormalities occur at or around the time of fertilization. The absence of a whole autosomal chromosome is invariably fatal.

The commonest whole chromosome disorder is Down's syndrome. This is caused by an extra chromosome number 21 – a state known as trisomy-21. Trisomy-13 and trisomy-18 also occur, and these are even more severe in their effects and are usually fatal early in life. Gross abnormalities of the sex chromosomes are also common. Klinefelter's syndrome, which is caused by one or more extra X chromosomes in males, features sterility, a eunuch-like body with underdeveloped genitals and enlarged breasts. The XYY configuration has effects varying from complete normality to antisocial tendencies and mental backwardness. Turner's syndrome affects only females. Girls with Turner's syndrome have only one X chromosome. They are of short stature and do not develop normal secondary sexual characteristics. They do not menstruate and are sterile. They often have other features such as webbing of the sides of the neck and various congenital abnormalities.

genetic linkage the greater than chance association of two or more characteristics in an individual, that occurs because the genes concerned are on the same chromosome.

genetic load the totality of abnormalities caused in each generation by defective genetic material carried in the human gene pool.

genetic mapping 1 the process of determining the location of the genes for various characteristics and diseases in the chromosomes. 2 the determination of the sequence of bases in the chromosomes corresponding to the genes. The human GENOME mapping project was completed at the beginning of the 21st century.

genetic marker a gene or DNA sequence that indicates the presence of a disease or a probable risk of developing it.

genetic pedigree a kind of family tree, drawn up to show by symbols the occurrence of a particular disorder or characteristic. Study of a pedigree can often demonstrate the mode of inheritance. The symbols indicate sex, mating, offspring, death without issue, the location of the person being studied (the propositus) and the individuals affected. Circles are used for females, squares for males. Lines indicate mating and offspring, and affected individuals are shown by blacked circles or squares. The propositus is arrowed.

genetics the branch of biology concerned with the structure, location, abnormalities and effects of the GENES. Medical genetics is mainly concerned with the expression of abnormal genes or gene combinations in the production of disease. Knowledge of such matters allows useful GENETIC COUNSELLING. William Bateson (1861–1926) was the English physiologist whose studies and publications led to his being known as the 'father of genetics'. Curiously, Bateson persistently opposed the chromosome theory of heredity.

genetic screening the use of AMNIOCENTESIS or CHORIONIC VILLUS SAMPLING before birth to obtain fetal cells or a small portion of fetal tissue on which chromosomal, and to some extent gene, studies can be made. A number of serious genetic disorders can, in this way, be diagnosed at a very early stage after conception so that the option of terminating the pregnancy can be considered.

-genic *combining form denoting* producing or creating.

geniculate 1 bent at an angle, like a knee. 2 capable of bending at an angle. The term comes from the Latin *geniculatus* meaning with bended knee.

geniculate body one of the four small oval prominences on the base of the brain lying in relation to the optic pathways and the THALAMUS. The lateral geniculate bodies contain important links (synapses) in the route of visual information from the eyes to the visual cortex at the back of the brain.

genioglossus one of the small extrinsic muscle of the tongue. It arises from centre of the back of the jaw bone (mandible).

genital pertaining to the GENITALIA.

genitalia a term usually implying the external organs of generation – the labia majora and minora and the clitoris in the female and the penis, scrotum and testicles of the male. Strictly, the genitalia also include all the other parts concerned with reproduction – the vagina, uterus, fallopian tubes and ovaries in the female; and the spermatic cords, vasa deferentes, seminal vesicles and prostate gland in the male.

genius a variously-defined term for a person of exceptional achievement or accomplishment, or for the capacity that allows such achievement. Genius is commonly manifested in the arts, literature, science or invention.

genome the complete set of CHROMOSOMES, together with the MITOCHONDRIAL DNA, containing the entire genetic material of the cell.

genomic pertaining to the GENOME.

genomic imprinting the concept, derived from an increasing body of compelling evidence, that the expression of some of the genes depends on whether they have been derived from the father or from the mother. It has been shown, for instance, that chromosomal deletion in chromosomes of parental origin may differ in their effect from the same deletion in the homologous chromosome of maternal origin. Many cancers feature the loss of a particular chromosome derived from a particular parent – usually the mother. The DNA of some genes is modified during the formation of gametes so as to have altered expression and be activated or inactivated.

genomics the study of the GENOME.

genotype 1 the total genetic information contained in a cell.
2 the genetic constitution of an individual organism. Compare PHENOTYPE.

genu a knee or any knee-like or bent anatomical structure.

genus the taxonomic category containing only species and set below the category of family. The generic name, in the Linnaean classification, is always written with a capital letter, the specific name in lower case. Thus, in *Homo sapiens*, *Homo* is the genus and *sapiens* the specific name.

geometric mean the average value of a set of n integers quantities, expressed as the nth root of their product. The geometric mean is used to determine the average of a skewed frequency distribution.

ger-, gero- *combining form denoting* old age or relating to the elderly.

geriatric medicine the medicine of old age. The branch of medicine concerned with the practical application of the science of GERONTOLOGY to the improvement of the quality of life of elderly people. Modern geriatrics is a growing and positive discipline much concerned with the abolition of conventional stereotypes. The emphasis is on the encouragement of maximal activity and achievement, regardless of age, and the avoidance, and when necessary, management, of conditions common in old age.

germ a popular term for any organism capable of causing disease. Germ plasm is living primitive tissue capable of developing into an organ or individual.

germ cells sexual reproductive cells. See OVA and SPERMATOZOA.

germ layer any one of three layers of tissue, the ectoderm, the mesoderm and the endoderm, into which the embryo differentiates.

germ line 1 the lineage of cells leading to the contemporary GERM CELLS.
2 often used loosely to refer to the cells of the ovary and testes that give rise, respectively, to the ova and spermatozoa, and to the ova and spermatozoa themselves.

gerontology the study of the biology, psychology and sociology of ageing. Gerontology is concerned with the changes that occur in the cells, tissues and organs of the body with age, with the natural limits of cell reproduction, the causes of natural cell death, the effects of life style and physical activity on longevity and the psychological and sociological effects of ageing.

gestalt a physical, mental or symbolic pattern or figure so arranged that the effect of the whole differs from, or is greater than, that of the sum of its parts. A unified whole, the full nature of which cannot be grasped by analyzing its parts.

gestalt psychology a school of psychology that held that phenomena, to be understood, must be viewed as structured, organized whole entities (gestalten). Thus the gestalt of a melody remains recognizable whether it be sung, played on a flute or heavily orchestrated. Gestalt theories have had an impact on the physiology of perception, but the philosophic view that psychological phenomena are irreducible gestalts no longer commands much support.

gestation period the period from fertilization of the egg (ovum) to the birth of the child. The human gestation period is 40 weeks, plus or minus 2 weeks.

GH *abbrev. for* GROWTH HORMONE.

GHRF *abbrev. for* GROWTH HORMONE releasing factor.

GHRIH *abbrev. for* GROWTH HORMONE release inhibiting hormone (SOMATOSTATIN).

giant cell a large multinucleate cell formed from the fusion of many MACROPHAGES.

gigantism excessive body growth. This is usually the result of an abnormal production of growth hormone by the pituitary gland in childhood, before the growing ends of the bones (the epiphyses) have fused. The excess hormone production is almost always due to a benign tumour – a pituitary adenoma. The height may exceed 2.4 m. Excess growth hormone after fusion of the epiphyses causes ACROMEGALY.

gingiva the gum.

gingival pertaining to the gums.

girdle the ring of bones at either the shoulder (shoulder girdle) or the pelvis (pelvic girdle) that, respectively, support the arms and the legs.

glabella the normally hairless area between the inner ends of the eyebrows immediately above the nose. From the Latin *glabellus*, hairless.

glabrous smooth. The term is applied to a hairless surface.

gland a cell or organized collection of cells capable of abstracting substances from the blood, synthesizing new substances, and secreting or excreting them into the blood (endocrine glands), into other bodily structures or on to surfaces, including the skin (exocrine glands). The simplest glands are single mucus-secreting goblet cells. Glands also produce digestive enzymes, hormones, tears, sweat, milk and sebum. LYMPH NODES are often miscalled 'glands'.

glandular 1 pertaining to, functioning as or resembling a gland or the secretion of a gland. 2 possessing glands.

glans the acorn-shaped bulb at the end of the penis or the small piece of erectile tissue at the tip of the CLITORIS. From the Latin word *glans*, an acorn.

Glasgow coma scale a numerical method of evaluating the level of coma by assigning numbers to the response to three groups of responses to stimulation – eye opening, best obtainable verbal response, and best obtainable movement (motor) response. The scores are added and a deteriorating total suggests the need for a change in management.

glenoid 1 of any smooth, shallow depression, especially in a bone, as in the glenoid cavity of the shoulder blade (scapula) with which the head of the upper arm bone (humerus) articulates.
2 the cavity of the scapula on the outer aspect.

glial tissue binding or connective tissue, especially in the nervous system. The glial cells of the brain, long believed to have only a supporting, nutritional and antipathogenic role, are now known to communicate with neurons and with one another, to have a function in the formation of synapses, and can determine which neural connections become stronger or weaker over time. It thus appears that glial cells are as important in thinking, learning and in the emotions as are neurons. Almost all intrinsic brain tumours are neoplasms of glial cells.

glio- *combining form denoting* glue-like; relating to GLIAL TISSUE.

GLP-1 a glucagon-like neuropeptide claimed to be an obesity mediator. Starved rats given the drug behave as if satiated and this state is reversed if given a GLP-1 antagonist.

gliosis propagation (proliferation) of nerve connective tissue (neuroglia) in the brain or spinal cord. This may occur as a repair process or as a response to inflammation.

globin a term for the collection of four polypeptides in the haemoglobin molecule.

globular protein any protein readily soluble in a weak salt solution. GLOBULINS are globular proteins. Also known as eublobulins.

globulins a group of blood proteins that include the family of IMMUNOGLOBULINS (Ig) or antibodies, comprising IgG, IgM, IgE, IgA, and IgD; factor VIII, the antihaemophilia globulin; antilymphocytic globulin; thyroxine-binding globulin; fibrinogen; prothrombin; vitamin D-binding globulin; and many others. Gamma globulins are the most strongly positively charged of the blood globulins and most of them are immunoglobulins.

glomerulus a microscopic, spherical tuft of blood capillaries, especially that within the BOWMAN'S CAPSULES of the kidney, through which urine is filtered. From the Latin *glomus*, a ball of thread.

glossal pertaining to the tongue. From the Greek *glossa*, a tongue.

glossopharyngeal nerves the 9th of the 12 pairs of cranial nerves arising directly from the brain and providing taste sensation to the back of the tongue and sensation in the lining of the throat. They also supply the CAROTID BODY and the CAROTID SINUS and a pair of muscles that assist in swallowing.

glottal pertaining to the GLOTTIS.

glottal stop a speech characteristic in which there is sudden interruption of the voice sound from a momentary complete closure of the vocal cords (GLOTTIS).

glottis the narrow, slit-like opening between the vocal cords and between the false vocal cords and the space between them. The vocal apparatus of the larynx.

glucagon one of the four hormones produced by the Islet cells of the PANCREAS, the others being insulin, somatostatin and a polypeptide of unknown function. The action of glucagon opposes that of insulin. It causes liver glycogen, a polysaccharide, to break down to glucose, thereby increasing the amount of sugar in the bloodstream. It can also mobilize fatty acids for energy purposes. Glucagon is a 20-amino acid peptide secreted by the alpha Islet cells.

glucocorticoids CORTISOL and other similar hormones produced by the outer zone (cortex) of the adrenal gland. The glucocorticoids suppress inflammation and convert AMINO ACIDS from protein breakdown into glucose, thus raising the blood sugar levels. Their effect is thus antagonistic to that of INSULIN.

glucogenesis the formation of glucose from GLYCOGEN.

gluconeogenesis the formation of glucose from non-carbohydrate sources, especially from AMINO ACIDS from protein. GLUCOCORTICOID hormones stimulate gluconeogenesis.

glucose grape or corn sugar. Glucose is a simple monosaccharide sugar present in the blood as the basic fuel of the body. Glucose is essential for life; a severe drop in the blood levels rapidly leads to coma and death. It is stored in the liver and the muscles in a polymerized form called GLYCOGEN. It is derived from carbohydrates in the diet, but in conditions of shortage can be synthesized from fats or proteins.

glucose-6-phosphate dehydrogenase an enzyme necessary for a biochemical pathway by which red cells obtain their energy. In its absence red cells are fragile and easily release their haemoglobin (haemolysis) under the action of a range of drugs.

glucose tolerance test a test of the body's response to a dose of glucose after a period of fasting. The blood sugar levels are measured at various intervals after the sugar is taken and the results plotted on a graph. A characteristic shape of the curve occurs in diabetes because the blood sugar levels rise to an abnormal height and take longer than usual to return to normal. A urine test also shows sugar, in diabetes.

glucuronic acid a substance formed from glucose that combines with many body waste products to form glycosides that are excreted in the urine.

glutamate a negatively-charged ion derived from the amino acid GLUTAMIC ACID and an important excitatory neurotransmitter in the central nervous system.

glutamic acid glutamate, an AMINO ACID present in most proteins. One of its salts, monosodium glutamate, is widely used as a seasoning and flavouring agent.

glutathione a tripeptide amino acid derivative that protects red cells from oxidative damage and, in the form of the enzyme glutathione peroxidase, plays an important role in detoxifying hydrogen peroxide and organic peroxides produced in the body. Glutathione also participates in the transport of amino acids from one cell to another.

glutathione synthetase one of the enzymes concerned in the synthesis of GLUTATHIONE from glutamate.

gluteal pertaining to the buttock region.

gluten the insoluble, glue-like protein constituent of wheat that causes stickiness in dough. Gluten consists of two proteins – gliadin and glutenin. Some people are sensitive to gluten and in these it causes the intestinal malabsorption disorder coeliac disease which is treated by a strict gluten-free diet. Gluten is found in wheat, oats, barley, rye and similar grain cereals.

gluteus maximus the largest of the three flat buttock or rump muscles. The gluteal muscles arise from the back of the pelvis and are inserted into the back of the upper part of the thigh bone (femur). They straighten (extend) the hip joint in rising from a stooping position and in running, climbing and going upstairs.

gluteus medius the buttock muscle lying between the GLUTEUS MAXIMUS and GLUTEUS MINIMUS.

gluteus minimus the smallest, thinnest and deepest of the three buttock muscles.

glycaemia the presence of glucose in the blood. This is essential for life but the amounts must remain within strict limits. See also HYPERGLYCAEMIA and HYPOGLYCAEMIA.

glycaemic index a measure of the effect a given food has on blood sugar levels, expressed in terms of a comparison with glucose, which has a value of 100. Fast-releasing foods that raise blood sugar levels quickly are high on the index, while slow-releasing foods, at the bottom of the index give a slow but sustained release of sugar.

glyco- *combining form denoting* sugar or glycine.

glycobiology the study of the role of carbohydrates in biological events and their association with disease processes and mechanisms. Major advances in the understanding of the roles of glycoproteins and glycolipids has, in recent years, elevated glycobiology into a discipline in its own right.

glycogen a polysaccharide formed from many molecules of the monosaccharide glucose and found in the liver and in the muscles. It is the primary energy store of the body as it breaks down readily to release molecules of glucose. Glycogen has been called 'animal starch'.

glycogenesis the formation of the polymer GLYCOGEN from many GLUCOSE molecules.

glycogenolysis the process of breakdown of GLYCOGEN to release molecules of GLUCOSE.

glycolysis the breakdown of glucose or other sugars under the influence of enzymes, with the formation of lactic acid or pyruvic acid and the release of energy in the form of adenosine triphosphate (ATP). The complex biochemical sequence by which glucose-6-phosphate is converted to pyruvate and ATP.

glycoproteins a class of proteins linked to carbohydrate units. They are called conjugated proteins and are of comparatively small molecular weight. Some, such as follicle stimulating hormone, luteinizing hormone and chorionic gonadotropin, lose their function if the sugar part is removed; others can continue to function even if deglycosylated. Some glycoproteins are cell adhesion molecules.

glycosuria sugar in the urine. This is one of the cardinal signs of diabetes.

glycosaminoglycans mucopolysaccharides, a range of polysaccharides containing amino sugars or monosaccharides in which the –OH group is replaced by an NH_2 group. Heparin is a glycosaminoglycan. All six classes contain substantial amounts of D-glucosamine and D-galactosamine.

glycosylated haemoglobin a normal chemical linkage between haemoglobin and glucose. The amount of glucose linked in this way is proportional to the average levels of glucose in the blood. This provides an independent check of the quality of blood sugar control in diabetes and is a useful monitoring aid.

gnath-, gnatho- *combining form denoting* the jaw.

GnRH *abbrev. for* GONADOTROPHIN-releasing hormone.

goitre enlargement of the THYROID GLAND from any cause.

Golgi apparatus a collection of stacked, flattened, cup-shaped sacs situated in the CYTOPLASM of cells near the nucleus and concerned with the movement of materials within the cell. The Golgi apparatus receives protein-containing vesicles from the endoplasmic reticulum, glycosylates them, sorts them into groups for different locations and transports them to other parts of the cell or to the cell membrane for export. (Camillo Golgi, 1843–1926, Italian microscopic anatomist).

Golgi, Camillo and Cajal, Santiago Ramòn Y Probably the greatest single advance in the technique of determining the nerve connections in the brain was the discovery by the Italian microscopic anatomist Camillo Golgi (1843–1926) of the use of silver salts to stain nerve tissue. This method brings out detail in the nerve cells never previously seen, and allows detail of the connections of nerve cells to be worked out. Golgi's method allows the investigator to pick out a few neurons from among the masses and to demonstrate all its branches and connections. Golgi also used metallic salts to demonstrate the intracellular organelles now known as the Golgi apparatus.

Once Golgi's staining method had been demonstrated, in the 1880s, it was taken up by the obscure Spanish histologist Santiago Ramòn Y Cajal (1852–1934) who improved it and set to work to use it to discover how the brain and nervous system was organized. Cajal soon became a master microscopist, in whose hands the technique proved wonderfully fruitful. In 1889 he lectured at the Berlin conference of the German Anatomical Society where he astonished the leading neurologists of the day with his demonstrations of brain structure. Cajal, encouraged, devoted the rest of his life to the study.

Golgi was none too pleased with some of Cajal's ideas, especially that the nervous system consisted entirely of nerve cells. Golgi was convinced, with other influential scientists, that nerve cells were connected by a network of non-nervous fibres – the reticularist theory. Eventually, Cajal was able to show that this theory was wrong, that the nervous system consisted only of nerve cells and their supporting tissues, and that nerve cells contacted each other by their own fibres. In 1904 Cajal published his enormous work *Histologie du Système Nerveux de L'homme et des Vertébrés* which established his reputation as the world's leading neuroanatomist, and which is still regarded as the major single work on the subject. In this, he showed that there was nothing random about the nerve connections in the brain but that these were elaborately structured. He demonstrated how the different regions of the brain differed in architecture and, to a limited extent, how they were interconnected.

In 1906 Cajal and Golgi shared the Nobel Prize in Physiology and Medicine.

Golgi tendon organ a tension-detecting sensor wrapped around collagen bundles in tendons that provides afferent nerve information about the state of tightness of the tendon.

gomphosis a non-moving joint, such as those between the bones of the CRANIUM.

gonadotrophic hormones PITUITARY GLAND hormones that stimulate the testicles and the ovaries to produce sperm (spermatozoa), eggs (ova) and sex hormones. Follicle stimulating hormone promotes the growth and development of the eggs ova and sperm, and luteinizing hormone prompts the production of sex hormones. Gonadotrophic hormones are used as drugs to treat infertility, delayed puberty and underdevelopment of the gonads.

gonadotrophin, chorionic see CHORIONIC GONADOTROPHIN.

gonads the sex glands. The gonads are the ovaries in the female, that produce eggs (ova) and the testicles in the male, that produce sperm (spermatozoa). Both also produce sex hormones.

good cholesterol an informal term for cholesterol carried in high density LIPOPROTEINS.

goose flesh skin in which the tiny erector pili muscles have contracted, causing the hairs to stand upright and the skin around each hair to form a small papilla. These muscles contract in response to cold or fear.

goose pimples see GOOSE FLESH.

GOT *abbrev. for* the enzyme glutamate oxaloacetate transaminase. This enzyme is released when heart and liver cells are damaged. Its detection and measurement provide helpful diagnostic information.

G proteins cell messengers that relay signals from over 1000 different cell membrane receptors to many different intracellular effectors such as enzymes and ion channels. G proteins have three subunits, alpha, beta, and gamma, each coded for by a different gene, selected from a total of 34 genes. G protein function is switched on and off by the binding and hydrolysis of guanosine triphosphate to the alpha subunit which is loosely attached to the others. Binding causes the beta and gamma fragments to separate as a dimer and to activate downstream effectors.

Graafian follicle a nest of cells in the ovary that develops into a fluid-filled cyst containing a maturing egg (ovum). One or more of these develops in each menstrual cycle, releasing one or more ova into the FALLOPIAN TUBE and leaving behind the CORPUS LUTEUM. (Regnier de Graaf, 1641–73, Dutch anatomist).

gracilis a long slender muscle lying on the inner side of the thigh.

graded potential a changeable electric charge on a membrane of variable duration and amplitude but that, unlike an action potential, has no refractory period or threshold.

graft a tissue or organ, taken from another part of the body or from another donor person, and surgically implanted to make up a deficit or to replace a defective part. To be successfully retained, a graft must quickly establish an adequate blood supply and must be able to resist immunological rejection responses.

granins a family of acidic, soluble secretory proteins that include various chromogranins and secretogranins. The granins occur in vesicles in neurons and neuroendocrine cells. They are prohormones and give rise to bioactive peptides and can be used as markers of sympathoadrenal activity and of secretion from normal and cancerous neuroendocrine cells into the bloodstream. Some granins are sensitive markers of particular diseases.

granulation tissue the tissue that forms on a raw surface or open wound in the process of healing. It consists of rapidly budding new blood capillaries surrounded by newly generated COLLAGEN fibrils secreted by cells called FIBROBLASTS, and many embedded inflammatory cells.

granulocyte a white blood cell (LEUKOCYTE) containing granules. Granulocytes include NEUTROPHILS, BASOPHILS, EOSINOPHILS and their precursors. MAST CELLS are also granulocytes. These cells are components of the immune system.

granulocyte colony stimulating factor a circulating hormonal substance that controls the growth of some of the white cells of the blood.

granulosa cells epithelial cells that surround the oocyte in developing ovarian follicles.

gravid pregnant. The term may be applied either to a woman or to a womb (uterus). From the Latin *gravid*, heavy or burdensome. See also PRIMIGRAVIDA.

gray a unit of absorbed dose of radiation equal to an energy absorption of 1 Joule per kilogram of irradiated material. 1 Gy is equivalent to 100 RADS. In radiotherapy, radiation is commonly applied to the area of the tumour in a dosage of around 2 Gy a day, five days a week for periods of 3–6 weeks.

greater omentum a large double fold of PERITONEUM attached to the greater curvature of the stomach and hanging down like an apron over the intestines. It contains many fat cells and is translucent and partially fragmented.

greenstick fracture a type of long bone break common in children in which the fracture is incomplete, the bone being bent on one side and splintered on the other.

grey matter that part of the CENTRAL NERVOUS SYSTEM consisting mainly of nerve cell bodies. The grey matter of the brain includes the outer layer (the cortex) and a number of centrally placed masses called nuclei. In the spinal cord, the grey matter occupies the central axis. The white matter consists of nerve fibres – axons of the nerve cells.

greenhouse effect the progressive earth-heating effect resulting from the transparency of the atmosphere to sun (solar) radiation at high frequencies and its relative opacity to energy re-radiated by the earth at a lower, less penetrative, frequency. Water vapour and carbon dioxide are the main elements concerned, and any increase in these, mainly from the burning of fossil fuels, enhances the heating effect. A rise in surface temperature could melt polar ice and cause widespread flooding.

grief the mental and physical responses to major loss of whatever kind, especially loss of a loved person. The mental aspects include unhappiness, anguish and pain, guilt, anger and resentment. The physical aspects are caused by overaction of the sympathetic part of the autonomic nervous system. This causes rapid breathing and heart rate, loss of appetite, a sense of a lump in the throat (globus hystericus), a fluttering sensation in the upper abdomen and sometimes severe restlessness. Grief follows a pattern of recognizable stages, some of which are: a sense of being stunned; refusal to accept the event; denial; a feeling of alarm; anger; a sense of guilt; and, eventually, consolation, adjustment and forgetting.

grinding-in a method of obtaining improved relationship of the opposing (occlusal) surfaces of teeth by inserting abrasive paste and moving the jaws relative to one another.

grippe a popular term for INFLUENZA.

gristle see CARTILAGE.

groin the area between the upper part of the thigh and the lower part of the abdomen. The groins slope outwards and upwards from the central pubic region and includes an obvious crease when the hip is flexed. The term is sometimes confused with loin, which is on the back between the lowest ribs and the back of the pelvis.

gross anatomy body structure visible with the naked eye, as distinct from HISTOLOGY which is the structure visible on microscopy.

growth hormone (GH) the hormone, somatotropin, produced by the pituitary gland, that controls protein synthesis and hence the process of growth. Excess growth hormone during the normal childhood growth period causes gigantism. Deficiency causes dwarfism. In adult life, excess causes ACROMEGALY. GH is secreted during periods of exercise and stress and for an hour or two after falling asleep. Growth hormone is also produced by breast tissue and in excess by breast cancers. It encourages cancer cells to metastasize.

growth hormone releasing factor (GHRF) a hormone used as a drug to test the function of pituitary growth hormone.

guanine one of the two purine bases of double-ring structure (the other being ADENINE) which, with the PYRIMIDINE bases form the 'rungs of the ladder', and the genetic code, in the double helix deoxyribonucleic acid (DNA) molecule. Guanine is also one of the ribonucleic acid (RNA) bases.

guanylyl cyclase an enzyme that catalyses the transformation of GTP to GMP.

guarding reflex tightening of the muscles of the abdominal wall when pressed upon by the examining hand in the presence of underlying tenderness from inflammation. Guarding makes examination difficult but is, in itself, of diagnostic significance.

gubernaculum any guiding structure such as the fibrous cord that extends from the fetal testicle in the abdomen down to the scrotum and forms a mould for the canal down which the testicle moves into the scrotum around the time of birth.

guilt a state of distress usually caused by the belief that one has contravened accepted moral, ethical, religious or legal standards of behaviour. Early conditioning in such matters remains powerful throughout life and guilt may be experienced even when early precepts have been long-since been abandoned as illogical. A deep, and seemingly inappropriate, sense of guilt is often a feature of psychiatric disorder.

gular pertaining to the gula or upper part of the throat (GULLET).

gullet the common term for the OESOPHAGUS.

GUM *abbrev. for* genitourinary medicine. This specialty has absorbed and replaced the former discipline known as venereology.

gustatoreceptor a TASTE BUD on the tongue.

gustatory pertaining to the sense of taste.

gustducin a protein that is released when the taste receptors in the mouth detect a bitter compound. This protein triggers a cascade of reactions that finally sends sensory messages to the brain that cause the experience of a bitter taste. Gustducin blockers have been developed in the expectation that a food additive might remove unpleasant flavours from foods. This may be expected to exacerbate the obesity pandemic in the Western world.

gut the intestine. The term is neither slang nor popular. A major gastro-enterology journal is called *Gut*.

gut-associated lymphoid tissue (GALT) patches of lymphoid tissue in the intestine comprising the Peyer's patches, solitary lymphoid nodules under the gut mucous membrane, and the vermiform appendix.

guttate 1 in the form of drops.
2 speckled or spotted as if by drops. Drop-shaped.

gynae- 1 *combining form denoting* woman or female.
2 informal *abbrev. for* GYNAECOLOGY.

gynaecologist a doctor specializing in the wide range of disorders of the reproductive organs and the breasts of women. Gynaecologists are also usually experts in the management of pregnancy and childbirth (obstetrics). From the Greek *gyne*, a woman. See also GYNAECOLOGY.

gynaecomastia the occurrence in the male of breasts resembling those of the sexually mature female. From the Greek *gynae*, a woman and *mastos*, a woman's breast.

gynandromorph a person having both male and female external genitalia and secondary sexual characteristics. A person displaying HERMAPHRODITISM.

gyrus one of the many lobe-like rounded elevations on the surface of the brain resulting from the infolding of the adjacent grooves (sulci). From the Greek *guros*, a circle.

Hh

habenular nucleus one of a pair of collections of nerve cells in the brain, situated on either side of the stalk of the PINEAL BODY, and having a strap-like form.

habilitation training of the disabled in needed skills.

habit one of many learned responses to a given stimulus that lead to regularly repeated behaviour performed with little thought. Habits are strengthened by repetition until they become automatic. Life would be intolerable without habits as they allow us to perform many essential repetitive tasks while devoting our thoughts to other things. Most habits are useful in this way; some, such as systematic working habits or the habit of eating in moderation, are of great value. Many are, however, harmful and may be difficult to break. Behaviour therapy includes some effective techniques for breaking undesired habits.

habit-forming tending to lead to physiological addiction.

habituation the development of a tolerance or dependence by repetition or prolonged exposure. From the Latin *habituare*, to bring into a condition.

haem- *combining form denoting* blood. From the Greek *haima*, blood.

haemagglutination clumping together of red blood cells.

haemagglutinin an antibody that causes HAEMAGGLUTINATION of red blood cells bearing the corresponding ANTIGEN.

haematemesis vomiting blood.

haematin the iron-containing portion of HAEMOGLOBIN. A complex of PORPHYRIN, iron and hydroxide ion.

haematinic pertaining to any substance that promotes blood production.

haematocrit 1 the proportion of the volume of cells to the total volume after the blood has been centrifuged.
2 a CENTRIFUGE used to separate the cells from the fluid part of the blood.

haematogenous originating in, or carried by, the blood.

haematologist a specialist in disorders of the blood.

haematoma an accumulation of free blood anywhere in the body that has partially clotted to form a semi-solid mass. Haematomas may be caused by injury or may occur spontaneously as a result of a bleeding or clotting disorder. In some sites, as within the skull, enlarging haematomas may be very dangerous. Infected haematomas may form abscesses.

haematoma auris 'cauliflower ear' resulting from repeated bleeding into the tissues of the ear, usually caused by boxing, with the formation of much internal scar tissue and distortion.

haematopoiesis the growth and maturation of the blood cells and other formed blood elements in the bone marrow. Haematopoiesis normally occurs in the flat bones, such as the pelvis, breastbone and shoulder-blades, but in times of extra demand may extend to other bones or even other tissues such as the liver. All haematopoiesis, including both red and white cells, develops from a single type of stem cell.

haematopoietic growth factors haemopoietins, factors including GRANULOCYTE COLONY STIMULATING FACTOR and granulocyte-macrophage stimulating factor that bring about the maturation of bone marrow stem cells to form normal blood cells.

haematoxylin a reddish or yellow dye used to stain biological material for microscopic examination.

haematuria blood in the urine.

haemo- *combining form denoting* blood.

haemoconcentration thickening of the blood from loss of plasma or water. This occurs in dehydration and after severe burns.

haemocytoblast the precursor cell from which any of the formed blood elements can be derived.

haemoglobin the iron-containing protein that fills red blood cells. Haemoglobin combines readily but loosely with oxygen in conditions of high oxygen concentration, as in the lungs, and releases it when in an environment low in oxygen, as in the body tissues. In health, each 100 ml of blood contains 12–18 g of haemoglobin. The various genetically induced abnormalities of haemoglobin are called haemoglobinopathies.

haemoglobin A the normal form of the haemoglobin molecule, containing two alpha chains and two beta chains.

haemolysis destruction of red blood cells by rupture of the cell envelope and release of the contained HAEMOGLOBIN.

haemolytic pertaining to HAEMOLYSIS.

haemolytic anaemia a form of anaemia arising from haemoglobin loss as a result of increased fragility of red blood cells.

haemophilia an X-linked recessive blood clotting disorder causing a life-long tendency to excessive bleeding.

haemophiliac a person suffering from HAEMOPHILIA.

haemopoiesis see HAEMATOPOIESIS.

haemoptysis coughing up blood.

haemorrhage an abnormal escape of blood from an artery, a vein, an arteriole, a venule or a capillary network.

haemosiderin one of the two forms in which iron is stored in the body. The bulk of the iron is stored as haemosiderin. See also FERRITIN.

haemostasis the act or process of stopping bleeding.

hair the filamentary keratin secretion of follicles in the skin. The outer layer, or cuticle, of each hair is made of overlapping flat cells arranged like roofing slates. Below this is the thick cortex of horny cells and the core of softer rectangular cells. The hair colour comes from pigment cells (melanocytes) of uniform colour present in differing concentration. Very curly hair comes from curved follicles.

hair cycle the repetitive sequence of growth and rest affecting the production of the hair follicles. The growth phase is known as anagen and this varies in length in different sites. The scalp anagen may last for several years. The rest phase is called the telogen. After about three months of telogen the hair is shed and a new anagen starts. Between these is the brief catagen during which the base of the hair becomes club-shaped.

hairiness, excessive see HIRSUTISM.

halitosis bad breath.

hallucination a sense perception in the absence of an external cause. Hallucinations may involve sights (visual hallucinations), sounds (auditory), smells (olfactory), tastes (gustatory), touch (tactile) or size (dimensional). Hallucinations should be distinguished from delusions – which are mistaken ideas.

hallucinogenic causing HALLUCINATION.

hallux the big toe.

hamate 1 hooked at the tip.
2 one of the bones of the wrist.

hamstrings the tendons of the three long, spindle-shaped muscles at the back of the thigh (the hams). These prominent tendons can be felt at the back of the knee on either side.

hamstring muscles the three large muscles on the back of the thigh – the semimembranosus, the semitendinosus and the biceps femoris. These muscles arise from the lower back part of the pelvis and are inserted, by way of the HAMSTRINGS, into the back of the bone of the lower leg (the TIBIA). Contraction of the hamstring muscles bends the knee and straightens the hip joint.

handedness the natural tendency to use one hand rather than the other for skilled manual tasks such as writing. Ambidexterity – the indifferent use of either hand – is rare. About 10 per cent of people are left-handed.

handicap any physical, mental or emotional disability that limits full, normal life activity. Handicap may be CONGENITAL or acquired as a result of injury or disease especially to the nervous or musculoskeletal systems.

hanging a method of judicial execution or of committing suicide. Judicial hanging by the drop method causes separation (disarticulation) of the upper neck vertebrae and severance of the spinal cord. Suicidal hanging usually causes death by strangulation.

hangover the state of general distress experienced on the morning after an evening of alcoholic over-indulgence. The symptoms include headache, depression, remorse, shakiness, nausea and vertigo. With the possible exception of stomach irritation, these are not caused by alcohol, most of which has already been metabolized. The breakdown products, such as acetaldehyde, and some of the other constituents (congeners) of alcoholic drinks are, however, toxic. Other factors such as smoking, dehydration, overeating, the recollection of indiscretion and the loss of sleep may contribute.

haploid having half the number of chromosomes present in a normal body cell. The germ cells, the sperm and eggs (ova) are haploid, so that, on fusion, the full (DIPLOID) number is made up. From the Greek *haploeides*, single.

haploidy the state of being HAPLOID.

haplosis the process of halving of the DIPLOID number of chromosomes by reduction division (MEIOSIS).

haplotype the entire set of allelic variants that may be found at any particular genetic location.

hapten an incomplete antigen that cannot, by itself, promote antibody formation but that can do so when conjugated to a protein. Most haptens are organic substances of low molecular weight. From the Greek *haptein*, to fasten.

haptic *adj.* 1 pertaining to the sense of touch. A haptic contact lens is a large lens fitting under the eyelids and covering the white of the eye as well as the cornea.
2 the part that touches. The haptic of an intraocular lens is the peripheral supporting part.

haptoglobin alpha$_2$-globulin, a plasma protein that binds free haemoglobin to form a complex too large to pass out of the kidneys into the urine.

hare lip a lay term for the appearance caused by a badly repaired CLEFT LIP.

Harvey, William William Harvey (1578–1657) was a successful London medical practitioner, who had been studying at Padua when Fabricius produced his influential book on veins there. Harvey never ceased to ponder on the anomalies between Galen's teaching and what he had learned at Padua. Finally, nearly thirty years later, after much thought and experimentation based on close observation, he decided, even at the cost of challenging Galen, to give the world the benefit of his findings. In 1628, under the title *Exercitatio anatomica de motu cordis et sanguinis* (*Movement of the heart and blood; an anatomical essay*), he outlined for the first time, in terms that would brook no argument, the way in which the blood was driven, by the pumping action of the heart, through the arteries to all parts of the body, and was returned, by way of the veins to the heart. The calculated volume moved in this way was such that Galen's idea – of blood being manufactured in the liver from ingested food – was nonsense.

One difficulty with Harvey's theory was that there appeared to be no channel for the blood to flow from the smallest visible branches of the arteries supplying organs, to the smallest branches of the veins draining blood from them. Harvey postulated large numbers of fine connections and, eventually, in 1661, this was shown to be correct when Marcello Malpighi, using the recently developed microscope, demonstrated the system of capillaries that links up all arteries and veins.

This idea of a continuous circulation was revolutionary and marked the beginning of modern physiology. Harvey saw clearly that the same blood was moved continuously round the body and was replenished with nutrients as it passed through the intestines. Harvey's idea, soon shown to be obviously correct, was, in its way, as important as that of Darwin.

The disciples of Galen did not readily admit, in the light of this incontrovertible evidence, that Galen had been wrong. A Paris professor of medicine, Guy Patin pronounced Harvey's findings absurd, paradoxical, impossible and harmful. Anatomy professor Jean Riolan stated stoutly that if current dissections no longer agreed with Galen, it was not because Galen was wrong but because the human body had changed in the meantime. For many, the attitude was 'I would rather err with Galen than be right with Harvey'.

Harvey's demonstration of the direction of blood flow in veins
The veins of the forearm can be made prominent by a moderately tight band tied briefly round the upper arm. The effect of the valves in these veins, in allowing the blood to flow only in the direction of the heart, can then readily be demonstrated. In Harvey's own words: 'But so that this truth may be more openly manifest, let an arm be ligated above the elbow in a living human subject as if for a blood-letting. At intervals there will appear, especially in country folk and those with varicosis, certain so to speak nodes and swellings not only where there is a point of division, but even where none such exists, and these nodes are produced by valves. If by milking the vein downwards with the thumb or a finger you try to draw the blood away from the node or valve, you will see that none can follow your lead because of the complete obstacle provided by the valve; you will also see that the portion of the vein between the swelling and the drawn-back finger has been blotted out, though the portion above the swelling or valve is fairly distended. If you keep the vein thus emptied and with your other hand exert a pressure downwards towards the distended upper part of the valves, you will see the blood completely resistant to being forcibly driven beyond the valve.'

hashish see CANNABIS.

HDL *abbrev. for* high density LIPOPROTEINS.

healing and repair

Anything as mobile as the human body is bound to suffer injury of all kinds, and, without a built-in system of healing and repair, the accumulation of physical damage, major and minor, would be disastrous. Surgery, which is implicitly based on this fundamental healing response of tissues, would be impossible. Structural damage from infecting organisms, although less visible, would, on aggregate, be equally dangerous and would often be more rapidly fatal than apparently grosser injuries. Failure to repair such microinjuries would soon lead to widespread and fatal tissue destruction. Happily, even in cases of severe immune deficiency, the natural processes of healing and repair continue as normal. Healing, whether of gross wounds or of microscopic tissue damage, is a dynamic series of events that involves remarkable action on the part of many body cells.

The area of the body most exposed to environmental injury is the skin surface. Healing and repair are so necessary here that we have evolved a system of continuous replacement of the surface layer – a system that continues to operate whether injury occurs or not. As the cells of the outer layer, the epidermis, are pushed out towards the surface, they die, become flattened and horny and are eventually cast off. The effects of minor surface injuries such as abrasions and shallow cuts, are thus eliminated and a normal surface restored. When necessary, the skin even responds to the threat of injury by an automatic protective thickening of this layer – the formation of calluses. Deeper injuries, however, which involve tissues below the level of the epidermis, need a more elaborate repair process.

Closed wound healing

Disruptive injury to deep tissues immediately brings into action an impressive series of changes. At first, all the small blood vessels around the injury constrict, but almost immediately they widen to a more than normal degree. As they do so the smallest vessels – the capillaries – become abnormally permeable, and proteins from the blood pass into the injury site. Large numbers of white blood cells migrate through the walls of the capillaries and congregate on and around the injured surfaces. This process is called inflammation and it is the body's basic response to injury of any kind. Torn blood vessels leak whole blood into the wound space and this blood soon clots so that the whole area between the surfaces becomes filled with semi-solid material. If the wound is clean and relatively free from organisms and foreign material, the white blood cells will be able to cope with the organisms and minor tissue debris present. The area will rapidly become sterile and the healing processes may proceed. If there is much contamination or foreign material, however, the inflammation may last for months and healing may be delayed. The basic surgical management of a wound is to open it widely, remove all foreign matter and devitalized tissue, and bring the edges closely together with deep stitches.

If the edges of a wound are kept together, surface healing is rapid. Within 24 hours the edges of the wound begin to thicken and the living cells of the epidermis enlarge and migrate across the wound plane. This is a rapid process and in about 48 hours the whole wound surface has acquired a new covering. This is called epithelialization. The healing of the deeper part of the wound still has a long way to go, however. At this stage the edges are held together only by delicate strands of the protein fibrin that forms the main part of the blood clot. It is now that the definitive healing cells, the fibroblasts, come into action. Fibroblasts are highly active cells that might be considered as mobile protein factories. They have oval nuclei and are of varying contour, but when active are usually star- or spindle-shaped.

Within two or three days of wounding they appear in their millions, probably from the surrounding tissues, where they have been lying at resting state. By about the tenth day, fibroblasts are the predominant cells present. Each one contains a large rough endoplasmic reticulum covered with enormous numbers of ribosomes – the cell organelles responsible for protein synthesis under the control of DNA. Fibroblasts are programmed to synthesize the structural protein collagen and each one becomes busy pouring out new, long, delicate collagen fibrils into the wound. Fibroblasts use the blood clot fibrin strands as scaffolding and actively move along them from one side of the wound to the other, generating collagen as they go. They do not move, as many cells do, by amoeboid action but by tiny adhesive feet that attach to the surface near the advancing edge of the cell and then move backwards underneath it.

The stage of collagen formation lasts for several weeks and during this time the fibroblasts produce a quantity of the protein that is enormous in relation to their size. The collagen fibres form bundles of ever-increasing thickness, that are randomly oriented between the tissue surfaces, and eventually build up into a massive collagen structure that binds the edges of the wound together. This structure is called a scar. Over the course of subsequent years the scar gradually changes in bulk and appearance, usually remodelling itself so that it becomes stronger and less conspicuous. Occasionally, unfortunately, collagen production is excessive and results in an ugly, heaped-up scar known as a keloid.

Open wounds

Wounds that are left open or that cannot be closed also show remarkable features in their healing. Soon after the injury, the surface of the wound becomes covered with fluid that oozes from the edges. This fluid is invaded with fibroblasts and with large numbers of buds of tiny blood vessels that grow out from the raw surfaces of the wound. These growing vessel buds produce a rough, granular appearance and were long ago given the name of granulation tissue. At the same time, epidermis from the wound edges begins to grow down the raw surfaces. After two or three days the edges of the wound begin to move towards each other. This process results in a great reduction in the size of the wound. Wound contracture does not occur by the production of new skin and the surrounding skin has to be stretched to allow it to occur. This sometimes causes trouble by limiting movements of joints.

The explanation of wound contraction remains obscure. Although much structural collagen is formed by the fibroblasts in the wound, collagen is not a contractile protein and does not cause wound contraction. It is believed that granulation tissue contains specially modified fibroblasts capable of acting in a manner similar to muscle cells, and pulling the edges of the wound together. Although almost all wounds that are left open will eventually contract and heal over, this process often leads to undesirable complications by shortening the surrounding skin and by producing ugly wide scars. For these reasons, wounds that cannot be closed are almost always treated by skin, or skin and muscle, grafting. Application of a graft prevents wound contracture as does the applications of steroid drugs.

Bone healing

The structure of bone is a scaffolding of collagen impregnated with mineral salts, especially salts of calcium and phosphorus. Bone contains many blood vessels which allow a constant interchange of materials. When a bone is broken a blood clot forms between the broken ends and, as in a soft tissue wound, many white cells collect in this clot. Fibrin that has formed in the clot connects the broken ends of the bone but, in relation to the forces acting on a fracture site by the weight of the parts concerned, provides no structural strength. If the fracture is to heal satisfactorily the broken bone must be immobilized so that very little movement can occur at the site. Again, as in soft tissue wounds, large numbers of fibroblasts invade the site, using the fibrin strands as pathways. Fibroblasts in bone are called osteoblasts and are either identical to or very similar to fibroblasts elsewhere.

The process of laying down collagen to re-establish the shape of a bone, requires a higher level of organization than the formation of scar tissue elsewhere, and another kind of cell, the osteoclast, is involved. Osteoclasts are large cells with multiple nuclei and are thought by some to originate from fibroblasts. They are, however, destructive rather than constructive, and more closely resemble macrophages. Osteoclasts secrete enzymes (collagenases) that break down collagen and release minerals. The function of the osteoclasts is to remove unwanted pieces and spicules of bone and prepare the site for new collagen deposition in an orderly manner.

The cooperative action of these two cells, the osteoblasts and the osteoclasts, will lead, in the course of a few weeks, to the replacement of the blood clot by a mass of semi-solid tissue called callus that bridges across the gap between the bone ends. Within this mass, new bone is being formed by a simultaneous process of collagen synthesis and mineralization. Inadequate immobilization will damage the callus bridge and may lead to the production of a false joint of gristle (cartilage).

Surfaces and adhesions

All body surfaces, inside and out, are covered with a non-stick layer. This is called an epithelium and it is an essential guard against the body's tendency to heal surfaces together. If the pads of a finger and thumb are sewn together, no healing will occur even if left for several weeks. As soon as the stitches are removed the surfaces will separate. But if, before stitching, the skin is removed over the areas to be brought in contact, healing will occur rapidly and will be secure, without stitches, in about two weeks. Loss of the epithelial covering of internal structures can cause much trouble because of the tendency for raw surfaces to heal together (adhesions). Such loss may result from surgical intervention or from certain disease processes that damage surface cells. Adhesions between abdominal organs can cause distortion and sometimes problems such as intestinal obstruction. Fortunately regeneration of damaged epithelium is normally rapid and effective and in the absence of persisting disease will usually restore normality.

Tissue regeneration

Organ regeneration does not occur in humans but some tissues have remarkable powers of regeneration. Perhaps the most impressive example of tissue regeneration is in the liver. Following surgical removal of a large part of the liver, the remainder will rapidly undergo regeneration so that in a matter of a week or so the bulk is made up. Often such regeneration is not merely a disorganized reproduction of cells but is an ordered process by which the structural and functional integrity of the organ is restored. In the continuing presence of damaging influences, such as alcohol, however, regeneration of liver tissue tends to be disorganized with nodules of new liver cells surrounded by scar tissue. This is called cirrhosis of the liver – a condition incompatible with normal liver function and often with life.

health

The term 'health' means 'wholeness'. It derives from the Old English word *haelp* meaning 'whole' and is related to the verb 'to heal' – that is, 'to make whole'. The Greeks did not consider health to be a single entity and used two separate words – *euexia* meaning 'a good habit of body', and *hygeia* meaning 'a good way of living'. They were also alive to the importance of mind, in this context. In the Platonic dialogue *Charmides,* Socrates argued that a physician who tried to heal the body without healing the mind, was a fool. One should, he said, attend first to the state of the mind, for health was dependent on quality of mind. Being well in body required that we should establish good habits of living. Socrates understood very well that we, as individuals, are to a large extent, responsible for the state of our own health.

The World Health Organization official definition of health reads: 'Health is a state of complete physical, mental, and social well-being and not merely the absence of disease and infirmity.' Unfortunately, although a positive attitude of this kind is clearly desirable, it is not at all easy to quantify health in this sense, and, for practical purposes, health professionals, especially community physicians and statisticians, must rely on negative indices – such as mortality rates and morbidity and disability figures – in trying to assess the health of a community.

From the individual point of view, however, some rough measure of quantification, is possible. Health implies 'fitness' – the ability to perform and to live a full life. Many people are free from organic disease but are so unfit that only the most sedentary of life-styles is possible to them. Such people cannot be considered healthy. Positive health is usually associated with a feeling of well-being and the absence of depression. It is almost always the result of high motivation for living. There is an important relationship between unfitness and ill-health. The kind of life-style that results in unfitness is very apt also to be the kind that leads, sooner or later, to actual disease. Although this does not necessarily follow, it is probable that the very unfit person, in addition to indulging in physical idleness, eats more than is necessary for his or her energy needs, takes an unsuitable diet, may drink too much alcohol, and may be a cigarette smoker.

The meaning of fitness

Fitness is essentially a matter of the amount of exercise that can be tolerated. This varies over wide limits. The concept of exercise tolerance is important as this allows useful comparisons of the state of fitness to be made. It is also important as a way of assessing heart and respiratory system disease. The exercise tolerance scale is a kind of fitness hierarchy with long-distance athletes at the top and bed-ridden cardiac cripples at the bottom. The scale may be calibrated in terms of the distance which can be travelled, at a brisk pace, before the subject is halted by severe breathlessness, or other effects. Most people are somewhere near the bottom of the scale.

People who are unfit but free from disease can, whatever their age, readily increase their exercise tolerance by taking exercise. Even people with a heart or respiratory condition can often also do so but close medical supervision is necessary. In the investigation of such cases doctors will always try to assess the exercise tolerance.

A patient with angina pectoris may be able to walk 50 metres before being stopped by pain. Later, if the condition gets worse, he may be able to go only 20 metres before the symptom occurs. In the most severe cases, simply being upright is sufficient to cause breathlessness. A useful idea of exercise tolerance is the performance on stairs in tall buildings. Fit men can easily run up several flights of stairs, two or three steps at a time. Some habitually do. Less fit men can only walk, one step at a time. Some instinctively avoid stairs, finding that even talking is difficult, while doing so, because of breathlessness.

Breathlessness and a fast pulse rate are the normal response to exertion. Their degree, however, for a given amount of exertion, varies widely between people of different standards of fitness. This is because of the many physiological and even structural changes that occur in the heart, lungs and muscles as a result of training. Because of their higher cardio-respiratory efficiency, athletes usually have slow resting pulse-rates. A long-distance runner may have a resting pulse-rate of as low as 40 beats per minute.

Above a certain heart-rate pumping efficiency drops. If a heart can work efficiently up to 160 beats per minute, an unfit person may only be able to double the useful output; the trained athlete may be able to quadruple it. Circulatory efficiency is not, however, simply a matter of heart-rate. The important parameter is cardiac output – the amount of blood that can be pumped in a given time, such as a minute. This is the product of heart-rate and stroke volume (the volume pumped per contraction). Stroke volume varies with the power and efficiency of heart muscle contraction with each beat. In a trained athlete, the heart output may rise from the resting value of 5 litres per minute, to as high as 35.

How to assess fitness

The degree of breathlessness on effort is a sensitive index of the degree of fitness and we should all be aware of this. Men or women well into middle age, who are not particularly concerned with athletic pursuits and who do not regularly play physical games, should be able to do the following: walk ten miles; walk at a brisk pace for a mile without embarrassing breathlessness; continue after this, at a more normal pace, without difficulty; run up a flight of 20 steps in seven bounds with no overt sign of distress; in an emergency, run 30 meters in under seven seconds.

These are by no means athletic standards but there are many who, because of a sedentary habit of life, and perhaps heavy smoking, would fail miserably. There are many who would be ill-advised even to try without a very gradual preliminary workup.

health education the inculcation of knowledge the possession of which can help to promote health and reduce the chances of disease. Health education is concerned with such matters as personal hygiene, cleanliness, exercise of body and mind, good diet, care of the skin and hair, and the avoidance of hazards such as smoking, excessive drinking and the abuse of drugs.

health food food claimed or believed to be more beneficial to health than the generality of nutriments. Numerous claims are made explicitly or implicitly for the medical benefit of various foods but many of these are of dubious authority. Some measure of control in the UK is provided by the Health Claims Initiative of the National Food Alliance and the Food and Drink Federation which produces guidelines for food manufacturers. In the USA the Food and Drug Administration (FDA) serves a similar, if more draconian function.

hearing and balance

In many of the simpler animals the functions of hearing and balance are served by separate mechanisms, often in different parts of the body. Hearing in vertebrates, however, evolved out of balancing mechanisms and in all, including humans, the ear is the organ both of hearing and of equilibrium. In man and most other mammals, the two functions are so intimately linked that they share a common nerve pathway to the brain. Disorders of balance are commonly associated with disorders of hearing.

Hearing has become an essential medium of social intercourse, so important that severe loss of hearing is often more damaging to human relationships than loss of vision. The atmosphere in which we live is seldom free from the periodic vibration that we interpret as sound. Much of this we can, and should, ignore. But often these vibrations convey important information to us that we neglect at our peril. Hearing has the advantage over sight that it can attract our attention without our active participation. Hearing, as the vehicle for music, is also a source of pleasure and satisfaction to millions. For these reasons we should be sensitive to the threats to our hearing offered by modern living and should understand how these threats arise and how they may be avoided.

Structure and function of the ear

The ear is conveniently divided into three parts – the outer, middle and inner ears. The outer ear consists of the pinna, the visible part, and the external auditory canal, or meatus, a tube about 2.5 cm long that ends in the ear drum. The pinna serves little more than a decorative function in humans and no longer has much sound-gathering power. The ear drum, or tympanic membrane is a delicate circular surface of stretched skin and fibrous tissue capable of free vibration in response to the most subtle changes in the pressure of the air in contact with it. The sensitivity of the drum to force is remarkable: the footstep of a mosquito upon it is clearly audible.

The middle ear lies between the ear drum and the outer bony wall of the inner ear. It is a narrow cleft crossed by a chain of three tiny articulated bones, the auditory ossicles, that link the drum to a window in the inner ear. The ear drum must be very free to vibrate, but the force of the vibrations must be transferred to fluid in the inner ear. This fluid has a much greater resistance to imposed movement (impedance) than the drum, and the purpose of the ossicles is to act as an impedance transformer, converting large-amplitude but low impedance vibrations at the drum to small-amplitude but high force vibrations at the window.

The middle ear is lined with mucous membrane containing blood vessels, and oxygen in the air in the middle ear is constantly being absorbed into the blood in these vessels. The resulting tendency to develop a partial vacuum and for the drum to be forced inwards is, however, prevented by the presence of a tube joining the middle ear to the back of the nose. This is the Eustachian tube through which air can pass into the middle ear to equalize the pressure on both sides of the drum. The front end of the tube opens each time we swallow.

The inner ear is the most complicated of the three parts. The hearing sensory apparatus is contained in a snail shell-like structure called the cochlea. The balancing mechanism consists of the saccule and the utricle, which bears the semicircular canals (see diagram). The whole combined apparatus, called, for obvious reasons, the labyrinth, is filled with fluid. The inner of the three auditory ossicles, the stapes or stirrup, has a footplate that fits neatly into the oval window of the outer wall of the labyrinth. Vibration of the stapes conveys the sound vibrations to the fluid in the labyrinth.

The cochlea contains a spiral membrane, the basilar membrane. This can be likened to a kind of elongated harp with strings (fibres), of different length and tension, free to vibrate in sympathy with vibrations in the cochlear fluid. Under the influence of different frequencies, different parts of the basilar membrane vibrate. High frequencies cause fibres near one end to vibrate; low frequencies cause vibration near the other end. Resting on the basilar membrane are many rows of hair cells. These are the transducers that convert vibrations into nerve impulses. Movement of these cells cause their protruding hairs to wipe against an overlying membrane and this causes the hair cells to fire, sending nerve impulses along the auditory nerve to the brain. Loudness (amplitude) is conveyed by a higher frequency of nerve impulse transmission, pitch (frequency) by the identity of the nerve fibres stimulated. In this way the brain is

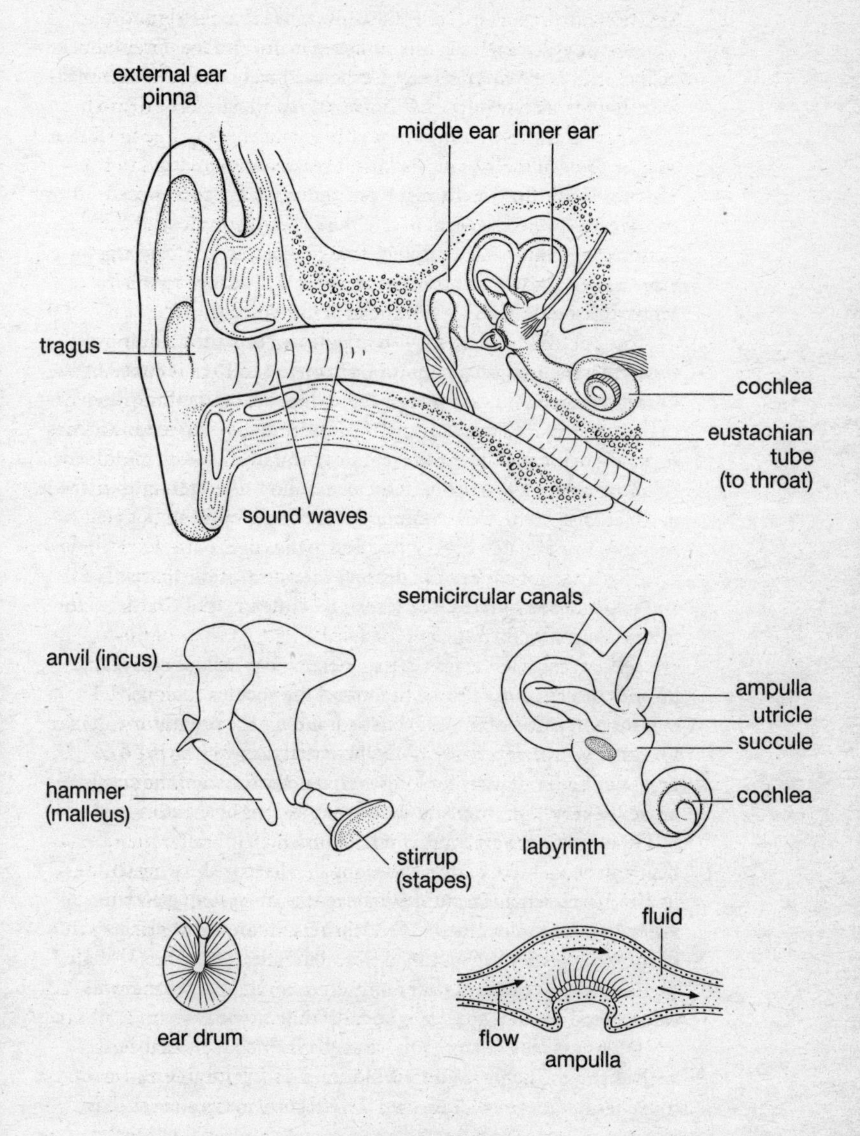

Fig. 7 **Hearing and balance**

informed of the amplitude and frequency of the sounds impinging on the ear drums. The quality of the sounds is inherent in the waveform, which includes harmonics. The hair cells of the inner ear are delicate and easily destroyed, especially by sustained loud noise or by high intensity impulsive noise, as in explosions.

Types of deafness

There are basically two distinct types of deafness – conductive deafness and sensori-neural deafness – and the causes are quite different. Conductive deafness is caused by disorders occurring between the exterior and the oval window in the outer wall of the inner ear; sensorineural deafness results from disorders of the inner ear.

Conductive deafness may arise from any cause that interferes with the transmission of vibrations from the air to the inner ear. Common causes include wax in the auditory canal, perforation of the ear drum, middle ear inflammation (otitis media), excessive sticky secretion in the middle ear so that movement of the ossicles is impeded ('glue ear'), damage to the ossicles from injury, and the condition of otosclerosis in which the footplate of the inner bone, the stapes, becomes fused by new bone formation to the edges of the oval window.

Most cases of conductive deafness are susceptible to treatment and many can be cured. Otosclerosis will often respond to delicate microsurgery on the stapes.

Sensorineural deafness is, in general, more serious as it usually implies destruction of some or most of the cochlear hair cells. The commonest cause of this is acoustic trauma – the progressive damage to hair cells from excessive noise. Damage can result either from noises of high intensity for long periods, or from sudden impulsive noises of very high intensity such as explosions or those caused by a blow on the ear. Very loud, sudden noises can literally shake the hearing mechanism to pieces, leaving the destroyed hair cells floating in the cochlear fluid. Whatever the cause, hair cell damage is permanent and irremediable. Hearing aids are of little or no value in the management of severe sensorineural deafness. Even the latest digital cochlear implants offer only crude sound perception and cannot readily allow speech to be fully understood.

Other causes of damage to hair cells include various drugs, such as the aminoglycoside antibiotics, aspirin, quinine and certain diuretics; Ménière's disease; and the effects of ageing (presbycusis).

Balancing mechanism

The part of the labyrinth concerned with equilibrium is known as the vestibular system. As in the cochlea, the principal feature of the balancing mechanism is the action of hair cells. In this case, however, the hair cells detect not fluid vibrations but fluid movement. Any change in the motion of the head, especially angular motion, causes the fluid in the vestibular system to move. And because of the position of the vestibular hair cells, movement can be detected in three different planes. Bodily acceleration or deceleration in any direction can also be detected. The detecting hair cells are situated in each of the three semicircular canals and in the utricle and saccule. Any movement of the fluid in which they are immersed will cause the corresponding hair cells to fire off impulses and these pass along nerve fibres of the same cranial nerve that carried impulses from the cochlea. The brain is thus informed of the plane in which movement of the head is occurring. Only a *change* in the rate of motion causes fluid movement. Movement at constant velocity does not cause hair cell stimulation.

heart the twin-sided, four-chambered controlled muscular pump that, by means of regular rhythmical tightening (contractions) of the chambers and the action of valves, maintains the twin circulations of blood to the lungs and to the rest of the body. The right side of the heart pumps blood through the lungs and back to the left side. The left side pumps the blood returning from the lungs through all parts of the body and back to the right side.

heart, arrest see CARDIAC ARREST.

heart massage see CARDIOPULMONARY RESUSCITATION.

heart sounds the sounds heard with a STETHOSCOPE applied over the heart. The most prominent sounds are caused by the closure of the heart valves. Heart abnormalities, especially valve disorders, cause additional sounds, called MURMURS. The timing and characteristics of these give much information about the state of the heart.

heart, stopped see CARDIAC ARREST.

heat-shock proteins a range of peptides that are well conserved throughout a wide range of organisms and are produced by the body in response to various stressors. They are found in higher than normal concentration in people who have suffered strokes and heart attacks, and in women with breast cancer.

heavy chains polypeptides forming the main part of the Y structure of an ANTIBODY, including the stalk and the inner edges of the two upper arms. Each heavy chain contains one variable region and three or four fixed domains. Attached to the outer side of the upper arm parts of the heavy chains are the two LIGHT CHAINS. The area between the free upper ends of the two pairs of chains is known as the N terminal. This is the antigen combining site and is the variable part of the antibody allowing for the enormous range of different specificities.

hedonism the philosophic and psychological proposition that pleasure, or gratification, is the only ultimate good, and that the pursuit of pleasure is the ultimate motivating force. The concept of 'pleasure' is, of course, susceptible to a variety of definitions.

Heidelberg man the origin of a fossil jaw found in a quarry at Heidelberg, Germany in 1907. The species is though likely to be that of *Homo erectus* who flourished some 400,000 years ago.

Heimlich manoeuvre a first aid procedure used to try to relieve choking caused by an inhaled foreign body such as a bolus of food. The upper part of the abdomen is encircled from behind, the hands are clasped together in front and the fists are suddenly and firmly forced upwards into the gap between the lower ribs so as to compress the air in the chest. (Henry Jay Heimlich, American surgeon, b. 1920).

helicase an enzyme that breaks the hydrogen bonds between the BASE PAIRS in DNA thus separating the two strands of the double helix in the process of replication. Helicase works in conjunction with single-strand binding proteins that attach to the outer side of each single strand preventing the two from rebonding so that two rows of free-ended bases are left as templates on which new complementary strands can be formed.

helicotrema a semilunar opening at the apex of the cochlea through which the fluid in the scala vestibuli and the scala tympani communicate so that sound vibrations can pass to the round window.

helix the folded margin of the outer ear. The form of the helix is determined by the underlying cartilage.

Helmholtz, Hermann Hermann Ludwig Ferdinand von Helmholtz (1821–94) was the son of a Potzdam schoolteacher who could not afford to give him a University education. So Helmholtz accepted a Government medical cadetship which obliged him, after graduating as a doctor at the age of 21, to serve for 10 years in the Prussian Army. Almost at once Helmholtz embarked on a series of brilliant experiments on the conversion of food into energy in which he showed that the work performed, and the heat produced, by the muscles exactly equalled the chemical energy contained in the food. He also showed that body heat was produced almost wholly by muscle contraction. Prior to this it had been generally believed that the work energy displayed by the body was derived from some mysterious 'vital force', emanating from the soul. On the basis of this research Helmholtz went on later to prove the fundamentally important principle of the conservation of energy.

His early work so impressed the government that he was released from his military obligations and became an instructor in anatomy in Berlin and, a year later, in 1848, was appointed Professor of pathology and physiology at Königsberg. Subsequently he held professorial chairs of anatomy and physiology at Bonn, of physiology at Heidelberg and of physics at Berlin.

His contributions to medicine and science were legion. He wrote the definitive textbook on the function of the eye, invented the ophthalmoscope and explained colour vision. He mastered physiological acoustics and wrote the still standard work on the nature and perception of musical tones. He did original work in neurology and was the first to measure the speed of the nerve impulse. He pioneered thermodynamics and meteorological physics. He did original work in electricity and inspired Hertz in his work on radio propagation. Not content with everything else, he was a brilliant mathematician and philosopher, producing work on non-Euclidean geometry, empiricist scientific philosophy and epistemology, especially on the metaphysics of perception. A great deal of his research was fundamental and germinal and was fruitfully followed up by such men as Pavlov (mental basis of the conditioned reflex), Kelvin (the atom), Kölliker (evolution), Clausius and Boltzmann (thermodynamics), Arrhenius (ionization in liquids) and de Vries (theory of mutations).

helminth a worm, especially a parasitic nematode or fluke (trematode). From the Greek *helmins*, a worm.

helper T cell a subclass of T lymphocytes which provides essential assistance to MACROPHAGES of the immune system. When organisms have been ingested by a MACROPHAGE, antigen is expressed on the macrophage surface at the site of the MAJOR HISTOCOMPATIBILITY COMPLEX (MHC) surface markers. Helper T cells bind to the combination of antigen and MHC and produce the CYTOKINE gamma-interferon. This diffuses to the MACROPHAGE and switches on the killing function so that the organism is destroyed. In the absence of helper T cell assistance the protective function of the immune system is seriously undermined. The HIV attacks and kills helper T cells thus causing acquired immune deficiency (AIDS).

hemeralopia inability to see as well in bright light as in dim. From the Greek *hemera*, day, *alaos*, blind and *ops* an eye.

hem see HAEM-.

hemi- *combining form denoting* half. From the Greek *hemi*.

hemianopia loss of half of the field of vision of one or both eyes.

hemiparesis muscle weakness on one side of the body.

hemiplegia paralysis of the right or left half of the body.

heparin a complex polysaccharide organic acid found mainly in lung and liver tissue. Heparin is thought to bind to THROMBIN and antithrombin in plasma thereby assisting in their combination and interfering with the cascade of reactions that end in blood clotting (coagulation). From the Greek *hepar*, the liver.

hepat-, hepato-, hepatico- *combining form denoting* liver.

hepatic pertaining to the liver.

hepatocellular pertaining to liver cells.

hepatocyte a liver cell. Hepatocytes are metabolically very active and are rich in MITOCHONDRIA, LYSOSOMES and ENDOPLASMIC RETICULUM. Each is surrounded by highly permeable capillaries.

hepatomegaly enlargement of the liver.

hepatosplenic pertaining to the liver and the spleen.

HER2 human epidermal receptor 2, a growth factor receptor that plays an essential role in cell proliferation and differentiation.

heredity the transmission from parent to child of any of the characteristics coded for in the molecular sequences on DNA known as the GENES. Heredity is mediated by way of the CHROMOSOMES which, essentially, consist of DNA. Of the 46 chromosomes in each body cell, 23 come from the mother and 23 from the father. The pattern of genes on the chromosomes is called the genotype; the resulting physical structure with all its characteristics is called the phenotype.

Hering-Breuer reflex a feedback system for controlling breathing. The stimulation of stretch receptors in the bronchi and lungs, occasioned by breathing in, eventually causes the DIAPHRAGM and the muscles between the ribs (intercostal muscles) to cease contracting. (Heinrich Ewald Hering, 1866–1918, German physiologist; and Josef Breuer, 1843–1925, Austrian physician).

hermaphroditism the rare bodily condition in which both male and female reproductive organs are present, often in an ambiguous form. Most hermaphrodites are raised as males but about half of them menstruate and most develop female breasts. The term derives from the mythical Hermaphroditos, who so inflamed the passions of a young woman called Salmacis that she prayed for total union with him. Her wish was granted and the two were fused into one body.

hernia abnormal protrusion of an organ or tissue through a natural or abnormal opening.

hetero- *combining form denoting* different. From the Greek *heteros*, meaning other.

heterochromatin a length of chromatin in the genome that is permanently highly condensed and whose DNA is not transcribed.

heterocyclic of organic compounds with one or more closed rings consisting of carbon atoms and atoms of other elements such as nitrogen, oxygen or sulphur.

heterograft a transplant of tissue taken from one species and grafted into another.

heterologous 1 derived from a different source.
2 of a transfusion or transplant from a different species.
3 of tissue not normally present at a particular site.
4 of parts of different organisms that differ in structure.

heteroploid having an abnormal number of chromosomes.

heterosexuality the common state of sexual orientation directed towards a person of opposite anatomical sex.

heterotopic occurring in an abnormal location, especially of normal tissue.

heterozygosity the state of possessing two different ALLELES for one or more genes. Loss of heterozygosity from all the abnormal cells in a lesion indicates monoconal proliferation and is a molecular marker of neoplasia.

heterozygous of a person carrying different genes at the same gene locus in corresponding chromosomes. The ALLELES are different. A single DOMINANT gene can manifest itself in a heterozygous person. Recessive genes are only manifest if both are present (HOMOZYGOUS).

HFEA *abbrev. for* Human Fertilization and Embryology Authority.

Hg the symbol for the element mercury. Blood pressure is measured in terms of the weight of a column of mercury measured in millimetres, as in 140/80 mm Hg (Systolic/Diastolic).

hiatus a gap or abnormal space.

hiccup repetitive involuntary spasms of the diaphragm causing inspirations, each followed by sudden closure of the vocal cords. In most cases the cause is unknown and it can be stopped by re-breathing into a small bag. Also called 'hiccough'.

hidr-, hidro- *combining form denoting* sweat or sweat gland.

hidrosis sweating. Excessive sweating is caller HYPERHIDROSIS.

high density lipoproteins see LIPOPROTEINS.

hilum a small gap or opening in an organ through which connecting structures such as arteries, veins, nerves or ducts enter or leave.

hindbrain the part of the embryonic brain from which the CEREBELLUM and the brainstem, with the nuclei of most of the cranial nerves, develop. Technically known as the rhombencephalon.

hinge region the part of the immunoglobulin (antibody) molecule between the Fab and Fc regions that allows flexibility.

hippocampus an infolded ridge of the surface of the brain (cerebral cortex) on either side. Each hippocampus forms a ridge on the floor of the LATERAL VENTRICLE. Disease or injury to the hippocampus causes memory defects. PET scanning in young subjects shows that the act of memorization is associated with a 5 per cent increase in blood flow to the hippocampus. From the Greek *hippos*, a horse, and *kampos*, a sea monster.

hirsutism excessive hairiness. Hirsutism in women may, rarely, be due to an excess of male sex hormone from an ovarian or adrenal gland tumour but is usually hereditary, ethnic or just unfortunate. Certain drugs can cause hirsutism.

hist-, histio-, histo *combining form denoting* tissue.

histamine a powerful hormone synthesized and stored in MAST CELLS and basophil cells from which it is released when antibodies attached to the cells are contacted by ALLERGENS such as pollens. Free histamine acts on H_1 receptors to cause small blood vessels to widen (dilate) and become more permeable to protein, resulting in the effects known as allergic reactions. It causes smooth muscle cells to contract. Histamine also acts on receptors in the stomach (H_2 receptors) to promote the secretion of acid. H_2 receptor blockers, such as cimetidine and ranitidine (Zantac) are widely used to control acid secretion.

histidine an essential amino acid and precursor of HISTAMINE.

histiocyte a fixed scavenging cell (phagocyte) found in connective tissue. Histiocytes are also known as reticuloendothelial cells and reticulum cells. A macrophage that does not migrate.

histochemistry the biochemistry of cells and tissues.

histocompatibility sufficient affinity between the genetic composition (genotypes) of donor and host to allow successful tissue or organ grafting.

histocompatibility antigens genetically determined glycoprotein groups situated on the outer membrane of all body cells to provide an identifying code unique for each person (except identical twins). These groups act as antigens and, in a foreign environment, provoke the production of antibodies against them. This is the basis of graft rejection. They are also called HLA antigens because they were first found on human white blood cells (leukocytes). There are three classes: class 1 proteins are found on the surface of most cells, even on red blood cells; class 2 proteins are found on the surface of B cells, antigen-presenting cells and some endothelial cells; class 3 proteins are components of the complement system. See also MAJOR HISTOCOMPATIBILITY COMPLEX (MHC).

histology the study of the microscopic structure of the body. All healthy tissues are identifiable microscopically and a knowledge of normal histology is an essential basis for the recognition of the specific microscopic changes occurring in disease. The microscopic study of diseased tissue and the identification of diseases by this means is called histopathology.

histolysis breakdown of the structure of body tissue.

histones strongly alkaline proteins associated with DNA. Histones may be released into the blood and appear in the urine when tissue breaks down in wasting illnesses.

histopathology the microscopic study of disease processes in tissues.

HLA human leukocyte antigen; the MAJOR HISTOCOMPATIBILITY COMPLEX.

HLA antigens see HISTOCOMPATIBILITY ANTIGENS and MAJOR HISTOCOMPATIBILITY COMPLEX.

hol- or **holo-**, *combining form denoting* wholly, entire. From the Greek *holos*.

Holley, Robert William the American biochemist Robert Holley (1922–93), born in Urbana, Illinois was educated at the University of Illinois, graduating in chemistry in 1942, and at Cornell Medical School where he received a PhD in organic chemistry in 1947. He was appointed to the staff of Cornell in 1948. In 1955 he moved to the California Institute of Technology and while there began to study nucleic acids. By 1958 Holley had succeeded in isolating the transfer RNAs for the amino acids alanine, tyrosine and valene from bakers yeast.

In 1962 he became professor of biochemistry at Cornell where his research team spent three years painfully isolating a gram of alanine transfer RNA from a large quantity of yeast. They then proceeded to work out the sequence of 77 nucleotides in the molecule of alanine transfer RNA. It was this work that earned him the 1968 Nobel Prize for Physiology or Medicine which he shared with Marshall Nirenberg and Har Gobind Khorana.

In 1966 Holley moved to the Salk Institute for Biological Studies in San Diego, California where he worked on growth factors of cells in culture.

holoblastic cleavage cleavage of a fertilized egg into two parts (blastomeres), each of which can form an entirely separate individual.

holocrine pertaining to a gland whose secretion is a breakdown product of the gland's own lining cells. A sebaceous gland is a holocrine gland.

holoenzyme a complete enzyme that includes its prosthetic coenzyme. An enzyme without its coenzyme is called an apoenzyme.

homeo- *combining form denoting* similar. From the Greek *homoios*, of the same kind.

homeobox genes a highly CONSERVED family of genes, found in a large range of different species, including insects. Homeobox genes are expressed early in embryonic development and are the determinants of body shape. These genes divide the early embryo into fields of cells each with the potentiality to develop into a particular part such as an arm, leg, tissue or organ. This is the key to the long-unexplained problem of how bodily configuration is determined by genetics. Mutations in homeobox genes cause severe bodily defects such as PHOCOMELIA.

homeostasis the set of physiological mechanisms by which constant conditions are maintained in the body. It is essential for the maintenance of health, and often even life, that the many variable elements such as temperature, heart rate, blood acidity, blood levels of many substances, hormone levels and blood pressure should be kept within prescribed limits. There are thousands of such parameters and all must be controlled. Some have natural limits, but in many cases control must be exercised by a process of automatic monitoring and self-regulating information feedback. Homeostatic control is most obviously exercised by the endocrine system under the general supervision of the pituitary gland, but other systems are extensively involved, especially the nervous system. Transducers of various kinds – chemical, pressure, orientational, positional – convert variations in these modalities into nerve impulse variations that are fed to the brain to effect an appropriate corrective response. Homeostasis is one of the central principles in physiology.

hom-, homo- *combining form denoting* same, like. From the Greek *homos*.

homograft a graft taken from a member of the same species as the recipient. Compare HETEROGRAFT.

homolateral situated on the same side. Ipsilateral. Compare CONTRALATERAL.

homologous 1 of corresponding structure, position, function or value.
2 having the same consecutive sequence of genes as another chromosome.
3 belonging to a series of organic compounds of which the successive members differ by constant chemical increments.
4 of transplantation in which the donor and recipients are of the same species.

homologous chromosomes chromosomes with the same genetic loci that occur in pairs, one derived from the mother and one from the father.

homologous genes genes from organisms of different species that code for the same enzyme or other product. Homologous genes need not have the same base sequences.

homology the important observation of the essential anatomical similarity between many different species of animal, including the human being. Homology provides strong support for the hypothesis that species had a common ancestor.

homonymous corresponding to the same side, as in homonymous hemianopia in which there is loss of half of the field of vision of each eye, the loss being either of both right halves or both left halves.

homophobia a slang term meaning fear of homosexuality.

homosexuality sexual preference for a person of the same anatomical sex. The term derives from the Greek *homos*, the same (not from the Latin *homo*, a man) and is applicable to women as well as to men. In 1991, differences in the size of a hypothalamic nucleus between heterosexual and homosexual men were reported; and in 1993 it was reported that a region of the long arm of the X chromosome was associated with male homosexuality. It is still not universally accepted, however, that homosexuality is genetically induced.

homozygous having identical gene pairs (ALLELES) at corresponding positions (loci) on the chromosome pairs. People who are homozygous for a quality or condition will always manifest it.

Hooke, Robert Robert Hooke (1635–1703) was born on the Isle of Wight. The son of a poor clergyman, he had to work as a waiter to pay his fees at Oxford University. There he met the great chemist and physicist Robert Boyle (1627–91) who at once recognized and encouraged his mechanical genius. Hooke became interested in every aspect of science and conducted many important and brilliant studies in physics – especially light and gravitation – astronomy, mechanical engineering, geology and biology. He invented the hair spring that made watches possible. He was appointed a member of the Royal Society in 1663 and became Secretary in 1677. Unfortunately, he was a quarrelsome, jealous, mean-minded and censorious man whose greatest satisfaction was to prove others wrong. His position in the Royal Society provided him with plenty of scope for his malice and he became notorious for his sustained attacks on other scientists. His unremitting persecution of Isaac Newton, for instance, finally drove the great mathematician and physicist into a mental breakdown.

The microscope, made famous by the Dutch biologist Anton van Leeuwenhoek (1632–1723), was further developed by Hooke who produced a compound instrument and used it to make many observations of nature. These, he published in the book *Micrographia* a collection of exceptionally beautiful drawings with a text in English. Among many other things, he describe the microscopic appearances of thin slices of cork which showed regular rows of geometric spaces which Hooke called 'cells'. Later studies showed that these were actually empty spaces – the dead remnants of living cells – but the observation was of fundamental importance and led others to look for, and find, a cellular structure in all living organisms.

hormones a term whose definition has expanded progressively over time with the growth of physiology 1 chemical substances produced by the ENDOCRINE and other glands or cells and released into the bloodstream to act upon specific receptor sites in other parts of the body, so as to bring about

various effects. Hormones are part of the control and feedback system of the body by which HOMEOSTASIS is achieved. The pituitary hormones are adrenocorticotropin, to prompt cortisone release from the adrenal cortex; follicle-stimulating and luteinizing hormones to produce sperm and egg maturation in the testis and ovary; prolactin for milk secretion in the breast; thyroid stimulating hormone for the thyroid; growth hormone for the bones and muscles; melanotropin for the pigment cells (melanocytes); and antidiuretic hormone for water reabsorption in the kidneys. Under these influences, each of the endocrine glands produces its own hormones.
2 any chemical mediator or carrier of information from whatever source to whatever destination. From the Greek *hormon*, to urge or stir up.

hospice a hospital specializing in the care of the terminally ill. Hospices are dedicated to providing the physical, emotional and psychological support and expert pain management needed to help the dying to accept the reality of death and to die in dignity and peace of mind.

host 1 an organism that provides a residence and nourishment for a parasite.
2 a person receiving a graft of a donated organ or tissue.

housekeeping genes constitutive genes, genes that provide the basic routine functions required by all cells for their sustenance.

hubris excessive arrogance and pride that tends to be the downfall of the possessor.

human beings – their place in nature

All scientists now agree that the two million or so species inhabiting the earth today, of which the human being is one, arose through a long history of gradual change that can be traced back to very simple life forms. The proposition that man is an animal is a statement of scientific fact – a reflection of the inescapable physical relationship between man and the many thousands of other animal species. In comparing man with closely allied species, it is hard to decide which are greater – the similarities or the differences. But if we exclude mental development and all its implications, it is clear that, on a purely physical level, the difference between man and closely related species are so minor that, in some particulars, only an expert can distinguish one from the other.

Comparative anatomy

This is the study of the relationship between the body structure of different animals. Even an elementary knowledge of comparative anatomy immediately illustrates the remarkable similarities between man and the other animals. A central element in comparative anatomy is the concept of homology – the relationship of body organs of different species that have the same evolutionary origin. In closely related species, corresponding structures are so similar as to leave no doubt as to their identity. But scientists such as Richard Owen (1804–92) showed clearly how even less obviously related structures in remote species such as the wings of a bird and the arms of a man, or the flipper of a seal and the hand of a man, also had common developmental and evolutionary origins. By the middle of the 19th century the idea of homology was becoming widely appreciated.

Man classified

To place man in the context of the animal kingdom it is necessary to consider how animals are categorized.

Animals can be divided into two very large groups – those with backbones (the vertebrates) and those without (the invertebrates). The vertebrates, in turn, can be divided into two very large groups – those that lay eggs that hatch to release the young, and those (the mammals) who produce milk to suckle and nourish their offspring.

The large group of mammals contains one relatively small group called the primates. Primates are an order of mammals with thumbs that can be bent across the palm so as to touch each of the other fingers; with fingernails rather than claws; and with four central biting teeth (incisors) in both upper and lower jaws. They have large brains and well-developed eyes that are directed forward. The majority live mainly in trees, and all produce a small number of young which are nurtured for a long time after birth. The primates include lemurs, lorises, pottos, galagos, tarsiers, marmosets, tamarins, Old and New World monkeys, apes and man.

Narrowing the definition further, we can say that, in terms of strict classification, man is also an ape. Apes are primates with no visible tail, with shoulder blades (scapulae) at the back rather than at the sides, and with a Y-shaped pattern of grooves on the surface of the grinding teeth (the molars). The apes consist of the gibbons, the gorilla, the orang-utang, the chimpanzees and man. Each of these groups is called a genus and each genus contains one or more species. The evolutionary lines of the present ape genera are thought to have separated from the line leading to man some time between 6 and 20 million years ago.

All the apes sleep at night and all are vegetarian, apart from man and the chimpanzees (the latter have been known to kill small animals for food). Chimpanzees and gorillas spend most of their time on the ground. Apes usually have single babies or sometimes twins and form families of varying size, in some cases of as many as 50 individuals. They have a life expectancy of about 30 years – much the same as that of early man. They are shy and retiring and will show aggressive responses only when threatened, although chimpanzees are often inquisitive and extroverted.

In the entire animal kingdom the intelligence of other apes is second only to that of man. Chimpanzees and gorillas are known to be able to master a sign-language vocabulary of as many as a hundred words, and chimpanzees commonly use simple tools, especially when motivated by the desire for food. They use sticks or twigs to extract termites from the ground and, in captivity, pile up boxes to stand on to reach otherwise inaccessible food. It seems probable that orang-utangs are equally intelligent but are reluctant to display it in the presence of humans. The other apes do not speak, but this is probably because their vocal equipment and brain connections are not adapted to speech, rather than because of any mental inadequacy. Primates communicate comprehensively with each other by means of signs, gestures and calls.

The brains of other apes closely resemble that of man. All the major features correspond in each case. The areas for voluntary movement, facial expression, bodily sensation, vision, hearing, smell, emotional reactions, balance, motor coordination, and so on, are the same, qualitatively, in both. In other apes, as in man, the sensory input to the brain areas concerned with emotion comes largely from the parts of the brain concerned with visual information processing. In less highly developed animals the connections come mainly from the parts concerned with smell appreciation.

The other apes suffer many of the same diseases as humans. Genetic disorders such as Down's syndrome and albinism are well known. Heart disease and strokes occur, and epilepsy has been observed. Infections, identical to those suffered by man, are commonplace.

Within the group of the apes is a small genus called *Homo*. Modern man, who is a member of this genus, has been given the specific name *sapiens*, meaning wise, judicious or sensible. Whether this specific name is the most appropriate one must remain an open question, but no one questions that the species *Homo sapiens* is the most highly evolved and dominant member of the group of apes. Other members of the genus *Homo* are long extinct. The earliest known species, *H. habilis*, is believed to have lived about 2 million years ago. Others were *H. erectus* (which included Peking man and Java man), and the sub-species *H. sapiens neanderthalis*. *H. sapiens* is believed to have appeared first about 200,000 years ago .

Physical anthropology

This is the study of the evolutionary biology of the species *Homo sapiens*. Its main purpose is to try to establish the process by which present-day man and his man-like ancestors (the hominids) evolved from early primates. Physical anthropology is an active science, constantly advancing as new fossil discoveries are made and advanced techniques are applied. Scientists use carbon dating methods to establish relative ages; biochemical studies to compare the physiology of man with other primates; detailed physical measurements of all parts of the body (anthropometry) to quantify comparisons; and genetic analysis especially the study of mitochondrial DNA, which is relatively unchanged through the generations, to help to determine human origins. Genetics have also been widely used in the studies of the more recent diversification of man. Various genetic markers such as the range of blood groups, tissue types and various gene mutations have proved valuable in tracing historic movement and intermarriage of human groups.

human evolution

Until about the middle of the 19th century almost all educated people believed that each new species was produced, fully developed, by a separate act of creation. Some believed that species appeared by a process of 'spontaneous generation', others attributed the creation of species to God, as related in the bible. But as the arranging and classification of the multitude of different species proceeded – especially following the work of the Swedish naturalist Carolus Linnaeus (1707–78) – people began to notice that there were close similarities between certain species, and the more enquiring began to wonder whether there might be some kind of relationship between them. By the beginning of the 20th century most of their questions had been answered.

As interest in the structure of the earth developed and became formalized in the science of geology, two important facts emerged: stratified layers of rocks had been formed at different periods in the past, and the fossils found in these different layers were the remains, or the impressions, of organisms that had been living during these periods. From these observations developed the science of palaeontology – the study of prehistoric life based on the evidence of the fossils.

The record of the rocks

Fossils had, of course, been recognized for thousands of years. Aristotle thought they were failed or abortive attempts at spontaneous generation from mud. Leonardo da Vinci and other Renaissance thinkers, however, recognized them for what they were

– the remains of once-living organisms. It was not always safe, in those days, to make too much of this obvious fact. The most impressive finding of the palaeontologists was the record of definite sequences of organisms, living at successive periods, but showing general resemblances to their predecessors. The older the rocks in which they were found, the simpler and more primitive were the life forms. This suggested the probability that modern living things came from earlier life forms by a gradual process of change – by evolution.

The earliest fossils of visible size date back almost 3000 million years. These have a cabbage-like structure and seem to be similar to modern aggregations of blue-green algae. In sediments laid down about 1500 million years ago, large numbers of single-celled organisms can be found. Sediments dating back about 590 million years contain the remains of multicellular soft-bodied animals such as worms, and in deposits made about 570 million years ago we find fossils of molluscs, jellyfishes, crustaceans and starfishes similar to those existing today.

Fossils of numerous species of fish and of strange jointed creatures called trilobites and large scorpion-like eurypterids are found in deposits dating back 400–500 million years. Those of insects and amphibians appear at about the same period. The first reptile appeared about 275 million years ago, giving rise to the dinosaurs, who dominated life on earth from 205–65 million years ago. The earliest mammals emerged at the same time as the dinosaurs, while the birds evolved from one group of dinosaurs around 175 million years ago.

Fossils of the primates – the group of mammals that contains lemurs, monkeys, apes and man – are found only in deposits dating from about 65 million years ago to the present day. Remains of recognizably humanlike creatures occur only in deposits laid down in the past few million years.

The evidence of embryology

There is a striking parallel between the stages of development found in the fossil record and that occurring during the development of every human being as an embryo in the womb. Each of us starts as a single fertilized cell. This divides repeatedly to form a hollow ball of cells similar to some of the most primitive organisms. The ball then folds in on itself to form a two-layered structure called a *gastrula*, which is again reminiscent of some simple organisms. From this, a three-layered embryo develops and each of the layers – the *ectoderm*, *endoderm*, and *mesoderm* – gives rise to particular organs and systems of the body. In the course of the development of these organs, the embryo passes through stages in which it clearly resembles, successively, the embryos of a range of organisms more primitive than ourselves. It is as if each of us repeats in our bodily development, the evolutionary history of man.

This concept was expressed in technical terms by the German biologist Ernest Haeckel (1834–1919) as 'ontogeny recapitulates phylogeny'. Ontogeny is the developmental history of an individual from the time of conception. Phylogeny is the sequence of events involved in the evolution of a species. Strictly speaking, this is not quite accurate. At no stage does a mammalian embryo resemble an adult fish, but at one stage it certainly resembles the embryo of a fish.

If the embryos of different classes of animals, such as birds, lizards or mammals, are compared at certain stages it is often impossible, by inspection, to say which is which. Haeckel was especially interested in this matter, and pointed out that the closer two animals resemble one another in bodily structure, the longer do their embryos remain indistinguishable. On the basis of this and similar evidence he concluded that we had '...sufficient definite indications of our close genetic relationship with the primates.'

The evidence of body structure

As knowledge of the structure of human bodies and of those of other species grew, it became apparent that structural resemblances were much closer than had been thought. When the anatomies of the vertebrates (animals with backbones) were compared, it was seen that they differed essentially only in the shape and size of individual parts rather than in possessing different parts. The similarities between even remote species were far greater than the differences. Many scientists contributed to this understanding, perhaps the most outstanding being the English comparative anatomist Richard Owen (1804-92), who showed the equivalence of many anatomical structures not only in terms of their position in the body but also in their developmental origins. Owen's concept of *homology* was illustrated by examples of organs in different species that not only shared the same essential anatomical structures (if not necessarily the same function) but also developed from the same germinal layer in the embryo.

Equally striking was the common finding that many animals possess vestigial structures that no longer serve any purpose but that are present, in functional form, in other species. Ostriches have rudimentary wings but cannot fly. Snakes have vestiges of hind-limb bones. Man has an appendix attached to the large intestine which corresponds to a blind-ended digestive structure in herbivorous animals, and a vestigial tail in the shape of the coccyx. Some marsupial mammals, that have been delivering live offspring for hundreds of millions of years, still show embryonic indications of the egg tooth with which the young of egg-laying species crack the egg-shell. All these observations strongly suggested common ancestry and some kind of evolutionary process.

The theories

A number of hypotheses were put forward to explain these suggestive facts. For a time, one of the most influential was that of the French naturalist Jean Baptiste Lamarck (1744–1829). Lamarck taught that animals and plants evolved by changing in response to changes in their environment. Giraffes, he suggested, acquired long necks by stretching up to reach the leaves at the top of trees. Tail-less mice could be produced by repeatedly cutting off the tails of mice and then letting them breed.

The trouble with Lamarck's idea was two-fold. One difficulty was to explain how acquired characteristics could be passed on. Lamarck suggested that bodily changes could somehow modify the sperm or the ova so as to pass on the new characteristic to the offspring. The second problem was that the theory, unfortunately, did not represent observable fact. Cutting off the tails of mice had no effect whatsoever on the length of the tails of the offspring. Lamarck's theory was eventually abandoned.

Almost all scientists believe that the real breakthrough in the understanding of evolution was made by Charles Darwin (1809–82) and Alfred Russel Wallace (1823–1913). Darwin and Wallace first outlined the theory in a joint communication to the Linnaean Society in London, in July 1858. But the real impetus to the spread of the new ideas came with the publication of Darwin's book *The Origin of Species by Means of Natural Selection, or the Preservation of Favoured Races in the Struggle for Life* in November 1859. This book made Darwin famous – some, at the time said infamous – and Wallace's contribution, which ranked with Darwin's, was largely ignored.

Darwin's finches

While on a voyage of scientific exploration in HMS *Beagle*, many years earlier, Darwin had visited the Galapagos Islands. There he had observed 14 species of finches, differing mainly in the shape and size of the bill, none of which were known to exist anywhere else in the world. Each species occupied its own island. Some of the finches were seed-eating, some insectivorous, and the beaks differed accordingly. These observations set Darwin thinking. It seemed highly probable to him that a single similar species of finch, a seed-eating bird common on the nearby mainland, must have colonized the islands a very long time before and that the isolated descendants had evolved into different forms in an environment in which they were not in competition with other birds.

Later, in 1838, Darwin read *An Essay on the Principle of Population* by Thomas Robert Malthus (1766–1834) an English clergyman. In this book Malthus suggested that human populations invariably grew faster than the available food supply and that environmental factors such as starvation, disease or war were necessary to limit them. Darwin's thoughts returned to the finches. It occurred to him that the first occupiers of the islands would have multiplied unchecked until they outstripped the available seed supply. Many would have died, but those who happened to be able to adapt to a different diet such as insects – perhaps by a naturally occurring variation such as a differently shaped beak – could then flourish and multiply until, they, in turn, outstripped their food supply. Purely by chance, some variations might prove to be well adapted to the current environment and thus survive to breed, while others might prove poorly adaptive, and disappear. It was a matter of the survival of the fittest. Over the course of millions of years this process, operating very slowly, could be seen inevitably to give rise to radical alterations in the characteristics of organisms that shared common ancestors in the past.

New species could evolve by the splitting of one species into two or more different species largely as a result of geographical isolation of populations. Such isolated populations would undergo different environmental pressures, experience different spontaneous changes and so could evolve along different lines. If isolation were to prevent interbreeding with other stock derived from the same ancestors, these differences might become great enough to establish a new species that cannot successfully interbreed except with its own kind.

This was the great idea of how changes in species occurred by 'natural selection' and Darwin saw that this principle alone was sufficient to explain evolution. Darwin never understood how inheritable changes occurred or how the necessary spontaneous variations in species (mutations) occurred, but this did not in the least detract from the power or persuasiveness of his theory. To most unbiased scientists of the time Darwin's idea was at once seen to be so persuasive as to be almost self-evident.

Objections to Darwin's theory

One of the difficulties faced, even in Darwin's time, by those who supported his theory was the problem of sterility between different species. It is one of the definitions of a species that its members are incapable of breeding successfully with members of other species. Mating and fertilization can, of course, occur between closely related species and new individuals can occasionally result but, at least in the case of animals, these are usually sterile or even more seriously defective. If species occurred by the selection of small chance changes, why should not close species be able to interbreed?

With our advantages in understanding genetics we now know that different species have incompatible genes and often even different numbers of chromosomes. As a result, the accurate matching up of chromosomes from the father with those from the mother cannot occur. Cells require matched pairs of chromosomes (homologous pairs) to function normally.

Another criticism of Darwin's theory is that evolution is not a steady, gradual process of change from one species to another as Darwin and his contemporaries believed. Instead, it appears to occur in sudden jumps. The fossil record suggests that new species appear to enter the record suddenly and to change little during their term of existence. Darwin believed this to be a false impression arising from insufficient data, but the difficulty is still far from being resolved.

Variation and mutation

We now understand, in considerable detail, how it happens that individuals within a species commonly differ in many features from one another and from their parents. We also know how it is that sudden, seemingly inexplicable changes in characteristics occur in individuals.

Variation within a species is commonplace, and one of the most striking examples is in humans. Differences in skin and hair colour, height, weight, body type, intelligence, and so on, abound. Modern man has enormous power over his environment and is able to interfere with many of the natural selective processes, but these processes are occurring, nevertheless. The lifetime of any individual, or even of many generations of people however, are insignificantly short periods in the context of evolution and changes are not apparent. This makes it difficult for us to perceive that we are subject to the same evolutionary forces we accept so readily as acting on other species.

Every group of organisms, man included, produces, over the course of very long periods, a range of mutant types. These mutations are changes in the genes brought about by environmental factors such as radiation. In the more complex animals mutations in the DNA of general body cells are not inheritable, but mutations in the eggs or sperm are passed on to offspring. Some of these changes are so damaging that the affected organisms do not survive long enough to breed and the mutation is lost. Many are of little positive value in assisting in the process of adaptation to changing environments.

A few, however, so change the individual that the power of survival is enhanced. If the characteristic caused by the mutation offers a substantial advantage to the individual and if the mutation is inheritable, a single such individual can, in the course of a number of generations, give rise to a very large number of individuals possessing the desirable new characteristic.

Major changes of this kind are very rare, and it is not surprising that we have recognized none in humans in the course of recorded history. Substantial evolutionary changes in species with a much shorter generation than humans, are, however, common. Although humans can insulate themselves from the effects of natural selection and to modify its operation by such advances as hygiene, nutrition, medical treatment and contraception, there is no reason to believe that we are immune from the same evolutionary processes as all other living organisms.

Creation and science

The 19th century saw a great popular spread of knowledge on palaeontology. Inevitably those who were concerned to promote conventional religious views of the creation of the world were much troubled by their inability to account for these facts.

One idea put forward was that successive creations of living forms had succumbed to successive catastrophes. It was still widely held, at that time, that the world had been created in the year 4004 BC – a date worked out by Bishop James Usher (1581–1656) from evidence in the bible. One difficulty with this chronology was the inescapable evidence of the time necessary for the formation of sedimentary rocks. Another was to account for the presence of the fossils. An ingenious, but not very plausible, suggestion put forward by some churchmen was that when God had created the world he had also created the fossils so as to provide a test for man's faith.

The evolution of man

The idea that humans evolved from a non-human species was widely rejected until the end of the 19th century. But as the evidence of comparative anatomy and of the fossil record accumulated, this conclusion gradually became irresistible. It soon became apparent to most scientists that both man and the apes had developed separately from a common ancestor. Few now believe that there is a 'missing link' between the apes and man. The paleoanthropological evidence of man's evolution is open to various interpretations and many 'facts' deduced by the experts have been challenged by others. There is, however, a large consensus of agreement about the broad lines of the evolution of man.

The earliest mammals on earth date back some 200 million years and are believed to have evolved from mammal-like reptiles of the subclass *Synapsida*. The *Synapsida* had begun to develop the essential mammalian characteristic of control and maintenance of body temperature and they also showed signs of differentiating teeth into front teeth for grasping and tearing prey and back teeth for chewing and grinding food. The *Synapsida* became extinct during the time of the dinosaurs but their inconspicuous descendants – tiny mouse-like creatures, the first mammals – were quietly biding their time during the 100 million years or so of dinosaur dominance, and survived when the dinosaurs fell. From these lowly beginnings, the mammals rose to dominance during the Cenozoic Era which began about 65 million years ago.

The success of the mammals was largely due to the extreme adaptability of a basic structure which allowed a large range of types to be evolved. Limbs could be modified into flippers, wings, and digging organs so that the habitat of mammals could be the surface of the land, the sea, the air or the subterranean environment. Thus mammals diversified into land animals, into marine forms like dolphins, whales, seals and walruses, into underground diggers such as the mole, and into flying and gliding animals like bats and flying squirrels.

Primate evolution

The order of mammals to which man belongs is called the primates. The primates, today, consist of over 160 species and are primarily tree dwellers. By about 45 million years ago primates similar to modern lemurs had developed. These had relatively large brains, forward vision and nails instead of claws. Ten million years later, a number of monkey-like primates appeared. This line included *Aegyptopithecus*, a small creature thought to be in the direct ancestral line of apes and humans. By about 20 million years ago the first true monkeys and apes appeared. Many fossils of these creatures have been found in Africa, where it is thought that the changes leading to man occurred. There is now general agreement that the evolutionary line that led to man probably diverged from the ape line between 20 million and 4 million years ago.

One of these early apes was *Dryopithecus*, thought by some to be a common ancestor of the modern apes and man. The identity and timing of the separation of the lines leading to apes and man is in dispute. One descendant of *Dryopithecus* is believed by some to be the earliest man-like creature. This is *Ramapithecus*, an ape-like animal that abandoned the trees as its main habitat. During this period, climates were warm, mild and rainy and supported many species of plant and animal life. In the grasslands of Africa some primates, such as *Ramapithecus* became adapted to a new way of life in which they began to try to conquer the ground environment, feeding mainly on roots, seeds and berries. A more popular candidate is the later genus *Australopithecus*, many fossils of which have been found in Africa. This creature lived between 5 million and 1 million years ago in what are now Kenya, Tanzania, Ethiopia and the Transvaal.

Australopithecus walked upright and had short canine teeth of little use in fighting. Its molar teeth were adapted for hard grinding of roots and berries. Its brain was larger than that of most apes and some specimens were about as tall as modern man. There is reason to believe that it could defend itself with simple weapons but probably did not use tools. For these and other reasons it is classified as being in the human family, *Hominidae*. There are three recognized species *Australopithecus africanus*, *A. robustus* and *A. boisei*. The latter was contemporary with early members of the genus *Homo*. Early tools found in conjunction with *Australopithecus* remains are probably the work of humans who shared the same sites.

The first humans

The evidence provided by fossil remains suggests that the first true humans appeared near the end of the Pliocene Epoch – between two and three million years ago. The period since then has seen enormous climatic variations. There have been a succession of ice ages in which the valleys of much of Europe and north America were covered with great sheets of thick ice. With so much of the earth's water frozen, the seas shrank and the ice provided land bridges between the continents, allowing migration. During these periods, plants and animals adapted to arctic conditions and the plains supported large herds of reindeer on which early man, who had developed skills as a hunter, was able to prey.

Between these long glacial periods were comparatively short periods – perhaps about 10,000 years – of warmth. Further south, especially in Africa, cool rainy periods, during which forests spread widely, alternated with dry periods leading to the formation of deserts and grasslands.

The first evidence of the development of behaviour distinguishing the species *Homo* from other animals appears in the form of tools and implements of stone and bone. These were used for hunting and for gathering food and as containers. There is plenty of evidence of tribal or family life and of the division of labour in the endless task of finding enough to eat.

The earliest human species is known as *Homo habilis* the specific name meaning 'handy, able or skillful'. More than 50 specimens of this type were found at Olduvai Gorge, East Africa between 1964 and 1981. This species was physically similar to *Australopithecus* but had a larger brain and survived after the latter became extinct. *H. habilis* used simple tools, including stone choppers and sharp-edged flint flake cutting implements. Some sites show evidence of butchery of quite large animals – up to the size of the hippopotamus – but it is not clear from this whether these had been hunted and killed or merely scavenged after death. Successive species of hominids appear to have followed one another without overlap, the most conspicuous differences being in the size of the brain as deduced from the volume of the cranial cavity of the skull.

Homo erectus

In 1891, a Dutch anatomist and palaeontologist Eugene Dubois (1858–1941) while serving as an Army surgeon in central Java found some fossilized bones at Trinil. These included the vault of a skull and a very human-like thigh bone. Dubois was convinced that the creature whose remains he had found walked erect and had characteristics intermediate between those of apes and man. Adopting a term first coined by Ernst Haeckel, he called it *Pithecanthropus erectus*. Subsequent discoveries of similar remains in other parts of Indonesia (Java man) and China (Peking man) and later in Africa and Europe (Heidelberg man) suggested that here was a species more evolved that *H. habilis*. Now known as *H. erectus*, this species lived some 1.5 million to 300,000 years ago. Some specimens have features close to those of *H. habilis*, others closely resemble modern man, *H. sapiens*.

H. erectus was an active big-game hunter who eventually learned to cook animal meat and to use the bones for various purposes. As well as eating meat, eggs, fish and small rodents, his diet consisted of roots, nuts, fruit and fleshy leaves. He preferred to live in open or lightly-wooded areas and used two-edged hand axes and various pounders and cutting flakes. There is no evidence that he buried his dead.

There remains considerable difference of opinion as to whether *H. erectus* was a direct ancestor of *H. sapiens* or whether the two species overlapped. There is also argument on how and where *H. erectus* evolved into *H. sapiens*. Most scientists now believe that the species of the genus *Homo* formed a continuous line of evolution.

The only major differences between *H. erectus* and *H. sapiens* are in the skull and the teeth. The limb bones are very similar, although those of *H. erectus* are usually more robust. The differences in brain size are significant. The cranial volumes of *H. erectus* range from 750–1225 ml, the average of 14 skulls from Java, China and Africa being 940 ml. The range for modern man is from about 1000–2000 ml with an average of about 1450 ml. So far as brain size is concerned, the largest brains of *H. erectus* were larger than the smallest brains of man today. The cranial volumes of the known samples of *Australopithecus* range from 440–520 ml. Those of *H. habilis* average 640 ml.

H. erectus was intelligent enough to be able to adapt to a range of habitats and to develop the use of fire for warmth, cooking protection and in hunting. This advance probably occurred about 500,000 years ago.

Neanderthal man

Homo sapiens can be said to have originated from *H. erectus* about 250,000 years ago. At that stage, however, *H. sapiens* did not particularly resemble modern man. He had heavy ridges above his eyes and a skull that sloped sharply back with little or no forehead. His teeth were large and protuberant. His brain size, however, was well within the the modern range.

Some 90,000–40,000 years ago *H. sapiens* developed a tool culture of which evidence was found in the Neander valley (Neanderthal) of West Germany, Le Moustier in France, Gibraltar, Italy and many other sites. Mousterial tools were made from large thin flakes struck from flint cores and used as blanks from which a range of cutting and scraping implements could be fashioned. These were made by the Neanderthals (sub-species *H. sapiens neanderthalis*). Mousterian tools allowed these people to cut and carve wood, to cut meat, to scrape hides, to make thongs.

The Neanderthalers were advanced peoples who lived in caves or in huts or even in rough stone-constructed dwellings, sometimes forming substantial societies. They constantly used fire, and killed and ate a wide variety of animals. They were probably capable of trapping game. They may have been cannibals, but performed rituals such as the formal burial of the dead. They were capable of looking after their aged members whose knowledge and experience they may have valued. It is likely that they communicated by speech.

Mysteriously, the Neanderthalers seem to have disappeared about 40,000 years ago when they were replaced by humans of modern type, exemplified by *Cro-Magnon man*, whose remains were first found in a rock shelter at Cro-Magnon in the Dordogne in 1868. The Neanderthalers, or at least some of them, probably gradually evolved into modern man. In some areas, however, they appear to have survived coincidentally with the more evolved humans, and to have been displaced or killed by emigrating moderns from other areas.

human cloning the production of a person genetically identical to another person by the insertion of a genome from a somatic cell into an ovum from which the DNA has been removed (somatic cell nuclear transfer). Human cloning is currently almost universally proscribed. At the present time it is also scientifically unfeasible. Because nuclear cloning bypasses the normal processes of gametogenesis and fertilization, it prevents the reprogramming of the clone's genome necessary for the development of an embryo into a normal human being. There is evidence that surviving cloned animals have serious abnormalities of gene expression.

Human Fertilization and Embryology Authority an organization set up by act of Parliament in 1990 to control research involving embryos. The HFEA regulates and inspects all UK clinics providing IVF, donor insemination or the storage of eggs, sperm or embryos. It maintains a register of persons whose gametes are used for assisted conception and licenses and monitors all human embryo research being conducted in the UK.

human genome the entire gene map of all the chromosomes that contains all the information needed to make a human being. The genome contains about 6000 million chemical bases forming about 30,000 genes. The completion of the sequencing of these bases, along the length of each chromosome, in the massive research programme known as the human genome project, was formally announced on 14 April 2003, but well before that date the information obtained had already had a major impact on science. As an index of the speed of progress in genomics, this announcement was made only a few days prior to the 50th anniversary of the publication of the structure of DNA by Watson and Crick.

human givens a term that is becoming popular with psychotherapists, psychologists, educationists and others concerned with the nature and pathology of the mind. It refers to a body of organizing but tentative ideas that are being developed from scientific advances in understanding of brain function and of the rich inheritance of innate knowledge patterns with which we are all born and which manifest themselves as emotional and physical needs.

human leukocyte antigen (HLA) the human MAJOR HISTOCOMPATIBILITY COMPLEX.

humanistic psychology a school of psychology that views people as individuals responsible for, and in control of, their destinies and that emphasizes experience as the source of knowledge. It suggests that we can acquire insight into the inner life of another person by trying to see things from that person's own point of view.

human social development

Early man was distinguished from most other species mainly by his use of tools. To begin with, he had little resources other than to use what was to hand – natural objects, especially stones. The use of stones as tools was so central to the culture of early man that most of his history has been designated the Stone Age. The Stone Age is the period during which primitive man emerged, began to fabricate and use chipped stone tools and weapons, and eventually developed some of the cultural elements of civilization. It is divided into three periods: the *Palaeolithic* period, or old Stone Age, dating from around 3 million years ago; the *Mesolithic*, or middle Stone Age, from about 12,000 years ago; and the *Neolithic*, or new Stone Age, that began about 10,000 years ago. These terms are derived from the Greek word *lithos*, meaning a stone.

Palaeolithic period

This very long period is also divided into three. In the Lower Palaeolithic man (*Homo erectus*) used crude flint stone choppers, picks, scrapers and bone flakes. In the Middle Palaeolithic, Neanderthal man had progressed to the stage of being able to prepare useful narrow flakes from flint cores, from which he could fabricate sharp scrapers and awls. Hand axes became a prominent part of his armamentarium. During the Upper Palaeolithic, which began about 40,000 years ago in the middle of the last Ice Age, *Homo sapiens sapiens*, who had largely displaced Neanderthal man, was able to produce sharp blades, knives and narrow chisels or engraving tools (burins). The use of a stone hammer to strike these chisels became commonplace, and with these, humans were able to fabricate needles, harpoons and other tools of wood and bone. These tools enabled them to sew skins together for clothing and drill small stones to make decorative beads. Other personal ornaments were made from mammoth ivory decorated with simple engraved patterns. Human and animal figures were carved from ivory or made from a mixture of bone and clay.

About 20,000 years ago *H. sapiens*, in Europe, made rapid progress. He made simple huts by scooping holes in the ground, surrounding them by upright poles and roofing them over with branches and thatch. He formed settlements for at least part of the year. The Upper Palaeolithic began in the New World around this time when people migrated across the Bering Straits from Siberia.

This period was dominated by a culture based on hunting. In Western and Central Europe, great herds of reindeer, horses and mammoths roamed everywhere and hunters became skilled and well organized, working in cooperative groups to drive and kill herds of game. Permanent communities grew up and, with plenty to eat, population densities rose rapidly. A number of cave, stone and ivory paintings, mostly representing animals, date from the Upper Palaeolithic period. Whether these paintings served a magical or ritualistic function or whether they merely indicate man's need to express himself remains obscure.

Mesolithic period

The Mesolithic period lasted from about 15,000–5,000 years ago. Around 10,000 years ago environmental temperatures rose and the thick sheets of ice and the tundra vegetation in the north gave way to forests. Game herds were replaced by animals less easy to hunt, and men had to adapt to the new conditions. This period saw an enormous expansion of mankind and many important technological developments. New tools and weapons of all kinds, including bows and arrows, handled utensils, fish-hooks, adzes and barbed harpoons were developed. There was much emphasis on the use of small flint chips (microliths) that were fitted, either singly or multiply, in handles, spears and arrows. Woodworking tools – axes and adzes – were used to make dugout canoes and paddles, as well as bows and arrows. Bows and arrows made it possible for hunters to bring down solitary game animals in the forests. Nets, traps and snares were used to catch fish and wild fowl. Plants became an increasingly important food source. Permanent settlements became less appropriate and population mobility increased.

In some areas of Europe, Mesolithic cultures appeared to be successful enough to persist for as long as two or three thousand years after cereal and live-stock farming had been introduced from the East. Many Mesolithic coastal sites were lost by the rising sea level that resulted from the melting of the ice.

Neolithic period

The Neolithic period began in South-west Asia about 9,000 years ago and reached Britain about 5,000 years ago. It featured the development of agriculture, the herding of livestock and the working of mines. The results of these activities were more stable populations and the emergence of more elaborate social structures. Neolithic tools are made of polished stone. The Neolithic gradually merged into the Bronze Age.

Man adapted in different ways to the ending of the Ice Ages. In some areas people chose to concentrate on certain reliable local resources such as the cultivation of edible wild grasses or the capture, herding and breeding of wild goats and sheep. This was particularly so in the Eastern Mediterranean areas – now Israel, Lebanon and Jordan. There is plenty of archaeological evidence of the transition, in these areas, from hunting to agriculture. The *Natufians*, who lived there in rock shelters, left evidence of the change from hunting and fishing to an agricultural economy – flint sickles that were used to cut wheat and straw, and picks that might have been used as plows. Animal husbandry became common and the records show that almost all the meat eaten in some of these areas was of sheep and goats. Gradually, the animals used in this way became domesticated and dependent on man for their survival.

Apart from Western Asia, which was probably the earliest area of neolithic domestication, several other such centres existed. Areas of early development of agricultural villages have been found in Mexico, China and South-east Asia. Some disputed sites also exist in tropical Africa. There is evidence that some of these centres of agrarian development spread to the surrounding regions. The primitive agricultural methods used soon exhausted the soil so that crops failed. This led to constant movement of cultivators to new land and a consequent extension to new areas, over the years, of their ideas and methods.

Prehistoric villagers often subsisted on only minimal diets and were prone to many diseases. Even so, their numbers increased rapidly. During the full development of the Neolithic period, woven mats were made, pottery was developed, huts were built, stone implements were polished. In some settlements large numbers of simple doorless dwellings were built close together. Villages soon extended as populations grew. The extension of land acquisition led to the concept of property and to the beginnings of the employment of labour by landowners. Within a few thousand years of the start of agriculture, there were clear signs of social stratification and the development of complex societies in which different trades and occupations emerged, a crude monetary system developed, and wealth could be accumulated and bequeathed. Early in this process certain inspired, or enterprising, individuals formed a priestly elite that presided over religious ceremonies and began to expound transcendental matters.

The discovery of metallic copper and lead in fires, around 5,000 years ago, was followed by deliberate smelting of these metals from ores. Later it was found that copper could be hardened by the addition of tin. With the production of bronze in this way, the stone age came to an end. The ability to work in bronze made it easier to produce both agricultural implements and weapons for use against both animals and man. The earliest known definite written records date from about the same time. Iron was first smelted about 3,000 years ago.

The most advanced cultures arose in what is now south Iraq, in the alluvial plains of the valleys of the Tigris and the Euphrates rivers, the land of Babel, or Babylonia. There, the *Sumerians* developed the first fully urban civilization and built many walled cities including Kish, Eresh, Lagash and Ur.

humeral pertaining to, or the region of, the HUMERUS.

humerus the long upper arm bone that articulates at its upper end with a shallow cup in a side process of the shoulder blade (scapula) and, at its lower end with the RADIUS and ULNA bones of the lower arm.

humoral 1 pertaining to extracellular fluid such as the blood plasma or lymph.
2 pertaining to B cell, antibody-mediated immunity.
3 pertaining to the aqueous and vitreous humours of the eye.
4 an obsolete term referring to the ancient medical theory, propagated mainly by Claudius GALEN), that the body contains four humours (blood, phlegm, yellow bile and black bile) and that health results from their correct balance.

humour the possession of, or the capacity to perceive, those things which excite laughter or the desire to laugh. Humour is one of the more mysterious characteristics of the human being and its nature has been endlessly argued. We laugh when we are painlessly surprised; when we perceive foolishness or qualities to which we consider ourselves superior; when we see the pompous deflated, the powerful threatened or the consciously superior mocked. Theories abound, none of them entirely convincing. Humour is, however, a valuable human attribute and its absence is a personality defect.

hunch back extreme curvature or angulation of the spine whether from congenital malformation, postural defect, crush fracture of a vertebra, spontaneous collapse of a vertebra from disease such as tuberculosis or exaggeration of the natural curve by bone weakening (osteoporosis).

hunger the symptoms of abdominal discomfort, pain, contractions of the stomach and craving for food induced by a drop in the level of sugar in the blood passing through the HYPOTHALAMUS of the brain.

hyaline translucent or glassy, as in hyaline cartilage. Of amorphous texture.

hyaloid HYALINE.

hyaluronic acid hyaluronate, a long polymer glycosaminoglycan, consisting of repeating disaccharide units, found in basement membranes, mature oocytes, skin, cartilage, the vitreous body of the eye and the synovial fluid of joints.

hyaluronidase an enzyme that breaks down proteins holding tissue planes together.

hybridization probing the use of a radioactively-labelled single-strand sequence of DNA to form a hybrid complementary pairing with an RNA strand and thus identify it; or the use of a labelled RNA strand to pair with and identify a DNA strand.

hydr, hydro- *prefix denoting* water. From the Greek *hydro*, water.

hydraemia excessive water in the blood.

hydrocephalus 'water on the brain' – an abnormal accumulation of cerebrospinal fluid within, and around, the brain. This occurs if the fluid, which is continuously secreted, cannot be normally reabsorbed, usually because of obstruction of the passages to the site of reabsorption, by a congenital abnormality or later acquired disease.

hydrochloric acid a strong acid, produced by the lining of the stomach, that breaks down connective tissue and cell membranes in the food, so that it can more easily be acted on by digestive enzymes. Hydrochloric acid also kills most of the bacteria ingested with the food.

hydrocortisone a natural steroid hormone derived from the outer layer (cortex) of the adrenal gland. Hydrocortisone has anti-inflammatory and sodium-retaining properties.

hydrogen bond a bond in which a hydrogen atom is shared by two other atoms. The hydrogen is more firmly attached to one of these (which is called the hydrogen donor) than to the other (which is called the hydrogen acceptor). The acceptor has a relative negative charge, and, as unlike charges attract each other, a bond is formed to the hydrogen atom. Hydrogen bonds are weak and easily broken but occur extensively in biomolecules. The link between the bases in the two chains of DNA are hydrogen bonds. Adenine links to thymine by two hydrogen bonds, and guanine links to cytosine by three hydrogen bonds.

hydrogen ion a free proton, which is the positively charged nucleus of a hydrogen atom.

hydrogen ion concentration the number of free hydrogen ions in a given quantity of fluid, such as blood. Hydrogen ion concentration (pH) determines acidity. In the blood it must be kept within the narrow limits of pH 7.37–7.45 or a fatal condition occurs.

hydrolases enzymes that promote the splitting of large molecules by attaching –OH from water to one moiety and –H from water to the other.

hydrolysis splitting of a compound into two parts by the addition of water (H_2O), the hydrogen atom (H) joining to one part and the hydroxyl group (OH) joining to the other. Hydrolysis is usually effected by a hydrolytic ENZYME.

hydrophobia violent and painful spasms of the throat muscles occurring as one of the principal symptoms of rabies. Literally, fear of water.

hydroxocobalamin vitamin B_{12}. This is the specific treatment for pernicious anaemia and is highly effective unless neurological damage has already occurred.

hydroxyapatite a principal ingredient in tooth enamel. A paste containing the compound has been used experimentally to seal small enamel cavities without drilling.

5-hydroxytryptamine see SEROTONIN.

hygiene the study of the promotion of health. Hygiene includes rules for personal conduct and cleanliness and Public Health measures such as preventive medicine. From the name of Hygieia, daughter of Aesculapius, the Greek God of medicine. Her sister, *Panacea*, sometimes called *Therapia*, provided healing.

hymen the thin, fringe-like ring of skin that partly occludes the lower end of the vagina in the virgin. Hymens vary considerably in thickness and extent and may even be imperforate, requiring a minor operation to allow menstruation. The hymen is usually torn during the first sexual intercourse and this may cause some bleeding. From the Greek *Humen*, the God of marriage.

hyoid bone the delicate U-shaped bone suspended from muscles in the upper part of the front of the neck, like a spar in rigging. It provides a base for the movements of the tongue and is usually fractured in manual strangulation. From the Greek letter *upsilon*, 'U' and *eides*, like.

hyper- *prefix denoting* above, beyond, over, excessive. From the Greek *huper*.

hyperactivity an unduly high level of restlessness and aggression with a low level of concentration and a low threshold of frustration, especially as resulting from minimal brain dysfunction. The term is usually applied to children and is sometimes called the 'hyperkinetic syndrome'.

hyperaemia an increase in the amount of blood in a part, organ or tissue as a result of widening (dilatation) of the supplying arteries.

hyperalimentation 1 nutritional intake in excess of normal.
2 total feeding by intravenous means (parenteral nutrition).

hypercatabolism unduly rapid breakdown of body tissues. This may occur in fevers.

hyperextension 'over-straightening' of a joint beyond its normal limits.

hyperflexion bending beyond the normal limits.

hyperglycaemia excessive levels of glucose in the blood. This is a feature of untreated or undertreated DIABETES MELLITUS.

hyperhidrosis excessive sweating. This is not usually the result of disease but may be caused by fever, overactivity of the thyroid gland and occasionally nervous system disorders. It is commonly the result of a stress reaction or other psychological upset.

hyperkalaemia excessive levels of potassium in the blood.

hyperkeratosis undue thickening of the outer layer of the skin so that a dense horny layer, such as a corn or callosity, results. This is a normal and essentially protective response to local pressure.

hyperkinesis excessive movement or activity. Hyperactivity.

hyperlipidaemia an abnormal increase in the levels of fats (lipids), including cholesterol, in the blood.

hypermetropia an inherent, dimensional eye defect in which neither distant nor near objects can be seen clearly when the eye is in a state of relaxed focus. Vision can be clarified by ACCOMMODATION and this is easy for the young, who are often unaware of hypermetropia. As the power of accommodation falls off with age, however, hypermetropia inevitably becomes manifest and convex spectacles will be needed.

hypernatraemia excessive sodium in the blood.

hyperosmotic of a fluid having a concentration of solutes great than that of the normal extracellular fluid.

hyperphagia overeating.

hyperpigmentation abnormally increased numbers of pigment cells (melanocytes) in a particular area of the body.

hyperpituitarism an abnormal excess of production of PITUITARY hormones, especially growth hormone.

hyperplasia an increase in the number of cells in a tissue or organ causing an increase in the size of the part. Hyperplasia is not a cancerous process. It is often a normal response to increased demand and ceases when the stimulus is removed. To be distinguished from HYPERTROPHY.

hyperpnoea abnormally rapid and deep breathing.

hyperpyrexia body temperature above 41.1°C (106°F). Hyperpyrexia calls for urgent treatment to lower the temperature, if permanent brain damage is to be avoided.

hypersensitivity an allergic state in which more severe tissue reactions occur on a second or subsequent exposure to an ANTIGEN than on the first exposure. A particular group of antibodies (IgE) is involved in many hypersensitivity reactions.

hypertelorism an abnormal increase in the distance between bodily parts, usually referring to an abnormal separation of the eye sockets (orbits) due to a much widened and enlarged SPHENOID bone. Such hypertelorism is a congenital condition and is sometimes associated with other developmental abnormalities and with mental retardation.

hypertension abnormally high blood pressure.

hypertonia increased muscle tension (tone).

hypertonic of a solution having a higher concentration of membrane-impermeable particles than normal extracellular fluid.

hypertrichosis see HIRSUTISM.

hypertrophy an increase in the size of a tissue or organ caused by enlargement of the individual cells. Hypertrophy is usually a normal response to an increased demand as in the case of the increase in muscle bulk due to sustained hard exercise. Compare HYPERPLASIA.

hypervariable regions the parts within the antibody and T cell receptor variable regions which show the greatest variability.

hyperventilation unusually or abnormally deep or rapid breathing. This is most commonly the result of strenuous exercise but the term is more often applied to a rate and depth of breathing inappropriate to the needs of the body. This results in excessive loss of carbon dioxide from the blood and sometimes a consequent spasm of the muscles of the forearms and calves.

hypervitaminosis one of a number of disorders that can result from excessive intake of certain vitamins, especially vitamins A and D. Overdosage with vitamin D can cause deposition of calcium in arteries and other tissues and kidney failure.

hypervolaemia an abnormal increase in the blood volume.

hypnagogic 1 causing sleep.
2 pertaining to the period during which a person is falling asleep. Of images, dreams or hallucinations occurring during this period.

hypno- *combining form denoting* sleep.

hypnopompic pertaining to the period during which a person is waking up from sleep. Of images, dreams or hallucinations occurring during this period.

hypnosis a state of abnormal suggestibility and responsiveness, but decreased general awareness often brought about by concentration on a repetitive stimulus. In the hypnotic state, the instructions of the hypnotist are usually obeyed, opinions apparently modified and hallucinations experienced. Many widely-believed myths are associated with hypnotism. It does not involve any kind of sleep; it is impossible without the full cooperation of the subject; and a hypnotized person will not perform actions that would normally be unacceptable. There is, however, inevitably some loss of personal will. Long-forgotten memories of obscure detail are not uncovered by hypnotism.

hypnotherapy treatment by HYPNOSIS.

hypnotic any drug or agent that induces sleep.

hyp, hypo- *prefix denoting* below, beneath, less than. Hypo is the exact opposite of HYPER. Hypoi also used to refer to position in the body, as in hypochondrium or hypogastric. In this sense, the body is to be understood as standing upright, so hypo indicates that the entity is nearer the soles of the feet than the part it qualifies. Hypo is, however, also used in the sense of 'deep to' as in hypodermic. From the Greek *hupo*. Compare HYPER.

hypoaesthesia less than normal sensitivity to any sensory modality.

hypocapnia reduced amounts of carbon dioxide in the blood, as after HYPERVENTILATION.

hypochlorhydria reduced amounts of acid in the stomach.

hypochondriac a person manifesting HYPOCHONDRIASIS.

hypochondrium the region of the abdomen immediately below the cartilages that join the lower ribs on either side to the breastbone. The term, literally, means 'below the cartilages'.

hypodermic under the skin, as in a hypodermic injection or injection needle.

hypogastrium the lowest of the 3 central regions of the ABDOMEN. The central area below the navel.

hypoglossal nerves the 12th and last of the pairs of nerves which arise directly from the brain (cranial nerves). They supply the muscles of the tongue and are necessary for talking and swallowing.

hypoglycaemia abnormally low levels of sugar (glucose) in the blood. Hypoglycaemia is dangerous as the brain is critically dependent on glucose and is rapidly damaged if this fuel is absent. Hypoglycaemia causes trembling, faintness, sweating, palpitations, mental confusion, slurred speech, headache, loss of memory, double vision, fits, coma and death. Behaviour is often irrational and disorderly and may simulate drunkenness. The commonest cause is an overdose of insulin in a diabetic. Diabetics are advised always to carry sugar lumps or glucose sweets for use as an emergency treatment of hypoglycaemia.

hypokalaemia abnormally low levels of potassium in the blood.

hypomania a sustained, but mild or moderate, degree of abnormal elation and hyperactivity.

hyponatraemia abnormally low levels of sodium in the blood.

hypophoria a latent tendency to vertical squint in which a covered eye becomes oriented downwards but resumes straight binocular FIXATION when uncovered.

hypophysis the PITUITARY gland. From the Greek *hupophusis*, an attachment underneath.

hypopigmentation a lower than normal concentration of pigment cells (melanocytes).

hypopituitarism abnormal underproduction of pituitary hormones.

hypoplasia underdevelopment of a tissue or organ as a result of a failure of production of a sufficient number of cells. Compare HYPERPLASIA.

hypopnoea abnormally slow and shallow breathing. Compare HYPERPNOEA.

hypotension abnormally low blood pressure.

hypothalamus the region of the undersurface of the brain immediately above the pituitary gland. The hypothalamus is the area in which the nervous and hormonal systems of the body interact. It receives information relating to hormone levels, physical and mental stress, the emotions and the need for physical activity, and responds by prompting the pituitary appropriately. Nuclei in the lower region of the hypothalamus contain neurons that secrete growth hormone releasing hormone, gonadotropin releasing hormone, somatostatin and other regulating substances.

hypothenar eminence the minor muscle bulk on the palm of the hand on the little finger side opposite the ball of the thumb (the THENAR eminence).

hypothermia below-normal body temperature. This may occur, especially in the elderly, as a result of prolonged exposure to low temperatures or may be brought about deliberately to reduce tissue oxygen requirements during surgery.

hypothesis a tentative proposition used as a basis for reasoning or experimental research, by means of which it may be rejected or incorporated into accepted knowledge. See NULL HYPOTHESIS.

hypothesis test a statistical procedure to determine the probability (p) that the observed result of a test or trial could have been obtained if the NULL HYPOTHESIS were true.

hypothyroidism underactivity of the THYROID GLAND. See MYXOEDEMA.

hypotonia a condition in which the muscles offer reduced resistance to passive movement.

hypoxaemia deficiency of oxygen in the blood. Hypoxaemia often causes CYANOSIS.

hypoxia deficiency of oxygen in the tissues. Local hypoxia can lead to GANGRENE; general hypoxia to the death of the individual. Hypoxia occurs mainly as a result of obstructive artery disease, ANAEMIA, certain forms of poisoning and suffocation.

hysteria a disturbance of body function not caused by organic disease but resulting from psychological upset or need. The affected person is apparently unaware of the psychological origin of the disorder. The term 'hysteria' has become politically incorrect and the condition is now usually referred to as a conversion disorder.

Hz *abbrev. for* Hertz, the unit of frequency equal to one cycle per second. (Heinrich Rudolph Hertz, 1857–94, German physicist).

Ii

-ia *suffix denoting* a disease or pathological condition, as in pneumonia.

-iasis *suffix denoting* a pathological condition produced by a specified cause, as in candidiasis, or relating to or resembling, as in elephantiasis.

-iatric *suffix denoting* a specified form of medical treatment or care, as in psychiatric.

-iatrics *suffix denoting* a class of medical treatment or study, as in paediatrics.

iatrogenic pertaining to disease or disorder caused by doctors. The disorders may be unforeseeable and accidental, may be the result of unpredictable or unusual reactions, may be an inescapable consequence of necessary treatment, or may be due to medical incompetence or carelessness. *Iatros* is the Greek word for a doctor.

-iatry *combining form denoting* medical treatment, as in podiatry or psychiatry.

-ic *suffix denoting* pertaining to, or characterized by, as in rheumatic.

ichthyo- *combining form denoting* fish.

-ician *suffix denoting* someone who practices or is a specialist in, as in physician.

icterus an alternative term for JAUNDICE. The Roman author Pliny, the Elder, believed that jaundice could be cured by gazing on the small yellow bird, the oriole. *Icteros* is the Greek word for a yellow bird.

id a Freudian term for that primitive part of our nature concerned with the pursuit of mainly physical and sexual gratification and unmoved by considerations of reason, logic or humanity. The id manifests the forces of the libido and the death wish, but is said to be the source of much of our psychic energy. Freud's choice of the term may have been a little prudish in its lack of specificity; *id* is a Latin rendering of the Greek *es* meaning it.

ideation the formation of ideas, thought, the use of the intellect.

idée fixe a fixed idea or obsession, often delusional, and having a marked effect on behaviour.

identical twins twins derived from the same egg (ovum) which, after the first division, has separated into two individuals. Identical twins thus have the same genetics.

identity the awareness of one's own nature and personality, especially in relation to a social context. An identity crisis arises when one feels unable to 'identify' with one's social environment.

ideogram a graphics representation of the G banding pattern of a CHROMOSOME.

ideomotor mental processes that immediately result in movement.

idio- *combining form denoting* personal, individual, originating within oneself, without external cause.

idiocy the state of a person with a mental age of less than three.

idiopathic of unknown cause.

idiosyncrasy a physiological or mental peculiarity.

idiot a person of severe mental deficit, incapable of coherent speech or of normal response to danger.

idiot savant a person, severely backward in most mental respects, who shows great precocity and ability in one particular direction, such as mental arithmetical calculation or data recall. The idiot savant often has an IQ of less than 50 and the particular skill is seldom of practical use.

idiotypes the molecular structure of the variable region of an IMMUNOGLOBULIN molecule that confers its antigenic specificity.

Ig *abbrev. for* IMMUNOGLOBULIN.

IgA immunoglobulin class A, an antibody class concerned with protection against virus and other infections in the mucous membranes of the body, especially in the respiratory and digestive systems.

IgD immunoglobulin class D, an antibody class probably concerned with the regulation of B LYMPHOCYTES. IgD is found almost exclusively on the surface of B lymphocytes.

IgE immunoglobulin class E, an antibody class concerned with immediate hypersensitivity reactions, such as hay fever (ALLERGIC RHINITIS). IgE has an affinity for cell surfaces and is commonly found on MAST CELLS.

IgG immunoglobulin class G. This antibody accounts for three quarters of the immunoglobulins in the blood of healthy people. It is widely distributed in the tissues and is the only immunoglobulin class that passes through the placenta to the fetus. It is concerned with protection against a wide range of infecting organisms. Note that the term gamma globulin refers to the whole class of immunoglobulins, not simply to IgG.

IgM immunoglobulin class M, an antibody class concerned especially with the breakdown of foreign cells and with preparing foreign material for PHAGOCYTOSIS.

ileocaecal pertaining to the ILEUM and the CAECUM. The ileocaecal valve is an internal fold of lining mucous membrane that prevents reflux of bowel contents from the caecum into the ileum.

ileocaecal valve a smooth muscle ring (sphincter) at the point at which the small intestine ends and the large intestine begins. The caecum is the dilated first part of the colon. This valve exercises some control over the movement of intestinal contents from the part concerned mainly with digestion and absorption and the part concerned mainly with water retention.

ileum the third part of the small intestine lying between the JEJUNUM and the start of the large intestine, the CAECUM. The contents of the ileum are of the consistency of a watery mud and almost all nutrients have been absorbed by about the middle of the ileum.

ileus failure, usually temporary, of the process of PERISTALSIS by which the bowel contents are moved onwards.

iliac arteries the two large arteries into which the abdominal AORTA divides in the lower abdomen, and which supply blood to the pelvic region and the legs.

iliacus the part of the ILIOPSOAS muscle arising from the inside of the back wall of the pelvis (the iliac fossa and sacrum) and being inserted into the top of the front of the FEMUR. The iliacus helps to flex the hip joint in walking.

iliac veins the three large veins on each side of the body that drain the pelvis and legs and accompany the ILIAC ARTERIES.

iliofemoral ligament a strong fibrous band running from the front of the pelvis to the top of the thigh bone (femur).

ilioinguinal pertaining to the iliac region and the groin.

iliopsoas a large muscle group arising from the inside of the back wall of the pelvis and lower abdomen and consisting of the ILIACUS and PSOAS major and minor muscles. These muscles are inserted into the front of the top of the thigh bone (femur) and act to bend the hip joint or to flex the trunk on the thigh as in sitting up from lying.

iliotibial tract a thickening in the sheet of tough fibrous tissue (fascia lata) that covers the outer side of the thigh. The iliotibial tract extends upwards from the prominence on the upper part of the main lower leg bone (tibia) (the lateral condyle) to the crest of the pelvis (the iliac crest).

ilium the uppermost of three bones into which the INNOMINATE bone of the pelvis is arbitrarily divided. Its most conspicuous feature is the iliac crest forming the upper brim of the pelvis on either side.

illusion a false sense perception from misinterpretation of stimuli. Most illusions are normal and harmless, but some are features of psychiatric conditions, especially depression. Compare DELUSION and HALLUCINATION.

imbecile an obsolete term for a mentally retarded person or, nowadays, a person with severe learning difficulty.

imbibition the taking up of fluid as by absorption into a gel.

immobilization the avoidance of movement of an injured or diseased part, especially a bone fracture, so that healing may take place. Effective fracture immobilization demands that the joint above and below the fracture should be unable to flex. Immobilization is achieved by means of slings, splints, plaster of Paris casts, cold-setting plastic casts and external steel bar fixators of variable design.

immortalization a change in a eukaryotic cell line that confers the ability to go on dividing and reproducing indefinitely. Immortalization implies the ability to continue to reform TELOMERES.

immune complexes combinations of ANTIGEN and ANTIBODY and sometimes COMPLEMENT proteins. These complexes may be deposited in the walls of small blood vessels where they react with complement, MAST CELLS, other white cells or PLATELETS to trigger local inflammation.

immune response genes genes, including all those within the MAJOR HISTOCOMPATIBILITY COMPLEX (MHC) that determine the total response to a given antigen.

immune surveillance a body mechanism by which early cancers are detected as being foreign and are attacked and usually destroyed. Immune surveillance is a T cell-mediated process without which cancer would be much commoner. See also CTLA4-IG.

immunity

The term 'immunity' has an absolute ring about it that is far from justified in the context of the human body. It is clear that humans are not immune from infection. We suffer respiratory infections such as colds, influenza, tonsillitis and pneumonia; skin infections such as herpes, impetigo and boils; intestinal infections such as dysentery, typhoid and salmonellosis; urinary infection; sexually transmitted infections, and so on. So the term has to be understood as implying only relative protection. In only a few instances is there an approximation to absolute immunity. Ironically, the term arose from such a case – the almost complete immunity from smallpox that followed a single attack or that followed vaccination with a similar organism. It was this unusual characteristic that made possible the complete eradication of smallpox.

But for the operation of an elaborate system of defences, the interior of the living human body would come close to being an ideal culture medium for infecting organisms. Organisms that cause human diseases have, by evolution, adapted almost perfectly to the conditions obtaining within the body – the temperature, supply of nutrients and other chemicals, the moistness and the darkness. When the body is no longer living, its suitability as a culture medium for organisms soon becomes only too apparent. The differences, in the response to infection, between the living and the dead body is a striking demonstration that a dynamic mechanism is operating in the former but not in the latter. This mechanism is known as the immune system.

In general, the body's ability to resist infection is determined, on the one hand, by the size of the dose of organisms to which it is exposed and by their virulence, and on the other, by the effectiveness of the body's resistance to attack. The outcome will depend on the relative strength of the opposing forces. In a healthy person, a minor assault by a small dose of organisms of low virulence is easily repulsed, while a large dose of highly virulent organisms might be overwhelming. In a person with impairment of the immune system – as from AIDS or other cause of immune deficiency – a small dose of an organisms of low virulence may lead to infection. In severe immune deficiency, the body becomes vulnerable to organisms that do not normally cause infection at all.

The nature of the defence

A primary requirement of an active defence system – a system that defends by attack – is that it should be able to distinguish friend from foe. As the immune system consists of no more than a large collection of separate and often free-moving cells of different kinds, the use of the term 'recognition' in this context is somewhat metaphorical. Immunological recognition is a chemical process concerned with the relationship of the molecular shape of receptors on the surface of the defending cells to molecular structures on the surface of invading organisms and other foreign material. This relationship becomes effective when the defender and the invader come into contact so that the fit between the surface molecules can be tried. The characteristically-shaped molecules on the surface of invaders are called antigens. For historic reasons, arising from less complete understanding than we have today, the term antigen is still often applied to the whole invader. Immunological recognition displays remarkable specificity. The immune response can, for instance, distinguish very small differences between antigens on closely similar strains of influenza virus. Such strains also have antigens in common. The immune system also displays memory. Once contact with a particular invader has been made, the response to subsequent contact with the same invader will be stronger, more rapid and more effective. Amplification of the response is said to have occurred.

The class of cells responsible for the final destruction of invaders, whether organisms or other foreign material such as a donated organ, are called phagocytes. *Phago-* means 'eater' and *cyte* means cell. There are two main varieties, the macrophages ('big eaters') and the microphages (small eaters, or polymorphs). Macrophages are large cells that live for years in various tissues of the body and accumulate gradually at sites of infection to which they are attracted by substances released when body tissues are injured. They can engulf invaders, flowing around them by amoebic action, and contain protein-splitting and other enzymes which they use to destroy them. Polymorphs are smaller, short-lived cells, attracted to the site of infection in enormous numbers. They, too, actively engulf bacteria and digest them. In the process, polymorphs are often killed. Pus consists largely of polymorphs, killed by bacterial toxins in the course of duty.

Phagocytes have limited inherent power of recognition of invaders but recognition is greatly enhanced when certain substances produced by the body have become attached to the surfaces of the invaders. These substances are called antibodies.

Antibodies

Apart from the phagocytes, the most important cells of the immune system are the lymphocytes. These small round cells are found both free in the blood circulation and tissue fluids and also packed tightly in millions in the lymph nodes found in the neck, armpits, groins and around the main blood vessels of the chest and abdomen. Lymphocytes also occur in enormous numbers in the spleen. There are two large classes of lymphocytes, the B lymphocytes (B cells) and the T lymphocytes (T cells).

The B cells produce antibodies, the T cells help them to do so but also directly attack certain invaders, such as viruses. Antibodies, or immunoglobulins, are proteins which attach to invading organisms or foreign substances and neutralize them so that they can be recognized and destroyed by phagocytes. Antibody production is a remarkable process. The B cells exist in millions of different genetically determined types. When an invader enters the body, helper T cells identify its antigens and select, from the range of B cells, the one or two capable of producing antibodies that will best fit the invader. Once these B cells have been selected they start to divide rapidly and produce large collections of identical cells (clones) called plasma cells which are the actual antibody factories. Plasma cells pour out antibodies at the rate of many thousands per second. B cells also produce clonal memory cells. When subsequent invasion by the same organism or material occurs, the response is much more rapid because the memory cells can immediately clone plasma cells to produce large quantities of the appropriate antibodies.

Active and passive immunity

There are five classes of immunoglobulins (Ig) – IgG, IgA, IgM, IgI and IgE. Each class has a particular general function and each may consist of any of thousands of different specific antibodies. The most plentiful immunoglobulin is IgG (gamma globulin) which is the class involved in defence against most bacterial infections and against many viruses and toxins. Human gamma globulin and other immunoglobulins, from pooled human plasma can, of course, be used therapeutically to help people cope with infections. This is called passive immunity and the method is valuable in providing protection against many infections such as hepatitis, chickenpox, measles, rubella, tetanus and poliomyelitis. It is especially useful for people with an inherent or acquired immune deficiency. One form of such deficiency is called agammaglobulinaemia – a congenital absence of gamma globulin.

A more generally useful and widely used method of conferring relative immunity is known as active immunization. To achieve this, organisms are either killed or so modified as to be harmless while still carrying the specific antigens that provoke an immune response. If such organisms are suspended in an injectable fluid and introduced into the body, the T and B cell responses result in the production of large quantities of immunoglobulins as well as memory cells. Such a response may be effective for a life-time, but more often, 'booster' doses are required, from time to time, to maintain the antibody levels (titres).

Immunization

The terms 'immunization' and 'vaccination' are interchangeable. The latter term is a historic curiosity arising from the central part the cow (Latin *vacca* a cow) played in the origins of immunization (Jenner's use of cowpox to protect against smallpox). Immunization against infectious disease has been one of the most successful enterprises in medicine and has been responsible for an enormous reduction in the incidence of human suffering and tragedy. In countries where acceptance of immunization is high, crippling and fatal diseases like poliomyelitis and diphtheria have been virtually eliminated and the incidence of many other less serious conditions greatly reduced. Immunization has been so successful that generations of parents have grown up who have never known the terrors of some of these diseases, and there is real concern among Public Health authorities that communities that fail to remember their histories may find them repeated.

Phagocytes

Phagocytes, large and small, are the ultimate scavengers of the body, engaged on an unremitting clean-up operation. Much of their work is concerned with picking up and digesting the bodies of foreign invaders – bacteria, viruses and other unwanted matter. But the macrophages are also concerned with the scavenging of some of the body's own cells that have become dangerous and have had to be destroyed. The task of recognizing cells that must be destroyed, whether foreign invaders or infected or cancerous body cells, is the principal function of the immune system. To understand how this is done it is necessary to look a little more deeply into this most complex of all the body systems.

The end-product of the activity of the immune system is the production of specific antibodies which latch on to invaders, such as bacteria, and inactivate them so that they can be destroyed by the phagocytes. Lymphocytes all look very much alike – small round cells with large, densely-staining nuclei – but in fact they fall into a number of radically differing groups. The production of antibodies is the function of one of the two classes of lymphocytes – the B cells. The other class is known as the T cells. In this usage, the term 'cell' is synonymous with 'lymphocyte'.

The T cells and their interactions

The T cells – so called because their early processing occurs in the thymus gland in the upper chest – fall into a number of distinct classes, the most important of which are the killer cells and the helper cells. Killer cells, often called cytotoxic cells, are aggressive lymphocytes that roam the body searching for, and destroying, cells bearing 'suicide notes'. These cells, asking to be killed, are cells that have been invaded by viruses or that have developed cancerous tendencies. And the suicide notes are short lengths of linked amino acids called peptides. Peptides can be thought of as segments of proteins (which are constructed from linked polypeptides). Proteins within the cell, such as the proteins of invading viruses, are constantly being chopped up into peptides and these peptides are transported to the cell surface and displayed, like recognizable flags, on the outside. Such peptides are antigens and occur in an enormous variety, depending on what has gone wrong in the cell.

Killer and helper T cells

Killer T cells arise from an unactivated form of T lymphocyte, the prekiller. These carry antigen receptors on their surfaces that are able to 'lock on' to the peptides. When this happens the now activated

T cell, able to recognize the specific peptide displayed by the damaged body cell, divides repeatedly. All the daughter cells (the members of the same clone) are now active killers able to deal with other cells that are putting out the same flags. Killer T cells destroy abnormal cells with enzymes.

The helper T cells also carry receptors for peptide flags, but they have a different, but equally important function. Again, the unactivated T cell – the pre-helper cell – has to be converted, this time into a helper cell. This occurs when macrophages, which have taken up antigens, especially bacterial toxins, put out specific flags on their surfaces. The unactivated T cells lock on to these and are converted to helper cells. These then divide to form large families of identical helper cells (clones) which select and latch on to the correct B cells. The helper cells then secrete powerful stimulating substances called interleukins which prompt the B cells to multiply and clone plasma cells that secrete the specific antibodies to tackle the antigens.

Diversity of T cell receptors

The antigen receptors on the surfaces of the T cells have two functions – to recognize 'self' and to identify the peptide 'flags' put out by abnormal cells. Both types are encoded for by genes which we inherit from our parents, but although the self-recognizing receptors come from only two genes in each individual, the genes for the variable part are inherited as many small and separate segments of DNA that undergo random recombination in the developing lymphocytes. The result is an enormous collection of genes producing an equal number of different T cell receptors. Many of these are exactly what are needed and do the job perfectly, binding to non-self peptides presented by own-body cells. Some, unfortunately, are of a type that can lock on to self-peptides, and lymphocytes that can do this would clone gangs of cells that would attack normal body cells.

The explanation of why this does not normally happen is that clones of T cells capable of attacking the body are automatically removed by the thymus early in life. All the antigens that might later be encountered on body cells are present in the thymus. T cells that do not have the receptors to bind to any sites in the thymus are useless and die in about three days. T cells that are found capable of attacking normal cells are destroyed. Only those that can form useful killer and helper lymphocytes are allowed to proceed into the blood circulation to perform their function. The T cells are also involved in the processes that lead to allergies. Children exposed early in life to certain environments normally thought to be unhygienic – such as cattle sheds and other farm situations are much less likely to develop asthma and other allergic disorders than children who have been reared in a clean environment. This occasions no surprise to those whose thinking takes into account evolutionary and historical factors.

How the immune system goes wrong

Many different diseases are caused by the phenomenon known as autoimmunity in which cells of the immune system appear to regard certain body cells as foreign, and attack and destroy them. Such autoimmune diseases include diabetes, myasthenia gravis, dermatomyositis, lupus erythematosus, scleroderma, Behçet's syndrome, certain thyroid diseases, rheumatoid arthritis, and possibly multiple sclerosis. The precise way in which autoimmunity occurs has been unclear for many years, but it now seems likely that a factor is the survival of the dangerous T cells that are normally destroyed by the thymus. This probably occurs because of the absence from the thymus of the particular antigens present on the cells attacked in autoimmune disease. Some experimental work has shown that foreign tissue implanted into the thymus can induce the production of lymphocytes that regard such tissue as 'self'. Grafts of the same foreign tissue have been permanently accepted.

Another form of immune disorder occurs when the body is infected by organisms whose antigens closely resemble those of certain body tissues. There is a streptococcus that commonly causes severe throat infections. This organism carries antigens closely similar to those occurring on the heart lining, the lining of the joints and in the kidneys. Infection by this organism promotes the production of antibodies in the normal way, but these antibodies not only attack the organisms but also lock on to the normal peptides in the heart, brain, joints and kidneys. The result may be rheumatic fever, rheumatic heart disease, St. Vitus' dance (rheumatic chorea) or glomerulonephritis.

Immune deficiency

This has been known for many years – long before the AIDS epidemic. Several forms are present from birth and some of these are due to a genetic inability to manufacture T cells. In other less severe forms, the genes for the globulin protein from which antibodies are made are missing or defective. One such is the disorder hypogammaglobulinaemia, in which there is a severe deficiency of gamma globulin (IgG). Failure of normal development of the thymus can cause severe early immune deficiency, as in the DiGeorge syndrome. Infants affected in this way show many of the features of AIDS – including widespread thrush infections, *Pneumocystis carinii* pneumonia and chronic diarrhoea. Vaccination with live vaccines is

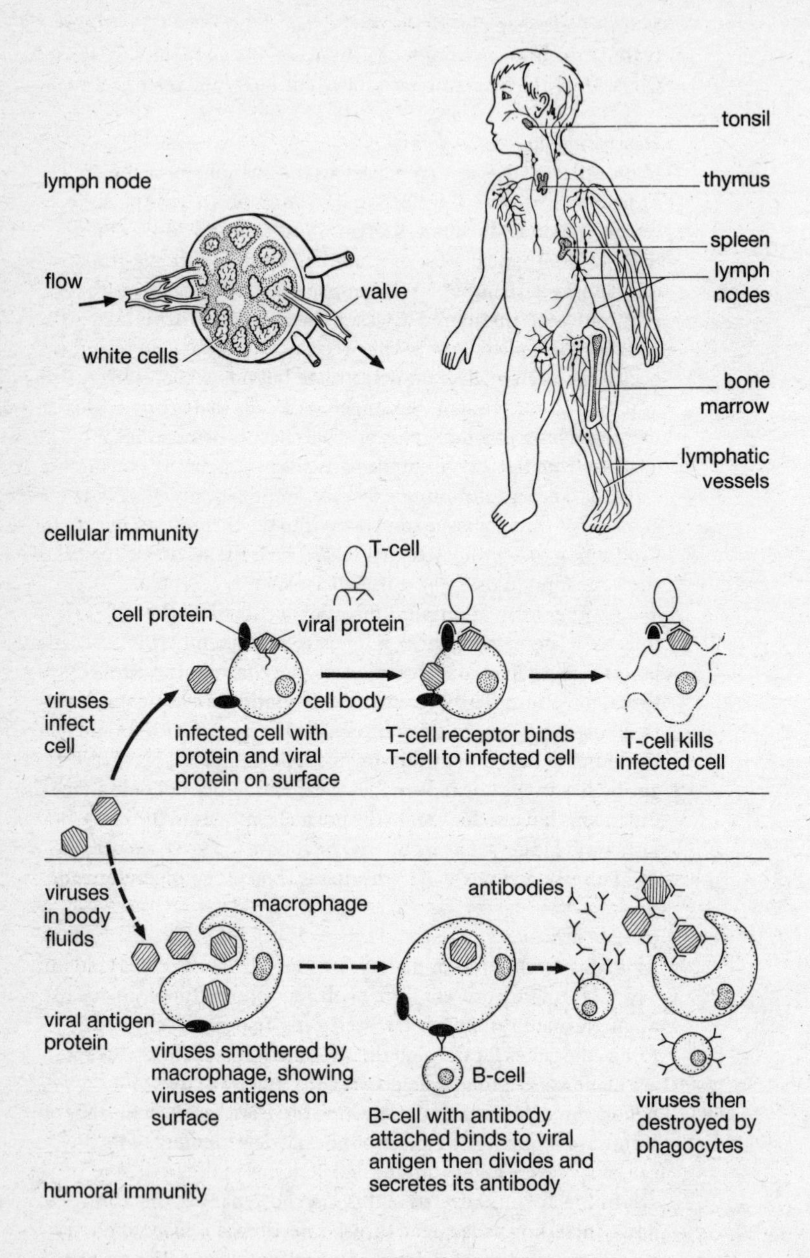

Fig. 8 **Immunity**

usually fatal in these children. Various other specific T cell disorders can occur, the best known being the acquired immunodeficiency syndrome (AIDS).

AIDS is caused by a specific virus, the human immunodeficiency virus (HIV), spread mainly by sexual contact. This is an RNA retrovirus that carries an enzyme, reverse transcriptase, that enables it to make DNA copies of itself in the helper T cells it invades. The virus binds to the surface of the T cell and injects its genetic material. This replicates rapidly in the T cell to form thousands of copies of the virus which then burst out of the cell and spread to other helper cells. The result is a profound deficiency of helper T cells. Some shortage of killer cells also occurs with reduced production of interleukin. B cells are not affected and some antibody production occurs although the only value of this appears to be to enable tests for HIV infection to be performed. The antibodies do not have any significant effect on the virus.

immuno- *combining form denoting* immune or immunity.

immunochemistry the science of the chemical processes underlying the antigen-antibody reaction and of the other chemical reactions involved in the operation of the immune system.

immunocompromised in a condition of diminished ability to resist infection, to reject foreign material gaining access to the body, and to react aggressively at a cellular level to early cancerous change.

immunodeficiency a state in which the body's immunological system of defence against infection, foreign material and some forms of cancer, is defective. Immunodeficiency may be of genetic origin, may be IATROGENIC or may be caused by infection with the human immunodeficiency virus (HIV) that has led to the development of AIDS.

immunofluorescence the detection and identification of antigenic material by observing, under the microscope, the fluorescence of known, specific, fluorescein-linked (conjugated) antibodies that have become attached to it.

immunogen any substance that can elicit an immune response. An antigen.

immunoglobulins ANTIBODIES, protective proteins produced by cloned B lymphocyte-derived plasma cells. There are five classes of immunoglobulins, the most prevalent being immunoglobulin G (IgG), or gammaglobulin which provides the body's main defence against bacteria, viruses and toxins. Immunoglobulins are Y-shaped protein molecules consisting of two inner heavy POLYPEPTIDE chains, forming a Y, and, attached to the outer side of the short arms of the Y, two light polypeptide chains. The heavy chains are held together and the light chains held to the arms, by disulphide bonds. The short arms of the Y, with the light chains, are called the Fab (fragment antigen binding) section of the antibody. The antigen combining site lies between the open ends of the light and heavy chains. See also IgA, IgD, IgE, IgG and IgM. Some immunoglobulins are prepared as drugs for the management of RHESUS incompatibility and antibody deficiency; for the prophylaxis of MEASLES and hepatitis A; to minimize fetal damage when a pregnant woman is exposed to RUBELLA; and to assist in the treatment of TETANUS and RABIES.

immunology the science and study of the many complex cellular and biochemical interactions involved in the functioning of the immune defences of the body and of the mechanisms that allow the body to distinguish 'self' from 'non-self'.

immunopathology the study of the role of immunological processes in the production of disease and in its diagnosis and treatment.

immunoprophylaxis prevention of disease by the use of vaccines.

impaction 1 the condition of being forced into and retained in any part of the body.
2 the situation in which the ends of a fractured bone are firmly driven into each other so that movement at the fracture site does not occur.
3 retention of an unerupted tooth in the jaw by obstruction by another tooth, especially a molar, so that its normal appearance is prevented.

impalpable unable to be felt.

imperforate having no opening. Used of a structure normally having an opening, as of an imperforate hymen or anus.

implantation 1 partial penetration of, and attachment to, the lining of the womb by the fertilized egg.
2 the introduction into the body of a donated or transferred tissue or organ or a prosthetic part, such as an intraocular lens.

impotence the inability to achieve or sustain a sufficiently firm penile erection (tumescence) to allow normal vaginal sexual intercourse. The great majority of cases are not caused by organic disease and most men experience occasional periods of impotence. It is often related to anxiety about performance and is usually readily corrected by simple counselling methods which prescribe sensual massage but forbid coitus. Organic impotence may be caused by some diseases including diabetes, multiple sclerosis, spinal cord disorders and heart disease.

impregnation the act or process of making pregnant, fertilizing or inseminating.

impression in dentistry, a negative mould of the teeth or other mouth structures, made in plastic, which is later filled with Plaster of Paris to provide a perfect copy of the anatomy.

imprinting 1 the rapid early development in young animals of recognition of the ability to recognize and to be attracted to others of their own species or to similar surrogates.
2 in genetics, changes that occur in a gene in passing through the egg or the sperm so that maternal and paternal alleles differ at the start of embryonic life.

in- *prefix denoting* not or in, into, within.

inanition a state of exhaustion or a bodily disorder arising from lack of any of the nutritional elements such as calories, protein, vitamins, minerals or water.

inborn of qualities or characteristics that are genetically determined rather than being acquired after conception.

inbreeding mating of the closely related. Inbreeding tends to promote similarities and to deny access to new genes. It increases the chances of offspring being HOMOZYGOUS for RECESSIVE genes and thus manifesting the effect but, in itself, has no inherent tendency to produce bad characteristics. Its reputation to do so arises from the observation of the undesirable effects of inbreeding in genetically disfavoured people.

incest sexual intercourse between close blood relatives, especially between brothers and sisters, fathers and daughters, or mothers and sons. The 'prohibited degrees' vary in extent in different legal systems. There is a strong social taboo against incest now thought to be based on social and psychological, rather than genetic, factors.

incestuous pertaining to INCEST.

incidence the number of cases of an event, such as a disease, occurring in a particular population during a given period. Incidence is usually expressed as so many cases per 1000, or per 100,000, per year. Compare PREVALENCE.

incisor one of the four central teeth of each jaw, with cutting edges for biting pieces off food. The incisors are situated immediately in front of the canine teeth.

incisura an indentation or notch in a structure.

inclusion bodies microscopically visible masses of virus material, or areas of altered staining behaviour, seen within cells in a number of virus infections such as RABIES, herpes infections, papovavirus infections and adenovirus infections.

incomplete abortion loss of part of the products of conception, usually the embryo or fetus, with retention in the womb of the placenta or some of the membranes. See also ABORTION.

incomplete dominance failure of one or other of two ALLELES to exert a dominant effect with the result that the PHENOTYPE has a form somewhere in between those of the two phenotypes that would be produced were either gene homozygous.

incontinence loss of voluntary control of one or both of the excretory functions.

incoordination the inability to accurately time and phase the various components of movement so that these tend to be effected separately causing clumsiness and lack of smoothness in MOTOR activity.

incubation period the interval between the time of infection and the first appearance of symptoms of the resulting disease. Incubation periods vary widely, from as little as a few hours in the case of CHOLERA to many weeks in some cases of RABIES.

incurable not able to be remedied by currently available medical means. The progress of medical science in the 20th century repeatedly showed that what is incurable today is often remediable tomorrow.

incus the middle of the three tiny bones (auditory ossicles) that form a chain across the middle ear linking the eardrum to the inner ear. The incus is anvil shaped.

index finger the finger adjacent to the thumb.

index of refraction a measure of the optical density of a transparent material such as glass or the cornea. It is the ratio of the speed of light through the material to its speed in a vacuum.

indicator a substance that undergoes an observable change, usually a change of colour, when a chemical alteration occurs in its environment. Indicators may demonstrate changes in acidity, the presence of various substances, such as sugar or protein in body fluids, or alterations in the concentrations of substances. Indicators are widely used in chemistry and in clinical medicine.

Indo-European language family a group of languages spoken by half the world's population and on every continent. The family includes many branches. The Germanic branch includes English, German, Dutch and the Scandinavian languages; the Italic or Romantic branch includes the Latin derivatives Italian, French, Spanish, Portuguese and Romanian; the Celtic branch includes Irish and Scottish Gaelic, Welsh and Breton; the Indo-Iranian branch includes Hindi, Bengali, Persian; the Baltic branch includes Lithuanian and Latvian; and the Slavic branch includes Russian, Polish, Czech and Bulgarian.

indole 2,3-benzopyrrole, an unpleasant-smelling product of protein breakdown that contributes to the odour of the faeces. In high dilution, indole has a pleasant smell and has been used in the perfumery industry. See also SKATOLE.

indolent of slow progression or taking a long time to heal. Causing little or no pain. Often used of skin ulcers.

inducer a molecule that causes a gene to be expressed by binding to a repressor protein so as to prevent it from acting to prevent expression.

induction the reasoning process by which conclusions are drawn from premises derived mainly from experience and observation. Induction often involves generalization and there is always the risk of error in arguing from the particular to the general (for example, 'all swans are white'). Contrast with DEDUCTION, in which conclusions follow logically from premises.

induction of labour the artificial initiation of the processes of birth, usually at a late stage in the pregnancy, before this has occurred spontaneously. This is done in cases in which the continuation of the pregnancy would be dangerous either to the mother or the baby.

induration 1 abnormal hardness of tissue as a result of a disease process or injury. 2 hardening of tissue.

industrial diseases diseases specifically caused by the effects of work processes or working conditions on health or by exposure to substances involved in work processes.

inebriety drunkenness.

inevitable abortion a separation of the placental attachment to the wall of the womb that has proceeded to such an extent that the death of the embryo or fetus cannot be prevented.

in extremis *adv.* near to death, in extreme danger of dying.

infantilism persistence of child-like characteristics of body and mind into adult life. Arrested development in an adult.

infant mortality the number of infants per 1000 live births who die before reaching the age of 1 year. Infant mortality is a sensitive index of the standards of public health in a society. The rate in Britain was about 150 in 1900. Today, in the best regions, it is as low as 8.

infection **1** the process by which organisms capable of causing disease gain entry to the body and establish colonies.
2 the state of injury or damage to part of the body resulting from this process.

infectious diseases diseases caused by organisms that can spread directly from person to person. Diseases requiring a transmission agent (vector) such as malaria, yellow fever and leishmaniasis, are usually excluded from this group. Common infectious diseases are chickenpox, diphtheria, food poisoning, gastroenteritis, glandular fever, hepatitis, influenza, measles, meningitis, mumps, rubella, tuberculosis and the sexually transmitted diseases.

infective capable of causing INFECTION.

infecundity infertility. Inability to bear children.

inferior situated below. An anatomical term referring to relationships in the upright body. The heart is inferior to the head, but no value judgement is implied. From the Latin *inferus*, below.

inferiority complex a concept of the Austrian psychiatrist Alfred Adler (1870–1937) indicating a general sense of unworthiness resulting from repressed perception of one's bodily defects. In popular usage the term simply implies a generally self-critical attitude or a boastful, self-exalting manner that compensates for feelings of inferiority.

inferior vena cava the largest vein in the body, carrying blood from the lower half of the body to the right atrium of the heart. See also SUPERIOR VENA CAVA.

infero- *combining form denoting* anatomically INFERIOR.

infertility the apparent inability of a particular couple to reproduce. The problem may rest either with the female or with the male or, rarely, with both.

infibulation female CIRCUMCISION with stitching together of the LABIA MAJORA so as to prevent sexual intercourse.

infiltration the movement into, or accumulation within, a tissue or organ, of cells or material not normally found therein. Cellular infiltration, as with LYMPHOCYTES, is often part of an immunological response.

inflammation the response of living tissue to injury, featuring widening of blood vessels, with redness, heat, swelling and pain – the cardinal signs 'rubor', 'calor', 'tumor' and 'dolor' of the first century physician Celsus. Inflammation also involves loss of function and is the commonest of all the disease processes. It is expressed by the ending '-itis'.

infra- *prefix denoting* below, beneath, INFERIOR to.

infraclavicular under the collar-bone.

inframandibular below the lower jaw.

infra–occlusion failure of a tooth to erupt fully so that the biting or grinding surface is unable to contact that of its fellow in the other jaw. Also known as infraclusion.

infraorbital lying below the ORBIT.

infrapatellar below the knee cap.

infra-red pertaining to electromagnetic radiation of wavelengths between 780 nanometers (nm) and 1 mm, being greater than those of visible light but shorter than those of microwaves. Heat radiation.

infrascapular below the shoulder blade.

infraspinatus muscle a muscle that runs from the back surface of the shoulder blade (SCAPULA) to the back of the upper part of the upper arm bone (HUMERUS). Its action is to rotate the arm outwards.

infraspinous situated below any spine or spinous process.

infrasplenic below the SPLEEN.

infrasternal below or deep to the breastbone.

infratemporal below the temporal region of the skull. Below the TEMPORAL FOSSA.

infratrochlear situated below the pulley for the tendon of the superior oblique eye-moving muscle.

infraumbilical below the navel.

infundibulum any funnel shaped bodily passages or structure, such as the stalk of the PITUITARY GLAND.

ingestion the process of taking food or other material into the stomach. Ingestion is followed by DIGESTION, ABSORPTION and, finally, ASSIMILATION.

inguinal pertaining to the groin.

inguinal ligament a slightly downward-sloping and downward-curving ligament that runs from a bony spine on the upper and outer front edge of the pelvis (anterior superior iliac spine) to the PUBIS. The ligament forms part of the tendinous attachment (aponeurosis) of one of the main abdominal muscles, the external oblique muscle. Also known as Poupart's ligament. (Francois Poupart, 1661–1709, French surgeon and naturalist).

inherent of a quality or part, existing naturally or intrinsically.

inheritance 1 the acquisition of a particular set of genes (GENOME) from the entire series of a person's forebears, by way of an equal number of genes from each parent.
2 the characteristics transmitted in this way.

inhibin a hormone that inhibits secretion of the FOLLICLE-STIMULATING HORMONE (FSH) by the pituitary gland. It is produced by the Sertoli cells of the seminiferous tubules in the male and by the granulosa cells of the ovaries in females.

inhibition arrest or limitation of a function or activity.

inhibitory synapse a synapse which passes an inhibitory signal to its post-synaptic neuron or neurons causing it or them to be less likely to have an action potential or to have reduced frequency of action potentials.

injury any permanent or semi-permanent disturbance of structure or function of any part of the body caused by an external agency. Such agency may be mechanical, thermal, chemical, electrical or radiational. The term may also be applied to damage caused by infecting organisms or to psychological trauma.

inlay dental restorative material that is cemented into a prepared cavity in a tooth.

innervation 1 the supply or distribution of nerve fibres to any part of the body.
2 the provision of nerve stimuli to a muscle, gland or other nerve.

innocent INNOCUOUS, non-malignant, BENIGN.

innocuous having no ill effect, harmless.

innominate artery a major, unpaired, artery that arises from the arch of the aorta, towards the right side of the body, and immediately divides into the right subclavian and right carotid arteries.

innominate bone the bone that forms each side of the pelvis. Each innominate bone is attached, behind, to the sides of the SACRUM and, in front, to the other innominate at the pubic junction (symphysis pubis). The innominate is nominally divided, for convenience, into the ilium, above, the ischium, below, and the pubis, in front. Bones were commonly named because of their resemblance to other things. The innominate bone bears little resemblance to any other shape, hence the term which means nameless.

innominate vein one of a pair of veins which drain the head and upper chest by way of the JUGULAR and SUBCLAVIAN veins. The two innominate veins join to form the SUPERIOR VENA CAVA.

inoculation immunization or vaccination. The procedure by which the immune system is stimulated into producing protective antibodies (IMMUNOGLOBULINS) to specific infective agents, such as viruses and bacteria by the introduction into the body of safe forms of the organism or of its ANTIGENIC elements.

inoperable referring to the stage in a disease, normally treated by surgery, beyond which surgery is not feasible or useful. The term is commonly applied to cancer that has spread widely. Many conditions once universally considered inoperable are now treated by surgery.

inorganic of chemical compounds, not having the structure of, or derived from, compounds found in living organisms. Not containing carbon.

inosculate to join by small openings.

inotropic influencing the force or speed of muscular CONTRACTILITY. Inotropic agents are used to improve the output of the heart in the treatment of heart failure and sometimes in acute circulatory failure (shock).

insanitary unhygienic.

insanity a legal rather than a medical term, implying a disorder of the mind of such degree as to interfere with a person's ability to be legally responsible for his or her actions. The term is little used in medicine but might equate to PSYCHOSIS.

insecurity the sense of concern and anxiety caused by uncertainty over any aspect of living, whether physical, social, spiritual or financial. Insecurity is believed by some psychiatrists to be a major cause of neurotic disorder, but feelings of insecurity are also often responsible for valuable creative achievement.

inseminate to introduce semen into the genital passage of a female whether by coitus or otherwise. Potentially to impregnate.

insensate lacking sensation, feeling or sensibility.

insensible lacking the power of feeling. Unconscious.

insensible perspiration movement of water through the skin by sweat glands at a rate below that of which the subject is aware.

insensible water loss loss of water by evaporation from the skin of perspiration so slight that the subject is unaware of it.

insertion mutation a mutation caused by the insertion into a DNA sequence of one or more nucleotides.

insidious of disease, occurring or progressing in an imperceptible manner so as to reach a harmful stage before being suspected.

insight 1 ability to appreciate the real nature of a situation.

2 awareness of the nature of one's own psychiatric symptoms with some appreciation of the possible causes or precipitating factors. People suffering from neurotic illnesses usually have considerable insight; those with psychotic disorders are often, by definition, deemed to be lacking in insight.

in situ in a normal position. The term is used also of cancer that remains at the site of origin and has not yet spread locally or remotely.

insomnia difficulty in falling asleep or in remaining asleep for an acceptable period. Insomnia is very common and is often caused by worry, tension, depression, pain or old age. Sleep requirements vary widely from person to person and those who sleep for apparently short periods seldom, if ever, suffer any harmful effects.

insomniac a person habitually suffering INSOMNIA.

inspiration the process of breathing in. Inhaling.

instincts complex, unlearned, inherited fixed action patterns or stereotyped behaviour shown by all members of a species. Instinctive responses are essential for survival and the physical basis for these patterns is 'hard-wired' into the brain. Much of social activity consists in the complex interplay of instinctive responses and education.

insulin a peptide hormone produced in the beta cells of the Islets of Langerhans in the PANCREAS. Insulin facilitates and accelerates the movement of glucose and amino acids across cell membranes, especially muscle cells. It also controls the activity of certain enzymes within the cells concerned with carbohydrate, fat and protein metabolism. Insulin production is regulated by constant monitoring of the blood glucose levels by the beta cells. Deficiency of insulin causes diabetes.

insulin resistance a state in which normal levels of insulin in the blood fail to produce the normal biological response. A feedback mechanism results in higher than normal levels of insulin and the blood sugar levels may be normal or raised. Insulin resistance is commonly associated with type II diabetes (non-insulin dependent diabetes), obesity and essential HYPERTENSION.

insulin shock HYPOGLYCAEMIA resulting from excessive insulin in the blood.

insult any injury, trauma, poisoning or irritation to the body.

integrator in neurology, a nerve cell with input from a number of dendrites that emits a signal that is the combined effect of the inputs.

integrins a family of linked polypeptide chains, alpha and beta, that mediate adhesions and other interactions between cells and the extracellular matrix and between cells and other cells. Integrins are expressed on endothelial cells, leukocytes, other cells and platelets, and act as receptors for fibrinogen, fibronectin, thrombospondin, von Willebrand factor, and vitronectin. See also DISINTEGRINS.

integument any outer covering, such as the skin or the outer membrane layer of an organ or the capsule of an organism or spore. When the term is used without qualification, the skin is implied.

intellection the act or process of performing a mental act.

intellectualization a DEFENCE MECHANISM in which a personal problem is analysed in purely intellectual terms, the emotional aspects being deliberately excluded.

intelligence

Most people are fairly sure that they know what intelligence is. But what they really mean is that they are able to recognize intelligence when they come across it – usually by comparing it with their own. Intelligence manifests itself by a general effectiveness in putting two and two together and coming up with four, in grasping the essence of difficult ideas, in learning and applying a new subject and in accumulating mental data. But to describe the manifestations of a thing is not to define it and there is still no general consensus of agreement among psychologists as to the definition of intelligence.

When people first began seriously to consider the nature of intelligence it seemed obvious that it was a matter of inherent brain power. Different people inherited brains of different quality and that was that. Intelligence obviously ran in families. Every now and then an exceptionally powerful brain turned up and the possessor became a genius. Sometimes people were born with very low-grade brains and these unfortunates were mentally deficient.

Further consideration, however, showed that this scheme was inadequate. The idea of innate brain power, independent of educational and environmental factors, did not fit with the observed facts. A mass of evidence showed that the development of intelligence depended largely on the input to the brain after birth. Children of highly intelligent parents only became intelligent if their total environment was conducive to the development of intelligence. And the kind of intelligence that developed was very much determined by the environment – meaning, in general, the type of education. It seems clear that two factors are necessary for the development of intelligence – the inheritance of good 'hardware' and the subsequent provision of plenty of good data.

Does inherent brain power exist?

The early idea that intelligence was a unitary power of the mind, independent of the mind's separate abilities, has also had to yield to more careful scrutiny. Numerous attempts to find this inherent power, so that raw intelligence can be defined, have failed. It seems that intelligence, on close inspection, resolves into a large complex of different abilities or skills that are present to different degrees. None of these – not even reasoning power, which, on examination resolves itself into the ability to perform one or other special skills – can be unequivocally selected as a central, innate entity we could call intelligence.

The list of mental skills is a long one and includes such things as verbal comprehension, word fluency, numerical ability, the ability to detect significant associations, the power of spacial visualization, speed of perception, the ability to acquire data, to memorize information, the power of recall, the ability to adjust to change, planning ability, the ability to chose between alternative courses of action, to resolve ethical dilemmas, and so on. The disparity in the degree to which these different skills may be present is remarkable. Two people can be generally acknowledged to be intelligent yet may have few abilities in common. One psychologist claimed that human intelligence comprises 120 distinguishable elementary abilities, each one involving an operation on something to produce a product.

Psychologists have been deeply divided on the question of how many distinguishable mental skills constitute intelligence and whether there is anything more than the totality of these skills. Some claim that there is no general factor in human intelligence. Most, however, subscribe to the view that there is such a factor while being a little vague as to its nature. Psychologists are also at variance in their definitions of intelligence. Fourteen well-known experts, when asked, produced different answers. These included such ideas as the ability to learn by experience, to adapt to changing environments, to think in abstractions, to perceive truth, to acquire skills, and so on. One expert, with seeming disingenuousness, defined intelligence as the ability to do well in intelligence tests.

A definition still fairly widely held is that of Charles Spearman (1863–1945), who was Grote Professor of mind and logic at University College, London. Spearman taught that there *is* a general ability, which he called 'g', and which is necessary for the performance of all mental tasks. Surrounding this, is a number of separate specific abilities, present to different degrees and capable of being separately measured. Spearman's concept has been widely discussed and many modifications suggested, but has not been seriously challenged, even by modern cognitive psychologists, who try to understand intelligence in terms of information processing. The nature of the basic quality 'g' has been a source of much argument. The psychologists H.A. Fatmi, R.W. Young and H.B. Barlow, writing in the scientific journal *Nature* and elsewhere, have made a strong case for the view that it is the capacity to detect new and non-chance associations.

Intelligence tests

The intelligence quotient (IQ) for children is the ratio of the mental age to the chronological age. When these are equal, the IQ is 100. The problem in assessing intelligence is to find fair, realistic and reliable ways of measuring the mental age. The difficulty relates closely to the difficulty of defining intelligence (see above) and of trying to determine factors for testing which are independent of educational and cultural influences. Intelligence tests compiled without due regard to these facts have, rightly, been condemned as being unfair to those candidates who do not share the educational and cultural background of the testers. Such criticisms have tended to bring all intelligence testing into disrepute.

The first formal tests of intelligence were devised by the French psychologist Alfred Binet (1857–1911), working with Theodore Simon at the request of the French government. The purpose was to determine which children were worthy to receive education. Binet originated the idea of IQ. These tests were subsequently repeatedly modified at Stanford University by Louis Terman and others, and the original Binet-Simon test (1908) became the Stanford-Binet tests. The current Stanford-Binet tests provide tasks for various ages from two to adulthood. Very young children are asked to make copies of objects, to string beads, build with blocks and answer questions on familiar activities. Tests for older children involve such things as detecting absurdities, finding what various pairs of words have in common, completing sentences with omitted words, explaining proverbs, and so on. Such tests are, of course, strongly educationally oriented and really test scholastic ability.

Stanford-Binet tests were largely superceded by the Wechsler tests, compiled by the New York psychologist David Wechsler. These are of two basic kinds – the Wechsler Adult Intelligence Scale (WAIS) and the Wechsler Intelligence Scale for Children (WICS). The tests involve progressively increasing difficulty. Each test has verbal and performance parts and these can be applied independently for those with language difficulties, or combined to give an overall score. The verbal parts test vocabulary, verbal reasoning, verbal memory, arithmetical skill and general knowledge. The performance sections involve completing pictures, arranging pictures in a logical order, reproducing designs with coloured blocks, assembling puzzles, tracing mazes, and so on. Again, these tests cannot be said to assess much more than the general educational level.

The level of definitive intelligence cannot be reliably assessed in infancy, and until about the age of five there are usually only the most general indications of whether adult intelligence is going to be high or low. An occasional child will show evidence of exceptional intelligence or of retardation but, until that age there is little indication in the average child. In most cases, however, from the age of about 12 onwards the results of intelligence tests are consistent with the level of later adult intelligence.

Intelligence tests are not, however, except in the most general way, accurate predictors of achievement. Scholastic achievement depends on a number of factors other than intelligence, especially the quality of instruction, parental expectations and a rich early educational environment. Achievement later in life is even less accurately predictable on the basis of intelligence tests and is determined by many other factors including quality of personality, physical appearance, opportunity, luck, and the possession of special skills. Good motivation can compensate for restricted intelligence and its absence can result in little effective use being made of high intelligence. All that notwithstanding, there is a general positive correlation between intelligence, as measured by tests, and material and professional success in life.

intemperance lack of restraint in personal indulgence in any activity, such as alcoholic consumption, likely to be harmful in excess.

inter- *prefix denoting* between, among, shared or mutual.

interarticular situated between articulating joint surfaces.

interatrial between the upper chambers of the heart.

intercalary occurring, or interposed, between parts.

intercellular among or between cells.

interclavicular between the collar bones.

intercondylar situated between two CONDYLES.

intercostal lying between adjacent ribs, as in the case of the respiratory INTERCOSTAL MUSCLES and the intercostal arteries, veins and nerves.

intercostal muscles voluntary muscles, situated between each pair of adjacent ribs, which, on contracting, raise the rib cage upwards and outwards so as to increase the volume of the chest and cause air to be forced in by atmospheric pressure.

intercourse 1 any form of human communication.
2 a popular term for SEXUAL INTERCOURSE, COITION or COPULATION.

intercristal between two crests.

interdental *adj.* 1 between the teeth.
2 a consonant pronounced with the tip of the tongue between the teeth, as 'th-'.

interdigital between the fingers or the toes.

interdigitation arranged in the manner of clasped fingers.

interface a surface forming a common barrier or boundary between two objects.

interferons a considerable range of antiviral protein substances produced by cells that have been invaded by viruses. Interferons are released by such cells and provide protection to other cells liable to be invaded, not only by the original virus, but also by any other infecting organism. They also modify various cell-regulating mechanisms and slow down the growth of cancers.

interleukins a range of CYTOKINES secreted by white cells of the immune system. Effector cells have surface receptors for the various interleukins.

interleukin-1 a powerful polypeptide hormone produced by MACROPHAGES and fibroblasts that acts on LYMPHOCYTES to increase their ability to respond to ANTIGENS. Interleukin-1 is also responsible for resetting the temperature regulating mechanism at a higher level and thus causing fever, for the induction of the release of ACUTE PHASE PROTEINS, and for promoting the absorption of bone by OSTEOCLASTS.

interleukin-2 a peptide chemical mediator released by helper T LYMPHOCYTES that stimulates clonal T cell and B cell division and proliferation. It is known as the T cell growth factor and is responsible for the activation on natural killer T cells.

interleukin-3 a CYTOKINE produced by T cells amd MAST CELLS that promotes the growth and differentiation of blood forming cells and the growth of mast cells.

interleukin-4 a CYTOKINE produced by HELPER T CELLS, MAST CELLS and bone marrow that has a wide range of actions. It promotes the proliferation of B cells, T cells, MAST CELLS and blood forming cells; it induces the formation of MAJOR HISTOCOMPATIBILITY COMPLEXES on B cells; and it is believed to promote class switching in B cells from one class of antibodies to another.

interleukin-5 a CYTOKINE produced by HELPER T CELLS and MAST CELLS that assists in the proliferation of activated B cells and eosinophil cells and in the production of IgM and IgA.

interleukin-6 a CYTOKINE produced by HELPER T CELLS, MACROPHAGES, fibroblasts and MAST CELLS that promotes the growth and differentiation of B cells, T cells and blood stem cells and the induction of ACUTE PHASE PROTEINS.

interleukin-7 a CYTOKINE produced by the bone marrow stromal cells responsible for the proliferation of B cell precursors, helper and cytotoxic T cells and activated mature T cells.

interleukin-8 a CYTOKINE produced by MONOCYTES that is the substance causing CHEMOTAXIS of T cells and NEUTROPHIL polymorph phagocytes.

interleukin-9 a CYTOKINE produced by T LYMPHOCYTES (T cells) that promotes the growth and proliferation of T cells.

interleukin-10 a CYTOKINE produced by helper T cells, B cells, MACROPHAGES, and the placenta that inhibits gamma-interferon secretion and mononuclear cell inflammation.

interleukin-11 a CYTOKINE produced by bone marrow stromal cells that promotes the induction of ACUTE PHASE PROTEINS.

interleukin-12 a CYTOKINE produced by T LYMPHOCYTES that inactivates natural killer cells.

interleukin-13 a CYTOKINE produced by T LYMPHOCYTES that inhibits mononuclear cell inflammation.

interlobar situated between lobes.

intermediate-filament proteins protein elements in the cytoskeleton of diameters intermediate between those of microfilaments and microtubules. They include keratins, vimetin, desmin, peripherin, syncoilin, alpha-internexin, nestin and synemin. The gene family has at least 65 members and more than 30 diseases are related to mutations of these genes.

intermenstrual 1 occurring between menstrual periods.
2 pertaining to the interval between menstrual periods.

intermuscular between muscles or muscle groups. Compare INTRAMUSCULAR.

intermuscular septa fibrous connective-tissue sheets that partition muscle groups in the limbs.

internal anal sphincter a muscular ring surrounding the lower end of the rectum at the innermost end of the short anal canal.

internal capsule a corridor within the brain for bundles of nerve fibres, especially the motor PYRAMIDAL TRACTS descending from the motor cortex. Each internal capsule lies on the outer side of the THALAMUS and CAUDATE NUCLEUS and on the inner side of the LENTICULAR NUCLEUS. Bleeding within an internal capsule is a common cause of paralytic STROKE.

internal carotid artery one of the two main divisions of the common CAROTID ARTERY. The internal carotid supplies blood to the main part of the brain (CEREBRUM) and associated structures, including the eye.

internal secretion any secretion absorbed directly into the blood rather than passed out on to an internal surface or to the exterior.

interneuron a nerve that connects other nerves. An internuncial neuron.

internuncial linking two neurons.

interorbital lying between the eye sockets (ORBITS).

interphalangeal between the bones of a finger, especially the finger joints.

interphase the resting stage between mitotic cell division when chromosomes are loosely coiled and cannot be seen by light microscopy. The interphase is divided into periods designated G1, S and G2.

interpupillary between the pupils of the eyes. Situated or occurring between the pupils. The interpupillary distance must be measured when glasses are being prescribed so that the lens centring will be correct.

intersegmental reflex a spinal REFLEX arc in which the input (sensory) and output (motor) nerves are connected by tracts running within the spinal cord between different segments of the cord.

intersex a person with bodily or psychological characteristics of both man and woman. A person of ambiguous sex. See also HERMAPHRODITISM.

interspinous situated between, or joining, spinous processes of the spinal column or elsewhere.

interstices small spaces or gaps between parts of an organ or between cellular structural elements of tissues.

interstitial pertaining to, or existing in, INTERSTICES.

interstitial fluid extracellular fluid lying is small spaces around and between cells.

intertrochanteric between the bony protuberances (greater and lesser trochanters) at the top of the thigh bone (femur).

interventricular foramen one of two holes (foramina) that connect the THIRD VENTRICLE of the brain to each LATERAL VENTRICLE. Also known as the foramen of Monro.

interventricular septum the membranous and muscular wall between the right and left ventricles of the heart.

intervertebral situated between adjacent vertebrae, as in the case of an INTERVERTEBRAL DISC.

intervertebral disc a disc-shaped, fibro-cartilage, shock-absorbing structure lying between the bodies of adjoining vertebrae in the spinal column. Each disc consists of an outer fibrous ring called the annulus fibrosus and an inner soft core called the nucleus pulposus.

intestine the part of the digestive system lying between the outlet of the stomach (the PYLORUS) and the ANUS. It consists, sequentially, of the DUODENUM, the JEJUNUM, the ILEUM, the wide, pouch-like caecum, that carries the APPENDIX, the large intestine, or COLON, the S-shaped SIGMOID colon, the RECTUM and anus.

intestinal tract the INTESTINE. The whole of the tubular structure of the digestive system stretching from the outlet of the stomach to the anus. Also known as the intestinal canal.

intima the innermost layer of a blood vessel or hollow organ.

intolerance a tendency to react adversely to stimuli of any kind or to drugs or foodstuffs.

intoxication 1 the action of a poison of any kind on an organism.
2 drunkenness or alcoholic poisoning.
From the Latin *intoxicare*, meaning to smear with poison.

intra- *prefix denoting* within, inside.

intra-abdominal lying or occurring within the cavity of the ABDOMEN.

intra-arterial within an artery.

intra-articular within a joint.

intracapsular within its capsule. In an intracapsular cataract operation the whole affected CRYSTALLINE LENS, including its capsule, is removed. This method is now almost obsolete.

intracardiac within one of the chambers of the heart.

intracellular within a cell.

intracerebral within the brain.

intracerebral haemorrhage bleeding inside the brain, usually from a small artery predisposed by ATHEROSCLEROSIS, but sometimes from rupture of a pre-existing small ANEURYSM. Intracerebral haemorrhage causes STROKE.

intracranial within the skull.

intractable resistant to cure.

intracutaneous within the skin. Intradermal.

intracytoplasmic sperm injection a method of in vitro fertilization (IVF) in which a single SPERMATOZOON is injected directly into an ovum which is then implanted into the womb. The method, which was introduced in 1993, allows women to become pregnant by partners who may be totally sterile as a result of low sperm counts, poor sperm mobility or even AZOOSPERMIA. Anxieties that the method may result in an unacceptably high number of defective babies seem to have been exaggerated. There are indications, however, that there is an increased risk of mild delays in development at 1 year in babies produced by this method of fertilization.

intradermal within the thickness of the skin.

intraepithelial within the epithelium.

intrahepatic within the liver.

intraluminal within any tubular structure. Within the lumen.

intramedullary 1 within the innermost tissue (medulla) of any organ.
2 within the bone marrow.
3 within the MEDULLA OBLONGATA.

intramural 1 within the walls of an organ.
2 within the substance of a wall.

intramuscular within a muscle. In an intramuscular injection the needle is passed deeply into the substance of a muscle before the fluid is injected.

intranasal within the nose.

intraocular within the eyeball.

intraocular lenses rigid, or flexible and folding, plastic optical lenses that are placed in the lens capsular bag after the cataractous and opaque contents have been removed either by emulsification or extrusion, and washout. Loss of the natural lens defocuses the eye markedly and additional optical power averaging 15 dioptres is required. This can readily be provided by an intraocular lens. Foldable lenses allow the whole cataract operation to be performed through a very short incision that may require no sutures.

intraocular pressure the hydrostatic pressure within the otherwise collapsible eyeball necessary to maintain its shape and allow normal optical functioning. Intraocular pressure must be adequate but not excessive as this may compress blood vessels within the eye and deprive important structures of blood. Damagingly raised pressure is called glaucoma.

intraoral within the mouth.

intraorbital within an ORBIT.

intraosseous infusion the process of supplying urgently needed fluid into the marrow cavity of a bone in a life-threatening condition in which normal access to the circulation is difficult, and delaying, or impossible.

intraparietal 1 INTRAMURAL.
2 within the parietal lobe of the brain.

intrapartum occurring during childbirth (parturition).

intraperitoneal within the cavity of the PERITONEUM.

intrapleural in or of the space between the two layers of the PLEURA. The space is normally occupied by a thin layer of lubricating fluid.

intrapulmonary within the substance of a lung.

intrarenal within a kidney.

intrathecal 1 within a sheath.
2 within the SUBARACHNOID SPACE.

intrathoracic within the chest (thoracic) cavity.

intratracheal within the TRACHEA.

intrauterine within the womb (uterus).

intravaginal 1 within the vagina.
2 within a tendon sheath.

intravascular within blood vessels within the lymphatics.

intravenous 1 within a vein.
2 into a vein. Intravenous injection of a drug achieves rapid action. It also permits the giving of irritating substances because these are rapidly diluted and dispersed in the blood.

intraventricular within a ventricle of the heart or brain.

intravesical within the urinary or other bladder.

intrinsic belonging to or situated within, the body or part of the body.

intrinsic factor a glycoprotein substance produced by the lining of the stomach (gastric mucosa) that complexes with vitamin B12 and promotes its absorption by the stomach, without itself being absorbed.

intro- *prefix denoting* into, inward or directed into.

introitus the entrance into any hollow organ or body cavity. The term is often used to refer to the entrance to the vagina.

intron a non-coding segment of a DISCONTINUOUS GENE. Introns are lengths of DNA interposed between coding segments (EXONS) in a gene and are transcribed into MESSENGER RNA but are then removed from the transcript and the exons spliced together. Introns do not contain biological information.

introspection examination, usually prolonged, of one's own thoughts, feelings, and sensations.

introversion 1 a physical turning in upon itself, as may occur with a hollow organ.
2 a directing of psychic energy in upon the self. See also INTROVERT.

introvert a person whose tendency of mind is to look inwards, to contemplate his or her own thoughts, feelings and emotions rather than to seek social intercourse. The introvert is often obsessive, anxious, hypochondriacal and solitary, more concerned with thought than with action. Compare EXTROVERT.

intuition knowledge apparently acquired without either observation or reasoning. The idea, although romantically attractive, wilts in the presence of modern psychological and physiological ideas. Few experts now believe that anything can come out of the brain that has not previously gone in, in however fragmentary a form. Intuition is probably the result of the synthesis of information from partly-conscious observations.

intumescence 1 the process of swelling or becoming engorged, as in the erection of the penis.
2 the condition of being swollen.

in utero within the womb.

invagination a folding into or ensheathing. The process of invagination occurs in the early development of the embryo when part of the BLASTODERM folds inward so that the hollow sphere becomes cup-shaped and double-walled.

invasive 1 involving entry to the body through a natural surface, usually referring to entry for diagnostic purposes.
2 having a natural tendency to spread, as of a cancer.

inversion mutation a mutation resulting from the removal of a length of DNA which is then reinserted facing in the opposite direction.

in vitro occurring in the laboratory rather than in the body. Literally, 'in glass'. Compare IN VIVO.

in vitro fertilization fertilization of an egg that has been withdrawn from the body, by sperm that have been obtained by masturbation. The procedure is done by adding semen to the eggs in a glass receptacle. A successfully fertilized ovum may then be artificially implanted into the womb (uterus) so that the pregnancy may continue. Fertilization can also be achieved by intracytoplasmic sperm injection. Recently cytoplasm containing known mitochondrial genetic defects has been eliminated by a process in which a fertilized nucleus is transferred into another ovum. Babies born from such a procedure can be said to have three parents.

in vivo occurring naturally within the body. Compare IN VITRO.

involucrum a sheath or covering of a part.

involuntary muscle smooth muscle, usually within an organ or blood vessel, that contracts under the influence of unconscious processes mediated by the AUTONOMIC nervous system rather than by the conscious will.

involution 1 decay, retrogression or shrinkage in size.
2 a return to a former state.
3 an infolding or INVAGINATION.

ion an electrically charged atom, group of atoms, or molecule. A positive ion is an atom that has lost an electron; a negatively charged ion is one that has gained an electron. See also IONIZATION.

ion channels protein ports in cell membranes that are specific for the passage of sodium, potassium, calcium and chloride ions in solution. Changes in the protein configuration, under the influence of various hormone molecule attachment, intracellular ion or other chemical concentration, or electrical potential, cause ion channels to open or close as required. Many diseases result from disordered function of ion channels.

ionic bond a chemical bond caused by the strong electrical attraction between ions of opposite charge.

ionization the state of an atom or group of atoms that has become positively charged by the loss of an orbital electron or negatively charged by gaining an electron. All body electrolytes such as sodium, potassium, calcium and magnesium become ionized in solution. Gases may be ionized by means of electrical discharges.

ionizing radiation radiation capable of causing ionization by breaking electron linkages in atoms and molecules. Such radiation includes alpha particles (helium nuclei), beta particles (electrons), neutrons, X-rays and gamma rays.

ipsilateral being located on, affecting or referring to, the same side of the body. HOMOLATERAL. Compare CONTRALATERAL.

IQ *abbrev. for* INTELLIGENCE QUOTIENT.

irido- *combining form denoting* the iris of the eye.

iris the coloured diaphragm of the eye forming the rear wall of the front, water-filled, chamber and lying immediately in front of the CRYSTALLINE LENS. The iris has a central opening, the pupil. It contains circular muscle fibres to constrict the pupil and radial fibres to enlarge (dilate) it.

iron an element essential for the formation of HAEMOGLOBIN. Lack of iron, or excessive loss leads to iron-deficiency anaemia.

irradiation exposure to any form of ionizing or other radiation either for purposes of treatment, as in radiotherapy, or to sterilize medical or surgical material and instruments.

irreducible incapable of being replaced or restored to a former state. Irreversible.

irremediable not able to be remedied or cured.

irritability 1 the state of being normally excitable or able to respond to a stimulus.
2 the state of abnormal excitability featuring an exaggerated response to a small stimulus.

ischaemia inadequate flow of blood to any part of the body. It is a serious disorder usually due to narrowing, from disease, of the supplying arteries.

ischaemic necrosis local tissue death (GANGRENE) due to an inadequate blood supply.

ischium the lowest of three bones into which the innominate bone, comprising one side of the pelvis, is divided. The ischial tuberosities are the bony prominences on which we normally sit.

Islets of Langerhans INSULIN secreting collections of cells lying in the INTERSTITIAL tissue of the PANCREAS. (Paul Langerhans, 1847–88, German medical student, later anatomy professor).

iso- *combining form denoting* equal or equivalent.

isochromosome an abnormal chromosome formed when, during the ANAPHASE of cell division, the CENTROMERE divides horizontally rather than longitudinally, thus producing a chromosome with two long arms and one with two short arms.

isogenic pertaining to individuals who are genetically alike, such as identical twins or closely inbred animals. Isogenic individuals are ideal donors to each other of organs for transplantation.

isolation the state of separation from other people of a person suffering from an infectious disease, or carrying infective organisms, so as to prevent spread of infection. Isolation is also used to protect immunocompromized people from organisms carried by healthy people (reverse barrier nursing).

isoleucine an essential AMINO ACID.

isomer a chemical compound having the same number of each type of atom (same percentage composition and molecular weight) as another compound, but having different chemical or physical properties.

isometric 1 of equal dimensions or length. 2 of muscular contraction, in which an increase in tension occurs without shortening.

isometric exercises muscular exercises in which muscle groups are pitted against each other so that strong tensing occurs without causing movement.

isotonic of a fluid that exerts the same OSMOTIC PRESSURE as another, especially as that of the body fluids. Body cells, such as red blood cells, can be immersed in an isotonic solution without being caused to change shape. 'Normal' saline is isotonic with blood.

isotope chemically identical elements whose atomic nuclei have the same number of protons but different numbers of neutrons. The number of protons determines the number of orbital electrons and hence the chemical properties. Radioactive isotopes are called radionuclides. From the Greek *iso-*, equal and *topos*, place. Isotopes occupy the same place in the Periodic table of the elements.

itching a tickling or irritating sensation causing a desire to rub or scratch. Itching is caused by the stimulation of certain nerve-endings in the skin, probably by certain enzymes called endopeptidases. Severe itching is called pruritus. The feeling as of ants moving under the skin is called formication.

-itis *suffix denoting* inflammation of.

ito cell one of numerous star-shaped cells lying in the spaces between the main liver cells (hepatocytes) and the blood channels (sinusoids) that store vitamin A.

IVC *abbrev. for* inferior vena cava, the largest vein in the body.

Jj

Jacob, Francois the French biochemist François Jacob was born in Nancy in 1920 and was educated at the University of Paris, obtaining an MD in 1947. During the war he had fought with the Free French forces and had been seriously wounded. In 1950 he became a research assistant at the Pasteur Institute in Paris where, in 1960 he became head of the Department of Cellular Genetics, a position he retained until 1991. From 1965 to 1992 he was also professor of cellular genetics at the Collège de France.

At the Pasteur Institute Jacob conducted research on the bacterium *Escherichia coli* studying how genetic material was transferred from one bacterium to another. By noting the order in which genes were transferred he was able to discover the gene sequence on the chromosome. After Watson and Crick had shown that DNA coded for proteins, Jacob was determined to discover how this was done. Working with colleagues he cultured *E. coli* in various different media and found that changes in the medium would result in changes in the amounts of different enzymes produced by the organisms. They concluded that an organism will produce more of a particular enzyme when it needs that enzyme. This led them to the discovery that the production of each protein requires three genes – a structural gene for the identity and order of the amino acids, a regulator gene that produces a substance that binds to and represses a third gene they called the operator gene. This binding prevents the formation of messenger RNA and thus of the protein concerned. When E. coli was grown on a medium containing lactose, the enzyme needed to metabolise lactose was plentifully produced because lactose binds to the repressor substance preventing it from binding to, and repressing, the operator gene.

For this important discovery, which was later shown to be of universal application, Jacob and his co-workers Jacques Monod and André Lwoff were awarded the Nobel Prize for Physiology or Medicine in 1965.

jamais vu a strong sense that one has never before seen what is currently being perceived, although logic contradicts it. There is also a sense of unreality and depersonalization.

James, William one of the major influences in the development of modern psychology was the American philosopher and scholar William James (1842–1910). His *Principles of Psychology* (1890) which took him twelve years to write and which is probably the best known psychological text ever written, can still be read with enjoyment and profit today. It is said of the brothers William and Henry that Henry, the novelist, wrote like a psychologist while William, the psychologist, wrote like a novelist. James was as much a philosopher as a psychologist and his book was immensely influential. He adopted no fixed position and although he founded an experimental psychology laboratory at Harvard, he was more interested in general observations on human nature, behaviour and experience. The *Principles* is a masterly exposition of a wealth of scholarly knowledge and personal experience concerning the human condition. One of James's most often quoted remarks is 'Sow an action, and you reap a habit; sow a habit and you reap a character; sow a character and you reap a destiny.'

jargon 1 technical or specialized language used in an inappropriate context to display status or exclusiveness.
2 the formulation of fluent but meaningless chatter by combining unrelated syllables or words.

jaundice yellowing of the skin and of the whites of the eyes (scleras) from deposition of the natural pigment, bilirubin, that is released when HAEMOGLOBIN is broken down. Bilirubin is normally excreted in the bile but cannot do so in certain liver diseases and in obstruction to the outflow of bile into the intestine. In such cases it accumulates in the blood causing jaundice.

Java man *Homo erectus* whose fossil remains were first found in Trinil, Java. There was much controversy over the apparent disparity between the low volume and primitive shape of the skull and the remarkably modern thigh bone (femur), but later finds resolved the argument.

jaw 1 the mandible, the U-shaped bone that articulates with the base of the skull high up in front of the ears. In biting and chewing (mastication) the mandible is pulled upwards by powerful muscles running down from the base and temples (temporal bones) of the skull.
2 the MAXILLA, or upper jaw.

jaw wiring the securing of the jaws in a proper relationship by means of malleable wire bound round the necks of the teeth. This is commonly done to splint single or multiple fractures of either jaw and is sometimes done to limit food intake in the treatment of severe obesity. Weight loss is rapid but old habits usually prevail when the wiring is removed.

jealousy in childhood syndrome emotional disturbance resulting from competition between siblings or occasioned by the arrival of a new baby. The condition features regression to more childish behaviour, temper tantrums, bedwetting or manifest anxiety and is managed by scrupulous parental fairness.

jejunal pertaining to the JEJUNUM.

jejunum the length of small intestine lying between the DUODENUM and the ILEUM and occupying the central part of the ABDOMEN. Much of the enzymatic digestion of food, and most of the absorption, takes place in the jejunum. Absorption occurs through the thin walls of millions of tiny finger-like processes on the lining, called villi.

Jenner, Edward smallpox, one of the few diseases to have been totally eliminated, was once the scourge of nations. 50 million people died from the disease in Europe in the 18th century and countless others were disfigured. Since it was known that an attack conferred immunity, attempts had been made to achieve this by exposing people to material from smallpox crusts. Many of these people had fatal attacks of smallpox.

As a young assistant medical practitioner Edward Jenner (1749–1823) had heard a milkmaid tell his principal that she could never get smallpox because she had contracted the mild disorder, cowpox, while milking. Many years later, while in practice in Gloucestershire, it occurred to Jenner to make practical use of this suggestion. In 1796 he scratched some material from cowpox pustules on the hand of a milkmaid into the arm of a healthy young boy. The boy developed cowpox. Two months later, Jenner, with what many thought criminal foolhardiness, inoculated the boy with smallpox. Nothing happened. In 1798 he was able to repeat the experiment and the same year he published his book *An Inquiry into the Causes and Effects of the Variolae Vaccinae.* Although his temerity was criticized, others tried and proved the method and soon Jenner's fame spread around the world. Soon, thousands of people, including the Royal Family, had been vaccinated. Parliament voted Jenner £30,000.

The medical Establishment was less enthusiastic than the lay public about Jenner's success. When he applied for a Fellowship of the London College of Physicians he was told he would have to be examined in his knowledge of the works of Galen. Quite rightly, Jenner refused and his application was rejected.

jerk 1 a sudden involuntary movement, usually of the head or a limb.
2 a reflex muscle or muscle group contraction in response to a sudden stretching by briskly tapping the tendon. A tendon reflex.

jet lag physiological disturbances from disruption of the normal circadian rhythms caused by rapid travel across time zones. This is the effect of change in the experienced periodicity of light and dark which quickly falls out of synchronization with the various periodic body functions. The result may be insomnia, tiredness during the day, a feeling of light-headedness, a sense of ill-being and reduced mental and physical performance. These effects are not restricted to air travel but are experienced by people living in the Arctic or Antarctic or in space or by people who have recently lost all vision. Jet lag tends to be worse when travelling east because of the difficulty in shortening the day; because the innate circadian period is longer than 24 hours, lengthening the day is easier.

jogging slow running conducted within the limits of the subject's ability to supply oxygen to the muscles for an indefinite period. Regular jogging increases the capacity for exertion, lowers blood pressure, improves the function and performance of the heart and may diminish the progress of arterial disease. Beginners should build up distance very gradually and should be equipped with suitable footwear.

joints

Joints are junctions between bones, whether movable or not. There are three kinds – fibrous, cartilaginous, and synovial. Fibrous joints are bones held firmly together with ligaments and allow little or no movement. Such joints occur in the pelvis and the vertebral column. Cartilaginous joints are somewhat more mobile because of the flexibility of gristle (cartilage). The most conspicuous cartilaginous joints are those between the ribs and the breast-bone (sternum). Synovial joints, such as those at the shoulder, elbow, hip and knee, are freely mobile. The bearing surfaces of synovial joints are covered with a thin layer of articular cartilage which is lubricated with fluid that exudes from the cartilage under pressure. Such joints are enclosed in capsules of tough fibrous tissue lined with a membrane, the synovial membrane, that secretes the lubricating synovial fluid.

Synovial joints are of several types. They include hinge joints (knee and finger), ball and socket joints (shoulder and hip), rotating joints (upper end of radius bone and between the upper two vertebrae) and sliding joints (wrist and feet). The range of movement of synovial joints is restricted by external ligaments and, in some cases also by internal ligaments.

joule a unit of work, energy and heat. A watt-second. The joule is being used increasingly to replace the CALORIE in nutritional contexts. The calorie is equal to 4.187 J. (James Prescott Joule, 1818–89, English physicist).

jugular pertaining to the throat or neck.

jugular veins the six main veins – the right and left internal and external jugulars and the front anterior jugulars – that run down the front and side of the neck, carrying blood back to the heart from the head. The internal jugulars are very large trunks containing blood at low pressure. The external and anterior jugulars are much smaller.

jugular venous pulse the visible movement of the skin caused by movement of blood in an internal jugular vein when the subject reclines with the neck at 45°. Examination of the pulse can provide information about the function of the right side of the heart. See also HEPATOJUGULAR REFLUX.

jumping genes see TRANSPOSONS.

junctional at the interface between two structures.

Jungian theory a body of psychoanalytic theory offered as an alternative to Freud's with its central emphasis on sex. Carl Gustav Jung (1875–1961) defined 'libido', more widely, as a general creative life force that could find a variety of outlets. He identified extraversion and introversion and suggested that people could be divided into four categories by their primary interests – the intellect, the emotions, intuition and the sensations. Like Freud, Jung was deeply concerned with symbols which he considered central to the understanding of human nature. He postulated the existence of a layered unconscious psyche, both personal and collective, the latter being common to all humankind. He proposed the concept of 'archetypes' – inherent tendencies to experience and symbolize universal human situations in distinctively human ways. Never very scientific, Jung later in life moved even further into the airy realms of metaphysical speculation about which no scientific comment is possible. Compare FREUDIAN THEORY.

junk food a popular term for highly refined and processed, readily assimilable and palatable food with a low level of roughage. Junk food has a high calorific value but is often low in vitamins and minerals and usually has a high content of saturated fats. Junk food encourages excessive intake and commonly leads to obesity. The full extent of the dangers of a largely junk food diet has not been established but few experts deny that the dangers exist.

junkie or **junky** a slang term for a narcotic, especially heroin, addict.

juvenile delinquency criminal behaviour by a young person. Juvenile delinquency has a peak incidence around fifteen or sixteen years of age and is commonly associated with peer pressures to conform, parental neglect and lack of social opportunity to direct energy into more acceptable channels. There is often a poor school record, with truancy and resentment of authority. Most delinquents eventually learn to conform to generally acceptable patterns of behaviour.

juxta- *combining form denoting* near to or alongside.

juxta-articular adjacent to a joint.

juxtamedullary situated in the cortex of the kidney near to the MEDULLA.

juxtapapillary near the head of the optic nerve (the optic disc) on the RETINA.

juxtaposition in apposition or side-by-side.

Kk

K the symbol for potassium; for temperature in the absolute scale; as **k** for kilo- as applied to many other units; and for the mass of large molecules in KILODALTONS.

kangaroo mother method a method of care for small babies in the first few months of life involving frequent breast feeding, maximal skin-to-skin contact of mother and baby, and early discharge from the maternity unit. The method implies careful selection after detailed examination.

kappa the tenth letter of the Greek alphabet, sometimes used to denote the tenth in a series.

karyapsis see KARYOGAMY.

karyo- or **caryo-** *combining form denoting* a cell nucleus. From the Greek karuon, a nut.

karyocyte a nucleated cell.

karyogamy the coming together and fusing of the nuclei of GAMETES.

karyogenesis the formation of a cell nucleus.

karyokinesis MITOSIS.

karyolysis destruction of a cell nucleus.

karyon the cell nucleus.

karyolymph the fluid in the nucleus of a cell.

karyomegaly an increase in the size of the nuclei of the cells of a tissue.

karyoplasm the PROTOPLASM of the cell nucleus. Nucleoplasm.

karyorrhexis fragmentation of the nucleus of a cell as seen in a dead cell.

karyosome a spherical mass of aggregated CHROMATIN material in a resting (interphase) nucleus.

karyotype 1 the individual chromosomal complement of a person or species. The genome. 2 the CHROMOSOMES of an individual set out in a standard pattern and obtained from a photomicrograph taken in METAPHASE that has been edited with software so that the separate chromosomes are arranged in numerical order. This is done for the diagnosis of chromosomal disorders, as in prenatal detection of fetal abnormality.

katabolism see CATABOLISM.

katal a unit of enzyme activity in the SI system. 1 international unit is equal to 16.6 nanokatal.

kb *abbrev. for* kilobase (1000 base pairs of DNA or RNA). This is a convenient measure of the length of a gene or other segment of nucleic acid.

K cells see KILLER CELLS.

Kekulé, August the famous German chemist Friedrich August Kekulé von Stradonitz (1829–96) was the first to suggest, in 1858, that atoms formed molecules by joining together in particular ways. The structural formulae that were produced on the basis of this idea rapidly extended the scope and rate of advance of organic chemistry. In 1861 Kekulé published a textbook in which, for the first time, organic chemistry was described as the chemistry of carbon.

The benzene molecule, containing six carbon atoms and six hydrogen atoms, was of considerable importance in organic chemistry because of the range of new synthetic dyes that were being developed from it. Unfortunately, benzene appeared to defy the known fact that carbon combined with four hydrogen atoms (was tetravalent). Its structure was unknown and this was holding up progress. One day in 1865, Kekulé was daydreaming on a bus, thinking casually of atoms. It seemed to him that he could see chains of atoms whirling in a dance. As he watched, he saw the tail of one chain attached itself to its own head, forming a spinning ring of atoms. Suddenly wide awake, Kekulé realized that this was the answer to the problem of the structure of benzene. The six carbon atoms were arranged in a closed ring with alternate double and single bonds and a hydrogen atom attached to each carbon. This also solved the valency problem. The idea was one of the major advances in all chemistry.

Kell blood group system a family of red blood cell ANTIGENS designated as the ALLOTYPES KK, Kk, kk and K-k-, believed to be on the short arm of chromosome 2. Antibodies to the K-antigen occur in about 10 per cent of people in England and can cause red cell breakdown (haemolytic) transfusion reactions. The group system is next in importance to the ABO and rhesus systems and is named after a woman whose serum contained the antibodies.

Kendall, Edward Edward Calvin Kendall (1886–72) was an American biochemist who took his doctorate in chemistry from Columbia University in 1910 and at once began research on the thyroid gland. The general idea of hormones had been introduced by Starling and Bayliss eight years before and it was known that overactivity of the thyroid was associated with increased metabolic activity throughout the whole body. It was also known that loss of the thyroid led to a severe slowing of all body functions. So, since it had such a widespread effect, it seemed likely that the thyroid acted by way of hormones. After four years of work on the iodine-containing protein thyroglobulin found in the gland, Kendall moved, in 1914, to Minnesota to become head of the biochemistry section of the Mayo clinic. There, in that same year, by systematically breaking down the thyroglobulin molecule and testing the fragments, he was able finally to isolate the active principle, thyroxine. This he found to be a fairly simple substance, related to the common amino acid tyrosine, but containing four iodine atoms in its molecule. Soon thyroxine was being widely used in the treatment of thyroid deficiency disorders.

Kendall taught physiological chemistry at the Mayo Foundation from 1921–51, but was not content to leave hormones alone. He and his colleagues had noted that a woman with rheumatoid arthritis was much improved while pregnant. Convinced that this was due to a hormonal effect, Kendal worked on the problem until, in 1934, he succeeded in isolating the corticosteroid hormones from the adrenal gland. This was a major breakthrough. With his associates he then developed a way of partially synthesizing cortisone and hydrocortisone, thus making these invaluable substances available for therapeutic use. In 1950 Kendall was awarded the Nobel Prize for Medicine or Physiology.

kerat-, kerato- *combining form denoting* CORNEA or horny KERATOSIS. From the Greek keras horn.

keratin a hard protein (scleroprotein) of cylindrical, helical molecular form occurring in horny tissue such as hair and nails and in the outer layers of the skin. Hair and nails consist almost wholly of keratin. Keratins are insoluble and cannot generally be split by PROTEOLYTIC enzymes.

keratinization the formation of, or conversion into, keratin. This normally occurs to a limited degree in the outer layers of the skin, but is especially prominent when skin is exposed to constant localized pressure. Corns and callosities are areas of keratinization. Also known as cornification or hornification.

ketoacidosis see KETOSIS.

ketogenesis the formation of acid KETONE BODIES, as in uncontrolled DIABETES, starvation or as a result of a diet with a very high fat content.

ketonaemia KETONES in the blood. Low levels are normal.

ketones a class of acidic organic compounds that includes acetone and aceto-acetic acid. Ketones have a carbonyl group, CO, linked to two other carbon atoms. They are formed in states of carbohydrate deficiency such as starvation or in conditions, such as diabetes, in which carbohydrates cannot be normally utilized. Acetone, aceto-acetic acid and beta-hydroxybutyric acid are called ketone bodies. Ketones are volatile substances and confer on the breath the sickly, fruity odour of nail-varnish remover.

ketonuria the presence of the ketone bodies acetone, aceto-acetic acid or beta-hydroxybutyric acid in the urine, usually in cases of untreated DIABETES.

ketosis the presence of abnormally high levels of KETONES in the blood. These are produced when fats are used as fuel in the absence of carbohydrate or available protein as in DIABETES or starvation. Ketosis is dangerous because high levels make the blood abnormally acid and there is loss of water, sodium and potassium and a major biochemical upset with nausea, vomiting, abdominal pain, confusion, and, if the condition is not rapidly treated, coma and death. Mild ketosis also occurs in cases of excessive morning sickness in pregnancy.

ketosteroid a steroid hormone to which an oxygen molecule has been attached, especially at the 17th carbon atom (C-17). An oxosteroid. A 17-ketosteroid is a steroid with a C–O (carbonyl) group at C-17. 17-ketosteroids are excreted in the urine as breakdown products of ANDROGENS and are present in excess in adrenal and gonadal overactivity.

kidney one of the paired, reddish brown, bean-shaped structures lying in pads of fat on the inside of the back wall of the ABDOMEN on either side of the spine, just above the waist. The kidneys filter the blood, removing waste material and adjusting the levels of various essential chemical substances, so as to keep them within necessary limits. In so doing, they produce a sterile solution of varying concentration known as urine. This passes down the ureters to the bladder where it is stored until it can be conveniently disposed of. The kidneys are largely responsible for regulating the amount of water in the body and controlling the acidity of the blood. Most drugs or their products are eliminated through the kidney. Kidneys control fluid and chemical levels by both filtration and selective reabsorption under the control of various hormones such as ALDOSTERONE from the adrenal gland, the ANTIDIURETIC HORMONE from the pituitary gland and PARATHYROID hormone from the parathyroid glands. Sodium, potassium, calcium, chloride, bicarbonate, phosphate, glucose, amino acids, vitamins and many other substances are returned to the blood and conserved. Proteins, fats and all the cells of the blood remain in the circulation. The kidneys produce ERYTHROPOIETIN, which stimulates the rate of formation of blood cells in the bone marrow. When blood pressure falls below normal the kidneys release the enzyme renin into the blood. This results in the formation of a further hormone, angiotensin, which rapidly causes blood vessels throughout the body to constrict and raise the blood pressure.

killer cells a subclass of large, granular LYMPHOCYTES that includes the natural killer (NK) cells and the killer (K) cells. These are important elements in the immune system and are the final effectors in the process by which damaged, infected or malignant cells are recognized and destroyed. At this stage in the process, killer cells act in conjunction with activated MACROPHAGES and cytotoxic T cells.

kilo- *prefix denoting* one thousand, as in KILOCALORIE.

kilobase a unit of measurement of the length of a DNA or RNA sequence equal to 1000 base pairs of DNA or 1000 bases of RNA.

kilocalorie or Calorie the amount of heat needed to raise the temperature of a kilogram of water by 1°C. This has been the standard nutritional unit of energy for years, but is now being replaced by the kilojoule. 1 kcal = 4.187 kJ. See JOULE.

kilodalton one thousand daltons. A dalton is roughly the weight of a hydrogen atom. The kilodalton is the standard unit used to represent the weight of large molecules such as proteins. It is normally abbreviated to K or Kd.

kinaesthesia perception of bodily movement, or of the sensation of movement. Compare PROPRIOCEPTION.

kinase see TRANSFERASE.

kinesiology the study of muscles and their effects on movements, especially in relation to physical therapy.

kinin one of a family of POLYPEPTIDES, released as a part of the inflammatory process, which increase the leakiness of small blood vessels and cause smooth muscle fibres to contract.

kinship the usually complex set of genealogical relationships and other ties between an individual and the members of his or her family. The links include those of genetic relationships, marriage, adoption and long cohabitation.

kissing the widespread human practice of pressing the lips against some part of the body of another person, especially the mouth. Kissing has social as well as sexual functions and these are usually kept apart. Some societies accept public kissing between adult males; others do not. Kissing can be dangerous, as, for instance, a means of transmission of infectious mononucleosis. It can be fatal when an adult with an oral cold sore kisses a child with eczema, producing the life-threatening Kaposi's varicelliform eruption.

kiss of life mouth-to-mouth or mouth-to-nose artificial respiration.

kleptomania a rare impulse disorder featuring recurrent stealing of things neither needed nor wanted. The object is the emotional relief of tension accompanying a successful theft rather than acquisition. This is often followed by strong guilt feelings. Only 1 person in 20 arrested for shop-lifting behaves in a manner consistent with the diagnosis. The cause is unknown.

Klinefelter's syndrome a male bodily disorder caused by one or more additional X (sex) chromosomes. Instead of the normal X and Y sex chromosomes, men with Klinefelter's syndrome have an XXY configuration. This has a feminizing effect. The penis and testicles are small and there may be female breast development and diminished sexual interest (libido). Homosexuality and transvestism are common. The diagnosis can easily be confirmed by chromosomal analysis. Hormonal and plastic surgical treatment can help. (Harry Fitch Kleinfelter, American physician, 1912–90).

knee the hinge articulation between the lower end of the thigh bone (FEMUR) and the upper end of the main lower leg shin bone (TIBIA). The knee cap (PATELLA) is a flat bone lying within the massive tendon of the thigh muscles and is not an intrinsic part of the joint.

knuckle the common name for a finger joint.

Koch's postulates a set of criteria to be obeyed before it is established that a particular organism causes a particular disease. The organism must be present in every case and must be isolated, cultured and identified; it must produce the disease when a pure culture is given to susceptible animals; and it must be recoverable from the diseased animal. (Robert Koch, 1843–1910, German bacteriologist).

koilo- *prefix denoting* hollow or empty.

koilorachic having a backward curve in the lumbar spine.

Kornberg, Arthur the American biochemist Arthur Kornberg, was born in Brooklyn, New York in 1918. He was educated at local schools and, on a State scholarship, studied pre-medical subjects at the College of the City of New York and then read medicine at the University of Rochester School of Medicine, graduating MD in 1941. After appointments at the Washington University School of Medicine and the Stamford University School of Medicine, Palo Alto, California, in 1959 he became a professor and head of the department of biochemistry at Stamford. He was an active professor until 1988 and then at the age of 70 became Emeritus Professor.

In 1956 Kornberg discovered the enzyme DNA polymerase. He had been working on the construction of artificial DNA to be built on a DNA template and this required him to find the enzyme that catalyzed the formation of the polymer from nucleoside triphosphates. While investigating the synthesis of coenzymes he did so and called it DNA polymerase. A year later he succeeded in synthesizing the circular DNA of a single-stranded virus (Phi X174) but this virus was not infective because its ring of DNA was broken. At that time, the enzyme ligase was not known, but when this was later found in 1966, the viral DNA ring was closed and the virus became active and could be reproduced in bacterial cells.

For the discovery and isolation of DNA polymerase Kornberg shared the 1959 Nobel Prize for Physiology or Medicine with Severo Ochoa who discovered the enzyme that catalyses the formation of RNA.

Krause's corpuscles spheroidal, laminated, delicately capsuled endings to nerve fibres, found in the skin and mucous membranes. They are receptors for cold stimuli. Also known as end bulbs of Krause. (Wilhelm Johann Friedrich Krause, 1833–1910, German anatomist).

Krebs cycle a cyclical sequence of 10 biochemical reactions, brought about by mitochondrial enzymes, that involves the oxidation of a molecule of acetyl-CoA to two molecules of carbon dioxide and water. Each turn of the cycle can result in the formation of 12 molecules of ATP per molecule of acetyl-CoA. ATP is the direct source of energy for all work performed in any cell. The Krebs cycle is one of the most important in all body biochemistry and occurs in all organisms that oxidise food totally to carbon dioxide and water. Also known as the citric acid cycle or the tricarboxylic acid cycle. (Hans Adolf Krebs, 1900–81, German-born English biochemist).

Krebs, Hans Adolf the German-born British biochemist Hans Krebs (1900–81), the son of an otorhinolaryngologist, was born in Hildesheim, Germany and studied at five German universities, taking a medical degree in 1925. He worked for a time in the Kaiser Wilhelm Institute for Cellular Physiology in Berlin and taught at the University of Freiburg, but with the rise of the Nazi party in Germany he decided, in 1933, to come to England. As a Rockefeller Research student he studied for a year at Cambridge University then worked as a demonstrator in biochemistry. In 1935 he moved to Sheffield University where he lectured in pharmacology. In 1945 he became professor of biochemistry at Sheffield. In 1947 he became a Fellow of the Royal Society, and in 1954 he was appointed professor of biochemistry at Oxford in which post he remained until he retired in 1967.

Krebs wasted little time in demonstrating his knowledge and skills as a researcher. While still working in Germany he investigated the important metabolic process of the degradation of amino acids in the body. He showed how, in a reaction known as deamination, nitrogen is removed to form ammonia and how ammonia and carbon dioxide are catalysed in the liver to form urea that is excreted in the urine. Krebs isolated various enzymes, showed that ketone bodies were produced in starvation and that ketones could be used by cells as fuel.

Krebs is best remembered, however, for his elucidation of the complex citric acid cycle (tricarboxylic acid cycle), commonly named the Krebs cycle, which is the final common pathway for the oxidation of fuel molecules – glucose, fatty acids and amino acids – by which cells are provided with energy. The Krebs cycle is also an important source of precursors of the storage forms of fuels and of many other molecules such as nucleotide bases and amino acids. It has been said that the function of the citric acid cycle is the harvesting of high energy electrons from carbon fuels.

For this notable advance he shared the 1953 Nobel Prize for Physiology or Medicine with Fritz Lipmann.

Kupffer cells fixed MACROPHAGE cells that line the fine blood sinuses (capillaries) of the liver and act as scavengers to remove senescent red blood cells, bacteria and other foreign material. (Karl Wilhelm von Kupffer, 1829–1902, German anatomist).

kyphoscoliosis an abnormal degree of backward curvature of the dorsal spine (KYPHOSIS) combined with curvature to one side (SCOLIOSIS).

kyphosis an abnormal degree of backward curvature of the part of the spine between the neck and the lumbar regions. Backward curvature is normal in this region and kyphosis is an exaggeration of the normal curve. It is commonly the result of bad postural habits in adolescence or of OSTEOPOROSIS. From the Greek *kyphos*, meaning bowed or bent.

Ll

labia the four lips of the female genitalia. The inner pair, the labia minora, surround the entrance to the vagina and the external opening of the urine tube (URETHRA) and join at the front to form a hood over the front of the head of the clitoris. The outer pair, the labia majora, are long, well-padded folds, containing muscle and fibro-fatty tissue, and covered with hair. They are normally closed and conceal the rest of the genitalia.

labial pertaining to the lips or labia.

labile liable to change. The term is applied to the emotions as well as to physiological change.

labiodental *adj.* a sound articulated by contact between with the lip and teeth, as in the sound 'f'.

labiomental pertaining to the lower lip and the chin.

labionasal pertaining to the lips and the nose.

labium the singular of LABIA.

labour the three stage process of delivering a baby and the PLACENTA by contractions of the muscles of the womb (uterus), of the DIAPHRAGM and of the wall of the abdomen. The first stage lasts from the onset of pains to full widening (dilatation) of the CERVIX, the second to the delivery of the baby and the third to the delivery of the placenta.

labyrinth any group of communicating anatomical cavities, especially the internal ear, comprizing the vestibule, semicircular canals and the cochlea.

laceration a wound made by tearing. An irregular wound of the tissues, as distinct from a clean cut (incised wound).

lachrymal an arbitrary and incorrect spelling of LACRIMAL.

lac operon a sequence of genes found in many bacteria that codes for the enzymes needed to break down lactose to glucose and galactose so that the sugars can be utilized. Studies on the lac operon have been important in genetic research. The lactose repressor protein controls the transcription of the lac operon.

lacrimal pertaining to the tears, to their production and to their disposal. Note that the spelling 'lachrymal' has neither logical nor etymological justification. The term comes from Latin *lacrima*, a tear.

lacrimal bone a small plate of bone, situated just inside the inner wall of the eye socket (orbit), with a shallow hollow to accommodate the LACRIMAL SAC.

lacrimal canaliculus one of four tiny tubes that carry tears from the inner corners of the four eyelids to the LACRIMAL SAC.

lacrimal gland the tear-secreting gland lying in the upper and outer corner of the bony eye socket (orbit) and opening by many small ducts into the upper cul-de-sac of the CONJUNCTIVA behind the upper lid. Lacrimal glands secrete during emotional weeping and when the eye is irritated. The eye is normally kept wet by tiny accessory lacrimal glands in the conjunctiva.

lacrimal sac the small bag lying under the tissues just inwards and below the inner corner of the eye. Tears drain into the lacrimal sac before being discharged down the nasolacrimal duct into the nose.

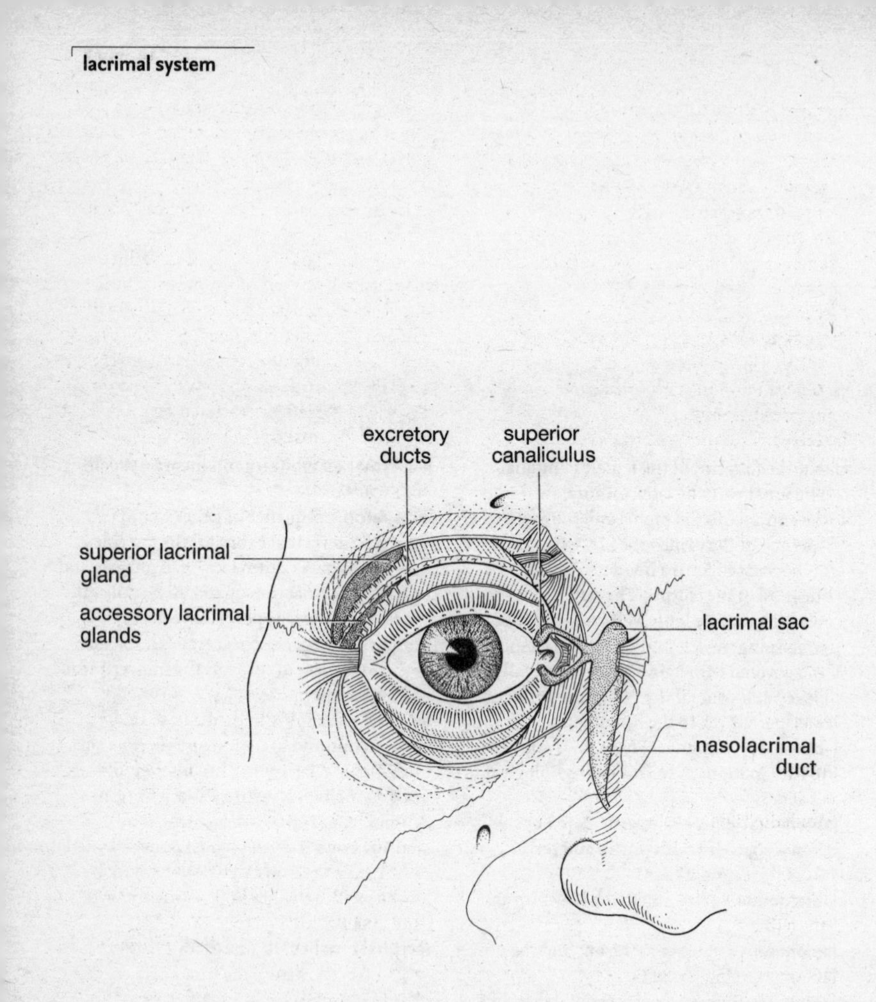

Fig. 9 **Lacrimal system**

lacrimal system the tear-producing LACRIMAL GLANDS and the drainage system – the CANALICULI and the NASOLACRIMAL DUCT – that carry surplus tears down into the nose.

lacrimation secretion of tears, especially excessive production as in weeping or in the presence of a foreign body or corneal ulcer. Compare EPIPHORA.

lacrimator tear gas.

lact-, lacto- *combining form denoting* milk, milk production or lactic acid.

lactalbumin any of a group of proteins contained in milk.

lactase an enzyme that brings about the HYDROLYSIS of LACTOSE to glucose (dextrose) and galactose. Beta-galactosidase.

lactate lactic acid in the ionized state.

lactate dehydrogenase (LDH) one of the cell enzymes released into the blood when heart muscle cells are damaged during a heart attack (myocardial infarction). A measure of the concentration of these enzymes can indicate the severity of the attack.

lactation the secretion and production of milk in the breasts (MAMMARY GLANDS) after childbirth.

lacteal 1 pertaining to milk.
2 a lymph vessel that absorbs and carries emulsified fat from the small intestine to the THORACIC DUCT and hence to the bloodstream.

lactic acid an acid formed when muscles are strongly contracted for long periods. Also formed from carbohydrates in the vagina by the action of DODERLEIN'S BACILLUS. Lactic acid is an ingredient in a range of drug formulations.

lactic dehydrogenase an enzyme that catalyzes the dehydrogenation of l-lactic acid to pyruvic acid. Dehydrogenation involves removing hydrogen atoms, usually two, from a molecule.

lactiferous able to produce, secrete, or convey milk.

lactoferrin an iron-binding protein found in milk and other body fluids and in neutrophil polymorph LEUCOCYTES in which its action helps to retard bacterial reproduction.

lactogenic promoting milk production (LACTATION).

lactoglobulin one of the proteins present in milk.

lactose the main sugar in milk. It is broken down by the digestive enzyme lactase (beta-galactosidase) to galactose and glucose.

lactosuria LACTOSE in the urine.

lacuna any empty space, missing part, cavity or depression.

lacus any small lake or collection of fluid.

laevocardia a reversal of the position of the abdominal organs (situs inversus) but with a normal position of the heart. Situs inversus is more commonly associated with a heart so placed that the apex points to the right instead of the left (DEXTROCARDIA).

laevulose fructose or fruit sugar, a monosaccharide found in honey and fruit. Combined with glucose, it forms the disaccharide cane sugar (sucrose).

lagging strand in DNA replication, the single strand forming a duplex in the direction away from the fork in the parental DNA. Replication on the lagging strand is discontinuous and can occur briefly in both directions.

-lalia *suffix denoting* a disorder of speech.

lalo- *combining form denoting* speech.

lalopathy any disorder of speech.

Lamarckism the discredited doctrine that species can change into new species as a result of characteristics acquired as a result of striving to overcome environmental disadvantages. It was claimed that such acquired characteristics became hereditary. (Jean Baptiste Pierre Antoine de Monet, Chevalier de Lamarck, 1744–1829, French naturalist).

Lamarck, Jean Baptiste de there were plenty of theories about evolution. For a time, one of the most influential was that of Jean Baptiste Lamarck (1744–1829). Lamarck was a French naturalist who taught that animals and plants evolved by changing in response to changes in their environment. Giraffes, he suggested, acquired long necks by stretching up to reach the leaves at the top of trees. Tail-less mice could be produced by repeatedly cutting off the tails of mice and then letting them breed. Lamarck's ideas of evolution by use and disuse were incorporated in his Zoological Philosophy (1809) and were accepted by most of his contemporaries.

The trouble with Lamarck's theory was two-fold. One difficulty was to explain how acquired characteristics could be passed on. Lamarck suggested that bodily changes could somehow modify the sperm or the ova so as to pass on the new characteristic to the offspring. The second problem was that the theory, unfortunately, did not represent observable fact. Cutting off the tails of mice had no effect whatsoever on the length of the tails of the offspring. Lamarck's theory was eventually abandoned.

lambdoid 1 resembling an inverted Y junction as in the Greek letter *lambda*.
2 pertaining to the SUTURE between the OCCIPITAL bone at the back of the skull and the PARIETAL bones on either side.

lamella any thin plate, layer or sheet, as of bone.

lamina any thin sheet or layer of tissue, especially the flat surfaces on the arch of a vertebra.

lamina cribrosa 1 the multiperforated plate of ethmoidal bone in the roof of the nose through which the fine fibres of the OLFACTORY NERVE pass.
2 the ring of perforations in the white of the eye, at the back of the globe, through which bundles of OPTIC NERVE fibres, and the central artery and vein of the RETINA, pass (lamina cribrosa sclerae).

laminar arranged in layers.

Landsteiner, Karl Karl Landsteiner (1868–1943) was an Austrian bacteriologist and immunologist who, in 1900, made a remarkable and important discovery by checking the effect of mixing blood serum from one person with the red blood cells from another. He found that while a particular serum would cause the red cells from one person to clump together, it might have no ill effects whatsoever on the red cells from another person. Landsteiner decided to check all the possible combinations and was soon able to show that human red blood cells fell into four groups, which he arbitrarily called A, B, AB and O. Human blood serum from any person with type A red cells could be mixed with these cells and no clumping would occur. But if the serum was mixed with cells from a person with type B cells, clumping always occurred.

We now know that people with A cells have antibodies in their serum to B red cells; people with B cells have antibodies to A cells; people with AB cells have no antibodies to red cells; and those with O cells have serum antibodies to both A and B. This means that group AB people can receive blood from anyone and group O people can donate blood to anyone. Transfused serum antibodies are so quickly diluted that they have no effect. It is the clumping of the transfused red cells that causes all the harm. And it was for this reason that earlier attempts at blood transfusion had been so often fatal that the procedures had been been prohibited. Landsteiner's work made transfusion safe and has saved countless lives. He was awarded the Nobel Prize in 1930.

Langerhans cell an antigen-presenting dendritic cell found in the skin. These cells pick up and process antigen in the skin and then move to the nearest lymph node where the antigen is presented to T cells. In this way the immune system may become sensitized to a contact allergen. (Paul Wilhelm Langerhans 1847–88, German physician, who described these cells when he was a medical student).

language a locally agreed system for the transmission of information by the articulation of sounds or the transcription of symbols.

language, origins of

There have been plenty of theories about the origins of language. Indeed, theorizing has been so popular that, as early as 1865, the *Société de Linguistique de Paris* included in its by-laws an article prohibiting discussion on the subject.

Body language has always been a means of communication and remains an important vehicle for the more developed animals. Some scientists believe that since the significance of certain bodily movements and gestures would have been well understood, speech may have originated in verbal imitation of such gestures – the jaw, lip and tongue unconsciously mimicking them, much in the way that a child will move its mouth during early attempts to write. This is the 'mouth-gesture' or 'TA-TA' theory.

Proponents of the 'BOW-WOW' theory hold that speech is purely imitative of natural sounds and arose when man discovered that he could make noises similar to those in nature. Onomatopoetic origins are, of course common and familiar, as when children call dogs 'bow-wows', cows 'moo-moos' and steam trains 'choo-choos'. A similar theory, the 'POOH-POOH' or 'OUCH' theory, suggests that language originated from the 'instinctive' cries made in response to emotions of various kinds. The 'YO-HEAVE-HO' theory would have it that language developed from the grunts and exclamations inseparable from hard labour. The trouble with these theories is the severe limitation they place on vocabulary.

One of the most fanciful ideas was the 'DING-DONG' theory of the German philologist Friedrich Max Müller (1823–1900). Müller was impressed with the concept of resonance and came to believe that there was a mystical harmony between sound and sense. The idea seems to have originated with the Greek philosopher Plato who also believed that there was a correspondence between names and their objects. None of these theories carry much weight nowadays.

It has been suggested that we might discover something about the origins of language by observing how children learn to speak. Some trials are said to have been based on this idea. The Greek historian Herodotus (485–425 BC) records that the Egyptian King Psammatichos isolated two infants in a mountain hut in the hope that they would begin to speak spontaneously. The King decided that the sounds they uttered were Phrygian – believed to be the original language of mankind. King James IV of Scotland is said to have conducted a similar experiment and concluded that the children spoke good Hebrew – the language then supposed to be aboriginal. We can now smile at the naive notion, implicit in these beliefs, that language is inborn rather than environmentally acquired. Even studies of the languages of contemporary primitive societies tell us nothing of origins. Far from being basic and providing clues to what original languages may have been like, such languages are often rich in vocabulary and complex in grammar and syntax.

Do other primates use language?

The answer to this is certainly yes. Although apes do not appear to have the necessary neurological equipment for the articulation of language, some, especially chimpanzees, can certainly understand and use it. The chimp Washoe was taught a vocabulary of 130 words in American Sign Language by the researchers Allan and Beatrice Gardner. Another chimp, Sarah, was able to converse with researcher David Premack using symbols consisting of pieces of plastic of different shape, size colour and texture. Sarah learned almost 130 'words'. These chimps went further and were able, without instruction, to combine symbols to produce new words. A refrigerator, for instance, was designated 'open-eat-drink'. Sarah was found capable of accurate performance of logical operations of the 'if this – then that' type.

There are good physical reasons why animals other than man cannot use speech. Vocal sounds made by chimpanzees and other animals appear to be mainly controlled by centres in the limbic area of the brain – the part responsible for emotional reaction rather than reason. Stimulation of these limbic centres produces the whole range of vocal utterances of which the animal is capable. In humans, vocal articulation is under the control of the cortex of the brain. Destruction by disease of the cortical speech areas in man – a common consequence of a stroke – will abolish speech. Such destruction has no effect on vocalization in lower primates.

It thus appears that, by an evolutionary step, the connections for speech have been transferred from the more primitive instinctive or reflex centres in the lower animals, to the cerebral cortex in man, making possible the voluntary and considered use of language. It is a moot point when this transfer occurred. We do not know, for instance, whether or not Neanderthal man was capable of articulate language. There is evidence that the vocal tract of the Neanderthals was unsuitable for speech. Some scientists go so far as to suggest that the appearance of language may have been the defining characteristic of *Homo sapiens* and that this might explain why *H. sapiens* did not, apparently, interbreed with his Neanderthal contemporaries.

Is there an innate element in language?

American linguist Noam Chomsky (b. 1928) believes so. He holds that the brain is structurally programmed for the organization of language. Without such programming he does not think it possible that children could learn a language, especially its complex grammatical and syntactical elements, so quickly. He holds that the rapidly-acquired competence to formulate an endless variety of sentences could never be explained by mere learning. It is essential, he believes, to postulate the existence of some kind of 'mental organ' operating on built-in sets of linguistic rules dictating what is possible and what is not. Chomsky's ideas, with their emphasis on the existence of linguistic universals at a physiological level, have been very influential.

Pidgins and creoles

These ideas have gained support from the study of *creoles* by the American linguist Derek Bickerton of the University of Hawaii. Creoles are a class of languages that arise spontaneously when people who speak mutually unknown languages are thrown into prolonged close contact as masters and servants. More than 100 creoles are known and their grammatical structures have been studied. The first step in the development of a creole is the use of a pidgin – a simple and inefficient language in which the servant group adopt some of the vocabulary of the masters and use it with some of the grammar of their own language. The children of pidgin speakers, however, typically speak a creole. In this they use the same vocabulary as the pidgin, but with a particular grammar which differs from anything of which they can have had experience. Extraordinarily, all creoles, whether old or recent, share a common, or at least similar, grammar. For instance, they all deal with tense in the same way.

When children are learning their native language, they make characteristic errors or non-conformities. A study of childrens' mistakes in different languages shows that they bear a strong resemblance to creole grammar. For instance, they do not use a change in word order to indicate interrogation, but rely solely on a rising intonation. Also like creole speakers, they often use a negative subject with a negative verb: 'Nobody don't love me'.

So there is a respectable body of opinion, by no means unopposed however, that holds that primitive man's language may have resembled a creole. Changing cultural forces would, of course, impose large changes on languages which would inevitably evolve, in every aspect, to meet contemporary needs.

Did all languages derive from a common source?

This view was, at one time, widely held, largely because of respect for biblical authority. In the words of the *New English Bible* translation of Genesis, Chapter 11: 'Once upon a time all the world spoke a single language and used the same words'. The story of the tower of Babel, that follows, describes how God confused the speech of man and 'made a babble of the language of all the world and scattered man all over the face of the earth'. This first language was long believed to be Hebrew and all the languages of the world were thought to derive from it.

There is no reason to believe that language derived from a single source and most scholars now agree that several different languages arose independently in different parts of the world among people who were geographically isolated from each other. No language is now believed to be, in any sense, basic or 'prior' to any other. Languages change constantly, by simplification and regularization, by loss of inflection, and by borrowing from other languages. Change may be slow or rapid, but is often extreme. Languages sometimes become almost unrecognizable in a matter of only a few hundred years. Children learn the language to which they are exposed from birth easily and naturally. There are no inherent differences in this respect between the different languages.

Language today

Today, some 6000 languages are spoken in various parts of the world and most of these also have regional dialects. It is estimated that about half of these will die out in the next 100 years. Languages are broadly classified into families – there are about 40 in the old world, the most important to the Western world being the Indo-European family to which most European languages and many Asian languages belong. Other families include Afro-Asiatic, Amerindian (e.g. North American Indian languages such as Chinook and Nootka), Austronesian (e.g. Fijian, Kuanua), the Dravidian (e.g. Tamil, Telugu), the Sino-Tibetan (e.g. Chinese, Tibetan), Finno-Ugrian (e.g. Finnish, Lappish, Magyar), Semitic-Hamitic (e.g. Hebrew, Arabic, and other North African languages) and the Altaic (e.g. Turkish, Mongolian).

Because of its commercial, academic and literary importance, English has now become one of the commonest languages spoken and read throughout the world. It is the native language of some 300 million people and is spoken by about 1000 million as a second national language. The number who have learned it as a foreign language is unknown but must be very large. Like all other living languages, it is constantly changing.

lanugo the short, downy, colourless hair that covers the fetus from about the fourth month to shortly before the time of birth. Similar hair sometimes grows on people with cancer, on those taking certain drugs and on girls with anorexia nervosa.

laparo- *combining form denoting* the flank, loin or abdominal wall.

large granular lymphocyte one of a range of larger-than-average lymphocytes with granules in the cytoplasm which function as natural killer and killer cells. The group also includes activated cytotoxic T cells.

large intestine the part of the intestine that extends from the end of the ILEUM to the ANUS. It starts in the lower right corner of the abdomen with the caecum, from which the APPENDIX protrudes, proceeds as the ascending COLON to the upper right corner, loops across to the upper left corner as the transverse colon then descends to the lower left corner as the descending colon. The intestine then swings down and centrally as the sigmoid colon and continues as the rectum and the anal canal. The main function of the colon is to reabsorb water from the bowel contents. The rectum is a temporary store for faeces.

laryng-, laryngo- *combining form denoting* the voice-box (LARYNX).

laryngeal 1 pertaining to the larynx.
2 produced in or with the larynx, as of a sound.

laryngology the branch of medicine concerned with the study and treatment of the LARYNX and its disorders. Laryngology is usually clinically associated with the study of the ear (otology) and of the nose (rhinology). An ear, nose and throat (ENT) specialist is called an otorhinolaryngologist (ORL).

laryngopharynx the lower part of the PHARYNX adjacent to the LARYNX.

laryngotracheal pertaining to the LARYNX and the TRACHEA.

larynx the 'Adam's apple' or voice box. The larynx is situated at the upper end of the wind-pipe (TRACHEA), just in front of the start of the gullet (OESOPHAGUS). At its inlet is a leaf-shaped flap of cartilage, the EPIGLOTTIS, that prevents entry of swallowed food. It has walls of cartilage and is lined with a moist mucous membrane and contains the vocal cords. These are two folds of the mucous membrane that can be tensed by tiny muscles to control their rate of vibration as air passes through them, and hence the pitch of the voice. The gap between the folds is called the glottis.

lassitude a disinclination to make an effort to achieve anything. Lassitude may indicate organic disease or depression, but is often due to boredom from lack of interests.

latent present but not manifest. Not yet having an effect.

lateral of, at or towards the side of the body. From the Latin military word *latus*, a flank or wing. Unilateral means occurring only on one side; bilateral means relating to both sides.

lateral inhibition a process in which the most active sensory nerve fibres in a bundle (i.e. those whose receptors are near the centre of an area of stimulus) inhibit action potentials in adjacent fibres from the periphery of the stimulus area. This increases the contrast between the most relevant and the least relevant information.

laterality 1 pertaining to one side.
2 a tendency to use or occur in one side rather than the other.
3 dominance of one hemisphere of the brain over the other.

lateral ventricles the fluid-filled cavities in each half of the brain (cerebral hemispheres) that communicate with the THIRD VENTRICLE by way of the interventricular foramen.

laterodeviation displacement to one side.

lateroflexion a bending to one side.

latissimus dorsi the broadest muscle of the back. It arises from the spines of the six lower thoracic vertebrae and from the FASCIA attached to the lumbar vertebrae and back of the pelvis and sweeps up and forward to be inserted by a short, broad tendon into the back of the top of the upper arm bone (humerus). It acts to pull the body upward in climbing and assists in heavy breathing.

latus the side of the body. The plural form is latera.

lavage washing out of a hollow organ or cavity, especially by irrigation. Stomach washout.

law of mass action the principle that the rate of a chemical reaction is proportional to the molar concentration of the reacting substances.

lazy eye a lay term applied both to an abnormal inturning of the eye (convergent strabismus or esotropia) and to the defect of vision often resulting from untreated squint in young children (amblyopia).

leading strand in DNA replication, the separated strand this is being converted to a duplex in the direction of the fork, or the opening of the fork, in the parent DNA. Replication is continuous along the leading strand. Compare LAGGING STRAND.

leaflet any small, leaf-like structure such as the cusps of a heart or other valve.

leaky mutant gene one of a pair of genes (ALLELE) that is less active than its normal partner. A leaky mutation produces only partial loss of a characteristic.

The way we behave is largely, but not entirely, a product of learning. Some of our more basic patterns of behaviour are the result of 'built in' neurological circuits. This is often called instinct and is fairly stereotyped. A particular stimulus will nearly always result in a predictable response. But one of the principal differences between humans and the less highly evolved animals is the extent to which human behaviour is determined not by instinct but by learned patterns.

Humans are learning animals, equipped by evolution to be the most efficient learning machines in existence. Learning implied the acquisition and storage of data, by any means and through any portal of sensory intake, and the integration of these data with already stored information so that the behaviour is subsequently modified. Data acquisition causes permanent internal changes but these are barely discernible. The easiest way to discover that learning has occurred is to observe changes in behaviour.

The idea of 'conditioning' is inherently distasteful to many people, as it seems to deprive humans of their freedom of will. But this is only one of a broader class of philosophical problems that arise from the undeniable fact that the more biological knowledge advances the more closely humans appears to resemble machines.

Pavlov

Modern thinking about learning started with the then startling findings of Ivan Petrovich Pavlov (1849–1936), a distinguished Russian physiologist, professor of physiology at the Institute of Experimental Medicine and the Military Medical Academy at St Petersburg. Pavlov knew that meat powder caused salivation when put in the mouths of puppies who had never tasted meat. This was an inherent, unlearned, response, unaffected by experience. Then Pavlov noticed that dogs who had already tasted the powder salivated in anticipation of it. This objective indication of a 'psychic' stimulation to salivation – a conditional response – interested Pavlov and he decided to study the effects of associated stimuli that would not normally occur in a dog's life – such as the ringing of a bell.

Pavlov soon found that dogs that salivated when allowed to see food while a bell was ringing soon salivated to the sound of the bell alone. This he called a conditional response and the process a conditional reflex. (Pavlov's translators got the word wrong and converted 'conditional reflex' into 'conditioned reflex'. This usage is now so well established as to be unalterable).

Pavlov found that for a conditioned reflex to be established the stimuli must occur together or the conditioning stimulus (the bell) must be applied very soon before the unconditioned stimulus (the food). If the food was presented before the bell, conditioning did not occur. If the bell was repeatedly rung without food being given, the reflex was gradually extinguished. Pavlov found that his dogs tended to give the same conditioned response to a number of similar stimuli, such as bells of different pitch but could be trained to discriminate between bells of only slightly different pitch. Other kinds of closely similar stimuli could also be distinguished.

In studying the limits to which closely similar stimuli could be distinguished, Pavlov found that animals straining at these limits became agitated and aggressive and tended to salivate strongly in response to all stimuli. Such animals lost the ability to develop normal conditioned reflexes. The implications of these extraordinary findings were not lost on Pavlov who spent the rest of his life exploring the possibility that much of human behaviour could be accounted for by such conditioned reflexes. His influence on physiological, psychological and philosophical thought was enormous. His principles, as applied to learning are now described as the classical conditioning theory.

Skinner

A further important theory of learning was that of the American behaviourist psychologist, Burrhus F. Skinner (1904–90) who suggested that a certain desired consequence will occur only if the subject makes the required response. The response, or action, is instrumental in obtaining the desired reward. According to Skinner, behaviour can result in reinforcement or punishment. Behaviour that leads to reinforcement is more likely to be repeated than behaviour that leads to punishment.

Reinforcement can be positive or negative. A positive reinforcer is any factor that increases the strength of the response that preceded it. A hungry cat trying to discover how to get out of a cage will try harder if some food is placed outside. A negative reinforcer is any factor whose removal will increase the strength of the preceding response. Primary reinforcers – such as the provision of food – produce their effect without previous association; conditioned reinforcers are those that act as a result of association with primary reinforcers. Money, for instance, is a conditioned reinforcer because we associate it with desirable elementary needs such as food, shelter and clothing.

Observational learning

A great deal of learning occurs in a vicarious way by our observation of the behaviour of other people and of its consequences. Such learning is especially efficient if the consequences of others' behaviour is painful to them. A child will learn to respect road traffic by seeing a dog run over and killed. Children with a natural, healthy fear of large domestic animals may lose their fear by observing other children petting them. Even the observation of fear in others can teach fear.

The concept of observational learning has wide implications. The observation of reinforcement or punishment being imposed on other people has been shown to affect the observer as if the reinforcement or punishment had been directly applied to him or her. One trial demonstrated that children who observed a subject being reinforced for aggressiveness later behaved more aggressively than another group that had observed the subject being punished for aggressiveness.

We are, of course, exposed to an environment providing an enormous mass of potential observational learning, and many factors will determine which elements will have the greatest effect. Clearly, the general nature of the environment is important in this respect, especially in childhood. Factors such as the moral attitudes and educational levels of the parents, social mores, peer group pressures and regular exposure to certain kinds of television programs cannot fail to have a major effect on the content of young minds.

Acquiring skills

Normal life involves the acquisition of a wide range of skills. Some are so basic to normal living that life would be severely restricted without them. Some of the more obvious include motor skills, such as articulating words, walking, running and making small, controlled movements as in writing; social skills, such as learning to relate successfully to others without aggression; arithmetical and other problem-solving skills; and musical and artistic skills. The acquisition of the more basic skills – such as standing, walking and speaking – appears to occur without any conscious volition, but the small child is, in fact, strongly motivated by the environment to practice these skills, and will do so until a high degree of efficiency is reached. Everyday observation of the behaviour of toddlers shows that this motivation operates even before the nervous system is sufficiently mature to allow them to be exercised.

Learning new skills is a slow process requiring repeated attempts. Improvement is not a smooth, gradual process but a series of leaps.

Often these are small, but, not infrequently, progress occurs in substantial jumps. These will not occur spontaneously but only if many attempts are made. This is why practice is so important in the acquisition of skills. A person trained in a skill has achieved some general advantage which, to a varying extent, can be transferred from one activity to another. Manual precision and control, acquired in learning to write, for instance, immediately confers improved efficiency in other delicate manual skills. Such skills can be analyzed and the components common to different activities recognized. Mathematical analyses have been made of the extent to which skills can be transferred in this way.

Ethical learning

Traditionally, questions of personal morality have been very much the province of the philosophers and the theologians, whose approach to these important matters has tended to be either speculative or dogmatic. But in the middle and late 20th century such questions began to be widely investigated as behavioural phenomena and have been the subject of much scientific study. There have been several different approaches.

The Swiss psychologist Jean Piaget (1896–1980) held that the sense of morality is based on nothing more than respect for an imposed system of rules. At first the child has no inherent morality, but ideas of right and wrong are soon imposed by parents and other environmental influences and these are, at first, accepted without question or even thought. Some time between the ages of about 6 and 12 the child gradually rationalizes the dogmatic system into a seemingly logical system of moral principles.

Some psychologists working in this field concentrate their attention on the significance of values and attitudes. Values are described as persistently-held beliefs or standards that relate to a person's goal in life and importantly determine behaviour. Attitudes are less profound and are manifested by whether our responses to various ideas, concepts, people and objects are favourable or unfavourable. When, in practice, attitudes are found to conflict with values, they tend to change.

The social-behavioural approach studies the influence of the social environment in determining moral awareness and the way social contexts affect moral behaviour. There is, unfortunately, no clear evidence of such direct causal relationships. It seems evident that, in most people, behaviour does not quite come up to the standards of the inner morality, nor is even always consistent with conscious ideas of how they should behave. More commonly, people behave in a manner that varies with the social context and this may either reinforce or weaken conformity with their ideas of how they should behave. Ethical behaviour would thus seem to be at least partly dependent on social learning.

The truth probably lies in a combination of all these theories. Moral character is an unstable general attribute, importantly determined by early influences, but currently modulated by the effects of different environments on our conduct. Good behaviour is an arbitrary, context-related, concept, probably requiring sound emotional, intellectual and informational integration.

learning disability a well-meaning euphemism for mental retardation. Other terms include developmental reading disorder and developmental word blindness. The condition should not be confused with DYSLEXIA which is a specific disorder. Young people with learning disability experience exceptional difficulty in acquiring an average standard of education. Learning disability is always apparent by the age of seven or, in severe cases, earlier. In spite of considerable research, the causes and nature remain obscure and controversial. There is no disagreement, however, that in mild cases the best treatment is intensive, individually-tailored, one-to-one instruction in reading and writing by an experienced remedial teacher. Behavioural and emotional problems, often secondary to the learning disability, also require appropriate skilled attention.

lecithinase an enzyme that breaks down lecithin.

lecithins a group of phosphoglycerides, plentiful in egg yolk, and occurring widely in the body, especially as a major constituent of cell membranes and in nerve tissue. Lecithins are important emulsifying agents and surfactants and are concerned, for instance, in the normal expansion of the fetal lung at the time of birth. Also known as phosphatidyl cholines.

lectins a group of proteins that bind firmly to specific small sugar (oligosaccharide) parts of glycoproteins and glycolipids. Most lectins in humans are cell membrane proteins with binding sites on the outside of the membranes, and are concerned in cell migration, tissue formation, fertilization, phagocytosis, agglutination, etc.

leiomyo- *combining form denoting* smooth muscle.

lemniscus a bundle of nerve fibres in the brain or spinal cord.

lens a regular transparent solid having convex or concave surfaces such that incident beams of light are bent so as to converge (convex lenses) or diverge (concave lenses). The CORNEA is the main light-bending lens of the eye. The internal CRYSTALLINE LENS effects fine adjustments to the focus by an alteration of curvature. Fixed lenses are used in spectacles or in contact with the corneas to correct inherent errors of eye focusing.

lens implant a tiny, plastic lens of high optical quality supported by delicate plastic loops, that is inserted into the eye at the end of a cataract operation to replace the opaque natural lens that has been removed.

lenticular pertaining to, or shaped like, a lens.

lentiform 1 lens- or lentil-shaped.
2 pertaining to the lenticular nucleus of the BASAL GANGLIA of the brain.

lenticulo-striate pertaining to the lenticular nucleus and CORPUS STRIATUM of the brain and their interconnections.

lentigo a local concentration of pigment-containing cells (melanocytes) in the skin. Lentigos (often called lentigenes) resemble freckles but occur as commonly on covered as on uncovered parts and do not become less conspicuous in winter time. They are harmless.

leper a person suffering from leprosy (Hansen's disease).

-lepsis *combining form denoting* a seizure.

leptin a hormone produced by fat cells which signals the state of repletion. Leptin is a protein of 167 amino acids coded for by the ob gene (for 'obesity'). Its receptor is expressed in the hypothalamus in an area known to be concerned with satiety and hunger. Leptin was at first thought to be the complete answer to appetite control but its action in humans has been found to be more complex than its action in rats. Only about one fifth of people respond to a high concentration of leptin by reducing food intake. Leptin provides a feedback signal from fat to the nervous system, stimulates the neurons that express proopiomelanocortin. When this molecule is split by proteolytic enzymes the anorexogenic peptide alpha MELANOCYTE-STIMULATING HORMONE is produced.

lepto- *combining form denoting* fine, soft, delicate, slender or weak.

leptomeningeal pertaining to the two soft inner layers of membrane that surround the brain, the PIA MATER and the ARACHNOID MATER, but not the DURA MATER.

leptomeningitis inflammation of the two fine inner layers of the MENINGES, the PIA and ARACHNOID maters.

lesbianism female homosexuality. Long-term stable lesbian relationships are common, as in male homosexuality. The term comes from the Greek female poet Sappho, who lived on the island of Lesbos with her followers during the 7th century BC.

lesion a useful and widely used medical term meaning any injury, wound, infection, or any structural or other form of abnormality anywhere in the body. Doctors would be at a loss without this term, but it is commonly wrongly regarded by lay people as implying some specific condition such as an adhesion. The word is derived from the Latin *laesio*, an attack or injury.

lethal injection an increasingly widely-used method of execution of condemned criminals. Since the method involves intravenous skills rarely available to non-medical people, doctors are commonly asked to advise or even participate – which few, if any, can contemplate without revulsion.

lethal mutation a mutation whose effect is so serious that the cell is killed. Lethal mutations will often result in the death of the organism concerned.

lethargy an abnormal state of apathy, sleepiness, drowsiness or lack of energy. Lethargy may be due to organic brain disease or to DEPRESSION. In Greek mythology the river Lethe flowed through Hades and the dead were required to drink its water so as to forget their past lives.

leuc-, leuco-, leuk-, leuko- *combining form denoting* white or LEUCOCYTE.

leucine one of the essential AMINO ACIDS.

leucoblast an immature white blood cell.

leucocyte any kind of white blood cell. The leukocytes include the neutrophil POLYMORPHONUCLEAR LEUKOCYTES ('Polymorphs'), EOSINOPHILS, BASOPHILS, LYMPHOCYTES and MACROPHAGES.

leucocytosis an increased concentration of white cells (LEUKOCYTES) in the blood other than one caused by one of the LEUKAEMIAS. Leukocytosis is an important indication that an inflammatory, usually infective process is occurring somewhere in the body.

leuconychia white discoloration of, or white patches in, the fingernails.

leucopenia an abnormal reduction in the number of circulating leucocytes.

leucorrhoea any whitish VAGINAL DISCHARGE containing mucus and pus cells. Leucorrhoea is purely descriptive and is not a specific disease.

leukaemias a group of blood disorders in which white blood cells reproduce in a disorganized and uncontrolled way and progressively displace the normal constituents of the blood. The leukaemias are a form of cancer (neoplasia) and unless effectively treated are usually fatal. Death occurs from a shortage of red blood cells (ANAEMIA), or from severe bleeding or from infection. The different types of leukaemia arise from different white cell types and have different outlooks.

leukotrienes powerful chemical agents released by MAST CELLS, basophil cells and

MACROPHAGES and involved in many allergic and other immunological reactions. They can be inhibited by corticosteroid drugs.

levator 1 any muscle that acts to raise a part of the body.
2 an elevator. A surgical instrument used to prize up a depressed piece of bone as after a fracture of ZYGOMA or skull.

Levi-Strauss, Claude while workers like Margaret Mead and Bronislav Malinowski went out into the field to gather material, the French philosopher, social theorist and anthropologist Claude Levi-Strauss (1908–) stayed at home to think about the implications of the mass of material accumulated by such workers, world-wide. From his scholarly studies of this material he has tried to detect the underlying structures of thought characteristic of primitive societies and to deduce from these the structural basis of all human thought. Levi-Strauss has been a formulator of theories on a grand scale. Applying the structuralist methods of linguistics to such matters as kinship, he showed that the cultural features of primitive societies were actually collections of codes that reflected the universal principles of human thought.
Since publication of his first major work *The Elementary Structures of Kinship* (1949, translated 1962) Levi-Strauss has exerted a strong influence on contemporary anthropology. Perhaps his greatest influence, however, arose from his studies of myth, articulated in the large four-volume work *Mythologiques* (1964–72, translated 1970–81). In this he demonstrates the systematic ordering of the codes of expression and argues that the real meaning of a myth lies not in its literary content or symbolism but in the underlying relationships of all its elements. This is accessible only by structuralist analysis.
Levi-Strauss's influence has extended far beyond anthropology into art, architecture, literature and sociology. His other writings include *Tristes Tropiques* (1955, translated 1964), the two-volume *Structural Anthropology* (1958, translated 1963 and 1976), and *The Savage Mind* (1962, translated 1966).

levocarnitine a nitrogenous muscle constituent necessary for the transport of long-chain fatty acids across the inner mitochondrial membrane. It is used as a drug to correct a deficiency of carnitine in people on dialysis.

Leydig cells cells lying between the seminiferous tubules in the testis that secrete testosterone. (Franz Leydig, 1821–1908, German zoologist).

libido sexual desire or its manifestations. In psychoanalytic theory, the term is used more generally to mean the psychic and emotional energy associated with instinctual biological drives.

lie the position or attitude of the fetus in the womb in relation to the long axis of the mother's body. Lie may be longitudinal (normal) or transverse.

lie detector a popular term for the polygraph – a collection of devices used to monitor and record various parameters of the body, such as the pulse rate, the blood pressure, the evenness and rate of breathing and the moistness, and hence the electrical resistance, of the skin. These vary with the state of the emotions and the results can be thought to cast light on significance to the subject of certain questions or statements. Emotional responses do not, however, necessarily indicate that the subject is lying or concealing the truth. Lie detection is a function of the interpreter, not the machine and it is the sensitivity, intelligence, imagination and experience of the operator that determines the forensic value of the procedure. This should always be challenged if lie detector evidence is used in court.

lienal pertaining to the SPLEEN.

lienorenal pertaining to the spleen and kidney.

life expectancy a statistical estimate of the number of years a person, of any particular age, is likely to live.

life support system medical equipment used to maintain respiration or the heart action, and possibly nutrition, in a person unable to survive without such support.

ligamentous 1 of the characteristics of LIGAMENTS.
2 pertaining to a ligament or ligaments.

ligaments bundles of a tough, fibrous, elastic protein called COLLAGEN that act as binding and supporting materials in the body, especially in and around JOINTS of all kinds. Ligaments are flexible but very strong and, if excessively strained, may pull off a fragment of bone at their attachment.

ligand a molecule or ION that binds to a central chemical entity by non-covalent bonds. A general term for any molecule that is recognized by a surface receptor.

ligand-sensitive channel a cell membrane channel that opens under the influence of the binding of a specific molecule to the channel protein.

ligase an ENZYME that promotes the linkage of chemical groups. Also known as a synthetase.

light chains the short polypeptide chains that form the outer parts of the upper arms of the Y-shaped ANTIBODY molecule. The light chains of all antibodies have one length (domain) that is the same for all (the constant domain) and one terminal domain that varies in its amino acid sequence (the variable domain). A domain is a single loop of about 110 amino acids. See also HEAVY CHAINS.

lightening the sense of relief felt, during the last three or four weeks of pregnancy, with the descent of the presenting part of the fetus, usually the head, more deeply into the pelvis so that the womb occupies a smaller volume of the abdomen. This reduces abdominal distention and occasions some easing of breathing and general discomfort. Lightening is usual in first pregnancies but may not be apparent later.

limbal pertaining to a LIMBUS.

limbic system a centrally situated, ring-shaped structure in the brain consisting of a number of interconnected nerve cell nuclei. The limbic system represents much of what constitutes the brain in the lower mammals and is concerned with unconscious and automatic (autonomic) functions such as respiration, body temperature, hunger, thirst, wakefulness, sexual activity and their associated emotional reactions. Diseases of the limbic system cause emotional disturbances, and these can include emotional lability, forced or spasmodic laughing and crying, aggression, anger, violence, placidity, apathy, anxiety, fear, depression and diminished sexual interest.

limbus an edge or distinct border, especially the margin of the CORNEA.

linea a line.

linea alba the central strip of white FASCIA running down the front wall of the abdomen from the bottom of the breastbone (STERNUM) to the PUBIS.

linear epitope a section of a protein, consisting of a sequence of amino acids, to which an antibody can bind. Compare CONFORMATIONAL EPITOPE.

linear regression a statistical method of predicting the value of one variable, given the other, in a situation in which a correlation is known to be significant. The equation is $y = a + bx$ in which x and y are, respectively, the independent and dependent variables and a and b are constants. This is an equation for a straight line.

lingual pertaining to, or pronounced with, the tongue.

lingual artery a branch of the external CAROTID artery that runs forward underneath the tongue to supply the tongue muscles.

lingual nerve a sensory nerve providing sensation to the inside of the jaw and tongue. It is a branch of the mandibular nerve – the lower of the three divisions of the great sensory nerve of the face, the trigeminal nerve.

lingula any tongue-shaped structure or process.

lingular tongue-shaped.

linin the delicate, thread-like material in the cell nucleus to which chromatin granules appear to be attached.

linkage 1 the location of genes on the same CHROMOSOME so that the characteristics they determine tend to remain associated.
2 the tendency of genes to remain together during recombination. This is proportional to their proximity to each other. Sex linkage simply implies that the particular gene is located on an X or a Y chromosome.
3 the force that holds atoms together in a molecule.

linkage disequilibrium the occurrence of combination of genes (linkages) in a population more often, or less often, than would be expected from their distance apart in the genome.

linkage map a CHROMOSOME map showing the relative positions of known genes.

Linnaean pertaining to the system of taxonomic classification and the binomial nomenclature widely used in biology, in which the name of the genus (generic name) is followed by the name of the species (specific name). (Carolus Linnaeus, or Carl von Linne, 1707–78, Swedish biologist).

linoleic acid the principle fatty acid in plant seed oils. An essential polyunsaturated fatty acid, interconvertible with LINOLENIC ACID and arachidonic acid and needed for cell membranes and the synthesis of PROSTAGLANDINS. It is plentiful in vegetable fats. Essential fatty acid dietary deficiency is rare.

linolenic acid an essential fatty acid. Like LINOLEIC and arachidonic acids it is polyunsaturated and found in vegetable oils and wheat germ.

lip-, lipo- *combining form denoting* fat or lipid.

lipaemia an increase in amount of emulsified fat in the blood, causing undue turbidity of the PLASMA.

lipase an enzyme that catalyzes the breakdown (hydrolysis) of fat molecules to glycerol and fatty acids.

lipectomy fat removal, usually by suction. A form of cosmetic plastic surgery designed to improve body contours. Small incisions are made in the skin and a blunt-ended metal sucker is passed through and moved around under the skin to suck out fat cells. Human frailty being what it is, the effect is usually temporary.

lipid profile a clinical chemistry assessment of the levels of fats in a patient's blood. The measurements include total cholesterol, total triglycerides, high- and low-density lipoproteins, and sometimes apolipoprotein E.

lipids see FATS.

lipochromes natural fatty pigments, such as carotene or LIPOFUSCIN.

lipofuscin a golden-brown pigment that occurs in granules in muscle and nerve cells in numbers proportional to the age of the individual. Also known as age pigment.

lipolysis lipid breakdown (HYDROLYSIS) into free fatty acids and glycerol under the influence of LIPASE. This is increased in DIABETES. Also known as adipolysis.

lipolytic pertaining to the breakdown of fat (LIPOLYSIS).

lipoproteins any complex of fats with protein. A conjugated protein consisting of a simple protein combined with a fat (lipid) group. The blood lipoproteins, which are the cholesterol carriers of the body, are classified by density, in accordance with the proportions of protein, as very low density (VLDL), low density (LDL) and high density (HDL). LDLs contain relatively large amounts of cholesterol. HDLs contain 50 per cent of protein and only 20 per cent of cholesterol. LDLs transport lipids to muscles and to fat stores and are associated with the arterial disease atherosclerosis and thus heart disease and stroke. HDLs are protective against these diseases because their main role is to transport cholesterol from the periphery back to the liver. They also carry paraoxanase enzymes that limit oxidative modification of LDLs necessary before cholesterol can be laid down in arterial walls. Blood concentration of HDL cholesterol shows a strong inverse correlation with the risk of coronary heart disease.

lipoxygenase an enzyme that catalyzes the production of leukotrienes from arachidonic acid.

lipping the formation of a curled edge at the bearing joint surface of a bone in OSTEOARTHRITIS and other degenerative bone disease.

lip-print the lip analogue of the fingerprint. Lip-prints are much less reliable as means of identification because, unlike fingerprints, they change with age.

lipreading a means of communication with the deaf. English speech involves more than forty distinct sounds but less than ten visibly distinguishable mouth patterns can be reliably identified. Other facial, bodily and contextual clues are, however, provided and a skilled lipreader can often discern or infer 60 per cent of spoken information.

lisp an anomaly in the production of 'ssssss' sounds (sibilants) in speech in which the tip of the tongue is protruded between the teeth instead of being placed high and close to the hard palate behind the upper front teeth. The lisp is largely under voluntary control and can be corrected by speech therapy.

-lith *suffix denoting* stone, as in FAECALITH.

lithiasis the formation of stones (calculi) anywhere in the body, but especially in the urinary and bile secretion and storage systems.

litho- *combining form denoting* stone.

lithotomy position the position in which a patient is placed for gynaecological operations or for any surgical procedure on the PERINEUM. The patient lies on his or her back with the knees up and the thighs spread wide. The feet and thighs are usually supported in slings.

litmus a powder derived from certain lichens that contains the natural dye azolitmin. This turns red in an acidic medium at pH below 4.5 and blue at an alkaline pH above 8.3. Paper strips impregnated with litmus form convenient indicators for checking urine acidity.

liver the largest organ of the ABDOMEN occupying the upper right corner and extending across the midline to the left side. It is wedge-shaped, with the thin edge pointing to the left, of a spongy consistency, reddish-brown in colour and moulded to fit under the domed DIAPHRAGM so that most of it lies behind the ribs. The liver receives chemical substances in the blood, especially in the nutrient-rich blood from the intestines (glucose, amino acids, fats, minerals and vitamins) and processes these according to the needs of the body. It takes up the products of old red blood cells and converts these into a pigment, bilirubin, which together with other substances, form the bile. It breaks down toxic substances into safer forms. Ammonia produced from protein breakdown is converted into urea, which is excreted in the urine. Alcohol and other drugs are altered to safer forms. To a remarkable degree, the liver is able to regenerate itself after disease, toxic damage or injury. But if this capacity is exceeded, functional liver cells form nodules and are replaced by inert fibrous tissue (cirrhosis) and the whole function of the body is severely affected.

liver failure the end stage of severe liver disease in which liver function is so impaired that it cannot meet the metabolic needs of the body. There is JAUNDICE, an accumulation of toxic substances in the blood such as ammonia, fatty acids and nitrogenous compounds causing a sweet musty odour in the breath, nausea and vomiting and brain damage with restlessness, disorientation, coarse tremor of the hands, sometimes aggressive outbursts, convulsions, weakness, coma and death.

livid black or bluish-black discoloration from accumulation of free blood in the tissues. Bruised or 'black-and-blue'. Post mortem lividity is the extensive bruising seen in the dependent areas of the body, indicating the position in the first 8–12 hours after death.

living will a document requesting and directing what should be done in the event of a person's later inability to express his or her wishes on medical management. The purpose is usually to try to ensure that exceptional measures are not taken to maintain life in the event of a terminal illness. The respecting of such a will has long been accepted in most States in the USA and has, since January 1998, also been a statutory right in Britain. The term refers to the fact that the writer's deposition may be enacted when he or she is still living. The Voluntary Euthanasia Society has recently produced a new draft will that also provides an opportunity for the patient to express the desire to be kept alive for as long as is reasonably possible.

LOA *abbrev. for* left occipitoanterior, a common position of the fetus in the womb, with the head down and the back of the head pointing to the front and a little to the left.

lobe a well-defined subdivision of an organ. Many organs, such as the brain, the lung, the liver, the pituitary, the thyroid gland and the prostate gland, are divided into lobes.

lobule a small LOBE or subdivision of a lobe.

lochia the discharge of blood, mucus and particles of tissue from the womb, mainly coming from site of the afterbirth (PLACENTA), during the first 2 or 3 weeks after birth. The discharge is red for the first 3 or 4 days and usually disappears by about the tenth day. Offensive-smelling lochia suggests infection and is a danger sign.

locked-in syndrome a state of total paralysis, except for eye movements, in which the victim remains conscious and able to communicate by eye movement codes. This nightmarish situation usually results from a basilar artery haemorrhage or thrombosis, or other damage, affecting the ventral pons with preservation of the dorsal tegmental area. This destroys almost all motor function, but leaves the higher mental functions intact. Compare PERSISTENT VEGETATIVE STATE.

locomotor pertaining to the function of voluntary movement.

loculated divided into small spaces, compartments or cavities.

locus the position on a chromosome at which the gene for a particular characteristic resides. The plural is loci. A locus can contain any of the ALLELES of the gene.

logarithmic growth phase the stage of growth in which cells are doubling in number during consecutive equal lengths of time. This rapidly leads to an enormous increase in the number of cells.

logo- *combining form denoting* word or speech.

-logy *suffix denoting* study, science, theory, thesis or creed.

loin the soft tissue of the back, on either side of the spine, between the lowest ribs and the pelvis. Compare GROIN.

longsightedness see HYPERMETROPIA.

long-term potentiation a persistent increase in effectiveness of some synapses when frequently activated.

lop-eared having ears with a folded down or drooping upper border.

lordosis an abnormal degree of forward curvature of the lower part of the spine, often associated with abnormal backward curvature of the upper part (KYPHOSIS). Lordosis is an exaggeration of the normal forward curve and often causes the buttocks to appear unduly prominent.

low-density lipoproteins the complexes by which cholesterol and other fats (lipids) are transported in the blood in conjunction with protein. High levels of low-density LIPOPROTEINS are associated with the serious arterial disease of atherosclerosis.

LSD *abbrev. for* lysergic acid diethylamide, a hallucinogenic drug derived from lysergic acid, once used in psychiatric research and treatment but now largely confined to illicit use. The drug is a powerful SEROTONIN antagonist and can induce a psychotic state with paranoid delusions that can last for months.

L-selectin an adhesion molecule that triggers off the interaction between leukocytes and blood vessel endothelium. It is shed from the surface membrane of haematopoietic cells and circulates in the blood as a functional receptor.

lumbar relating to the LOINS and lower back.

lumbar puncture passage of a needle between two vertebrae of the spine, from behind, into the fluid-filled space lying below the termination of the spinal cord. Lumbar puncture is usually done to obtain a sample of cerebrospinal fluid for laboratory examination in the investigation of disorders of the nervous system. It also allows antibiotic drugs, anaesthetic agents and radio-opaque substances to be injected.

lumbo- *combining form denoting* the LOINS or lumbar.

lumbosacral pertaining to the region of the LUMBAR vertebrae and the curved central bone at the back of the pelvis (SACRUM).

lumbrical 1 pertaining to the LUMBRICAL MUSCLES.
2 pertaining to, or resembling, an earth worm, especially in reference to the intestinal parasite Ascaris lumbricoides.

lumbrical muscles the four small intrinsic muscles of the hand lying between the METACARPAL BONES and acting on tendons of another muscle, the flexor digitorum profundus, to straighten the fingers and bend the joints between the fingers and the palms. Similar muscles, with similar actions, occur in the feet and assist in walking.

lumen the inside of any tube, such as a blood vessel, an air passage (bronchus) or the intestine.

luminal pertaining to a LUMEN.

luminous emitting or reflecting light.

lumpectomy a minimal operation for breast cancer in which no attempt is made to remove more than the obvious lump. Supplementary treatment with radiation or chemotherapy is then given.

lunacy a legal term for psychotic disorder, no longer used by doctors. The origins of the word derive from the old belief that madness was caused by the full moon. From the Latin *luna*, the moon.

lunate bone one of the bones of the wrist (carpal bones). The lunate is the middle of the three bones in the row nearest to the forearm.

lung the paired, air-filled, elastic, spongy organ occupying each side of the chest and separated by the heart and the central partition of the chest known as the mediastinum. Each lung is surrounded by a double-layered membrane called the PLEURA. The function of the lungs is continuously to replenish the oxygen content of the blood and to afford an exit path from the blood for carbon dioxide and other unwanted gases. The right lung has three lobes and the left two. An air tube (BRONCHUS) and a large artery and vein enter each lung on its inner aspect and these branch repeatedly as they pass peripherally. The smallest air passages end in grape-like clusters of air sacs, the alveoli, the walls of which are very thin and contain the terminal branches of the blood vessels. In this way the air comes into intimate contact with the blood so that interchange of gases can readily occur.

lung cancer an inaccurate term usually referring to cancer of the lining of one of the air tubes (bronchi). The medical term is bronchial carcinoma. This tumour accounts for more than half of all male deaths from cancer and the incidence in women is rising rapidly. In most cases it is caused by cigarette smoking.

lung disorders these include actinomycosis, anthracosis, baggasosis, bird fancier's lung, bronchial asthma, bronchitis, bronchopneumonia, emphysema, farmers' lung, haemothorax, laryngotracheobronchitis, legionnaires' disease, lung abscess, lung cancer, obstructive airway disease, pigeon fancier's lung, pneumonia, pneumothorax, pulmonary embolism, pulmonary fibrosis, pulmonary hypertension, pulmonary oedema, respiratory distress syndrome, sarcoidosis, silicosis, tracheitis and tuberculosis.

luteal pertaining to the CORPUS LUTEUM, its functions and its hormones, especially progesterone.

luteal phase the phase of the menstrual cycle following ovulation when the CORPUS LUTEUM is active in the ovary.

luteinizing hormone a hormone released by the PITUITARY GLAND that stimulates egg production (ovulation) from the ovary, in the female, and testosterone from the testicle, in the male. In men, rising blood levels of testosterone inhibit secretion of luteinizing hormone, while, in women, rising levels of oestradiol (estradiol) prompt an increased secretion of luteinizing hormone in the middle of the menstrual cycle.

luteoma a collection of multiplied CORPUS LUTEUM cells in the ovary sometimes occurring in the last 3 months of pregnancy. The luteoma occasionally secretes male sex hormones. It is not a true tumour and regresses after the baby is born.

luxation dislocation.

lymph-, lympho- *combining form denoting* LYMPH or lymphatic tissue.

lymph tissue fluids drained by the lymph vessels and returned to the large veins. Lymph varies in character in different parts of the body. Lymph from the tissues contains large numbers of white cells, mainly LYMPHOCYTES, and is usually clear. Lymph from the intestines is milky, especially after a meal, because of the large number of fat globules which it contains. Fat-laden lymph is called CHYLE.

lymphadenopathy any disease process affecting a LYMPH NODE. Also known as lymphadenosis.

lymph glands the incorrect term for LYMPH NODES. These are not glands, although commonly so described, even by doctors.

lymph nodes small oval or bean-shaped bodies, up to 2 cm in length, situated in groups along the course of the LYMPH drainage vessels. The nodes have fibrous capsules and are packed with lymphocytes. The main groups of lymph nodes are in the groins, the armpits, the neck, around the main blood vessels in the abdomen, in the MESENTERY and in the central partition of the chest (the MEDIASTINUM). Lymph nodes offer defence against the spread of infection by producing ANTIBODIES, and become involved in the spread of cancer. They can become cancerous, forming LYMPHOMAS.

lymphoblast an immature LYMPHOCYTE.

lymphocytes specialized white cells concerned in the body's immune system. Several different types can be distinguished. B lymphocytes produce antibodies (IMMUNOGLOBULINS) and are divided into the plasma cells that secrete the immunoglobulins and memory cells that act when the event that stimulated antibody selection recurs. T lymphocytes help to protect against virus infections and cancer and are divided into helper cells, suppressor cells, cytotoxic cells, memory cells and mediators of delayed hypersensitivity. There are also large granular lymphocytes. These are the KILLER CELLS (K cells) and the natural killer cells (NK cells).

lymphocytosis a abnormal increase in the number of LYMPHOCYTES circulating in the blood. A form of LEUKOCYTOSIS.

lymphoid pertaining to LYMPH or lymphatic tissue.

lymphokine-activated killer cells cells artificially produced by incubating lymphocytes from cancer patients with INTERLEUKIN-2 (IL-2). The result is a class of natural-killer cells whose targets are not restricted to cells carrying particular antigens. They offer potential as treatment modalities.

lymphokines CYTOKINES produced by lymphocytes. Lymphokines attract MACROPHAGES to the site of foreign material and activate them to kill organisms, cause other T cells to clone and provide other cells with protection against virus invasion. They include INTERFERONS, CHEMOTACTIC factor, transfer factor and INTERLEUKIN-2.

lymphomas a group of cancers of lymphoid tissue, especially the lymph nodes and the spleen. There are two kinds. If certain large, irregular, multinucleated cells, called Reed-Sternberg cells, are present, the disease is called Hodgkin's Lymphoma. If not, it is called a non-Hodgkin's Lymphoma. Ninety per cent of non-Hodgkin's lymphomas are of clonal masses of B cells, 10 per cent of T cell origin. They vary considerably in their degree of malignancy and have many features in common with certain leukaemias. There is tiredness, loss of weight and sometimes fever. At a certain stage there may be pressure on various structures of the body. This may cause paralysis by compression of the spinal cord, difficulty in swallowing from pressue on the oesophagus, difficilty in breathing, obstruction of the bowel causing vomiting, and obstruction of the lymph vessels causing lymphoedema. Treatment depends on the cell type and on the extent of spread. In some cases, no treatment is needed and often patients are watched for years without intervention. But when treatment is required, radiotherapy is often best and may be curative.

lynching a manifestation of mob violence in which a person, conceived to be guilty of some offence, is killed, usually by hanging, without due process of law.

lyo-, lyso- *combining form denoting* loosened or dispersed.

Lyon hypothesis see X-INACTIVATION.

Lyonization see X-INACTIVATION.

lyophilic 1 readily dissolving.
2 of a colloid, quickly dispersing because of an affinity between the dispersed particles and the dispersing medium.

lyophilized freeze dried.

lysin any substance capable of causing LYSIS, especially a specific antibody that brings about a complement fixation reaction.

lysis the destruction of a living cell by disruption of its membrane. Haemolysis is lysis of red blood cells. This will occur if the cells are placed in plain water.

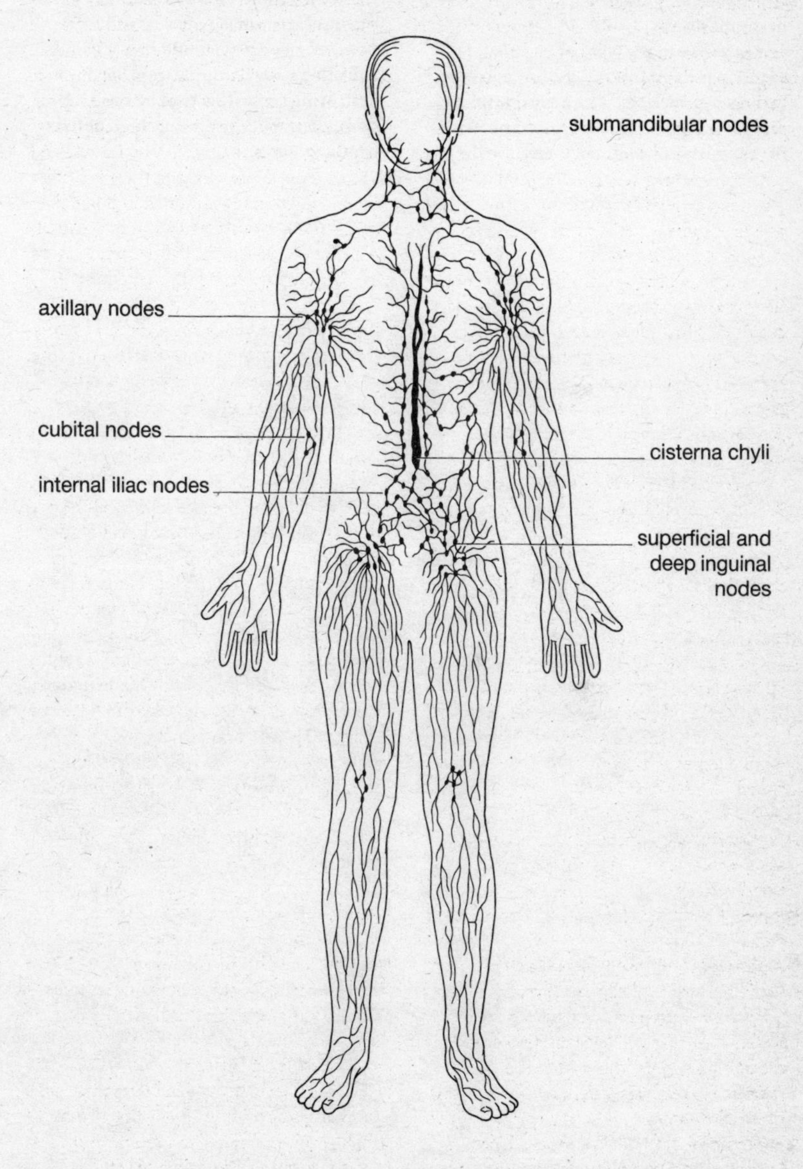

Fig. 10 **Lymph nodes**

lyso- *combining form denoting* LYSIS or decomposition.

lysosome one of the types of ORGANELLE found in cell cytoplasm. Lysosomes contain various hydrolytic enzymes capable of digesting large molecules (macromolecules), the products of which can then leave the lysosomes. Injury to lysosomes may release enzymes that can damage the cell.

lysozyme an enzyme found in tears, milk and other body fluids and capable of destroying certain bacteria by breaking down their walls by the digestion of their peptidoglycans.

Mm

macerated softened by prolonged contact with liquid.

macrobiotics a non-scientific system of diet based on the yin-yang (opposing, light–dark, male–female) principle. The principles of sound nutrition are well established on demonstrable scientific grounds that do not involve this idea. Over-enthusiastic adherence to a macrobiotic dietary could lead to ill-health.

macrocephalous having an abnormally large head.

macrocheilia abnormally large lips.

macroencephaly the state of having an abnormally large brain.

macrocrania having an abnormally large skull in proportion to the size of the face.

macrodactyly abnormally large fingers or toes.

macrodontia abnormally large teeth.

macrogamete the mammalian ovum, which is much larger than the male sex cell, the microgamete or spermatozoon.

macroglia one of the two forms of neurological connective tissue (glia).

macroglobulin see IgM.

macroglossia enlargement of the tongue.

macrognathia an abnormally large jaw.

macromastia abnormally enlarged breasts.

macromelia abnormal enlargement of one or more limbs.

macromolecule a very large molecule, such as a protein or other long polymer. DNA is a macromolecule, as is the polysaccharide glycogen.

macronormoblast an abnormally large form of any of the precursor series of cells that lead to mature red blood cells.

macrophage an important cell in the immune system. Macrophages are scavenging cells, large PHAGOCYTES derived from blood MONOCYTES, and are found all over the body, especially in the liver, lymph nodes, spleen and bone marrow. Some are stationary within the tissues (fixed macrophages), others are free and move about, being attracted to sites of infection. Connective tissue (fixed) macrophages are called histiocytes; those in the liver are called Kupffer cells; and those in the nervous system are called microglial cells. Macrophages are amoeboid and ingest foreign material and bacteria, which they destroy. In order to bring about an immune response, most ANTIGENS must first be processed by macrophages so that their antigenic elements can be presented to LYMPHOCYTES on the macrophage surfaces. The term derives from the Greek *macros*, large and *phageo* to eat.

macrophage activation factor a LYMPHOKINE that prompts a MACROPHAGE into action.

macroscopic visible to the naked eye.

macula any small flat spot.

macula lutea the yellow spot in the centre of the RETINA on which the image of the point of greatest visual interest falls when something is observed. The macula is the most sensitive part of the retina and is devoid of blood vessels. Here, the concentration of colour-sensitive cones is maximal and the visual resolution is greatest. The full visual acuity is possible only by the use of the centre of the macula – the fovea.

magic the use of spells, thoughts or ritual to attempt to influence unrelated events or to summon supposed supernatural forces. Historically, there has been considerable overlap between religion, magic and science and these have been separated only with great difficulty and, even today, incompletely. The reduction, by science, of the status of magic to childish superstition has been one of the triumphs of human achievement. Magical thinking is a feature of early childhood and of the mental processes of many adults.

major histocompatibility complex (MHC) cell surface protein markers, coded for by a large cluster of genes on chromosome 6, that control the activities of cells of the immune system. The MHC molecules indicates the tissue type and are important in organ donation. They have, however, wider functions in the immune system. Infected cells used their MHC sites to signal the fact to helper T cells and cytotoxic T cells so that they can be attacked. There are two classes of MHC. Class I MHC molecules are present on virtually all body cells other than red blood cells; class II MHC molecules occur on antigen-presenting cells such as MACROPHAGES and B cells. Cytotoxic T cells (CD8) bind to MHC class I, while helper T cells (CD4) bind to MHC class II. MHC variations have been used extensively in human population studies.

major histocompatibility locus a region on chromosome 6 that contains a large number of genes coding for lymphocyte cell surface ANTIGENS that determine the tissue type and immunological responses of the individual.

major immunogene complex a gene region that codes for the HISTOCOMPATIBILITY ANTIGENS, for the LYMPHOCYTE surface antigens, for the control factors of the plasma cells that produce antibodies, and for the proteins of the COMPLEMENT system.

mal- *combining form denoting* disease, bad, abnormal or defective.

malabsorption one of a number of disorders in which there is a failure of movement of some of the elements of the diet from the small intestine into the bloodstream so that malnutrition may occur in spite of an adequate diet.

-malacia *suffix denoting* softening.

malady any disease, disorder or illness.

malaise a vague general term for feeling unwell. Although included in the list of symptoms of most diseases, the term has no diagnostic value.

malar bone see ZYGOMA.

malar relating to the cheek bone (zygoma) or to the prominence caused by the cheek bone. From the Latin *mala*, the cheek or cheekbone, and perhaps also the Latin *malum*, an apple.

maleruption abnormal positioning, or failure of appearance, of teeth.

male symbol ♂ – the universally used biological symbol for the male. The symbol is the zodiac sign of Mars the Roman God of war, epitomizing maleness, and represents a shield and a spear rather than an erect penis.

malformation any bodily deformity or structural abnormality resulting from a defect in development or growth.

malignant a term usually applied to cancerous tumours but also used to qualify unusually serious forms of various diseases tending to cause death unless effectively treated. The term is opposite in meaning to benign and derives from the Latin *malignus*, evil.

malingering a pretence to be suffering from a disease, or the simulation of signs of disease, so as to gain some supposed advantage such as avoidance of work or of presumed danger, or to obtain money by fraudulent claims for compensation.

Malinowski, Bronislaw the Polish-born British anthropologist Bronislaw Malinowski (1884–1942) was the main founder of modern social anthropology. After studying physics, mathematics, psychology and sociology he was inspired by a reading of Fraser's *Golden Bough* to undertake a research project in Australia in 1914. Prevented by the outbreak of war, he was restricted to the Trobriand Islands near New Guinea where for four years he lived among the islanders studying their lives and work in great detail. In 1922 he published his findings in the book *Argonauts of the Western Pacific*. This was followed by *Crime and Custom in Savage Society* (1926) and *Sex and Repression in Savage Society* (1927). The quality of this work led to his appointment in 1927 as the first Professor of Social Anthropology at the London School of Economics.

He later did fieldwork in Africa, the Oaxaca Valley, Mexico and elsewhere and is acknowledged as the originator of modern methods of ethnographic fieldwork. Malinowski was appointed to a visiting professorship at Yale University in 1939. His later publications include *A Scientific Theory of Culture* (1944), *The Dynamics of Culture Change* (1945) and *Magic, Science and Religion* (1948). Much of what he wrote was in support of his belief that the origins of social and cultural institutions are to be found in the basic physiological and psychological needs of the human being.

malleolus either of the two bony protuberances on either side of the ankle. The inner (medial malleolus) is a process on the lower end of the TIBIA, the outer (lateral malleolus) is a process on the lower end of the FIBULA.

malleus the outermost and largest of three small bones of the middle ear, the auditory ossicles. From the Latin *malleus*, a hammer.

malnutrition any disorder resulting from an inadequate diet or from failure to absorb or assimilate dietary elements. The term is now often used to describe the effects of an ill-chosen, even if calorifically adequate, diet or of excessive food intake. Relevant diseases include sprue, malabsorption, coeliac disease, Crohn's disease, anorexia nervosa, vitamin deficiency, beri-beri, pellagra, scurvy, xerophthalmia, rickets and kwashiorkor.

malocclusion a poor physical relationship between the biting or grinding surfaces of the teeth of the upper jaw and those of the lower. Malocclusion is readily correctable by orthodontic treatment.

Malpighian corpuscle a tiny spherical body, one of millions in the kidney, consisting of a bundle of CAPILLARIES enclosed in a capsule which is attached to a tubule for the drainage of urine. (Marcello Malpighi, 1628–94, Italian anatomist).

malposition 1 abnormal location of any part of the body.
2 during pregnancy, an abnormal lie of the fetus in the womb.

maltase an enzyme that splits MALTOSE.

Malthusian theory the theory that populations tends to increase faster than the means of their subsistence so that starvation, poverty and misery are inevitable unless populations are controlled by disease, famine, celibacy, 'vicious practices' (contraception), infanticide or war. The theory was proposed in *An Essay on the Principle of Population* (1798). (Thomas Robert Malthus, 1766–1834, English theorist).

maltose a disaccharide sugar consisting of two linked molecules of glucose. It is produced by the enzymatic splitting, during digestion, of starches and glycogen.

mamillary bodies paired, rounded, breast-like swellings on the underside of the HYPOTHALAMUS of the brain just behind the stalk of the PITUITARY GLAND.

mammary gland the breast. This is rudimentary in the male but developed and capable of the function of long-term milk production in the female. The word *mamma* is both Greek and Latin for breast and may derive originally from the sound made by hungry babies.

mammography a method of X-ray examination of the breasts using low-radiation (soft) X-rays and specially designed apparatus to reveal density changes that might imply cancer. Mammography is used in cases of suspected breast cancer and as a screening procedure on groups of women. It cannot be relied on to exclude cancer and does not distinguish between benign and malignant tumours, but tumours that cannot be felt may be detected.

mancinism the condition of left-handedness.

mandible the lower jaw bone. The head of the mandible, on either side, articulates with a hollow on the underside of the temporal bone, just in front of the ear. This is called the temporo-mandibular joint. The mandible is pulled upwards by powerful masticatory muscles. In dislocation of the mandible, the heads slip forward out of the hollows in the temporal bone and the mouth remains wide open until the dislocation is reduced by downward pressure on the back teeth.

mania a state of physical and mental overactivity featuring constant compulsive and sometimes repetitive movements and unceasing loquacity. The manic phase of a manic depressive illness. From the Greek *mania*, raving madness.

-mania *suffix denoting* exaggerated feeling for, compulsion towards, or obsession with.

manubrium the shield-shaped upper part of the breastbone (STERNUM). The inner ends of the collar bones (clavicles) and of the first and second ribs articulate with the manubrium.

mapping the process of determining the order of GENES, and their functions, on the CHROMOSOMES. The human genome project, currently under way, is designed to map the entire collection of the human chromosomes, with incalculable potential benefit to human-kind, and a huge increase in responsibility for the application of the knowledge.

marasmus a state of wasting or emaciation from starvation, usually in infants. There is weakness, irritability, dry skin and, unless rapidly corrected by feeding, retardation of growth and of mental development.

marijuana see CANNABIS.

marital counselling the process of analysis of relational problems between spouses and the giving of advice on how they may be relieved.

marker 1 a trait, condition, gene, or substance that indicates the presence of, or a probable increased predisposition to, a medical or psychological disorder.
2 a gene whose location on a chromosome is known so that it can be used as a point of reference for MAPPING new mutations.

masculinization see VIRILIZATION.

masochism the achievement of sexual arousal or gratification by the experience of physical or mental pain or humiliation. Masochism is said to derive from a partly repressed sense of guilt which inhibits orgasm but which can be assuaged by punishment so that orgasm becomes possible. (Leopold von Sacher-Masoch, 1835–95, Austrian pornographic novelist).

mass a fundamental property of a quantity of substance related to the amount of matter present. Mass is not the same as weight, which is the force acting on a mass as a result of gravitation.

massage stimulation of skin and muscle by rubbing, kneading, stroking, pummelling or hand-hammering with therapeutic intent. Massage has little physical effect but the psychological and symbolic effect of human touch can be deeply soothing and can relieve symptoms, especially those of undue muscle tension.

mass movement the shift of an appreciable proportion of the colonic faecal contents into the rectum in a single process occasioned by contraction of a long segment of colon. Mass movement is quickly followed by the desire to defaecate.

masseter a short, thick, paired muscle in each cheek running down from the cheekbone (zygomatic arch) to the outer corner of the jawbone (mandible). The masseters act to raise the lower jaw and compress the teeth together in the act of chewing.

mast- *combining form denoting* breast or breast-like.

mastatrophy shrinkage of the breasts.

mast cell a connective tissue cell found in large numbers in the skin and mucous membranes and in the lymphatic system. The mast cell plays a central part in allergic reactions. It contains numerous large granules – collections of powerfully irritating chemical substances such as HISTAMINE; SEROTONIN; heparin; the proteases tryptase and chymase; CYTOKINES; PROSTAGLANDINS; and LEUKOTRIENES. In people with allergies, the antibody (immunoglobulin), IgE, remains attached to specific receptors on the surface of the mast cells. When the substance causing the allergy (the ALLERGEN) contacts the IgE, the mast cell is triggered to release these substances and the result is the range of allergic symptoms and signs. Mast cells closely resemble blood basophil cells, and the latter also carry receptors for IgE.

mastication chewing.

mastoid bone a prominent bony process which can be felt behind the lower part of the ear. This is not a bone in its own right but a protuberance on the lower, outer aspect of the TEMPORAL BONE. The mastoid process is honeycombed with air cells and these communicate with the middle ear. Infection can spread to the air cells from the middle ear, causing a mastoiditis. The term refers to a fanciful breast-like appearance.

mastopathy any disease or disorder of the breast (mammary gland).

mastoptosis sagging of the breasts.

masturbation self-stimulation of the genitals, with sexual fantasizing, in order to reach orgasm. Over 90 per cent of males and 75 per cent of females are believed to masturbate at one time or another. Average frequency varies from three or four times a week in adolescence to once or twice a week in adult life. There is evidence that frequent male masturbation between the ages of about 20 and 50 substantially reduces the risk of prostate cancer. An equal frequency of orgasm by sexual intercourse does not, apparently, provide the same advantage.

maternal mortality the number of women who die each year, from causes associated with pregnancy or childbirth, for every 1000 total births. Deaths during pregnancy from causes unrelated to pregnancy are excluded but causally related deaths are counted, even if they occur months or years after the pregnancy. Maternal mortality is a useful index of the standards of medical care in a community.

matrix the scaffolding or ground substance of a tissue which supports the specialized functional cells.

maxilla one of a pair of joined facial bones that form the upper jaw, the hard palate, part of the wall of the cavity of the nose and part of the floor of each eye socket. The maxillae bear the upper teeth and each contains a cavity called the maxillary antrum or sinus.

maxillary sinus the mucous membrane-lined air space within each half of the maxillary bone. The maxillary sinuses drain into the nose. Also known as the maxillary antrum.

maxillofacial pertaining to the MAXILLA and the rest of the face. Maxillofacial surgery is much concerned with the treatment of facial injuries and fractures of the facial skeleton.

McClintock, Barbara the American geneticist Barbara McClintock (1902–92), the daughter of a physician, was born in Hartford, Connecticut and was educated at Cornell University, College of Agriculture, New York, taking a PhD in 1927. Her work on botany had stimulated her interest in genetics and, in 1936, when a new department of genetics was set up in the University of Missouri she accepted the position of assistant professor. In 1941 she moved to the Carnegie Institute, Cold Spring Harbor, New York, where she continued to work for the rest of her life.

At Cold Spring Harbour McClintock carried out a long series of studies on the genetics of maize extending over many years concentrating on genes coding for pigment and on mutations that led to pigment changes. She used X rays to cause mutations and genetic rearrangements, and studied the ways in which chromosomal damage was repaired. In the course of this work she noted that the mutation rates throughout the generations were variable. She concluded that the genes that cause the pigmentation were subjected to control by other genes. These she called controlling elements. One of these was situated near to the pigment gene and could switch it on or off. The other was more remote on the same chromosome and controlled the rate at which the pigment gene was switched. McClintock went on to make a surprising discovery – that the two controlling genes could, as one generation succeeded another, move to different parts of the same chromosome or even to other chromosomes to control other genes. McClintock reported these findings in 1951 but they were ignored for ten years until they were independently confirmed in bacteria and fruit flies by other workers.

This discovery of what came to be called 'jumping genes' answered a number of previously puzzling questions in Mendelian genetics and earned her the Nobel prize for Physiology or Medicine in 1983 when she was more than 80 years old.

Mead, Margaret the American anthropologist Margaret Mead (1901–78) was one of the most effective popularizers of the discipline of cultural anthropology and her books were read all over the world. She graduated BA at the age of 22 and before proceeding to her doctorate carried out some notable fieldwork in Samoa. This led to the publication of the book *Coming of Age in Samoa* (1928). She graduated Ph.D. 1929 at Columbia University and in 1930 published *Growing up in New Guinea*. Her work was greatly concerned with cultural sexual matters – a fact that contributed to its popularity with the lay reading public, who were interested to learn that there were some societies in which adolescence was not necessarily fraught with sexual problems. In 1835 she published *Sex and Temperament in Three Primitive Societies* which was concerned with sexual roles and aggression. Her later publications included *Male and Female* (1949) and *Growth and Culture* (1951). In general she was concerned to show that human characteristics, especially the differences between male and female, are the result of cultural conditioning rather than heredity.

Throughout her whole professional life Margaret Mead was closely associated with the American Museum of Natural History in New York. Latterly, her work has attracted adverse criticism from other anthropologists. Even so, probably more than any other worker in the field, she made anthropology accessible to the public.

meat substitutes non-animal protein food products, derived from soya beans, wheat gluten, yeast or other sources, and usually flavoured and textured to resemble natural muscle protein. These products are a reasonably effective substitute for animal protein but may not contain all the essential AMINO ACIDS.

mean arterial pressure an average of the blood pressure over the cardiac cycle, roughly equal to the diastolic pressure plus one third of the difference between the diastolic and the systolic pressure (i.e. the pulse pressure).

meatus any passage or opening in the body.

mechanoreceptor a sensory receptor that responds preferentially to physical deformation such as stretching, twisting, compressing or bending.

mechanosensitive channel a cell membrane ion channel that is opened by stretching of the membrane.

Meckel's diverticulum a small pouch-like sac that protrudes from the interior of part of the small intestine (ILEUM) in about 2 per cent of people. Normally harmless, it sometimes becomes infected and causes a condition indistinguishable from appendicitis. Meckel's diverticulum may also lead to twisting (volvulus) or infolding (intussusception) of the bowel. (Johann Friedrich Meckel II, 1781–1833, German anatomist).

meconium the thick, greenish-black, sticky stools passed by a baby during the first day or two of life, or before birth if the fetus is deprived of an adequate oxygen supply (fetal distress). Meconium consists of cells from the lining of the fetal bowel, bowel mucus and bile from the liver. Once feeding is established meconium is replaced by normal stools.

media the middle wall of an artery or vein. The media is composed of smooth muscle and elastic fibres and is the thickest of the three layers. Also known as the tunica media.

medial situated toward the midline of the body. Compare LATERAL.

median 1 situated in or towards the MEDIAN PLANE of the body.
2 in statistics, the middle value when observations are ranked in order of magnitude.

median plane the vertical plane that divides the body into right and left halves.

median nerve one of the two major nerves of the arm, supplying most of the muscles and providing sensation in the two-thirds of the hand on the thumb side.

mediastinitis inflammation of the MEDIASTINUM.

mediastinum the central compartment of the chest, flanked on either side by the lungs, and containing the heart, the origins of the great blood vessels, the TRACHEA and the main BRONCHI, the OESOPHAGUS and many lymph nodes.

medicine 1 the branch of science devoted to the prevention of disease (hygiene), the restoration of the sick to health (therapy)

and the safe management of childbirth (obstetrics). Medicine is a scientific discipline but the practice of medicine involves social skills and the exercise of sympathy, understanding and identification, not normally demanded of a scientist.
2 medical practice not involving surgical operative intervention. In this sense, medicine and surgery are distinguished.
3 any drug given for therapeutic purposes.

medicolegal pertaining to both MEDICINE and law.

Mediterranean diet a diet featuring a high intake of vegetables, legumes, fruit, nuts, cereals and olive oil; a moderately high intake of fish; a low to moderate intake of cheese and yoghurt; a low intake of other dairy products, saturated fats, meat and poultry; and a moderate and regular intake of wine taken with meals. Research has shown that close adherence to the traditional Mediterranean diet is associated with a significant increase in longevity. The diet was found to reduce deaths both from coronary heart disease and from cancer.

medulla 1 the inner part of an organ, especially of the kidney, the adrenal and the shaft of long bones. Compare CORTEX.
2 the MYELIN layer of nerve fibres. See also MEDULLA OBLONGATA.

medulla oblongata the part of the BRAINSTEM lying below the PONS and immediately above the spinal cord, just in front of the CEREBELLUM. The medulla oblongata contains the nuclei of the lower four CRANIAL NERVES, the vital centres for respiration and control of heart-beat, and the long motor and sensory tracts running down to and up from the spinal cord. Disease or injury to the medulla is always serious, often fatal.

medullated MYELINATED.

mega- 1 *prefix denoting* one million.
2 *combining form denoting* large.

megabase a unit of measurement of the length of a segment of DNA equal to 1,000,000 base pairs. Usually contracted to Mb.

megakaryocyte an unusually large bone marrow cell that releases many small fragments of its CYTOPLASM as the blood PLATELETS essential for clotting (blood coagulation).

megalo-, -megaly *combining forms denoting* abnormal enlargement.

megalomania a delusion of power, wealth, omnipotence or grandeur.

meibomian gland one of the 20–30 glands that occupy each eyelid, lying parallel to each other and perpendicular to the lid edge and opening on to the lid margin. The meibomian glands secrete an oily fluid that prevents adhesion between the lids and forms an outer layer on the tear film over the cornea and conjunctiva so as to retard drying. (Heinrich Beibom, 1638–1700, German professor of Medicine, History and Poetry).

meiosis the process in the formation of the sperm (spermatozoa) and eggs (ova) in which chromosomal material undergoes recombination (meiosis I) and the chromosomes are reduced to a single set of 23 (haploid number) instead of the normal 23 pairs (meiosis II). This allows the restoration of the normal number when the spermatozoon fuses with the ovum. See also MITOSIS.

Meissner's plexus a network of nerve fibres, from the PARASYMPATHETIC NERVOUS SYSTEM, lying in the wall of the intestines between the mucous membrane lining and the muscle layer. These fibres control and coordinate the movements and changing contractions of the intestines. (Georg Meissner, 1829–1905, German anatomist).

melan-, melano- *combining form denoting* black.

melancholia DEPRESSION.

melanin the body's natural colouring (pigment) found in the skin, hair, eyes, inner ears and other parts. In body cells, melanin is bound to protein. It is a complex POLYMER formed from the amino acid TYROSINE (4-hydroxphenylalanine) by oxidation via dopa and dopaquinone.

melanin concentrating hormones small peptides formed in the pituitary and brain that regulate skin colour by altering the concentration of melanin.

melanocortin 4 receptor (MC4R) a cell receptor that is strongly stimulated by the alpha melanocyte-stimulating hormone – a peptide that produces a sense of fullness after eating. Mutations of the gene that codes for MC4R are commonly associated with severe uncontrollable overeating and pathological obesity. Such mutations can be found in nearly 6 per cent of people with a life-long history of obesity.

melanocyte a pigment cell of the skin. A cell carrying MELANIN or capable of producing melanin.

melanocyte-stimulating hormone the PITUITARY GLAND hormone that promotes the synthesis of MELANIN in MELANOCYTES. Also known as melanotropin. Alpha melanocyte-stimulating hormone is an anorexogenic peptide produced by the enzymatic cleavage of proopiomelanocortin (POMC) that is a powerful agonist of the MELANOCORTIN 4 RECEPTOR.

melanoderma abnormal darkening of the skin from excessive numbers of MELANOCYTES.

melanoma any benign or malignant tumour of MELANOCYTES.

melanonychia blackening of the nails with MELANIN.

melanosis abnormal pigmentation of the tissues from excessive deposition of MELANIN. Melanosis coli is pigmentation of areas of the COLON due to an accumulation of MACROPHAGES containing melanin or a similar pigment.

melatonin a hormone synthesized from serotonin in the pineal gland and elsewhere. Melatonin production has a strong circadian rhythm, being secreted mainly in the period between about 2100 hours and 0800 hours. Bright light suppresses melatonin secretion and exogenous melatonin can alter the timing of the body clock. For these reasons it has been proposed as a means of combatting jet lag. Other methods have been found more generally useful.

membrane fluidity the ability of lipid molecules to move sideways within their own single-molecule-thick layer.

membrane potential the difference in millivoltage between one side of a membrane and the other.

membrane-protein ion channels voltage-controlled and gated cell membrane pores through which currents of sodium and potassium ions pass in water solution. The movement of these ions through the controlled channels underlie all electrical activity of the nervous system, including the brain and are thus responsible for all behaviour, thinking, and emotional activity. Voltage-gated ion channels have a central pore domain surrounded by voltage-sensor regions.

memory

When we try to imagine the nature and physical basis of human memory we are apt to think of mechanical analogies such as digital computers, tape recorders, holograms, and so on. These analogies can be helpful but they are often misleading. The processes involved in the human memory are more complex than they seem, certainly much more complex than those involved in the processing and storage of data on a hard disc.

Aristotle (384–322 BC), who spent much time thinking about memory, came to the conclusion that it was situated in the heart, not the brain. Medieval ideas of memory were, as today, based on analogies with the technology of the time. Then, the most advanced technology was hydraulic, and memory was conceived as a flow of fluid through pipes, controlled by valves. Descartes (1596–1650) also subscribed to a hydraulic analogy, the fluid being 'animal spirits' which were driven by the pineal gland to different parts of the brain where they encountered physical traces left by the object we wish to remember.

Memory is a function of the brain's ability to store and retrieve information. The term is also used for the actual information store. Memory in this sense is not, however, as in a computer, a discrete and recognizable part of the brain in which data is stored. Certain known parts of the brain are concerned in the temporary storage, registration, processing and recall of information that is to be permanently stored, but there is no single part corresponding to the long-term store – no unique part corresponding to the disc drive on a computer.

Memory of facts and events is called declarative memory; memory of how to do things is called procedural memory; and short-term memory storage used in day to day activities such as temporarily remembering a phone number is called working memory.

Input

The sensory experience that must precede the storage of data in memory is very complex, and many different classes of sensation must be stored – visual, auditory, olfactory, gustatory, tactile. In addition, data to be memorized may take many forms – pictorial, verbal, numeric, and so on. Data is never presented to us in simple form, but always as part of a complex matrix in which it is embedded. The context of the data is likely to form part of the memory; indeed daily experience shows us how important context and associations are for effective memorizing. A single item of information conveyed to us by way of speech will be accompanied by a considerable context of other data – the appearance of the speaker's face, its spacial relationship to other things, the quality of the voice, the indications of emotion, and much other information. There is no reason to believe that all this material is stored in the same place and good reason to believe that it is distributed to those parts of the brain known to be concerned with the different sensory functions.

Physical basis of memory

Complex perceptions of this kind cannot all be separately stored at each moment of perception. Even the immense storage capacity of the human brain would soon be used up if this were happening. So some kind of analysis and selection is necessary to determine what should be recorded, and it seems likely that only significant changes in familiar contexts are registered. But such analysis and selection cannot be performed unless the data are temporarily stored so that they can be operated upon. This implies the existence of a short-term store or 'buffer' which is constantly in use.

Buffers

The idea that we have only one temporary memory store of this type does not fit the facts. Individual experience suggests that there are several short-term buffer stores for different modalities of memory – speech sounds, non-verbal sounds, touch, vision and so on. There is almost certainly a short-term store that codes for the articulation of a sequence of words that we are about to use, in spoken conversation. It is common to forget what we were about to say.

Again, we obviously have a short-term memory for small items of new data that have to be briefly remembered. This buffer has to be constantly refreshed. We can remember a new telephone number if we repeat it internally until we dial it. But if we are addressed or engage in other mental activity, the number will be lost as new incoming data displaces the current contents of the buffer. There is, also, a strict limit to the length of such an item of new data that can be held in this short-term buffer. Most people can readily hold a seven- or eight-digit number, but not a twelve digit number. The buffer can also be emptied by a blow or an electric shock to the head. All these facts suggest that short-term memory involves some kind of dynamic neuronal circuit, possibly of circulating nerve impulses.

There is reason to believe that each recall of a long-term memory of an event is not a recall of the original event but of the previous recall of it. And because recall is susceptible to error, memories may readily be falsified. Common experience indicates that this is so, and false memories may easily be induced by suggestion.

Mnemonics

We all carry in our brains a very large quantity of data, recorded, somehow, via short-term memory and preserved, often for a lifetime, in permanent storage. This mass of data is highly organized in terms of meaning and association and the better the organization the more accessible it is to us. The efficiency of retrieval is heavily dependent on clues, mnemonics and, in particular, cues. Similarly, efficient registration for long-term storage demands good organization and strong association. A scholar with a deep grasp of his or her subject assimilates new data on that subject with the greatest of ease, so long as it can be related to existing stored data. Entirely new matter, unrelated to any previous experience, is much more difficult to memorize.

Interface circuits

The exact sites in the brain through which sensory data must pass to be stored in memory are known. These interface circuits – through which long-term memory is recorded and recalled – are contained in two large structures on the inner surfaces of the temporal lobes of each cerebral hemisphere – two massive collections of nerve cells known as the amygdala and the hippocampus. The amygdala is connected to all the sensory areas of the cortex by two-way pathways. The hippocampus also has extensive connections. Destruction of these two nuclei leads to profound loss of memory.

Basis of long-term memory

The exact physical basis of long-term memory remains uncertain, but several hypotheses have been put forward. The sheer size of the data-base in relation to the size of the brain implies that the unit of information – whether it be a binary digit (bit) or some other code – must be very small. This has led some scientists to suggest a protein molecule as the basis. It would be naive, however, to think of long-term memory storage as based on two kinds of protein molecules corresponding to the presence or absence of a spot of magnetization on a computer disc store – binary 1 or 0. For one thing we have no reason to suppose that the nervous system operates on the binary system. For another, the life-time of protein molecules is very much shorter than the known length of human memory.

There is a good deal of evidence that the sites of memory are the same areas of the brain where the corresponding sensory impressions are processed – the various parts of the outer layer (the cortex). It now seems almost certain that, in memory recall, the amygdala and the hippocampus engage in a kind of feed-back dialogue with the appropriate part of the cerebral cortex – playing back the kind of neurological activity that occurs during sensory experience.

This being so, it seems likely that long-term memory store takes the form of the interconnection of many nerve cells in a particular way – the development of new and relatively permanent circuits, each representing an item of memory. There is evidence that particular connections – activation of the nerve-nerve junctions known as synapses – occur as a result of repeated stimulation of the kind that occurs during sensory experience. The way in which repeated stimulation leads to permanent link-up is also beginning to be understood. There is also evidence of the surprising fact that the brain can grow new nerve cells. For several reasons, this hypothesis seems more plausible than the idea of a kind of bit-mapping with protein molecules representing the pixels.

memory cells B lymphocytes that retain information about previous challenge, as by infective organisms, so that antibodies can be more rapidly produced in response to a subsequent infection with the same agent.

menarche the onset of MENSTRUATION. Compare MENOPAUSE.

Mendel, Gregor the Austrian monk and biologist Gregor Johann Mendel (1822–84) derived a completely new theory of heredity by a study in the artificial fertilization of peas. This research was a model of experimental science, almost unique for its time. In the course of seven years, Mendel grew some 28,000 garden pea plants in garden beds and pots and recorded seven of their characteristics. One of these – the seventh in his list – was plant height. In this trial he produced a large number of apparently pure bred peas by careful self-pollination, guarding them against insect pollination. The peas were of two types, tall and dwarf. The dwarf plant seed produced only dwarf plants. In the case of the seeds from the tall plants, however, some produced tall plants but some produced dwarf plants. Mendel noticed that for every dwarf plant there were three tall plants. Evidently there were two kinds of tall pea plant – those that bred true (i.e. passed on their own characteristics) and those that did not. Mendel then cross-bred true-breeding tall plants with dwarfs. The outcome, in every case, was a tall plant. He then self-pollinated these hybrid plants. The result was of the greatest interest. One quarter of the offspring were true-breeding dwarfs, one quarter were true-breeding tall plants, and the remainder were of the non-true-breeding tall variety.

Mendel's findings for the other characteristics, such as colour, were the same. His conclusion was a very important one – when plants with different characteristics were cross-bred, the outcome was not a blend of features. Instead, characteristics distributed themselves among the offspring in accordance with simple arithmetical ratios. Mendel was a mathematician and recognized the ratio 1:2:1 as corresponding to the terms of the binomial series. If the letters A and a are envisaged as corresponding to two contrasted characteristics of the plant, the distribution of the characteristics will be A + 2Aa + a. Mendel was also able to show that different pairs of contrasted characteristics could be inherited independently of each other without interacting in any way.

Tallness was a dominant characteristic, dwarfness was a recessive characteristic. Mendel used these terms and showed how they were related mathematically. Ironically, the mathematics made his paper appear boring to the ordinary reader and this probably accounts for the scant attention it received. Mendel had little appreciation of the fundamental importance of his work. Although he had read Darwin's *Origin of Species* he seems not to have grasped the relevance of his work to Darwin's great theory. His paper was concerned only with the rules for the hybridization of plants and said nothing about the wider implications. History was to show the critical importance of Mendel's great discovery.

Mendelian disease a disease caused by a single mutated gene. Also known as a 'single gene disorder'. Although some 5,000 such diseases are known, nearly all of them are comparatively rare. Some of the more common are cystic fibrosis, sickle cell disease, thalassaemia, haemophilia, Marfan's syndrome, phenylketonuria, one form of retinitis pigmentosa, Duchenne's muscular atrophy, Huntington's disease and adult polycystic disease.

menhidrosis monthly sweating, sometimes with blood in the sweat, as a form of vicarious menstruation. Also known as menidrosis.

meninges the three layers of membrane that surround the brain and the spinal cord. The innermost, the PIA MATER, dips into the brain furrows (sulci). The intermediate ARACHNOID MATER bridges over the furrows, leaving the subarachnoid space which contains cerebrospinal fluid and many blood vessels. The outer later, the dura mater, is a tough, fibrous protective covering attached, in the skull, to the overlying bone and forming a tube to enclose the spinal cord.

meningovascular pertaining to the MENINGES and to the blood vessels supplying the brain.

menopause the end of the reproductive period in women when the ovaries have ceased to form GRAAFIAN FOLLICLES and produce eggs (ova) and menstruation has stopped. The menopause usually occurs between the ages of 48 and 54. There is reduced production of oestrogen hormones by the ovaries and this may cause accelerated loss of bone bulk (OSTEOPOROSIS) and thinning and drying of the vagina (atrophy) with difficulty and discomfort in sexual intercourse. The common menopausal symptoms (hot flushes, night sweats, insomnia, headaches and general irritability) have not been proved to be due to oestrogen deficiency, but these usually settle if oestrogens are given. The term is derived from the Greek *meno*, a month, and *pausos*, cessation.

menstrual disorders these include amenorrhoea, oligomenorrhea, dysmenorrhea, menstrual irregularity, mittelschmerz, menorrhagia, polymenorrhea, metrorrhagia and premenstrual tension (pmt).

menstrual period the sequence of days, usually 3–7 but averaging 4 days, during each menstrual cycle when the lining of the womb (the endometrium), together with a quantity of blood, is being shed via the cervix and the vagina. The volume lost varies between 50 and 150 ml. Menstrual blood is dark in colour and does not clot.

menstruation the periodic shedding of the lining (ENDOMETRIUM) of the womb (uterus) at intervals of about 28 days causing bleeding through the vagina of 3–7 days duration in the non-pregnant female. The purpose of menstruation is to renew the endometrium so that it is in a suitable state to ensure implantation of a fertilized egg (ovum).

menstruation taboos the long-held unfounded superstition that any contact, especially sexual contact, with a menstruating woman is polluting or will have dire consequences. This is but one of the many manifestations of male sexism throughout the ages

mental pertaining to the mind.

mental deficiency see MENTAL RETARDATION.

mental retardation intellectual ability so much below average as to preclude the performance of most forms of work or other social functions. Mentally retarded people usually require supervision and guidance if they are to avoid distress or danger. There are degrees of mental deficiency. So far as INTELLIGENCE QUOTIENTS (IQs) can be measured in the mentally retarded, the mildly defective have IQs from 70 down to about 55; the moderately defective have IQs from 54–40; and the severely defective have IQs below 40.

mento- *prefix denoting* the chin.

mentum the chin.

mercy killing EUTHANASIA.

Merkel's discs Merkel's corpuscles, small cup-shaped touch receptors in the epidermis, each consisting of a single sensory nerve fibre ending in contact with an epithelial cell. (Friedrich S. Merkel, German anatomist and physiologist, 1845–1919).

mes-, meso- *prefix denoting* middle, MEDIAL, intermediate or connective.

mesencephalon the middle section of the embryonic brain.

mesentery the complex, double-layered folded curtain of PERITONEUM that encloses the bowels and by which they are suspended from the back wall of the abdomen. Blood and lymphatic vessels run to and from the intestines between the two layers of the mesentery.

mesoappendix the small MESENTERY of the APPENDIX.

mesocolon the folded membrane of PERITONEUM by which the colon is suspended from the inside of the back wall of the abdomen.

mesoderm the intermediate of the three primary germ layers of the developing embryo, lying between the outer ECTODERM and the inner ENDODERM. Mesoderm develops into the bones and muscles, the heart and blood vessels and most of the reproductive system.

mesolithic period the middle period of the stone age when the use of flint implements had become widespread, and hunting had not yet been replaced by herding and animal breeding. In Europe the mesolithic period started about 8000 BC.

mesomorph a person of powerful musculature and large, strong bones.

mesothelium lining cells originating in the primitive MESODERM of the developing embryo. Mesothelium occurs in the PERITONEUM, PLEURA and PERICARDIUM as well as elsewhere in the body.

messenger 1 pertaining to ribonucleic acid (RNA) that carries the coded information for protein synthesis from the DNA to the site of protein synthesis, the RIBOSOMES.
2 a HORMONE or other effector capable of acting at a distance from its site of production. A second messenger is a hormone, produced within a cell and operating on internal structures, when another hormone acts on the outer cell membrane.

messenger RNA commonly written as mRNA, this is the molecule that reads the genetic code from DNA. Before this can happen the double helix must separate into two single strands. One of these carries the same sequence as the mRNA and is called the coding strand. The other is called the template, or antisense, strand and it is this strand that directs the synthesis of the mRNA by complementary base pairing. In RNA the base uracil replaces thymine. The messenger RNA molecule then leaves the cell nucleus and passes out through a nuclear membrane pore to the site of protein synthesis. There the appropriate amino acids are selected and placed in the right order by TRANSFER RNA which, using its anticodons, reads the code on the messenger RNA.

met-, meta- *prefix denoting* beyond, after, following, with, next to, transcending, changed or transformed.

meta-analysis an attempt to improve the reliability of the findings of medical research by combining and analyzing the results of all discoverable trials on the same subject. In crude terms the advantages are obvious: trials that find against a hypothesis will cancel out the effect of those that find for it. Pooling of raw data is not, however, without statistical hazard and it has become apparent that meta-analysis can introduce its own sources of inaccuracy. The method is currently undergoing refinement.

metabolism the totality of the body's cellular chemical activity, largely under the influence of enzymes, that results in work and growth or repair. The 'building-up' aspects of metabolism are known as anabolic and the 'breaking-down' as catabolic. Metabolism involves the consumption of fuel (glucose and fatty acids), the production of heat and the utilization of many constructional and other biochemical elements provided in the diet, such as AMINO ACIDS, fatty acids, carbohydrates, vitamins, minerals and trace elements. The basal metabolic rate is increased in certain disorders, such as hyperthyroidism, and decreased in others. Anabolism can be artificially promoted by the use of certain steroid male sex hormones (androgens or anabolic steroids).

metabolite any substance involved in METABOLISM either as a constituent, a product or a byproduct.

metacarpal bone one of the five long bones situated in the palmar part of the hand immediately beyond the CARPAL bones of the wrist and articulating with the bones of the fingers (PHALANGES).

metacarpus the five bones of the palm of the hand.

metacentric of a chromosome in which the CENTROMERE is at or near the centre.

metamorphosis major alterations in structure and appearance occurring in an organism, such as the human embryo, in the process of its development from egg (ovum) to baby.

metaphase stage of MITOSIS or MEIOSIS during which the chromosomes are aligned around the equator of the cell and are visible on microscopy. The stage at which the banding pattern of the chromosomes is apparent.

metaphor a figure of speech in which a word or phrase accepted in one context is deliberately used in another, for the purpose of emphasis or illustration. This is an important device in human thought and expression and one of the mechanisms by which language is enlarged and enriched. Popular and useful metaphors such as 'the bed of a river' and 'understanding' become incorporated into the language.

metaphysics the branch of philosophy concerned with speculation about matters, such as being, knowing, reality, or the existence of God or the ultimate nature of matter, which cannot be directly observed and which are thus beyond the reach of science.

metaphysis the growing part of a long bone. The metaphysis lies between the growth plate (the EPIPHYSIS) and the shaft (the diaphysis).

metaplasia an abnormal change in the character or structure of a tissue as a result of changes in the constituent cells. Metaplasia often involves a change of cells to a less specialized form and may be a prelude to cancer.

metastasis 1 the spread or transfer of any disease, but especially cancer, from its original site to another place in the body where the disease process starts up. Metastasis usually occurs by way of the bloodstream or the lymphatic system or, in the case of lung disease by coughing and re-inhalation of particles to other parts of the lung.
2 the new focus of disease, so produced.

metatarsal one of the five long bones of the foot lying beyond the TARSAL bones and articulating with the bones of the toes (PHALANGES).

metatarsus the five bones of the foot.

metazoa all the members of the animal kingdom that consist of more than one cell. All animals more complex than the one-celled protozoa.

metencephalon the part of the hindbrain of the embryo from which the CEREBELLUM and the PONS develop.

metenkephalin one of the several endogenous opium-like substances having a sequence of five amino acids. See also ENDORPHINS.

-meter *suffix denoting* a measuring instrument.

metr- *combining form denoting* the womb.

micelles tiny aggregates of fatty acids and monoglycerides formed by the detergent action of the bile acids on digested fats so that these materials can be made soluble and absorbed into the LACTEALS of the intestinal VILLI.

micro- *combining form denoting* very small, abnormally small or pertaining to microscopy.

microarray a collection on a chip of thousands of biological probes such as DNA single-strand sequences, protein-detecting molecules, or any other biologically-identifying material, that can be used to survey a specimen for the presence of a target gene or substance. Binding to the target substance can be made to cause fluorescence to occur at the unique spot or spots on the array and the chip can then be read by a scanner and the result displayed on a computer monitor. See also DNA MICROARRAYS.

microbes microscopic organisms but especially bacteria or viruses capable of causing disease. The word is almost synonymous with, but slightly upmarket from, the term 'germs'.

microbicide any agent capable of killing MICROBES. The term has, however, recently been applied specifically to any intravaginal, topical gel preparation for use by women to reduce their likelihood of acquiring HIV infection during sexual intercourse with husbands or others. The first candidate microbicide, nonoxynol-9, was shown to damage vaginal epithelial cells and actually encourage infection. Research for a safe and effective microbicide continues and includes the development of agents blocking CD4 binding sites, inhibitors of GP41-medicated fusion, prevention of HIV take up by dendritic cells and conventional pharmacological attack on HIV replication.

microcephaly abnormal smallness of the skull. Microcephaly often reflects poor brain development and is usually associated with MENTAL RETARDATION.

microcheilia abnormally small lips.

microcyte a red blood cell of very small diameter.

microgamete 1 the male sex cell or spermatozoon. The OVUM is the macrogamete.
2 the motile male sex cell of the malarial parasite.

microglia neurological connective tissue MACROPHAGES. Compare MACROGLIA.

microglossia having an abnormally small tongue.

micrognathia having an abnormally small upper or lower jaw.

micromastia having abnormally small breasts.

micrometastases small collections of cancer cells, discernible only on microscopic examination, occurring in the lymph nodes of people with cancer. They may occur in treated cancer patients and identify a high-risk group among those previously deemed to be free of the disease.

micromelia having abnormally small arms or legs. Compare PHOCOMELIA.

micronutrients dietary substances necessary for health but required only in very small quantities. Vitamins and minerals.

microphthalmos the state of having an abnormally small eyeball. In severe degrees there is usually a gross defect of vision.

microRNA very short segments of RNA involved in RNA INTERFERENCE and recently shown to be important elements in the development of plant structure. Many genes that code for microRNA have been found in many species including *H. sapiens*. There is much current speculation into their role in human development.

microsomal enzyme system a collection of enzymes in the smooth endoplasmic reticulum of liver cells that modify molecules to make them more POLAR and less lipid-soluble.

microtubules cytoplasmic tubular filaments made from the protein tubulin. Microtubules contribute to the cytoskeleton and assist in the movements of organelles within the cell.

microvilli the millions of tiny, hair-fine, finger-like protrusions on the surface cells of EPITHELIUM which greatly increase the effective surface area so as to facilitate absorption. Microvilli occur especially on the secretory and absorptive surfaces, and are formed by extensions of the cell membrane. They have a central core of actin filaments bound together with the protein VILLIN.

middle ear the narrow cleft within the temporal bone lying between the inside of the ear drum and the outer wall of the inner ear. The middle ear is lined with mucous membrane, contains the chain of three auditory OSSICLES and is drained into the back of the nose by the EUSTACHIAN TUBE. It is a common site of infection, which gains access by way of the tube. Middle ear infection is called OTITIS MEDIA. Also known as the tympanic cavity.

mid-life crisis a psychological upset, sometimes affecting people in middle age, who feel that their lives have become meaningless and devoid of satisfaction. This may lead to injudicious sacrifice of established achievement and a damaging change of lifestyle.

midwifery the nursing speciality concerned with the conduct of antenatal care, labour and childbirth. Midwifery differs from OBSTETRICS to the extent that it is concerned primarily with the normal. Complications and undue difficulties are managed or supervised by doctors specializing in obstetrics.

milk the secretion of the breast (MAMMARY GLAND) of any mammal. Cow's milk differs from human milk, mainly in the composition of the fats. Human milk fats contain a higher proportion of unsaturated fatty acids that provide more resistance to bowel organisms than those in cow's milk. Human milk also contains maternal antibodies that provide the baby with protection against many organisms, until it is able to produce its own.

milk ejection reflex a reflex stimulated by suckling and mediated by oxytocin in which milk is moved from the breast alveoli into the milk ducts.

millivolt one thousandth of a volt. Electrophysiological processes mainly operate in the millivolt range, usually below 100 millivolts.

mind and body

Questions of how people think and feel, form opinions and beliefs, how they develop and mature mentally, how they change, and above all, how and why they act the way they do, are the province of psychology. Although modern psychology is essentially concerned with mental function, this can, in the present state of science, only be assessed by the observation of behaviour. In this context, the term 'behaviour' is used in a very wide sense and includes all detectable responses including speech. One preliminary difficulty is to decide exactly what we mean by the word 'mind'.

The history of psychology is inextricably linked with, and often inseparable from, that of other reaches of human thought and speculation – philosophy, theology, demonology and magic. While some of the early ideas now seem ridiculous, they were plausible in the light of what was then known. The study of the history of human ideas is, in itself, a source of information about the mind.

The soul and the mind

For many centuries, a clear verbal distinction was made between the mind and the soul, the latter being believed to be the immaterial essence of a person, the actuating and animating cause of life, and the immortal part of man. The soul was an entity distinct from the body – the spiritual as distinct from the physical part. The soul was also often regarded as the seat of the emotions or sentiments. Thus, although man was regarded as consisting of three parts – body, mind and soul – no clear logical distinction was made, in popular thought, between the mind and the soul. The Greek philosopher Aristotle (384–322 BC) believed that the soul had the power to think and called the thinking soul the mind (*nous*). He also held that the mind seemed to be some kind of independent substance implanted within the soul and incapable of being destroyed.

Historic ideas of the mind

Early Western thinkers, prompted by the Greek physician Galen (*c.* 130–201), conceived the world as compounded of four elements – earth, air, fire and water. Corresponding to these, in the human body, were the four humours – black bile, yellow bile, blood and phlegm. Our emotional and temperamental natures were, it was believed, determined by the relative proportions of these humours. If there was any preponderance of one humour, the resulting characteristic state of mind might be, respectively, melancholic, irascible, sanguine and phlegmatic.

The concept of the humours persisted throughout the Middle Ages, being gradually modified with time. In 16th century England all forms of psychological upset were described as melancholy. During that period, all scholarship and philosophy were concerned with the support of religious faith and, in accordance with theological principles, attributed severe emotional disturbances and madness to demonic possession. This idea often had painful consequences for the unfortunate sufferer whose attendants were often thereby justified in using violence to drive out the evil spirit.

Descartes and dualism

The central figure in the 17th century emergence of scientific and philosophical thought was the Frenchman René Descartes (1596–1650). Descartes was well aware of the mind-soul difficulty and stated that, in his definition of mind as being a substance distinct from matter, he used the term 'mind' rather than 'soul' since the word 'soul' was ambiguous. Subsequent philosophers tended to ignore this ambiguity.

In his *Discourse on Method* (1637) Descartes wrote 'I am a substance whose whole nature or essence is to think, and which for its existence needs no place, nor depends on any material thing.' This idea was incorporated into the concept which came to be described, after his time, as 'Cartesian dualism'. There were, he held, two fundamentally different kinds of substance – material substance which could be seen and touched, and thinking substance which had no dimensional qualities. The human body was of the former substance, the mind – which included thoughts, emotions, desires and volitions – was of the latter.

Cartesian dualism – the mind–body concept – had a dominant effect on the idea of mind and was accepted by most philosophers until well into the 20th century. It is, unthinkingly, held by many people to this day and is only now, with the growth of knowledge of the relationship between body function and its psychological concomitants, beginning to be seriously challenged. The concept of dualism raises a number of grave difficulties, some of which were recognized by Descartes himself. One of these is how to explain the close causal link between mind action and body action – how an immaterial substance can cause changes in a material substance, as occurs, for instance, when a man decides to walk, and does so. Descartes, somewhat disingenuously, tried to get round this difficulty by suggesting that the mind acted on the body through the pineal gland, which is situated in the middle of the head.

Ironically, Descartes' contemporary, the philosopher Baruch Spinoza (1632–77) came much closer to present-day ideas of the mind than any subsequent thinker. Spinoza held that mind and body were two modes (modifications) of the single substance that is all Nature. The mind was the idea of the body and vice versa.

Modern ideas of the mind

The Cartesian notion that the mind is a separate, non-physical entity is gradually being replaced by ideas more in keeping with modern physiological knowledge. An emotion such as fear, for instance, whether occasioned by perception of danger or by the memory of such danger, is invariably associated with the release into the blood of hormones such as adrenaline, that cause the heart to beat rapidly, the breathing to deepen, the palms to sweat, the hair to rise, and so on. A drug that blocks the action of adrenaline can abolish the emotion, although the purely intellectual awareness of the danger persists. An injection of adrenaline can induce a like emotion even in the absence of any external cause of fear. This implies that at least some mental experience is so inextricably linked up with bodily function as to be inconceivable apart from it. Many physiologists now believe that *all* mental experience is a product of neurological activity and that the mind is an epiphenomenon of brain and bodily activity.

mineralocorticoids hormones from the outer layer (cortex) of the adrenal gland that promote retention of sodium and excretion of potassium in the urine. ALDOSTERONE is the most powerful mineralocorticoid. Compare GLUCOCORTICOIDS.

minerals chemical elements required in the diet, usually in small amounts, to maintain health. Apart from iron and calcium, deficiency is comparatively rare. The essential minerals are calcium, iron, magnesium, copper, selenium, phosphorus, fluorine, potassium, sodium and zinc.

mio- *prefix denoting* narrowing, reduction or diminution.

miosis constriction of the pupil.

miotic a drug that constricts the pupil.

miscarriage a spontaneous ABORTION. Spontaneous ending of a pregnancy before the fetus is mature enough to survive even with the best supportive care.

misfolded proteins see CHAPERONES.

missense mutation a mutation caused by a change in a nucleotide sequence that changes a codon specifying a particular AMINO ACID into one that specifies a different amino acid.

mitochondria one of the class of important tiny elements (organelles) in the cytoplasm of nucleated (eukaryotic) cells. They may be rod-shaped, spherical, branched or ring-shaped. Mitochondria have double-layered walls, the inner layer being deeply infolded to form compartments. They contains genes and RIBOSOMES and are the site of cell respiration. Their many functions include the Krebs cycle, metabolism of fatty acids, amino acids and steroids, pyruvate oxidation, and the production of energy in the form of adenosine triphosphate (ATP).

MITOCHONDRIAL DNA is transmitted only from the mother and, apart from mutations, remains unchanged through the generations. Mitochondria are believed to have developed from bacteria that colonized primitive eukaryotic cells more than a billion years ago, thereby providing them with aerobic metabolism.

mitochondrial DNA a small circular DNA molecule of which all MITOCHONDRIA in cells have several copies. It contains 16,569 base pairs and 37 genes. Being present in the cytoplasm of the cell, is transmitted exclusively by the mother. A mature ovum contains about 100,000 copies of mitochondrial DNA; sperm contain none. The genome has been completely sequenced. Its main function is to code for enzymes needed by the mitochondrion. The sites of common mutations, especially deletions, are known as a number of diseases are caused by these defects in the mitochondrial DNA. See MITOCHONDRIAL DNA DISEASES.

mitochondrial DNA diseases diseases caused by mutations in MITOCHONDRIAL DNA (mtDNA) or by nuclear DNA mutations affecting components in mitochondrial processes. Because both mitochondrial and nuclear genes may be involved, inheritance may be maternal or Mendelian. They are all rare. Pure mitochondrial diseases caused by point mutations include Leber's hereditary optic neuropathy; ragged red fibres; mitochondrial encephalomyopathy with lactic acidosis and stroke-like episodes; and neurogenic weakness with ataxia and retinitis pigmentosa. Major rearrangements of mitochondrial DNA cause the Kearns-Sayre phenotype; chronic progressive external ophthalmoplegia; Pearson's syndrome; and excessive ageing.

mitogen 1 any agent that promotes cell nuclear division (MITOSIS). Mitogens are important in genetic research and technology. Pokeweed mitogen, derived from the plant *Phytolacca americana*, is a powerful mitogen of B LYMPHOCYTES.
2 any substance that non-specifically causes lymphocytes to proliferate.

mitosis the division of a cell nucleus to produce two daughter cells having identical genetic composition to the parent cell. First the long strands of CHROMATIN replicate and coil up to form dense chromosomes with the two copies (chromatids) joined at the CENTROMERE so that they appear X-shaped. At the same time, the envelope of the cell nucleus disrupts (prophase). Then two sets of strand-like microtubules (the spindle) appear, radiating from each end of the cell to the centre, the metaphase plate, and the chromosomes align themselves on the plate with the centromeres at the equator (metaphase). The copies of each chromosome (chromatids) now separate and move to opposite poles of the spindle (anaphase). Finally, the cell separates into two, the chromatin uncoils and the nuclear envelope of each reforms (telophase).

mitotic index the percentage of cells in a population that are actually undergoing MITOSIS. The mitotic index is a measure of the reproductive or growth activity of a tissue.

mitral valve the valve on the left side of the heart lying between the upper and lower chambers (ATRIUM and VENTRICLE). It has two cusps and is said to resemble a Bishop's mitre.

mitral valve disorders these include mitral stenosis, mitral incompetence and mitral valve prolapse.

mittelschmerz pain or discomfort in the lower abdomen felt by women at the time of OVULATION, between menstrual periods. The cause is uncertain but is thought to be either stretching of the Graafian follicle or irritation of the PERITONEUM from blood or follicle fluid. The term is German for 'middle pain'.

mixed venous oxygen saturation (SvO_2) the balance between oxygen delivery and extraction, and a measure of the adequacy of oxygen delivery to the tissues. Tissue hypoxia may occur even when heart rate, blood pressure and central venous pressure are normal. Knowledge of the SvO_2 has been shown to be valuable in any condition or state in which tissue hypoxia is possible. The normal range is 65–75 per cent.

M line a narrow dark band in the centre of the H zone of the sarcomere and at right angles to the long axis of the sarcomere.

MMR vaccination a protective active immunization against measles, mumps and rubella that should be routine for all children unless a strong medical reason contraindicates it. Concern was raised in 1998 that MMR vaccination could lead to an intestinal disorder that allows the absorption of otherwise non-permeable peptides capable of causing autism and other developmental problems. The evidence was reviewed by the Joint Committee on Vaccination and Immunization and no case was found for abandoning a vaccine of proved effectiveness and safety.

mobilization the process of relieving stiffness or restoring the full range of movement in a joint or a person, usually after illness or injury or after prolonged forced immobility. Local mobilization after fractures, joint or joint capsule disorders or other injuries may require physiotherapy or even manipulation under anaesthesia.

modality 1 a type or mode, especially of sensation, of the senses or of medical treatment.
2 a quality that denotes mode, mood or manner.

molality the number of moles (see MOLE 1.) of solute in 1000 g of solvent.

molar one of the 12 back grinding teeth. From the Latin *mola*, a grindstone.

molarity the concentration of a solution expressed in terms of the mass of dissolved substance in grams per litre divided by its relative molecular mass.

molar solution a solution of such concentration that the number of grams of it dissolved in 1 litre is equal to its relative molecular mass.

mole 1 the basic unit of amount of substance; the amount that contains the same number of entities as there are atoms in 0.012 kg of carbon-12. The entity may be an atom, a molecule, an ion, and so on. From the German *molekül*, a molecule.
2 a coloured (pigmented) birth-mark (naevus). Hairy moles may be disfiguring but are never dangerous. Moles seldom undergo malignant change, but any alteration is size, shape or colour should be reported at once. From the Latin *mola*, a millstone.

molecular biology the study of cellular phenomena, and especially genetics, at a molecular or chemical level. As knowledge has progressed the term has become more and more synonymous with 'biology', for, today, biology aims at nothing less than a full understanding of the functioning of living things in terms of the nature and interactions of their molecules.

molecular genetics a now out-dated term with its implication that a serious study of genetics is possible at other than a molecular level. Even those whose interest in genetics is limited to heredity can no longer pursue their discipline without becoming involved in MOLECULAR BIOLOGY.

molecular weight a term now superceded by relative molecular mass, the sum of the relative atomic masses of all the atoms in the molecule.

Mongolian spot a bluish-black pigmented birthmark (NAEVUS) occurring on the buttocks or lower part of the back, especially in coloured children. Mongolian spots are caused by a local accumulation of the normal skin pigment (melanin). They have usually disappeared by the age of about 4.

monitoring close surveillance or supervision, especially of people liable to suffer a sudden and dangerous deterioration in health. Monitoring involves checks of various parameters such as pulse rate, temperature, respiration rate, the condition of the pupils, the level of consciousness, the degree of appreciation of pain and various blood gas concentrations such as oxygen and carbon dioxide. Long-term instrumental display of the electrocardiogram is also common in monitoring and this may also be recorded for diagnostic purposes, using computer analysis.

mono- *combining form denoting* single or alone.

monoamine oxidase MAO, one of a group of enzymes found in brain cells, in peripheral adrenergic and dopaminergic nerve endings and in the intestinal wall and liver. These enzymes play an important part in the breakdown of the neurotransmitters NORADRENALINE, DOPAMINE and SEROTONIN (5-hydroxytryptamine). Non-selective inhibition of these enzymes will elevate mood. Selective MAO-A enzymes act on serotonin; selective MAO-B inhibitor drugs act on phenylethylamine in the glial cells of the brain and elsewhere. The selective MAO-A inhibitor drugs are used to treat DEPRESSION and anxiety; selective MAO-B inhibitors are used to treat Parkinsonism.

monoblast the precursor of the MONOCYTE normally found only in the bone marrow.

monochromasia complete absence of any perception of colour. Monochromasia is very rare. There is an absence or severe deficiency of cones in the RETINA and VISUAL ACUITY is poor. A person with this defect is called a monochromat.

monoclonal of a group of cells derived from a single cell, all having the same GENOTYPE. Of a single CLONE of cells.

monoclonal antibodies ANTIBODIES (immunoglobulins) produced by hybrid B lymphocyte tumours (myelomas). The type of antibodies produced depend on the selection of the B cell. These can be fused to cultured mouse, or even human, myeloma cells to form immortal tumours (hybridomas) – clones of cells that continue indefinitely to generate large quantities of the particular antibody produced by the B cell. The availability of quantities of almost any desired antibody has major diagnostic, therapeutic and research implications and monoclonal antibody production is one of the most important biotechnological advances of our time. Monoclonal antibodies can be made that will seek out and recognize cancers anywhere in the body, and this offers a number of intriguing possibilities for treatment.

monocular vision vision with one eye only, even if both are capable of seeing. In such a case, vision is suppressed in one eye, often to avoid seeing double.

monocyte a large white blood cell with a round or kidney-shaped nucleus. There are no granules in the CYTOPLASM. The monocyte migrates to the tissues where it becomes a MACROPHAGE.

monocytopenia a reduction in the number of MONOCYTES in the circulating blood.

monolayer a sheet of cells only one cell thick.

monomer one of the chemical groups many of which are repetitively linked together to form a POLYMER.

mononuclear cell a cell with a single spherical or near-spherical nucleus, as distinct from one with a lobed nucleus as in the case of the polymorphs. The mononuclear cells include LYMPHOCYTES, MONOCYTES and immature GRANULOCYTES. An unhelpful and obsolescent designation dating from a period prior to the detailed differentiation of cells of the immune system.

mononuclear phagocyte system blood MONOCYTES and tissue MACROPHAGES.

monoplegia paralysis of a single muscle or group of muscles or of a single limb.

monorchism having only one testicle in the scrotum. This is usually due to failure of descent of one testicle from the abdomen and occurs in about 1 boy in 50. An undescended testicle remains sterile and is much more prone to cancer than normal.

monosaccharide the simplest form of sugar. Monosaccharides are classified by the number of carbon atoms in the molecule. They may thus be trioses, tetroses, pentoses, hexoses, etc. The commonest monosaccharide in the body is GLUCOSE, which is a hexose, with six carbons.

monosomy the absence of one complete AUTOSOMAL chromosome of a pair. This is a lethal condition. Compare TRISOMY as in DOWN'S SYNDROME.

mons a mound or rounded eminence, especially the mons pubis (or veneris), the hairy mound of fatty tissue covering the junction of the pubic bones in adult females.

moon–faced pertaining to the full-cheeked, hamster-like appearance caused by excessive doses of corticosteroid drugs or by excessive production of the natural adrenal cortical hormone in Cushing's syndrome. Also known, inelegantly, as 'cushingoid'.

morbid anatomy the branch of pathology dealing with the visible structural changes caused in the body by disease and injury and discernible at postmortem examination.

morbidity the state of being diseased or suffering.

morbidity rate the number of cases of a disease occurring in a given number (usually 100,000) of the population. The annual morbidity figure for a disease, in a particular population, is the number of new cases reported (incidence) in the year.

Morgan, Thomas Hunt the most important early work on genetics was by the American biologist Thomas Hunt Morgan (1866–1945), who selected for study the tiny fruit fly *Drosophila*. This fly bred quickly and easily and its cells contained only four prominent chromosomes. By studying successive generations of *Drosophila*, Morgan was able to show that de Vries was right about spontaneous mutations, and also that various groups of characteristics were often inherited together. This was called *linkage*, and it showed that particular sets of genes always occurred on a particular chromosome. Morgan, however, found an important exception to the rule. It became apparent that, from time to time, during the process of cell division, pairs of chromosomes were exchanging segments with each other. This was called 'crossing over'. Morgan's work progressed to the point where he was able to show the approximate position of certain genes on the chromosome, and he showed that the further apart two genes were located on a chromosome, the more likely were their characteristics to be separated by crossing over.

Morgan's work on *Drosophila* culminated in his publication, in 1926, of *The Theory of the Gene*, which contained gene maps for *Drosophila* and placed genetics on a new and sound foundation.

morning sickness vomiting or retching of pregnancy. A common symptom occurring from the 6th to the 12th week. It usually occurs soon after waking but seldom has any effect on health or on the pregnancy and almost always stops before the 14th week. The extreme case is called hyperemesis gravidarum.

moron a person of a mild to moderate degree of learning difficulty (mental retardation). A person with an IQ of between 50 and 70. The term is not used in medicine. From the Greek *moros*, stupid.

morpheme the smallest element of speech that conveys either factual or grammatical information. Compare with phoneme which is a speech sound that serves to distinguish one word from another.

morphogenesis the origin and development of the form and structure of the body.

mortality rate death rate. The ratio of the total number of deaths from one or any cause, in a year, to the number of people in the population. Crude mortality is the number of deaths in a year per thousand total population. The age-specified mortality rate is the number of deaths occurring in a year in people of a particular age or in a particular age-range. See also INFANT MORTALITY.

morula an early stage in the development of the embryo at which it consists of a solid spherical ball of apparently identical cells.

mosaic gene see DISCONTINUOUS GENE.

mosaicism the state in which two or more genetically different types of cell occur in the same individual. Although the cells are all derived from the same fertilized egg, they do not all possess the same number of chromosomes. In about 1 per cent of cases of DOWN'S SYNDROME there are two different cell lines, one normal and the other with an additional chromosome 21 (trisomy 21). The effect of mosaicism varies with the proportion of cells containing abnormal chromosomes. Compare CHIMERA.

motion sickness nausea or vomiting induced by any sustained, repetitive, passive movement of the body in any vehicle of transportation. There is abdominal discomfort, pallor, sweating, salivation, depression, nausea, vomiting, apathy, loss of appetite and sometimes a loss of the will to live. The cause is unknown, but it is related to repetitive stimulation of the inner ear balancing mechanisms. The word nausea derives from the Greek word *naus*, a ship.

motor 1 causing movement.
2 carrying nerve impulses that stimulate muscles into contraction or cause other responses such as gland secretion. From the Latin *movere*, to move.

motor cortex the part of the surface layer of each hemisphere of the main brain (cerebrum) in which voluntary movement is initiated. These areas can be mapped out to show which parts of them are responsible for movement of any particular parts of the body. The map resembles a distorted and inverted human figure.

motor endplate the point of junction of a motor nerve fibre and a muscle fibre. The motor endplate is a modified area of the muscle fibre membrane at which a synapse occurs. A motor nerve axon ending may have up to 50 synaptic knobs (boutons) but a single muscle fibre has only one endplate. The neurotransmitter is acetylcholine.

motor neuron, motor neurone a nerve carrying MOTOR impulses. The upper motor neurons have their cell bodies in the surface layer (CORTEX) of the brain and their axons running down the spinal cord. The lower motor neurons have their cell bodies in a column at the front of the spinal cord and their axons running out of the cord in the spinal nerves.

motor proteins the protein classes of myosins, kinesins and dyneins. Also known as molecular motors. There are more than 40 myosins, they function in muscle contraction and some other processes, and move along filaments of the protein actin. Kinesins are concerned in transport of proteins, organelles and vesicles along microtubules and chromosome segregation. Dyneins are very large molecules that also move along microtubules to power cilia and flagella. All the motor proteins are powered by energy from ATP binding, hydrolysis and release. Molecular proteins convert chemical energy (ATP) into kinetic energy.

mouth-to-mouth resuscitation maintenance of an oxygen supply in a person unable to breath spontaneously by periodic inflation of the lungs by blowing into the mouth or nose. This is done 16–20 times a minute and is verified by watching the chest rise and fall. Also known as the 'kiss of life'.

mucin a glycoprotein that is the main constituent of MUCUS. The term is also used as a generic name for the substance used as a drug formulated with xylitol as artificial saliva.

muco- *combining form denoting* MUCUS or MUCOUS MEMBRANE.

mucopolysaccharide POLYSACCHARIDEs containing amino sugars or that are polymers of MONOSACCHARIDES in which one of the –OH groups is replaced by an NH_2 group. Mucopolysaccharides are important structural materials in the body forming the ground substance of many connective tissues in which fibrous proteins are embedded. Now more commonly called glycosaminoglycans.

mucopurulent containing MUCUS and PUS.

mucous membrane the lining of most of the body cavities and hollow internal organs such as the mouth, the nose, the eyelids, the intestine and the vagina. Mucous membranes contain large numbers of goblet-shaped cells that secrete mucus which keeps the surface moist and lubricated.

mucus a slimy, jelly-like material, chemically known as a MUCOPOLYSACCHARIDE or GLYCOPROTEIN, produced by the goblet cells of MUCOUS MEMBRANES. Mucus has important lubricating and protective properties. It prevents acid and enzymes from digesting the walls of the stomach and intestines. It traps fine particulate matter, including smoke, in the lungs. It lubricates swallowing and the transport of the bowel contents. It facilitates sexual intercourse.

Müllerian duct an embryonic structure that, in females, develops into the ducts of the reproductive system. The Müllerian duct degenerates in the male under the influence of the protein Müllerian inhibiting substance. (Johannes Peter Müller, German physiologist and comparative anatomist, 1801–58).

multigene family a collection of genes, usually but not necessarily situated together, that have either a common or associated function or have similar nucleotide sequences.

multigravida a woman who has had at least two pregnancies.

multimetric protein a protein consisting of more than one discrete polypeptide strands.

multineuronal pathway a chain of neurons linked by synaptic connections.

multipara a woman who has carried at least two babies to a viable stage or who has delivered more than one live baby.

multiple births the production of more than one individual in a single parturition. Twins occur in about one in every 84 births in Britain, and about one in 300 pregnancies results in identical twins. Multiple births may occur as a result of the near-simultaneous release and fertilization of more than one ovum, or from the fertilization of a single ovum which then separates into two or more parts from each of which a new individual develops, all having identical genetic constitution. Non-identical (dizygotic or fraternal) twins may be of the same or of different sex and are no more alike than any other two siblings. Each has his or her own PLACENTA and set of membranes. Identical (monozygotic) twins share a placenta and often compete for nourishment so that they are of unequal weight when born. Multiple pregnancies are shorter than single pregnancies. Triplets and higher number births are biologically rare. Triplets have a natural frequency of about one in 7000 births. They are more common as a result of the use of anti-infertility drugs that induce multiple ovulation.

multiple personality a rare psychiatric dissociative disorder in which a person appears to have two or more distinct and often contrasting personalities at different times, with corresponding differences in behaviour, attitude and outlook. This condition is quite distinct from schizophrenia.

multiple pregnancy a pregnancy with more than one fetus.

mummification drying and shrivelling of the whole or part of the body.

mural on the wall of a hollow organ or structure.

murder the killing of a human being in circumstances that are neither accidental nor lawful but with prior intent to kill (with malice aforethought).

murine pertaining to mice and rats.

murmur a purring or rumbling sound of variable pitch heard through a stethoscope especially over the heart or over a narrowed or compressed artery. Murmurs are caused by turbulence in blood flow and often imply disease such as heart valve narrowing or incompetence.

muscarinic 1 producing the effects of post-ganglionic cholinergic stimulation of the parasympathetic. Having an effect similar to that of the mushroom poison MUSCARINE.
2 of an acetylcholine receptor that responds to muscarine. Compare NICOTINIC.

muscle

Movement of the body, or of any part of it, is possible only as a result of the action of muscles. The movement results from muscle contraction – shortening. A contracting muscle does not change its volume – only its length, so when it shortens it becomes thicker. Many muscles act across joints so that when the muscle shortens the joint bends. Others change the diameter of body cavities or tubes so that the contents are moved. The most striking example of this kind of muscle action is the heart. Some, such as those of the tongue, effect movement by their action on each other. Muscles can only work by shortening and exerting a pull or, if circularly arranged, a squeeze. Even when we seem to be pushing something, the muscle action involved is a pulling one. Contraction involves the conversion of chemical energy into movement (kinetic) energy.

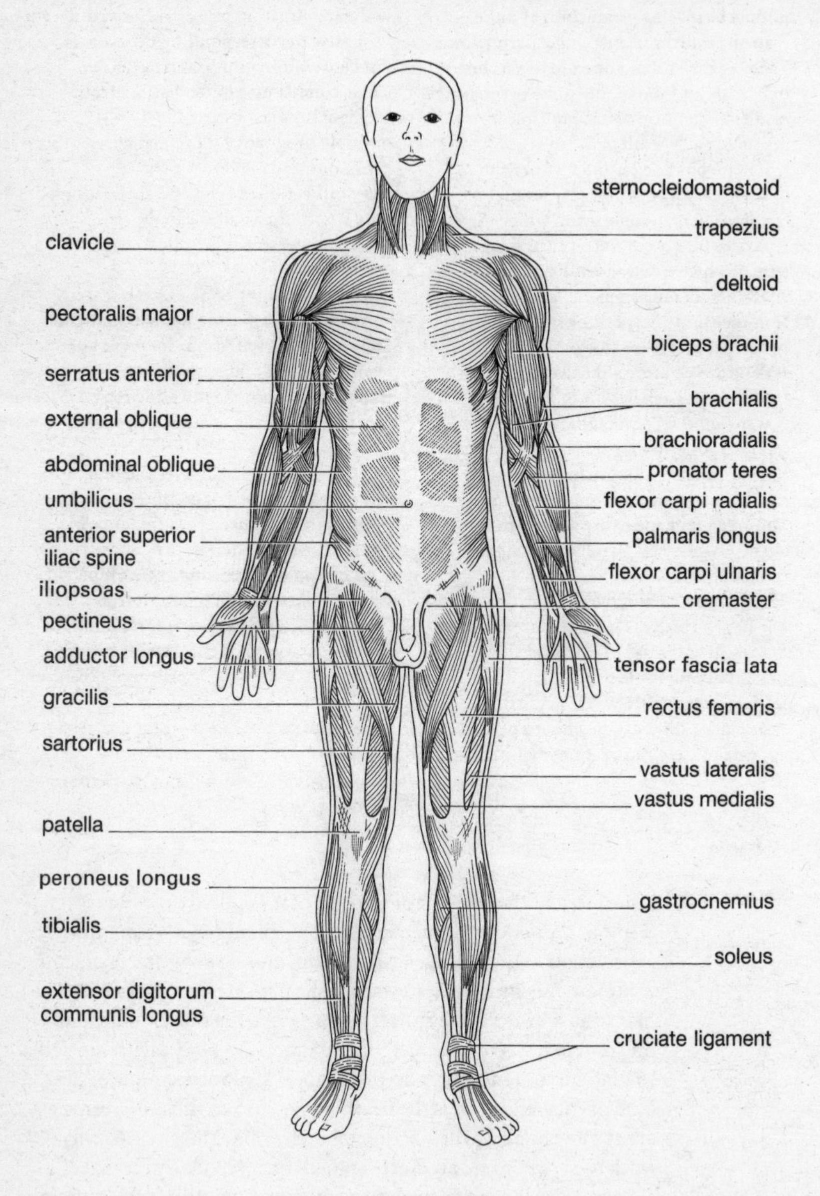

Fig. 11 **Muscle (front view)**

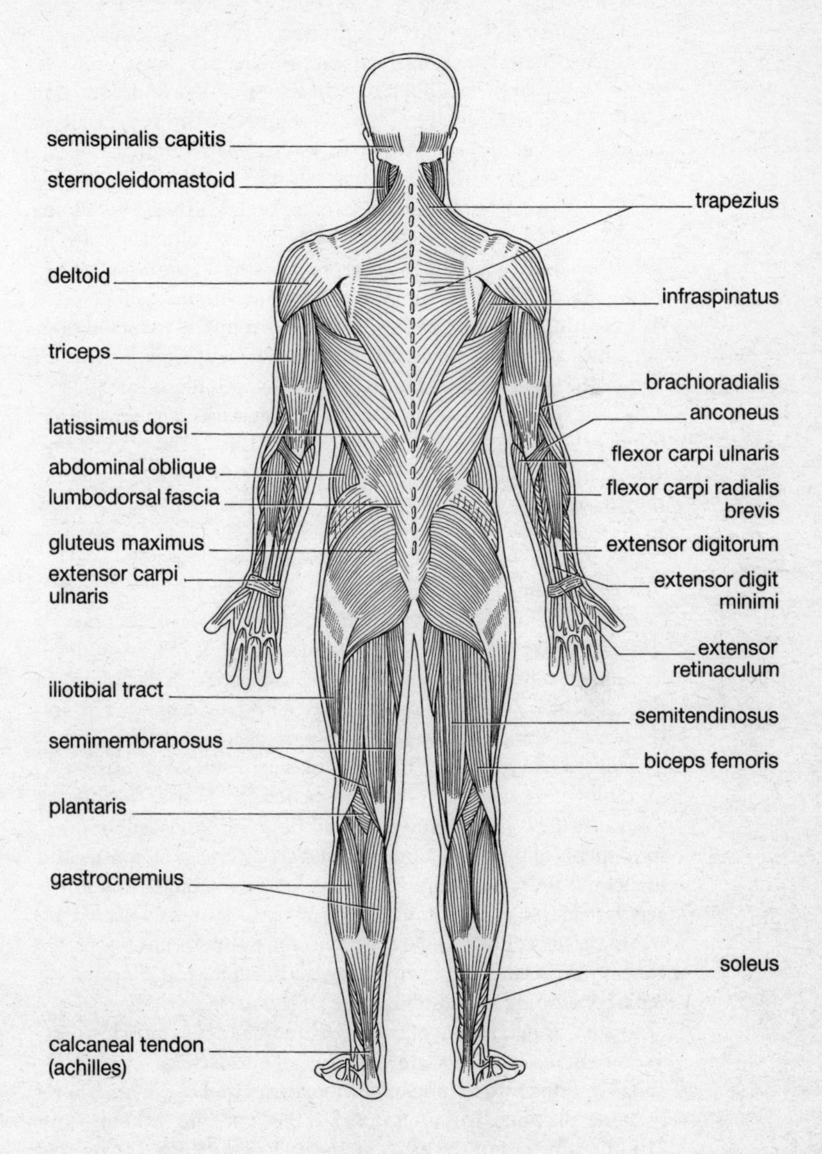

Fig. 12 **Muscle (rear view)**

On average, muscle constitutes about 40 per cent of the weight of males and about 36 per cent of the weight of females. Because of minor anatomical variations, the exact number of voluntary muscles in the body is uncertain but it is in the region of 660. Muscle is about as efficient as an internal combustion engine (around 25 per cent) in converting chemical energy into work. A reasonably fit man can exert about half a horsepower – at least for a time.

The most conspicuous muscles of the body are those of the limbs. Limb muscles work in opposing groups, those on one side of a joint being balanced by a comparable group working on the other side. As one group contracts, the opponents relax but without losing tension. The coordinated action of these opposing groups allows stability and smoothly controlled movement. While movement is taking place in a limb, the rest of the body must be appropriately braced and this may involve simultaneous contraction of muscles on either side of joints so as to stabilize them. The control of this complex and ever-changing interaction is provided by a compact, real-time computer – the cerebellum – connected to the voluntary centres in the brain and to the muscles by way of the peripheral nerves.

The main voluntary muscle groups

Several muscles act on the shoulder joint. The powerful shoulder-pad muscle, the deltoid, raises the whole arm while other muscles simultaneously act to fix the shoulder-blade on which the upper arm bone articulates. The elbow joint is flexed by the biceps that crosses the front of it, and extended by the triceps that crosses behind it. The biceps also rotates the forearm. Among the most impressive examples of muscle control and interaction are those of the forearm that cause the complex movements of the hand. Almost all the movements of the fingers and the hand are mediated by contraction of forearm muscles operating through long fine tendons that run across the front (flexors) and the back (extensors) of the wrist.

In the lower limb, the powerful hip and buttock muscles are concerned mainly with moving and stabilizing the hip joints in walking, running and climbing. The large muscle group that originates from the front of the thigh (the quadriceps muscles) has a common insertion in a large tendon that encloses the kneecap and is inserted into a bump on the top of the front of the main lower leg bone (the tibia). This group extends the knee as in kicking. The opposing group, the hams, on the back of the thigh, are inserted into the back of the tibia, by the hamstrings, and bend (flex) the knee. Two of the main calf muscles are inserted into the back of the heel bone by way of the Achilles tendon. Others act on the foot, by way of long fine tendons analogous to those of the forearm muscles that act on the hand.

The most important of our voluntary muscles are those that maintain breathing. These are the intercostal muscles that pull the ribs upwards and outwards, so increasing the side-to-side dimension of the chest, and the high-domed diaphragm that simultaneously flattens. The resulting increase in the volume of the chest causes air to be sucked into the lungs. When these muscles relax, the lungs collapse by elastic recoil.

The abdominal contents are less well protected than those in the chest, but the abdominal wall contains strong sheets of muscle that can contract quickly to form a firm protective barrier. The same muscles can strongly compress the contents of the abdomen, helping us to empty the bladder and rectum and, in childbirth, to help to push out the baby. The spine (vertebral column) is surrounded by a strong mass of longitudinal muscle that supports it, maintaining the proper relationship of the bones, and allowing bending and twisting in any direction. Most back trouble is caused by weakness of these muscles.

Much of our muscle action is somewhat less energetic. All the subtleties of facial expression and the indications of the emotions are achieved by the contraction of muscles under the skin, acting with great precision to alter the configuration of the mouth, the eyebrows, the eyelids, the nostrils and the cheeks. Delicate and finely controlled muscles in the voice box (larynx) act through levers on the vocal cords, tensioning and loosening them so as to change the pitch of the voice and prolong or interrupt the sounds produced. Tiny, but rapidly-acting, muscles in the middle ear act to tense the minute chain of bones that connect the ear drums to the inner ear, so as to protect us from unduly loud noises. Tiny circular muscles constrict the pupils of the eyes; radial muscles widen them.

Biting and chewing are made possible by muscles arising from the base of the skull and the outside of the temple bones and attached to the jaw bone. These muscles of mastication are capable of pulling the jaw bone powerfully upwards and rocking it sideways so that food can be cut and torn by the front teeth and ground by the molars.

The fine structure of muscle

Muscles consist of bundles of thousands of individual muscle fibres bound together by collagen. Each fibre is a single, cylindrical cell, up to a foot long and up to one tenth of a millimetre in diameter, containing a bundle of several hundred smaller sub-fibres called myofibrils. Myofibrils, in turn, contain bundles of smaller protein filaments arranged along them in a repeating pattern. Each unit of this repeating pattern is called a sarcomere and it is in the sarcomeres that the real action occurs. Sarcomeres in adjacent fibers lie accurately in line, and it is this that gives voluntary muscle fibres their striped appearance. The actual structure of sarcomeres is far too fine to be seen with anything less than a powerful electron microscope. Sarcomeres are made up of thick filaments of the protein myosin, between which run thin filaments of the protein actin, like interlacing fingers. The thick filaments are linked together in groups so as to leave spaces between into which the free ends of the thin filaments, also connected together in groups, can move.

Muscle contraction

Muscle contraction requires a lot of energy and this is provided by many mitochondria lying within the muscle fibre alongside the sarcomeres. Contraction of the muscle fibre does not occur by shortening of the myosin and actin filaments as might be supposed. In fact, the contraction results from sliding of the thin filaments more deeply between the thick filaments. This is brought about by the action of short protrusions on the myosin filaments called cross bridges which bind, briefly, to the thin filaments and, acting much in the manner of the oars of a rowing boat, pull the two sets of filaments together. The links between the myosin and the actin filaments then break. This action is performed repeatedly so as to pull the thick and thin filaments as far as possible into each other. The minute shortening movement achieved by a single sarcomere is multiplied many times by the number of sarcomeres in each fibre and the muscle fibre, as a whole, shortens to about two-thirds of its resting length. Fibres cannot contract partially; they do so completely or not at all.

Muscles are made to contract by electrical nerve impulses entering them from motor nerves. Like the muscles, these nerves are also made up of bundles of fibres. One motor nerve fibre is connected to a group of muscle fibres, and each time an electrical pulse (nerve impulse) reaches the end of the nerve fibre all the connected muscle fibres contract fully but briefly. Fibres can contract as frequently as 50 times a second. By varying the frequency of contraction and the number of fibres involved, the nervous system can effect the most delicate gradations of muscle power.

Muscles and exercise

Muscles are highly responsive in another sense. Intense, sustained muscular work builds up the bulk of muscles, and disuse soon leads to wasting (atrophy). When muscles are built up by exercise, the number of muscle fibres does not increase, but each fibre enlarges by an increase in the number of its myofibrils. This hypertrophy increases the power of the muscle. In addition, the blood supply to exercised muscles increases so that the fuels from which energy is derived can be more efficiently delivered. The build up of simple substances, such as amino acids, into complex proteins such as myosin or actin is called anabolism and, as everyone is now aware, this process can be assisted, not without danger, by taking certain anabolic steroid drugs.

Different kinds of exercise affect muscles in different ways because muscles contain different kinds of fibres that can be selectively improved. Periods of intense, maximum-effort exercise increase the number of myofibrils in the fast-acting fibres that are used only during such work. Prolonged exercise of moderate intensity, such as jogging, increases the number of mitochondria in the other muscle fibres. It follows that athletes must work hardest at the kind of activity in which they wish to excel.

muscle disorders these include cardiomyopathy, claudication, compartment syndrome, cramps, dermatomyositis, fibroids, muscle spasm, muscular dystrophy, myasthenia gravis, myocarditis, myoma, tetany and trichinosis.

musculature the muscle system of the body or the intrinsic muscles of a part of the body.

musicians' overuse syndrome severe, disabling pain and loss of speed, accuracy and agility from functional disturbance in the muscles used in playing musical instruments. This occurs from overwork or excessive practising, especially if there is faulty technique. The affected muscles become swollen and there may also be some loss of feeling in the part. Treatment involves strict limitation of playing time and the use of well-designed supports for instruments. It may sometimes be necessary to give up playing for weeks or months.

mutagen any agent capable of changing the structure of DNA without immediately killing the cell concerned. Any surviving MUTATION may be perpetuated to all descendants of the cell. Mutagens include ionizing radiation such as ultraviolet light, X-rays, gamma rays and cosmic rays and a wide range of chemical substances including the tars in cigarette smoke.

mutant any organism or cell with a gene or genes that have suffered a MUTATION.

mutation any persisting change in the genetic material (DNA) of a cell. Mutations most commonly involve a single gene but may affect a major part, or even the whole of, a chromosome or may change the number of chromosomes (genomic mutation). A nonsense mutation is one that alters the sequence of bases in a CODON so that no amino acid is coded. Many mutations have an unfavourable effect on the cell concerned and are not passed on, but non-lethal mutations are replicated in daughter cells. Mutation in a cell in the GONADS that gives rise to a SPERMATOZOON or an egg (OVUM), will be passed on to a clone of sperm or eggs and one of these may take part in fertilization so that the mutation is passed on to every cell in the body of the future individual, including the GERM CELLS. New mutations occurring in the sex cells (germ line mutations) may thus lead to hereditary abnormalities. Mutations in body cells (somatic mutations) cannot do this but can cause cloned abnormalities including cancers. See also FRAME SHIFT MUTATION, INSERTION MUTATION, INVERSION MUTATION, LEAKY MUTATION, LETHAL MUTATION, MISSENSE MUTATION, POINT MUTATION, NONSENSE MUTATION.

mutation rate the number of instances of a particular gene mutation occurring in a population in one generation.
mutator protein an antiviral factor found in human cells that converts one codon in DNA so that it codes for the amino acid uracil instead of cytosine. Uracil replaces cytosine in RNA. When HIV infects cells it makes a DNA copy of its own RNA to insert into the host DNA. The mutator protein offers the possibility of developing drugs that could interfere with the replication of infecting viruses.
mutism inability or refusal to speak. There are many possible causes including deafness from birth, mental retardation, severe DEPRESSION, SCHIZOPHRENIA, HYSTERIA, brain tumour and HYDROCEPHALUS.
my-, myo- *combining form denoting* muscle.
myalgic encephalomyelitis (ME) see CHRONIC FATIGUE SYNDROME.
myco- *combining form denoting* fungus. From the Greek *muces*, fungus.
mycology the science or study of fungi. Medical mycology is limited to the study of fungi that infect or affect humans and to those from which useful drugs can be derived.
mycosis any disease caused by a fungus.
mydriasis widening (dilatation) of the pupil of the eye, usually as a result of instillation of a mydriatic drug, such as ATROPINE or CYCLOPENTOLATE.
myel-, myelo- *combining form denoting*
1 the spinal cord.
2 the bone marrow.
3 MYELIN.
myelin the fatty, white material forming a sheath around most nerve fibres and acting as an insulator. See also DEMYELINATION.
myelinated possessing a MYELIN sheath.
myelin sheath see MYELIN.
myelocoele the narrow central canal of the spinal cord.
myelocyte an immature white blood cell normally found in the bone marrow.
myeloid pertaining to the bone marrow.
myo- *combining form denoting* muscle.
myocardial infarction (MI) a heart attack. The death and coagulation of part of the heart muscle deprived of an adequate blood supply by coronary artery blockage.
myoclonus a sudden, brief, involuntary muscle contraction usually causing a jerk of a limb. This occurs most commonly as a normal phenomenon in people half asleep but myoclonic contractions are a feature of epilepsy and of many other brain diseases.
myoglobin the muscle cell equivalent of the haemoglobin of the blood. Myoglobin acts as a temporary oxygen store from which oxygen is drawn as the muscle requires it.
myokymia one of a range of conditions featuring involuntary, fine, twitching or rippling of muscle fibres. The common eyelid twitch, or fasciculation, is an example of myokymia.
myology the science and study of muscle.
myometrium the muscle wall of the womb (uterus).
myopathy any disease or disorder of muscle. The myopathies include congenital or acquired conditions such as the muscular dystrophies, inflammatory muscle disorders and metabolic and drug-induced disorders.
myopia short-sightedness. A condition in which the optical power of the eye is too great in relation to the distance from the lens to the RETINA. Only diverging rays from near objects focus sharply. Myopia is corrected by weakening (concave) lenses. The derivation is uncertain but probably relates to 'muscular eye' – a reference to the tendency of myopes to screw up their eyelids so as to stop down the optics and improve distance vision.
myringo- *combining form denoting* the eardrum.
mythopoiesis the internal fabrication of mythic events or false memories that may subsequently be revealed or acted out in multiple personalities, trances, 'demonic possession', seemingly psychic phenomena or conviction of their reality.
myxo- *combining form denoting* mucus.
myxoid of a mucin-rich constitution.

Nn

naevus any coloured growth or mark on the skin present at birth. A birthmark. From Latin *naevus*, a spot or blemish.
nail a protective and functional plate of a hard, tough protein, KERATIN, lying on the back surface of the last PHALANX of each finger and toe.
nanism DWARFISM due to arrested development.
nano- *prefix denoting* one thousand millionth (one billionth).
nanogram one billionth of a gram.
nanometre one billionth of a metre.
nanoparticles particles of very small size that are being exploited increasingly in medicine as in other sciences. Nanoparticles have recently been developed as delivery vehicles for drugs and for gene therapy. One unexpected finding emerging from this research is that nanoparticles of cerium oxide only 5 nanometres in diameter inserted into nerve cells in culture appear to increase the three week life of these cells to about six months. It is speculated that they are having a useful effect in mopping up damaging FREE RADICALS.
nanosecond one billionth of a second.
nape the back of the neck.
narcissism possession of an exaggerated and exhibitionistic need for admiration and praise and an overweening conviction of one's own merits and attractiveness. Narcissus, a character in Greek mythology, was a youth who fell in love with his own reflection in a pond.
narco- *combining form denoting* numbness, narcosis or stupor. From Greek *narke*, numbness.
narcoanalysis psychoanalysis carried out while the patient is in a drowsy state induced by drugs such as sodium amytal or thiopentone (thiopental). The vogue for narcoanalysis seemed to have passed, but there is a suggestion of renewed interest in the USA.
narcosis a state of unconsciousness that may range from sleep to deep, irreversible coma. In most cases narcosis is caused by a drug.
narcotic a drug which, in appropriate dosage, produces sleep and relieves pain. Overdosage of narcotics may cause coma and death. Most narcotics are derived from opium or are synthetic substances chemically related to morphine.
nares the plural of NARIS.
naris a nostril. Either of the two external openings of the nasal cavity.
nasal pertaining to the nose.
nasal bones the two variably-sized, flat and roughly rectangular plates of bone forming the bridge of the nose. Each nasal bone is attached to the FRONTAL BONE above, the MAXILLA to the side and to the ETHMOID internally.
nasal conchae see TURBINATES.
nasal septum the thin, central partition that divides the interior of the nose into two passages. The septum consists of a thin plate of bone, behind, and a thin plate of cartilage in front. Both are covered with MUCOUS MEMBRANE. Deflection of the septum to one side (deviated septum) is common and usually harmless.

nasion the centre of the junction (suture) between the nasal and the frontal bones. The centre of the bridge of the nose.

naso- *combining form denoting* nose or nasal.

nasolacrimal canal the bony groove or conduit from the lacrimal bone through the MAXILLA that carries the NASOLACRIMAL DUCT.

nasolacrimal duct the membranous passage that carries the overflow of tears from the LACRIMAL SAC to the interior of the nose, opening just under the lower TURBINATE. Blockage of the nasolacrimal duct causes a persistently watering eye.

nasopharynx the space at the back of the nose, above and behind the soft palate. Normally this space is continuous with the space at the back of the mouth, but in swallowing it is shut off from the oropharynx by the soft palate pressing against the back wall. On the back wall of the nasopharynx are the openings of the EUSTACHIAN TUBES and, in childhood, the ADENOIDS.

natal 1 pertaining to birth.
2 pertaining to the buttocks.

nates the buttocks.

natriuresis excretion of sodium by the kidneys.

natriuretic peptide a PEPTIDE present in the blood and raised in quantity in people with long-term (chronic) heart failure. It arises in the upper chambers of the heart, probably as a result of stretching, and causes blood vessels to widen and the output of urine to increase.

natural antibodies antibodies in the serum to red blood cell antigens that are not present in the body of the same individual. Natural antibodies are not derived from previous exposure to antigens. They determine the blood groups.

natural childbirth a term used to encourage the concept that having a baby should be a normal and natural process rather than a medical or surgical event, operation or emergency. A clear understanding of what is involved and informed instruction on the nature and cause of pain in labour, together with exercises in relaxation and cooperation, has made labour less difficult and painful, and more rewarding, for millions of women. It has not, however, made childbirth either easy or painless.

natural immunity the ability to resist infection that does not depend on prior experience of the invading organism and the resultant production of antibodies or amendment or selection of LYMPHOCYTES. Natural immunity is a general and non-specific resistance to infection possessed by all healthy individuals. Also known as natural resistance.

natural killer cells a class of large, granular lymphocytes that bind directly to cells bearing foreign ANTIGENS and kill them. Natural killer cells do not require prior exposure of the immune system to the antigen and kill their victims by programmed cell death (apoptosis).

natural law the idea that there exists a 'built-in' body of ethical or moral principle common to all humankind and perceived without instruction. Human experience may seem to support the idea, but an equally strong case can be made for the proposition that the illusion of a natural law may arise from forgotten early conditioning and from the notion of justice induced by personal experience of injustice.

natural selection the Darwin-originated principle that individuals of a species happening, by normal genetic rearrangement or by mutation, to possess inherited characteristics with survival value relative to a particular environment are more likely to survive long enough to reproduce and increase the numbers having these characteristics. Instances of natural selection occur quickly enough in some organisms such as bacteria and moths to be readily observed. (Charles Darwin, 1809–82, English naturalist).

nature versus nurture the perennial argument as to whether heredity or environment is more influential in determining the outcome of any individual's development. It is now apparent that, in many particulars, the two are so intimately inter-related in their effects as to be almost inseparable.

naturopathy a system of folk medicine that claims that all disease can be cured by restriction to a largely vegetarian diet free from all contaminants and drugs. Such a regimen, if possible, might well promote health but there are many causes of disease other than dietary and many environmental hazards are unavoidable.

nausea the unpleasant feeling of sickness that often precedes vomiting. From the Greek *naus*, a ship.

navel the depressed scar in the centre of the abdomen left when the UMBILICAL CORD drops off and the opening into the abdomen heals. The medical term is umbilicus.

navicular the outermost of the wrist (CARPAL) bones on the thumb side in the nearest row. The bone is roughly boat-shaped, hence the name, which is derived from the Latin *navis*, a boat. Also known as SCAPHOID.

Neanderthal man the sub-human sub-species *Homo sapiens neanderthalensis* that inhabited much of Europe, North Africa and parts of Asia from about 125,000 to 40,000 years ago. The first remains were discovered in 1856 in a cave in the Neander valley near Dusseldorf. Much argument persists as to whether these were a direct ancestor of man or a separate blind-ended branch of evolution.

neolithic period the last phase of the stone age that immediately preceded the bronze age. In the neolithic period stone implements were ground and polished, animals were domesticated and bred, agriculture was practised and pottery made.

near point the shortest distance from the eye at which fine detail can be sharply perceived. Except in short-sighted (myopic) people, the near point moves progressively further away with age. Reading glasses will generally be needed when the near point exceeds about 40 cm.

near-sightedness see MYOPIA.

neck any narrowing or constriction in a body or part. A cervix.

neck, broken see BROKEN NECK.

necr-, necro- *combining form denoting* death.

necrobiosis natural death of cells and tissues occurring in the midst of healthy tissue. Natural cell death as opposed to death from disease or injury.

necromania an abnormal desire for the company of a dead body or dead bodies, often the body of a loved one.

necrophilia the desire for, or practice of, sexual intercourse with a dead person. This somewhat limiting propensity is described as a psychosexual disorder (paraphilia) in which arousal is possible only if the object of sexual interest is dead.

necrophobia an abnormal fear of death or of corpses.

necropsy an autopsy, or postmortem examination, of a body.

necrosis the structural changes, such as those of gangrene, that follow death of a body tissue. The most obvious changes are in the cell nuclei which become shrunken and condensed (pyknosis) and no longer take a basic stain. Cell CYTOPLASM becomes more homogeneous and spaces (vacuoles) develop.

necrotic pertaining to the death of tissue (NECROSIS).

needle exchange programmes an effective public health measure to reduce the prevalence of conditions such as AIDS and hepatitis B that are often spread among intravenous drug abusers by sharing needles. About half of new HIV infections are now caused by needle sharing. New sterile syringes and needles are supplied free of charge to discourage the practice. Many studies have shown that this is a valuable measure that does not encourage illegal drug usage but the USA, almost alone among developed countries, persistently banned it.

negative feedback a control entity in which part of the response to the stimulus, acting in opposition to it, is applied to the stimulus. In other words, in a system with an input and an output, a proportion of the output signal, which must be in opposite phase to the input signal, is carried back to join and modify the input signal, reducing the effect of distortional changes caused by the system. The feedback signal is usually much smaller than the input stimulus, but exerts a powerful stabilizing and linearizing effect on the whole system. Negative feedback, long used to improve the stability and quality of electronic circuits, is now known to be an important principle in a wide range of biological and physiological systems.

nemat-, nemato- *combining form denoting* thread-like, as in NEMATODE worms. The term derives from the Greek *nema*, a thread.

neo- *combining form denoting* new.

neologism 1 a newly coined word or phrase. 2 a meaningless word used by a psychotic person.

neonatal pertaining to a new-born baby.

neonate a new-born baby.

neonatologist a doctor specializing in the management, assessment, diseases and intensive care of newborn babies, especially those that are of low birth weight and those with congenital abnormalities. After the first 4 weeks of life, care may pass to a general paediatrician.

neoplasia the process of tumour formation.

neoplasm a collection of cells, derived from a common origin, often a single cell, that is increasing in number and expanding or spreading, either locally or to remote sites. A tumour. Neoplasms may be BENIGN or MALIGNANT. The term literally means a new growth.

neoplastic pertaining to a NEOPLASM.

nephr-, nephro- *combining form denoting* the kidney.

nephritic 1 pertaining to the kidneys. 2 pertaining to kidney inflammation (nephritis).

nephrology the study of the structure, function and disorders of the kidney.

nephropathy any disease of the kidney involving observable change.

nerve a pinkish-white, cord-like structure consisting of bundles of long fibres (axons) of nerve cells and fine blood vessels held together by a connective tissue sheath. Individual fibres are usually insulated with a layer of white fatty material called myelin. The larger nerves contain both MOTOR and SENSORY fibres. Twelve pairs of nerves arise directly from the brain. These are called cranial nerves and carry impulses subserving smell, eye movement, vision, facial movement and sensation, all other sensation in the head, hearing, taste, movements of the soft palate, tongue and neck muscles, and control of the heartbeat and the secretion of stomach acid. Thirty one pairs of nerves emerge from the spinal cord. These control all the other muscles of the body and carry impulses for sensation from all parts of the body to the spinal cord and thence to the brain.

nerve deafness hearing loss resulting from damage to the transducers in the inner ear (Organ of Corti) or to the acoustic nerve connections to the brain, rather than from mechanical interference with the transmission of sound vibrations to the inner ear (conductive deafness).

nerve growth factor a peptide substance that stimulates growth and differentiation of NEURONS in the sympathetic and sensory nervous system. Nerve growth factor has been found effective in promoting the healing of corneal ulcers due to loss of the sensory innervation of the cornea. Corneal transparency has been restored by this means.

nerve impulse the wave-like progression of electrical depolarization that passes along a stimulated nerve fibre. The nerve impulse results from a movement of positive and negative ions across the membrane of the fibre.

nervous breakdown a popular and imprecise term used to describe any emotional, neurotic or psychotic disturbance ranging from a brief episode of hysterical behaviour to a major psychotic illness such as schizophrenia.

nervous habit see TICS.

nervous system the controlling, integrating, recording and effecting structure of the body. The nervous system is also the seat of consciousness, of the intellect, of the emotions and of all bodily satisfaction. The central nervous system consists of the brain and spinal cord. The peripheral nervous system consists of the massive ramification of nerves running to every part of the body outside the brain and cord.

neur-, neuro- *prefix denoting* nerve.

neuralgia pain experienced in an area supplied by a sensory nerve as a result of nerve disorder that results in the production of pain impulses in the nerve.

neural networks artificial electronic or software systems that can simulate some of the neurological functions including a crude form of vision. In conjunction with expert software systems neural networks are expected to prove important in medicine in the future.

neurasthenia a state of constant fatigue, loss of motivation and energy and often insomnia and muscle aches associated with general and persistent unhappiness. In the present state of knowledge, and in the absence of any evidence of a cause, the state described as neurasthenia is considered not to be of organic origin and, in particular, to have nothing to do with nerve function.

neurilemma an outer covering of flattened cells that surrounds the MYELIN sheath of the nerve fibres (axons) of the larger peripheral nerves. Neurilemma also covers the axons of non-myelinated nerve fibres. Also known as the sheath of Schwann.

neuroanatomy the study of the structure of the nervous system and its relation to function. A knowledge of neuroanatomy is a prerequisite for the diagnosis of neurological diseases and for the accurate location of the LESION causing the disorder.

neuroblast a cell in the embryo that gives rise to nerve cells.

neurocyte a nerve cell (NEURON).

neurogenic of a lesion caused by interruption of the nerve supply.

neuroglia the network of branched cells and fibres that forms the supporting connective tissue of the central nervous system. Certain brain tumours arise from neuroglial cells.

neuroglobin an oxygen-transport material in the brain analogous to haemoglobin in the blood and myoglobin in the muscles. Neuroglobin is a small protein of 151 amino acids. It is thought to have a neuroprotective effect in cerebral ischaemia, but its full function remains unelucidated.

neuroleptic 1 capable of bringing about emotional quietening without impairing consciousness. Capable of modifying abnormal psychotic behaviour.
2 any drug having these effects.

neurologist a doctor trained in neurology, who specializes in the ANATOMY and PHYSIOLOGY of the nervous system and in the diagnosis and treatment of its disorders. Neurologists are learned diagnosticians and do not engage in operative treatment.

neurology the medical speciality concerned with the nervous system and its disorders. See NEUROLOGIST. Compare NEUROSURGERY.

neuromodulator a substance that, while not affecting the rate of firing or conduction of nerve impulses, can change the effect on a nerve of other neurotransmitters. Neuromodulators can control neurotransmitter synthesis or the amounts of neurotransmitter released in response to other stimuli. Adenosine is an example of a neuromodulator.

neuron, neurone the functional unit of the nervous system. A neuron is a single cell having a very long, fibre-like extension, called an AXON, and one or many short extensions called DENDRITES. The axon may be 100,000 times as long as the diameter of the cell body – some are as long as 1 m. Nerve impulses are moving zones of electrical depolarization and these travel outwards along the axon from the cell body. Incoming impulses travel to the cell body along the dendrites. Neurons interconnect with each other at specialized junctions called SYNAPSES, situated mainly between the end of an axon of one neuron and the cell body or the dendrites of another. Many neurons receive as many as 15,000 synapses, some more. Most synapses are interneurons connecting with other nerve cells, rather than with muscles or glands.

neuropathology the study of disease changes occurring in the tissues of the nervous system.

neurophysins a group of proteins found in the rear lobe of the PITUITARY GLAND and thought to be carriers for the hormones OXYTOCIN and VASOPRESSIN and to be concerned in their storage.

neurophysiology the study of the function of the nervous system.

neuropsychiatry the branch of medicine concerned with the effects on mind and behaviour of organic disorders of the nervous system. Neuropsychiatrists must be well versed in two formerly quite distinct disciplines – neurology and psychiatry.

neuroradiology the speciality concerned with the diagnosis of neurological disease by X-ray and associated methods of examination.

neurosis any long-term mental or behavioural disorder, in which contact with reality is retained and the condition is recognized by the sufferer as abnormal. Attempts have been made to prohibit the term as pejorative and insulting but these have failed mainly because of a more complete and humane understanding of the subject and of the plight of neurotic sufferers. A neurosis essentially features anxiety or behaviour exaggeratedly designed to avoid anxiety. Defence mechanisms against anxiety take various forms and may appear as PHOBIAS, OBSESSIONS, COMPULSIONS or as sexual dysfunctions. In recent attempts at classification, the disorders formerly included under the neuroses have, possibly for reasons of political correctness, been given new names. The general term, neurosis, is now called anxiety disorder; hysteria has become a somatoform or conversion disorder; amnesia, fugue, multiple personality and depersonalization have become dissociative disorders; and neurotic depression has become a dysthymic disorder. These changes are helpful and explanatory but ignore the futility of euphemism. Psychoanalysis has proved of little value in curing these conditions and Freud's speculations as to their origins are not now widely accepted outside Freudian schools of thought. Neurotic disorders are probably best regarded as being the result of inappropriate early programming. Cognitive behaviour therapy seems effective in some cases.

neuroticism the state of a person persistently and excessively prone to anxiety and to a preoccupation with self rather than with the external world. Neuroticism often involves hypochondriasis. See also NEUROSIS.

neurotransmitters a range of small-molecule chemical substances released by EXOCYTOSIS from a nerve ending on the arrival of a nerve impulse. Neurotransmitters are specific to particular neurons. They interact with receptors on adjacent structures to trigger off a response, either excitatory or inhibitory. The adjacent structure may be another nerve, a muscle fibre or a gland. The main neurotransmitters are acetylcholine, glycine, glutamate, dopamine, noradrenaline, adrenaline, serotonin, histamine and GABA (gamma-amino-butyric acid). With the exception of the adrenalines all the neurotransmitters are AMINO ACIDS or derivatives of amino acids.

neutraceuticals foods that are said to benefit health. See HEALTH FOOD.

neutral 1 neither acid nor alkaline. Having a pH of 7.

2 in the context of pharmacological INSULIN the term refers to soluble insulins that have not been formulated so as to prolong their action. Neutral insulins include those described as Rapid, Actrapid and Velosulin.

neutrophil a white blood cell of the granulocyte group, with a multilobed (polymorph) nucleus and numerous granules in the CYTOPLASM that stain neither red with eosin nor blue with basic dyes. Neutrophils are the major circulating PHAGOCYTES of the granulocyte group. Compare EOSINOPHIL and BASOPHIL.

neutrophil exudation the movement of neutrophil polymorphs out of the blood capillaries into the tissue spaces. The polymorphs squeeze through narrow spaces in the capillary walls by amoeboid action.

neutrophilia an increased number of NEUTROPHIL white cells (leucocytes) in the blood. This is often an indication of an infection somewhere in the body.

newton the unit of force required to accelerate a mass of 1 kg by 1 m per second per second. 1 N is equal to 100 000 dynes. (Sir Isaac Newton, 1642–1727, English mathematician, alchemist and physicist).

niacin nicotinamide, one of the B group of vitamins. Nicotinic acid. Niacin is present in liver, meat, grains and legumes. It is a constituent of coenzymes involved in oxidation–reduction reactions. Deficiency causes PELLAGRA. Niacin is being used in the treatment of high blood cholesterol levels.

nicotinamide see NIACIN.

nicotine a highly poisonous alkaloid drug derived from the leaves of the tobacco plants *Nicotiana tabacum* and *Nicotiana rustica*. Large doses are fatal. Very small doses are obtained by inhaling the smoke from burning tobacco and this is done for the sake of the desired slight stimulant and mood-elevating effect and to alleviate nicotine withdrawal symptoms. Nicotine increases the heart rate and raises the blood pressure by narrowing small arteries. This effect can be dangerous. Nicotine, in the doses acquired by smokers, is comparatively harmless but the other constituents of tobacco smoke are responsible for an enormous burden of human disease. Nicotine is dispensed in the form of dummy cigarettes, skin patches and chewing gum so that people who wish to stop smoking may still, for a time, continue to enjoy the perceived advantages. The drug is also used as an insecticide.

nicotinic having the effects of acetylcholine and other nicotine-like substances on autonomic ganglia and the neuromuscular junctions of voluntary muscle. Compare MUSCARINIC.

nicotinic acid see NIACIN.

nicotinic receptors ACETYL CHOLINE receptors that also respond to NICOTINE. These receptors are at nerve-muscle junctions and in the autonomic nervous system.

night blindness moderately reduced to severely defective vision in dim light. Night blindness (nyctalopia) occurs in many people with no objectively discernible eye disorder, but is common in short-sighted people, in those with vitamin A deficiency and in the early stages of degenerative diseases of the RETINA including RETINITIS PIGMENTOSA.

nightmare a frightening dream occurring during rapid eye movement (REM) sleep often connected with a traumatic prior event such as an assault or a car accident. Nightmares may be caused by withdrawal of sleeping tablets. The Anglo-Saxon word *maere* means an evil male spirit or incubus intent on sexual intercourse with a sleeping woman, but nightmares seldom have a sexual content.

night sweat drenching perspiration occurring at night or during sleep. Night sweats may be a feature of any feverish illness but do not indicate any particular diagnosis.

night terrors sudden attacks of severe panic occurring during deep non-REM sleep and associated with very high heart rates, rapid respiration and often screaming. There is a sense of suffocation, imprisonment in a small space or a conviction of impending death. Night terrors occur most often around the age of 5.

night waking see SLEEPWALKING.

NIH *abbrev. for* the National Institutes for Health. This is an American government-sponsored body, the principal interest of which to the general public is the excellent website (www.nih.gov/) providing reliable medical information.

nipple the conical or cylindrical projection from the breast that is surrounded by a darker areola. The female nipple is larger than that of the male and is perforated by the ducts of the milk-secreting segments of the breast.

nitric oxide nitrogen monoxide (NO), one of the eight oxides of nitrogen consisting of a single nitrogen atom and a single oxygen atom. In 1987 nitric oxide was found to be an important physiological mediator, a relaxant of smooth muscle in the walls of blood vessels that was derived from the inner lining (endothelium) of blood vessels. Later it was shown that nitric oxide was far more than simply an endothelium-derived relaxing factor (EDRF). Three different enzymes synthesize nitric oxide, from endothelium, nerves and macrophages and the NO produced has actions all over the body. Nitric oxide is involved in controlling blood pressure; in the phagocytic action of MACROPHAGES; in inhibiting PLATELET aggregation and hence blood clotting; in limiting the development of the principal arterial disease (atherosclerosis); in controlling the heart action; in relaxing the smooth muscle in the air tubes of the lungs and the walls of the intestine; in a range of brain functions; and in promoting penile erection.

nitric oxide synthase one of three enzymes that catalyze the synthesis of NITRIC OXIDE in blood vessel ENDOTHELIUM, nerve fibres and MACROPHAGES. These enzymes are coded for, respectively, on chromosomes 7, 12 and 17. They act by splitting off a nitrogen atom from the amino acid L-arginine which is then combined with an oxygen atom from molecular oxygen to form NO.

nitrogen an inert, colourless and odourless gas constituting about 80 per cent of the atmosphere. The element is present in all proteins and occurs in the urine in the form of urea. Under pressure, considerable nitrogen will dissolve in the blood. The release of gaseous nitrogen in the blood in bubbles that can block small arteries is the chief danger in far too sudden decompression in divers (also known as the bends).

nitrogen balance the difference between the amounts of nitrogen taken into and lost by the body. Nitrogen is taken in mainly in the form of protein and is mainly lost in urea in the urine.

nitroso-redox balance the balanced interaction between the production of nitric oxide and that of superoxide. Upset of this balance plays a fundamental role in cell and organ failure.

NMDA receptors receptors for the NEUROTRANSMITTER n-methyl D-aspartate. This is a glutamate transmitter. Abnormally high levels of NDMA are thought to lead to neural dysfunction.

noble savage an entirely fictitious and romantic notion, mainly promulgated by Jean Jacques Rousseau (1712–78) in his novel 'Emile', that those who dwell in natural surroundings and are untouched by civilization, remain innocent and noble. Rousseau believed that the child should not be exposed to the corrupting influence of civilization but should be allowed to develop his or her own natural and inherent virtue.

noci- *combining form denoting* pain or injury.

nociceptors nerve endings selectively responding to painful stimuli. Stimulation of nociceptors causes the sensation of pain.

noct-, nocti- *combining form denoting* night.

nocturia passing urine during the night. One definition, that of the International Continence Society, stipulates two or more voids during the night. Normally, the stimulus of a filling bladder is insufficient to awake a person and prompt him or her to get up to urinate, but rapid or excessive filling, or undue bladder awareness, will do so.

nocturnal emission spontaneous ejaculation, with orgasm, occurring during sleep, often at the climax of an erotic dream. The phenomenon is experienced from time to time by most males with restricted sexual opportunities who do not regularly masturbate. Also known as a 'wet dream'.

nodes of Ranvier narrow gaps between the ends of the segments of myelin that insulate single nerve axons. (Louis Antoine Ranvier, 1835–1922, French pathologist).

nodule a small, solid knot-like lump of tissue occurring anywhere in the body. Nodules in the skin are easily felt. The term implies nothing about the nature of the lump.

noesis a psychological term for the process by which information is derived. The cognitive process. Cognition. From the Greek *noesis*, understanding.

nomenclature a system of names used in a science or other discipline.

non- *prefix denoting* not.

nonpolar of a molecule or molecular region in which the chemical bonds share electrons equally so that there are no externally operative electric charges.

non-REM sleep sleep during the intervals between perios of rapid eye movement during which there is no dreaming and the tone is high in the postural muscles.

nonsense codon one of the three nucleotide triplets (codons), UAG, UAA or UGA that mark an end point to a particular protein synthesis. U is the base uracil; A is the base adenine; and G is the base guanine.

nonsense mutation a POINT MUTATION which changes a CODON that specifies an amino acid into a termination codon – one that marks the position where translation of a messenger RNA sequence should stop. The result is a gene with a segment lopped off. Such a gene will code for a protein that may have missing amino acids and may thus be functionally defective.

non-specific 1 not attributable to any definite causal organism.
2 of a drug, having a general, as distinct from a particular, effect.

non-specific ascending pathways sensory input neuronal chains for general sensory information that are used by various modalities of sensation.

non-specific immune response an immediate protective responses of the immune system that does not require previous exposure to the invader.

non-verbal communication transmission of information from person to person without the use of words, as by gesture, bodily attitude, expression, exclamation, and so on.

noradrenaline norepinephrine, an important adrenergic NEUROTRANSMITTER released by POSTGANGLIONIC adrenergic nerve endings and secreted by the MEDULLA of the adrenal gland. Noradrenaline acts chiefly on alpha-adrenergic receptors and causes constriction of arteries and a rise in the blood pressure. This is a SYMPATHOMIMETIC action. One of the catecholamines.

normal distribution Gaussian distribution, a distribution which when expressed graphically is bell-shaped. The distribution to which many frequency distributions of biological variables, such as height, weight, intelligence, etc correspond.

normoblast a nucleated red blood cell precursor showing the features of normal red cell development, as distinct from those of the MEGALOBLAST.

normothermia a body temperature within normal limits. The term is used mainly in contexts in which hypothermia is a possibility or a risk.

nose a term used both for the externally visible part and for the internal nasal air passages. The nose is the normal entry route for inspired air, which is warmed, moistened and cleaned. Chemical particles in the air stimulate the nerve endings of the olfactory nerves in the roof of the nasal cavity, giving rise to the sensation of smell.

noso- *combining form denoting* disease. From the Greek *nosos*, disease.

nosocomial of disease pertaining to, or acquired in, a hospital. The term is used especially to refer to infections more likely to occur in hospital than out of hospital. From Greek *nosokomion*, a hospital.

nostril one of the paired openings into the NOSE that contains hairs which can trap gross particulate matter in the inhaled air.

nostrum a medicine, especially a patent, secret and often QUACK remedy.

notochord a rod-like structure, present in early development, derived from the MESODERM and giving rise to the spine. In the adult, the notochord is represented by the pulpy centres (nucleus pulposus) of the intervertebral discs.

noxious harmful to health.

nuchal translucency test a method of ultrasound detection of fetuses with a chromosomal abnormality, especially Down's syndrome. The examination consists of the assessment of small collections of fluid at the back of the neck and spine of the fetus that increase translucency. About three-quarters of cases of Down's syndrome can be detected in this way.

nuclear envelope the double membrane, with perforations (pores), surrounding a cell nucleus. The outer membrane extends into the endoplasmic reticulum. The pores allow transport of macromolecules in both directions.

nuclear pores openings in the NUCLEAR ENVELOPE allowing chemical messenger communication between the nucleus and the cytoplasm of the cell.

nuclear radiation radiation and particles coming from the nuclei of radioactive atoms during radioactive decay and nuclear reactions. Alpha particles are the nuclei of helium atoms, beta particles are high-speed electrons, X-rays and gamma rays are electromagnetic radiations of wavelength progressively shorter than visible light. Nuclear radiations differ in penetration, beta particles being least penetrative and the gamma rays most. Ionizing radiations can dislodge linking electrons from molecules, such as those of DNA, and cause damaging changes. In general, radiations are most destructive to cells most rapidly dividing.

nuclease any one of several enzymes that break down NUCLEIC ACIDS.

nucleate possessing a NUCLEUS.

nucleic acid DNA or RNA. A very long polymer molecule made up of MONOMERS of either deoxyribonucleotides or ribonucleotides, joined by PHOSPHODIESTER BONDS. Nucleic acids constitute the chromosomes of almost all living cells and, by virtue of the order of the contained purine and pyrimidine BASE PAIRS, manifest the genetic code.

nucleolus a small, dense rounded body found in the nucleus of most cells. The nucleolus generates RIBOSOMES and is the site of the transcription of ribosomal RNA. The size and number of nucleoli in a cell nucleus vary with the amount of protein synthesized by the cell.

nucleoside a molecule compounded of a purine or pyrimidine base attached to a sugar (ribose or deoxyribose). The genetic code in DNA and RNA depends on the order of the nucleosides. A nucleoside is a NUCLEOTIDE without the phosphate group.

nucleosome the structural subunit of CHROMATIN consisting of about 200 BASE PAIRS and a barrel-shaped core of eight histone protein molecules (an octamer).

nucleotide a molecule formed from the bonding of a purine or a pyrimidine base with a sugar and a mono-, di- or tri-phosphate group. Compare NUCLEOSIDE. Four different nucleotides may polymerize to form DNA. They are 2′-deoxyadenosine 5′-triphosphate; 2′-deoxyguanosine 5′-triphosphate; 2′-deoxycytidine 5′-triphosphate; and 2′-deoxythymidine 5′-triphosphate. These lengthy names are commonly abbreviated to dATP, dGTP, dCTP and dTTP. Even this is too clumsy when printing out the sequence of nucleotides in a length of DNA. In that case they are abbreviated to A, G, C and T (for adenine, guanine, cytosine and thymine). In RNA the sugar is not 2′-deoxyribose, but ribose itself. Also one of the RNA bases differs from that in DNA. Thymine is replaced by uracil. So the nucleotides of RNA are adenosine 5′-triphosphate; guanosine 5′-triphosphate; cytidine 5′-triphosphate; and uridine 5′-triphosphate. These are abbreviated to ATP, GTP, CTP and UTP or simply A, G, C and U.

nucleus **1** of a body cell, the central structure consisting of the tightly bundled genetic material DNA surrounded by a nuclear membrane.

2 of an atom, the central core of protons and, except in the case of hydrogen, neutrons which is surrounded by a rapidly moving cloud of electrons, widely separated from it. The forces which bind together the protons and neutrons are immensely powerful and it is these forces which are released in an atomic explosion. From the Latin *nucleus*, a nut or kernel.

nucleus pulposus the pulpy core of the INTERVERTEBRAL DISC, that is surrounded by the ANNULUS FIBROSUS. Degeneration of the annulus and/or undue vertical stress may lead to some of the nucleus pulposus being squeezed (prolapsed) through the annulus. This usually occurs at the back of the disk and the prolapsed material may press on the nerve roots entering and leaving the spinal cord, causing severe pain and sometimes muscle weakness. The imprecise lay term for this process is 'slipped disk'.

nuclide an artificially produced radioactive isotope of an element. Many of these are used in medicine as tracers or for RADIOTHERAPY.

null cells a group of large granular LYMPHOCYTES that falls into neither the T cell nor the B cell category. The group includes the NATURAL KILLER CELLS and the KILLER CELLS.

null hypothesis the assumption that one variable has no effect on another variable, or that only one hypothesis can possibly account for a phenomenon. The assumption that there are no differences in two populations in matters relevant to the current investigation. In a clinical trial of a new treatment the null hypothesis might be that the proportion of patients improved by it was the same as the proportion improved by the existing standard treatment. See HYPOTHESIS TEST.

nullipara a woman who has never given birth to a viable child.

null mutation a mutation that eliminates the function of the affected gene. In many cases the null mutation is a complete deletion of the gene.

nummular coin-shaped.

nutation nodding of the head, usually involuntary.

nutrient anything that nourishes. Any physiologically valuable ingredient in food.

nutrition 1 the process by which substances external to the body are assimilated and restructured to form part of the body or are consumed as a source or energy. 2 the study of the dietary requirements of the body and of the amounts of water, carbohydrates, fats, proteins, vitamins, minerals and fibre needed for the maintenance of health.

nutritionist a person who specializes in the study of NUTRITION and especially in the applications of the principles of nutrition in the maintenance of health and the treatment of disease.

nyctalopia inability to see well in conditions of poor illumination. NIGHT BLINDNESS. From the Greek nyktos, night, alaos, blind and ops, the eye.

nymphomania excessive desire by a woman for copulation. The concept of nymphomania is largely a fiction, engendered in less liberal days by male wish-fulfilment fantasy or by puritanical and censorious contemplation of healthy female sexuality.

nystagmus persistent, rapid, rhythmical, jerky or wobbling movement of the eyes, usually together. The movement is usually transverse and most commonly of a 'sawtooth' pattern with a slow movement in one direction followed by a sudden recovery jerk in the other.

Oo

obesity excessive energy storage in the form of fat. This occurs when food intake exceeds the requirements for energy expenditure. Obesity is a hazard to health and longevity and increases the risk of high blood pressure (hypertension), diabetes, various cancers, osteoarthritis, foot trouble and depression.

obesity mediators see GLP-1, LEPTIN, OB GENE and OREXIN.

ob gene a gene on chromosome 7 that codes for the cytokine LEPTIN which is produced in adipose tissue and which exercises control over food intake and energy expenditure. Ob is an abbreviation of 'obesity'.

objective the lens in a microscope nearest to the object being examined.

obligate able to survive only in a particular environment. Used especially of certain parasites.

ob receptor one of a family of LEPTIN receptors on the choroid plexus of the brain through which the cytokine appears to enter the brain to perform its function of switching off appetite.

obsession a compulsive preoccupation with an idea or an emotion, often unwanted or unreasonable, and usually associated with anxiety.

obstetrician a doctor specializing in the conduct of childbirth and possessing the skills, knowledge and experience required to ensure that this is achieved with the minimum risk to mother and baby. From the Latin *obstetrix*, a midwife.

obstetrics the branch of medicine concerned with childbirth and with the care of the woman from the onset of pregnancy until about 6 weeks after the birth, when the reproductive organs have returned to normal.

obtundity a state of reduced consciousness.

obturation closure of an opening.

obturator any device, object or anatomical structure that closes or obstructs an opening or cavity.

obturator foramen a large opening on each side of the pelvis below and in front of the socket for the hip bone (the ACETABULUM).

obturator muscles the obturator externus and obturator internus muscles that cover the outer surface of each side of the front of the pelvis. The tendons of these muscles pass behind the hip joint and are inserted into the top of the femur. They rotate the thigh outwards.

Occam's razor a principle in science and philosophy that one should try to account for an observed phenomenon in the simplest possible way and should not look for multiply explanations of its different aspects. For instance, a range of symptoms and signs occurring together should always, if possible, be attributed to a single disease rather than to several different diseases occurring simultaneously. (William of Occam, ca. 1290–1349, English philosopher).

occipital bone the curved, shield-shaped bone forming the lower rear part of the skull.

occipital lobe the rear lobe of the main brain (cerebral hemisphere). The occipital lobe is concerned with vision.

occiput the back of the head.

occlusion 1 closing off or covering of an opening, or obstruction to a hollow part. 2 the relationship of the biting surfaces of the teeth of the upper and lower jaws. 3 the deliberately covering of one eye for periods of weeks or months in the treatment of AMBLYOPIA in children.

occult concealed or hidden, especially of traces of blood in the faeces or sputum which can be detected only by special tests.

ochre codon one of the three CODONS that causes termination of protein synthesis. A stop codon. It is the triplet UAA (uracil, adenine, adenine).

ochre mutation any mutation that changes a codon to the stop OCHRE CODON.

Ockham's razor see OCCAM'S RAZOR.

ocular 1 pertaining to the eye. 2 the eyepiece of an optical device such as a microscope.

oculist 1 an OPHTHALMOLOGIST. 2 an ophthalmic OPTICIAN (optometrist).

oculomotor nerve the third of the paired cranial nerves arising directly from the brainstem. This nerve supplies four of the six small muscles that move the eye, the muscle that elevates the upper lid and the circular muscles of the iris. Paralysis of an oculomotor nerve causes the pupil to be enlarged, the lid to droop and the eye to be unable to turn inwards.

odont, odonto- *combining form denoting* teeth.

odontoblast a specialized connective tissue cell, lying in the outer surface of the dental pulp, that produces DENTINE.

odontoid process a strong tooth-like process projecting upwards from the front arch of the second vertebra of the neck (the axis bone) around which the first vertebra rotates to allow the head to turn to either side.

odontology the study of the structure, growth, function and diseases of the teeth.

-odynia *suffix denoting* pain.

oedema excessive accumulation of fluid, mainly water, in the tissue spaces of the body. Oedema may be local, as at the site of an injury, or general. It often affects specific organs, such as the brain or the lungs.

oedipus complex the Freudian belief that much psychiatric disorder, especially the 'psychoneuroses', are caused by the persisting effects, including unresolved guilt feelings, of the child's unconscious wish to kill the parent of the same sex and to have sexual intercourse with the parent of the opposite sex. The notion was one of the central tenets of Freudian dogma but is no longer widely held. Freud derived the term from the name of the swollen-footed, mythical hero of Sophocles' tragedies who was nailed up by his feet as a baby (hence the swelling) but who survived to kill his father and marry his mother. See also FREUDIAN THEORY.

oesophagus the gullet. The oesophagus is a muscular tube, about 24 cm long, extending from the throat (pharynx) to the STOMACH. Just above the stomach it passes through the DIAPHRAGM. In swallowing, food is carried down by repetitive controlled contractions of the muscular walls, known as PERISTALSIS. Immediately above the stomach the wall of the oesophagus shows an increased tendency to contract, thus forming a muscle ring known as the cardiac SPHINCTER. This normally closes after swallowing, to prevent regurgitation of the stomach contents.

oesophagus disorders these include heartburn, pharyngeal pouch, hiatus hernia, reflux oesophagitis, oesophageal varices.

oestrogen one of a group of steroid sex hormones secreted mainly by the ovaries, but also by the testicles. Oestrogens bring about the development of the female secondary sexual characteristics and act on the lining of the uterus, in conjunction with progesterone, to prepare it for implantation of the fertilized OVUM. They have some ANABOLIC properties and promote bone growth. Oestrogens are used to treat ovarian insufficiency and menopausal symptoms, to limit postmenopausal osteoporosis, to stop milk production (lactation) and to treat widespread cancers of the PROSTATE gland. They are extensively used as oral contraceptives.

oestrogen receptors chemical groups to which oestrogens bind and which are commonly found in substantial quantities on breast cancers. The oestrogen receptor status of a woman with breast cancer is important because the prognosis is much worse in women with high levels of oestrogen receptor alpha than in those with oestrogen alpha negative cancers. The status can easily and cheaply be determined by gene expression micro array analysis. 70–80 per cent of breast cancers are oestrogen receptor alpha positive.

-oid *suffix denoting* like.

Okazaki fragments short segments of DNA, 1000–2000 bases long, that later join up to form continuous lengths of DNA. Okazaki fragments occur in replicating DNA in both prokaryotes and eukaryotes. They form up on the 'lagging' strand during replications and join by ligation. (Reiji Okazaki, Japanese geneticist).

olecranon the hook-shaped upper end of one of the forearm bones (the ulna) that projects behind the elbow joint, fitting into a hollow on the back of the lower end of the upper arm bone (humerus) and forming the point of the elbow. The olecranon prevents over-extension of the elbow.

olfaction the sense of smell or the act of smelling.

olfactory pertaining to the sense of smell.

olfactory nerves the nerves of smell and the first of the 12 paired CRANIAL NERVES arising directly from the brain. The olfactory nerves lie on the floor of the front section of the cranial cavity and run forward, close together, over the roof of the nose. From their terminal bulbs many tiny, hair-like filaments pass down through perforated bony plates (cribriform plates) into the upper part of the nose. Tiny particles of odorous material dissolved in the nasal secretions can be detected by the olfactory nerve endings with great sensitivity and discrimination.

oligo- *combining form denoting* little, few or an abnormally small quantity of.

oligodendrocyte a connective cell of the central nervous system (glial cell) that participates in the formation of the myelin sheaths of nerve cell axons.

oligomenorrhoea abnormally infrequent menstrual periods. Often the interval between periods exceed 40 days.

oligophrenia mental deficiency.

oligospermia an abnormally low concentration of sperm in the seminal fluid. Fertile semen contains about 100 000 000 sperm per ml. Semen with less than about 20 000 000 sperm per ml is likely to be infertile.

oliguria an abnormally small output of urine. In health, the urinary output varies from 700 ml to 2 l. Oliguria is usually caused by inadequate fluid intake or increased fluid loss in sweating or diarrhoea. A more serious cause of oliguria is KIDNEY FAILURE, either acute or, less often, following long-term kidney disease.

olive 1 a smooth, oval swelling on each side of the upper part of the MEDULLA caused by the underlying olivary nucleus. These nuclei connect with each other and with the CEREBELLUM.
2 the smooth, elliptical tip of a vein stripper, used in the treatment of VARICOSE VEINS.

-oma *suffix denoting* a tumour.

omega-3 marine triglycerides fish oils containing the omega 3 polyunsaturated fatty acids eicospentaenoic and docosahexaenoic acids. A third type of omega-3 fatty acid, alpha linoleic acid, is derived mainly from vegetable oils. There is good evidence that fish oils are protective against coronary heart disease and have value after heart attacks. They are also used to treat patients with abnormally high blood fat (triglyceride) levels who are at risk of heart attacks.

omentum one of two double folds of PERITONEUM, the greater and lesser omenta, that hang down like aprons from the liver and stomach over the coils of small intestine. The omenta usually contain fat and are often effective in sealing down and localizing areas of inflammation of the peritoneum (PERITONITIS). Also known as the epiploon.

omni- *prefix denoting* all.

omnivorous eating food of both animal and vegetable origin, as in the case of most people.

omphal- *combining form denoting* the umbilicus. From the Greek *omphalos*, the navel.

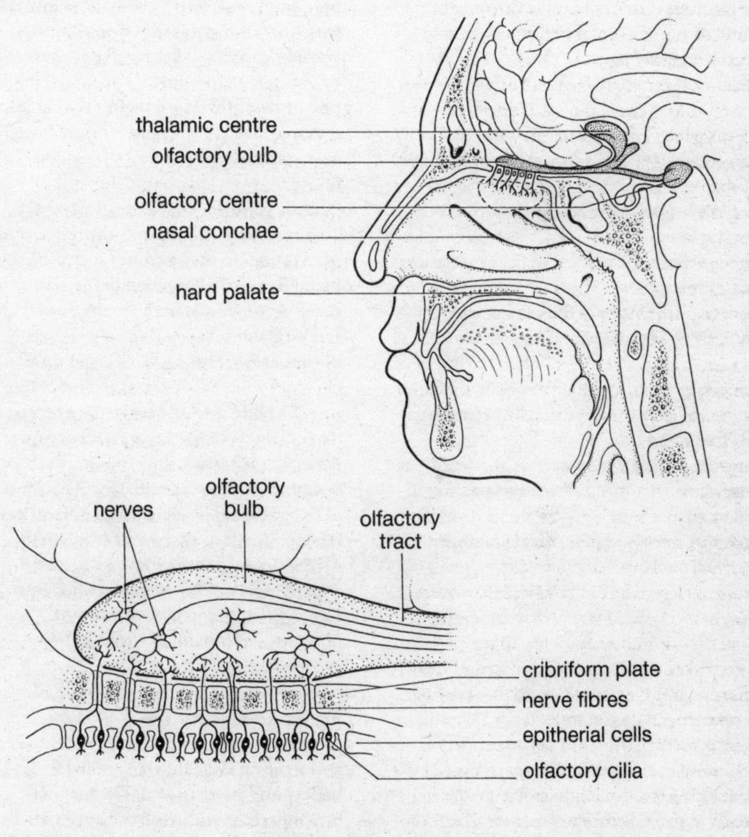

Fig. 13 **Olfactory nerves**

onanism an obsolescent term for MASTURBATION or COITUS INTERRUPTUS.

oncho- *combining form denoting* a swelling, mass or tumour. From the Greek *onkos*, a lump.

onco- *combining form denoting* a tumour.

oncogenes genes that contribute to cancerous changes in cells. Oncogenes are mutations of normal cell genes and must work together to cause cancer. Similar or identical genes are found in viruses known to be able to cause cancer. If one of the three virus genes – gag, pol or env – is replaced by an oncogene, such as ras, the virus becomes capable of causing cancer. The normal ALLELES of the oncogenes are called proto-oncogenes.

oncogenesis the process of the origination of a tumour.

oncology the study of the causes, features and treatment of cancer. An oncologist is a cancer specialist.

oncolysis destruction or breakdown of a tumour, either spontaneously or as a result of treatment.

ontogeny the development of an individual organism of a species from fertilization to maturity.

onycho- *combining form denoting* a finger- or toenail or claw.

oo- *combining form denoting* an egg (ovum).

oocyte a cell in the OVARY that undergoes MEIOSIS to produce an OVUM. In meiosis the 46 chromosomes are reduced to half the normal number (HAPLOID), so that the full complement can be restored by a haploid contribution from the sperm. Primary oocytes develop in the ovaries of the fetus but only a fraction of these will ever give rise to OVULATION. Secondary oocytes divide to form the mature OVUM but the second maturation division occurs only after the ovum has been fertilized by a sperm.

oogenesis the production of egg cells (ova) in the ovaries and their preparation for release and fertilization. Oogenesis starts in the fetal ovary with the formation of OOGONIA. These divide by MEIOSIS to form OOCYTES.

oogonia the precursors of OOCYTES in the OVARY derived from primordial female germ cells that have migrated to the site of the ovaries.

oophor- *combining form denoting* the OVARY.

oophoron obsolete term for the OVARY. From the Greek roots *oon*, an egg and *phoros*, bearing.

open reading frame in RNA, a sequence of base pair triplets (codons) with no introns, which is translatable into a protein.

operable capable of being effectively treated by a surgical procedure.

operant conditioning a method of behaviour therapy in which a response is reinforced or suppressed, whenever it occurs, by immediate reward or punishment.

operator gene the first gene in an OPERON. The operator gene controls the transcription of the genes in the operon.

operculum a covering membrane, flap or lid of tissue, especially in the brain, or over an erupting tooth.

operon a row of consecutive genes on a chromosome that operates as a functional unit. The structural genes in the operon are preceded by two regulatory sites occupied by regulatory genes, the promoter and the operator. These are essential for the expression of the operon. The genes in an operon have related functions that occur sequentially. All the genes in the operon are turned on and off together. All are transcribed into one large segment of MESSENGER RNA.

ophthalm-, ophthalmo- *combining form denoting* the eye.

ophthalmologist a doctor who specializes in OPHTHALMOLOGY.

ophthalmology the combined medical and surgical speciality concerned with the eye and its disorders. The practice of ophthalmology involves a mastery of ophthalmic optics, of the structure, function and diseases of the eyes, of the associated neurological systems concerned with vision, of the range of general conditions that affect the eyes and of the microsurgical skills and techniques used in the treatment of many ophthalmic conditions.

-opia, -opsia *combining form denoting* a specified form of vision.

opis-, opistho- *combining form denoting* back or backwards.

opisthotonus spasmodic, powerful contractions of the back and rear neck muscles causing the body to arch backwards so that the heels approximate to the head. This is a feature of tetanus, severe meningitis, strychnine poisoning and some brainstem disorders.

-opsis *combining form denoting* something resembling that specified.

opsoclonus a rare ophthalmic disorder featuring involuntary jerky (saccadic) eye movements in random directions. Opsoclonus should not be confused with NYSTAGMUS in which the movements are periodic and uniform. The disorder most commonly affects infants with neuroblastoma. In adults it may be related to other malignancies or to neurological infections. It is often associated with myoclonus and ataxia.

-opsy *combining form denoting* examination.

opt- *combining form denoting* the eye or vision.

optical activity the property of some substances in solution of causing of a ribbon-like beam of polarized light projected through the solution to twist through a small angle. Some substances cause rotation of the beam to the left (levorotatory), some to the right (dextrorotatory). Some optically active drugs are more potent in one form than in the other; a few are almost inert in one rotatory form.

optic chiasma the junction of the two OPTIC NERVES lying under the brain. In the chiasma optic nerve fibres from the inner half of each RETINA cross over. Those from the outer half do not. Thus fibres from the inner half of each retina run out of the chiasma in close association with fibres from the outer half of the retina on the other side. The two optic tracts so formed run into the brain. This arrangement ensures that input to both eyes from the right field of vision causes signals that pass to the left half of the brain, and vice versa.

optic disc the small circular area at the back of the eye at which all the nerve fibres from the retina come together to form the optic nerve.

optician 1 a person qualified to test vision, examine the eyes and prescribe glasses and contact lenses (ophthalmic optician).
2 a person qualified to fit spectacle frames, to make the measurements necessary to ensure that spectacle lenses are properly centred in the frames, and to fit contact lenses (dispensing optician).

optic nerves the 2nd of the 12 pairs of CRANIAL NERVES which emerge directly from the brain. The optic nerves are bundles of about one million nerve fibres originating in the RETINAS and connecting these to the brain. Each nerve passes through a channel in the bone at the back of the eye sockets (orbits) to reach the inside of the skull and to join with, and partially cross over, its fellow in the OPTIC CHIASMA.

optic tract the part of the nerve pathway for visual impulses lying between the OPTIC CHIASMA and the first set of connections (SYNAPSES) in the brain – the lateral geniculate body. If an optic tract is destroyed half of the field of vision of each eye is lost. The field loss in each eye is on the side opposite to the damaged tract.

optimal most favourable or desirable.

optometrist the American equivalent of ophthalmic OPTICIAN.

oral pertaining to the mouth.

oral sex any sexual activity in which the mouth is substituted for the vagina.
One consequence of this practice is the confusion between the strains of the herpes virus responsible for oral and genital herpes. Formerly, these were, respectively herpes simplex 1 and herpes simplex 2.
The matter is no longer so simple.

oral stage the first and most primitive of Freud's proposed stages of psychosexual development, in which the mouth is said to be the focus of the libido and the primary source of satisfaction. Such ideas no longer command widespread respect.

orbicular 1 circular or spherical.
2 circular and flat.

orbicularis oculi the flat, circular, SUBCUTANEOUS muscle that surrounds each eye and that, on contraction, causes the eye to 'screw up'.

orbicularis oris the SPHINCTER muscle surrounding the mouth used to close the mouth and purse the lips.
orbit the bony cavern in the skull that contains the eyeball and OPTIC NERVE, the muscles that move the eye, the LACRIMAL GLAND, a quantity of fat and various arteries, veins and nerves.
orchi- *combining form denoting* the testicle.
orexin one of a pair of centrally-acting neuropeptides produced by the lateral hypothalamus where the sensation of hunger is mediated. Rats given orexin will eat about ten times the normal amount of food and rats starved for 48 hours have more than twice the normal concentrations of orexin. There are also specific receptors for the two orexins. These facts are being exploited in the design of drugs that can both stimulate and reduce appetite.
organ any part of the body consisting of more than one tissue and performing a particular function.
organelle any one of the bodies forming the internal functional components, or 'little organs', of the cell. The organelles include MITOCHONDRIA, the GOLGI APPARATUS, the ENDOPLASMIC RETICULUM, RIBOSOMES, LYSOSOMES and the CENTRIOLES.
organic 1 pertaining to animals or plants, rather than to non-living matter.
2 pertaining to an organ of the body.
3 caused by a pathological change in bodily structure rather than by a purely mental process.
4 of a chemical compound (other than carbon dioxide or its salts or carbon monoxide), containing carbon.
5 of food, grown without the use of artificial fertilizers.
organic chemistry the chemistry of carbon compounds.
organism any living animal or plant.
organ of Corti a spiral structure on the inner surface of the basilar membrane of the COCHLEA that contains a large series of sensory receptors that respond to sound vibrations of different frequencies and stimulate appropriate nerve impulses subserving hearing. (Marquis Alfonso Corti, 1822–88, Italian anatomist).
organon organum a system of rules, principles or methods used in scientific investigation.
orgasm the sequence of bodily, and especially genital, processes, occurring at the climax of sexual intercourse, and involving the pleasurable release of heightened muscle tension followed by the decline of TUMESCENCE in erectile tissue. Most sexual activity becomes focused on the achievement of orgasm – an inherent property making for species survival.
orgasmolepsy sudden widespread bodily relaxation and transitory loss of consciousness occurring during an ORGASM.
origin the point or area of attachment of a muscle that remains mainly fixed when the muscle contracts. Compare INSERTION.
Origin of Species the popular abbreviation of the title of Charles Darwin's epoch-making book in which he proposed a scheme of biological evolution. The full title is: *On the Origin of Species by Means of Natural Selection, or the Preservation of Favoured Races in the Struggle for Life.*
oropharynx the part of the PHARYNX at the back of the mouth and extending down to the top of the OESOPHAGUS.
orphan genes isolated solitary genes situated remotely from their usual cluster site.
ortho- *combining form denoting* straight, upright, corrective or normal.
orthodontics the dental speciality concerned with the correction of irregularities of tooth placement and in the relationship of the upper teeth to the lower (occlusion). Teeth can readily be permanently moved by sustained pressure using braces, springs, wires and harnesses.
orthognathous having the lower jaw correctly aligned with the upper so that it neither protrudes nor recedes.
orthopaedics the branch of surgery concerned with correction of deformity and restoration of function following injury to, or disease or congenital abnormality of, the skeletal system and its associated ligaments, muscles and tendons. The term derives from the Greek for a 'straight child', not 'straight foot', as is commonly thought.

orthophoria a state of perfect alignment of the two eyes even when they are deprived of the assistance of the binocular reflexes that maintain alignment. In orthophoria, the eyes remain in alignment even when one is covered. See also ESOPHORIA and EXOPHORIA.

orthopnoea difficulty in breathing when lying down.

orthoptics a discipline, ancillary to ophthalmology, concerned mainly with the management of squint (strabismus) in childhood and the avoidance of AMBLYOPIA.

orthosis an appliance worn on the body to reduce or prevent deformity or to provide support, relieve pain and facilitate movement.

orthostatic pertaining to the erect posture.

orthotics the branch of medicine dealing with the use of supportive mechanical devices. See also ORTHOSIS.

os a bone or a mouth.

-ose *suffix denoting* characterized by or full of.

-osis *suffix denoting* a process or activity, an increase in, a disease or pathological process, or a non-inflammatory process.

osmole the standard unit of osmotic pressure. The osmole is equal to the molecular weight of the dissolved substance expressed in grams divided by the number of particles or IONS into which each molecule of the substance dissociates in solution.

osmolality the property of a solution that depends on its concentration in osmolal units. See OSMOLE.

osmolar having a concentration of 1 OSMOLE per litre.

osmolarity the concentration of a solution in OSMOLAR units.

osmoreceptors cells in the HYPOTHALAMUS that monitor the concentration of solutes in the blood. An abnormal increase, as occurs in dehydration, stimulates the release of VASOPRESSIN from the PITUITARY gland. This acts on the kidney to reduce water loss.

osmosis the automatic movement of the fluid part of a solution through a membrane, separating two quantities of the solution, in such a direction as to dilute the solution of higher concentration. The membrane is permeable to the liquid but not to the dissolved substance. Such a membrane is said to be semipermeable and membranes of this kind occur widely in the body. Osmosis is an important principle on which much of physiology is based.

osseo- *combining form denoting* bone.

osseous composed of, containing or resembling bone.

ossicle a small bone, especially one of the tiny bones, the auditory ossicles, in the middle ear that link the ear-drum to the inner ear. From the outside-in, the auditory ossicles are the malleus, the incus and the stapes.

ossification the process of conversion of other tissues into bone. Most bone forms from CARTILAGE but some is laid down by other connective tissue (membranous bone). Ossification may also occur in tissues that have been the site of disease such as long-term inflammation.

ost-, osteo- *combining form denoting* bone.

osteoblasts bone matrix-building cells. They are found on the outer surfaces of bone and on the surfaces of internal bone cavities. Osteoblasts are continuously active remodelling bone that has been broken down by OSTEOCLASTS.

osteoclasts large multinuclear cells that scavenge damaged bone. They work in conjunction with OSTEOBLASTS in promoting repair of bone fractures. Osteoclasts release calcium from bone under the influence of parathyroid hormone when the level of blood calcium drops.

osteocytes bone cells produced by OSTEOBLASTS.

osteogenesis bone formation.

osteogenin an extract from bone that has been used experimentally to induce muscle, cartilage and other soft tissue to transform into bone. Osteogenin is a glycoprotein that can transform muscle flaps placed in moulds into cancellous bone matching the exact shape of the mould.

osteoid resembling bone.

osteology the science and study of bones.

osteolysis reabsorption or demineralization of bone.

osteomalacia bone softening as a result of defective mineralization, usually occurring because of defective calcium absorption from vitamin D deficiency. Osteomalacia is commoner in women than in men. The softened bones may distort or bend under the body weight. The condition is treated with vitamin D.

osteoporosis a form of bone atrophy involving both the COLLAGEN scaffolding and the mineralization. It is thought to be due to predominance of reabsorption of bone over natural bone formation. It is commonest in women after the menopause and tends to be progressive, giving rise to the risk of fractures from minimal trauma.

osteoprogenitors unspecialized stem cells that develop into OSTEOBLASTS.

osteoprotegerin (OPG) a secreted protein which inhibits osteoclast differentiation. The cytokine TRANCE binds to osteoprotegerin.

ostium an opening or mouth.

ot-, oto- *combining form denoting* the ear.

OTC *abbrev. for* over the counter. This refers to drugs or other remedies that may be purchased from a pharmacist without a doctor's prescription.

otic pertaining to the ear.

otocyst the precursor, in the MESODERM of the head of the embryo, of the membranous LABYRINTH of the inner ear.

otolith one of the many tiny calcareous particles found in the utricle and sacculus of the inner ear. These move under gravitational and accelerative forces causing stimulation of hair cells and the production of nerve impulses that provide the brain with information about the position and movement of the head.

otology the surgical speciality concerned with the study of the structure, function and disorders of the ears and the treatment of ear disease.

-otomy *suffix denoting* a surgical incision into.

otorhinolaryngology the surgical speciality concerned with the diseases of the ear, nose and throat. Also known as otolaryngology.

ototoxicity damage to ear function by the poisoning effects of drugs or other agents. These can affect both hearing and balancing mechanisms.

oval window the oval opening in the outer wall of the inner ear in which the footplate of the inner of the three AUDITORY OSSICLES, the stapes, is free to vibrate.

ovari- *combining form denoting* the OVARY.

ovary one of the paired female gonads, situated in the pelvis, one on each side of the womb (uterus), just under and inward of the open ends of the FALLOPIAN TUBES. Ovaries are almond-shaped and about 3 cm long. They are the site of egg (OVUM) formation and release one or more ova each month about 14 days before the onset of the next menstrual period. This is called ovulation. See also OOCYTE, OOGONIA.

overdose a quantity of a drug well in excess of the recommended dose.

overpopulation probably the greatest threat facing humankind today. Populations do not grow linearly – by addition; they grow exponentially – by multiplication. This means that it is not only populations that increase; the rate of growth of populations also increases – an alarming fact. Thirty years ago world population was rising by about one million every eight days; today, there are a million more people every four days. We can take little comfort in the knowledge that people are having fewer babies; the real trouble is that every year there are far more people to have babies. The present rate of growth of about 1.7 per cent on today's world population of 6.4 billion means an annual increase of over 100 million. This is a far greater increase than the higher rate of growth produced on the smaller total population of a few decades ago. Most of the growth is taking place in the less well developed countries and they have an average rate of increase of 2.4 per cent. A general increase in life expectancy compounds the problem.

overt obvious.

oviduct the FALLOPIAN TUBE.

ovulation the release of an OVUM from a mature Graafian follicle in the OVARY. Ovulation occurs about half way between the beginning of consecutive menstrual periods, usually about 14 days before the expected date of onset of the next period.

ovum the female gamete or ovum which, when fertilized by a spermatozoon, can give rise to a new individual. The egg is a very large cell, compared with other body cells, and contains only 23 chromosomes, half the normal number (haploid). Like most other cells ova contain many mitochondria each containing many copies of mitochondrial DNA. This DNA is not present in sperm.

oxidative phosphorylation the process of cellular respiration occurring within the MITOCHONDRIA and responsible for the production of adenosine triphosphate (ATP). In this process energy derived from the oxidation of hydrogen, to form water, is transferred to ATP.

oxidative stress the widespread effects of oxygen FREE RADICALS on any part of the body.

oxidoreductases a group of enzymes that promote oxidation-reduction reactions. The group includes some enzymes also known as dehydrogenases and oxidases.

oxy- *combining form denoting* sharp.

oxycephaly a skull abnormality causing the head to assume a conical or peaked appearance. Oxycephaly is due to premature closure of the irregular junctions on the side of the skull between the bones of the vault (the coronal or lambdoidal sutures). Also known as tower head.

oxygen a colourless, odourless gas, essential for life, that constitutes about one fifth of the earth's atmosphere. Oxygen is required for the functioning and survival of all body tissues, and deprivation for more than a few minutes is fatal. The respiratory system captures oxygen from the atmosphere and passes it to the blood by means of which it is conveyed to all parts of the body. Oxygen is needed for the fundamental chemical process of oxidation of fuel to release energy. This series of reactions is known as oxidative phosphorylation and involves the synthesis of the universal energy carrier ATP (ADENOSINE TRIPHOSPHATE) in the inner membranes of the MITOCHONDRIA of the cells.

oxyhaemoglobin the bright red, oxygenated form of HAEMOGLOBIN in the red blood cells, in which a molecule of oxygen has attached itself reversibly to the iron atom in the haemoglobin molecule. Oxyhaemoglobin transports oxygen from the lungs to the tissues.

oxyntic cells cells in the glands of the lining of the stomach that secrete hydrochloric acid and INTRINSIC FACTOR. Also known as parietal cells.

oxytocic 1 hastening delivery of a baby by inducing contraction of the womb muscles. 2 a drug that hastens childbirth.

oxytocin an OXYTOCIC hormone produced by the pituitary gland. The hormone promotes contraction of the womb and the letdown of milk during suckling. It is used as a drug in obstetrics to bring on labour at term and to augment slow labour. There is recent evidence that the hormone may have a role to play in promoting trust between humans.

ozone a gas consisting of molecules in which three atoms of oxygen are linked together. Concentrated ozone is a blue explosive liquid. Even in low concentrations the gas is poisonous and highly irritating. The ozone layer in the stratosphere, between 10 and 50 km above the earth's surface, is produced continuously by the action of ultraviolet radiation from the sun and forms a protective barrier, cutting down the intensity of the ultraviolet component in sunlight. Without the ozone layer we would suffer serious biological effects from solar radiation, including a large increase in the incidence of skin cancer. Atmospheric ozone is broken down by the catalytic action of chloro-fluoro-carbons (CFCs) and other substances. Recent studies have shown that ozone is involved in the oxidative stress production of atherosclerosis in arteries. Cholesterol is converted by ozone to 5,6-secosterol which is cytotoxic and induces the formation of foam cells in the presence of low-density lipoproteins.

Pp

p21 an anticancer protein of the class of cyclin-dependent kinase inhibitors. Cyclin-dependent kinases are enzymes that have been described by Elizabeth finkel as 'the engines that drive the cell-cycle past its various control checkpoints.' Inhibition of these enzymes provides a way of controlling rapidly-reproducing cells such as cancer cells. High levels of p21 are found in ageing cells.

p53 a gene that induces production of p21. A tumour suppressor gene, the absence or mutation of which can greatly increase the probability that cancer will develop. When DNA damage, as from anticancer drugs, occurs in normal cells the expression of p53 is increased. The gene may then act to protect the cell by halting the cell cycle so that the DNA damage can be repaired, or, if the DNA damage is too severe, p53 can kill the cell by APOPTOSIS so that the defective DNA is not passed on. Mutations in p53 have been found in a range of cancers including those of the breast, colon, ovary, bladder and oesophagus. They have also been found to be associated with failure to respond to anticancer drugs.

pabulum any substance that provides NOURISHMENT.

pachy- *combining form denoting* thick.

pachydermatous abnormally thick-skinned.

Pacinian corpuscle a specialized encapsulated ending of a sensory nerve occurring in the deep layers of the skin. Pacinian corpuscles respond to touch and heavy pressure. (Filippo Pacini, 1812–83, Italian anatomist).

paed-, paedo- *combining form denoting* child.

paediatrics the medical speciality concerned with all aspects of childhood diseases and disorders and with the health and development of the child in the context of the family and the environment.

paedomorphism retention by the adult of juvenile characteristics.

paedophilia recurrent sexual urges towards a prepubertal child by a person over the age of 16 and at least five years older than the child. If manifested, these urges, which constitute criminal sexual molestation, usually involve fondling of the genitals and oral intercourse. Vaginal or anal intercourse is relatively infrequent except in cases of incest. Active paedophilia is not a medical or psychiatric condition; it is a crime universally condemned. The Internet has become a vehicle both for the promulgation of paedophilic graphics and for the promotion of paedophilic contacts.

pain an unpleasant or distressing localized sensation caused by stimulation of certain sensory nerve endings called nociceptors, or by strong stimulation of other sensory nerves. Nociceptors are stimulated by the chemical action of substances, such as prostaglandins, released from local cell damaged by injury or inflammation. Whatever the site of nerve stimulation, pain is usually experienced in the region of the nerve endings. Referred pain is pain experienced at a site other than that at which the causal factor is operating. Pain impulses pass to the brain via a series of control 'gates' analogous to those in computers and these can be modulated by other nerve impulses. Pain commonly serves as a warning of bodily danger and leads to action to end it. Pain is best treated by discovering and removing the cause. It is a complex phenomenon with many components – somatic, emotional, cognitive and social. From the Latin *poena*, punishment. See also ENDORPHINS.

palaeolithic period the old stone age dating from about 2.5 million to about 10 000 years ago. The palaeolithic was followed by the mesolithic (middle stone age) and neolithic (new stone age) periods. The palaeolithic was characterized by the use of chipped stone implements of flint and quartzite; the neolithic by ground tools.

palaeontology the study of rock fossil plants and animals.

palatal 1 pertaining to the palate.
2 of phonation, produced with the front of the tongue near or against the hard palate.

palate the partly hard, partly soft partition that forms the roof of the mouth and separates it from the nose. The hard palate consists of a plate of bone, part of the MAXILLA, covered with mucous membrane. The soft palate, attached to the back of the hard palate, is a small flap of muscle and fibrous tissue enclosed in a fold of mucous membrane. It can press firmly against the back wall of the PHARYNX, sealing off the opening to the nose during swallowing and when one is blowing out through the mouth.

palatine pertaining to the palate.

pallor undue paleness of the skin, observable also in certain mucous membranes, such as the CONJUNCTIVA and the lining of the mouth. Transient pallor may result from constriction of the blood vessels near the surface. Longer-term pallor may be due to ANAEMIA or lack of exposure to sunlight. Permanent pallor may be caused by ALBINISM.

palmar pertaining to the palm of the hand. Compare PLANTAR.

palpable able to be felt.

palpation examination by feeling with the fingers and hands.

palpebral pertaining to the eyelids.

palpitation abnormal awareness of the action of the heart, because of rapidity or irregularity. Irregularity is most commonly due to extrasystoles each of which causes a brief sense of stoppage.

palsy an obsolete term for PARALYSIS, retained for historical reasons in a few cases such as those of BELL'S PALSY and CEREBRAL PALSY.

palliative care the application of specialist knowledge to the relief of severe suffering, physical or mental, in people in whom such relief cannot be achieved by cure of disease or disorder. Palliative care is now a recognized specialty mainly exercised in hospices and in large hospitals with cancer units. It is not, however, necessarily limited to those with no prospect of recovery.

pan- *combining form denoting* all, or the whole of.

panacea a mythical universal remedy or cure-all. See also QUACKERY.

panagglutinin an AGGLUTININ capable of agglutinating red blood cells of any blood group.

pancreas a dual function gland situated immediately behind the STOMACH with its head lying within the loop of the DUODENUM, into which the duct of the pancreas runs. The pancreas secretes digestive enzymes capable of breaking down carbohydrates, proteins and fats into simpler, absorbable, compounds. It is also a gland of internal secretion (an endocrine gland). The endocrine element consists of the Islets of Langerhans, specialized cells that monitor blood and produce four hormones – INSULIN, GLUCAGON, SOMATOSTATIN and a pancreatic polypeptide of unknown function.

pancreatic juice the clear, alkaline fluid that passes along the pancreatic duct into the DUODENUM and contains enzymes that are able to break down proteins, carbohydrates and fats into simpler and absorbable substances.

pandemic a world-wide EPIDEMIC.

panic attack the episode characteristic of the PANIC DISORDER.

panic disorder a condition featuring recurrent brief episodes of acute distress, mental confusion and fear of impending death. The heart beats rapidly, breathing is deep and fast and sweating occurs. Overbreathing (hyperventilation) often makes the attack worse. These attacks usually occur about twice a week but may be more frequent and they are especially common in people with AGORAPHOBIA. The condition tends to run in families and appears to be an ORGANIC disorder with a strong psychological component.

pantothenic acid one of the B group of vitamins and a constituent of coenzyme A which has a central role in energy metabolism. Deficiency is rare.

papilla a small, nipple-like projection.

papillary muscles the finger-like muscular processes arising from the floors of the VENTRICLES of the heart to which are attached the strings (chordae tendineae) that tether the cusps of the atrioventricular valves.

papilloma a benign tumour of skin or mucous membrane in which epithelial cells grow outward from a surface around a connective tissue core containing blood vessels. Papillomas may be flat or spherical with a narrow neck (pedunculated) and occur on the skin, in the nose, bladder, larynx or breast. Most skin papillomas are WARTS.

Pap smear a popular term for the cervical smear test. A method of screening women for the earliest signs of cancer of the neck (cervix) of the womb. A wooden or plastic spatula is used gently to scrape some surface cells off the inner lining of the cervix and these are then spread on a microscope slide, stained and examined by a pathologist skilled in cytology. (George Nicholas Papanicolaou, 1884–1962, Greek-born American anatomist).

papule any small, well-defined, solid skin elevation. Papules are usually less than 1 cm in diameter and may be smooth or warty. From the Latin *papula*, a pimple.

para- *prefix denoting* alongside, beyond, apart from, resembling or disordered.

paracrine pertaining to the chemical transmission of information through an intercellular space. Compare ENDOCRINE and EXOCRINE.

paradigm 1 a human being's mental model of the world, which may or may not conform to that of others but is often stereotypical. 2 In the philosophy of science, a general conception of the nature of scientific operation within which a particular scientific activity is undertaken. Paradigms are, of their nature, persistent and hard to change. Major advances in science – such, for instance, as the realization of the concept of the quantum or the significance of evolution in medicine – involve painful paradigmic shifts which some people, notably the older scientists, find hard to make.

paraesthesia numbness or tingling of the skin. 'Pins-and-needles' sensation.

paralysis temporary or permanent loss of the power of movement of a part of the body (motor function). Paralysis may be due to damage to the nerve tracts or peripheral nerves carrying motor impulses to the muscles to cause them to contract or may be due to disorders of the muscles themselves. In the former case, the damage is most commonly within the brain or the spinal cord. Paralysis of one half of the body is called hemiplegia. Paralysis of the legs and lower part of the body is called paraplegia. Paralysis of all four limbs is called quadriplegia.

parametric of data that are normally distributed, the distribution curve being symmetrically bell-shaped. Data that are not normally distributed are said to be skewed and the bell shape of the curve is distorted.

parametric test a test or trial that depends on the assumption that the data involved are normally distributed.

parametrium the connective tissue lying between the two layers of the BROAD LIGAMENT of the womb and separating the CERVIX from the bladder.

paramnesia a memory disturbance in which fantasy is recollected as experience.

paranoia a delusional state or system of delusions, usually involving the conviction of persecution, in which intelligence and reasoning capacity, within the context of the delusional system, are unimpaired. HALLUCINATIONS or other mental disturbances do not occur. Less commonly there may be delusions of grandeur, of the love of some notable person, of grounds for sexual jealousy or of bodily deformity, odour or parasitization.

paraphilia any deviation from what is currently deemed to be normal sexual behaviour or preference. Thus, paraphilia may include bestiality, exhibitionism, fetishism, homosexuality, masochism, paedophilia, sadism, transvestism and voyeurism.

paraplegia paralysis of both lower limbs. Paraplegia caused by damage to the spinal cord is also associated with paralysis of the lower part of the trunk and loss of normal voluntary bladder and bowel control. In spastic paraplegia there is constant tension in the affected muscles often leading to fixed deformities. In flaccid paraplegia the muscles remain limp. Paraplegia may be hereditary or caused by birth injury, injury or disease of the brain or spinal cord or sometimes the peripheral nerves, as in alcoholic or dietary polyneuropathy, or by HYSTERIA or old age.

parapsychology the attempted study, by scientific methods, of a range of real or imagined phenomena not explicable by science. The subjects of parapsychology include EXTRASENSORY PERCEPTION, telepathy, clairvoyance, spoon-bending and the movement of objects without physical force (telekinesis). The history of science has been a long and painful struggle to escape from the realms of magical thinking and superstition and many scientists are concerned at the possible dangers of conferring a kind of respectability and plausibility on matters which they consider to be without scientific basis.

parasite an organism that lives on or in the body of another living organism, and depends on it for nutrition and protection. Ectoparasites live on the surface, endoparasites live inside. Parasites do not contribute to the host's welfare and are often harmful. Human parasites, which cause thousands of diseases, include viruses, bacteria, fungi, protozoa, worms, flukes, ticks, lice, bugs, some burrowing flies and leeches.

parasitology the study of organisms which use other organisms as their living environment. Medical parasitology is concerned mainly with the larger, usually visible, parasites of humans such as the various worms and the external parasites (ectoparasites). Bacteria, viruses and protozoa, although parasites, are so important as to require separate disciplines and are not normally included in medical parasitology.

parasuicide an apparent attempt at suicide, usually by self-poisoning, in which the intent is to draw attention to a major personal problem rather than to cause death.

parasympathetic nervous system one of the two divisions of the AUTONOMIC NERVOUS SYSTEM. The parasympathetic system leaves the central nervous system in the 3rd, 7th, 9th and 10th CRANIAL NERVES and from the 2nd to the 4th SACRAL segments of the spinal cord. Parasympathetic action constricts the pupils of the eyes, promotes salivation and tearing, slows the heart, constricts the BRONCHI, increases the activity of the intestines, contracts the bladder wall and relaxes the SPHINCTERS and promotes erection of the penis. See also SYMPATHETIC NERVOUS SYSTEM.

parasympathomimetic any effect similar to that produced by stimulation of the parasympathetic nervous system, especially the effect of cholinergic drugs.

parathyroid glands four, yellow, bean-shaped bodies, each about 0.5 cm long, lying behind the THYROID GLAND, usually embedded in its capsule. The parathyroids secrete a hormone, parathyroid hormone (parathormone or PTH), into the blood if the level of calcium in the blood drops. This hormone promotes the release of calcium from the bones, controls loss in the urine and increases absorption from the intestine, thus correcting the deficiency in the blood. Maintenance of accurate levels of blood calcium is more important, physiologically, than the strength of the bones. Secretion of abnormal quantities of PTH from a parathyroid tumour can lead to bone softening. Underaction of the parathyroids causes a dangerous drop in the blood calcium.

parenchymal pertaining to, or resembling, the functional elements of an organ or tissue, rather than to its structural parts (matrix).

parenchyme parenchyma, the functional tissue of an organ, as distinct from its purely structural elements.

parenteral of drugs or nutrients, taken or given by any route other than by the alimentary canal. Parenteral routes include the intramuscular and the intravenous.

parenteral nutrition intravenous feeding. This is required when the normal (enteral) route cannot be used. Early attempts at intravenous feeding via peripheral veins invariably led to severe thrombophlebitis within a matter of hours because of the strong sugar solutions used. A central venous cannula had therefore to be used. Developments in design of cannulas and new feeding solutions, with calorie-rich lipids in place of strong sugar concentrations, amino acids and weaker carbohydrates may, it is hoped, allow safe peripheral vein feeding.

parenting the process of caring for, nurturing and upbringing of a child.

paresis WEAKNESS or reduction in muscle power, as compared with complete PARALYSIS.

parietal pertaining to the wall or outer surface of a part of the body. From Latin *parietem*, a wall or partition.

parietal bone one of two large bones of the vault of the skull lying between the FRONTAL and OCCIPITAL bones on either side and forming the sides and top.

parietal cells large cells in the lining of the stomach that secrete hydrochloric acid. Also known as oxyntic cells.

parietal lobe the major lobe in each half of the brain (cerebral hemisphere) that lies under each parietal bone.

parotid glands the largest of the three pairs of SALIVARY GLANDS. Each parotid is situated over the angle of the jaw, below and in front of the lower half of the ear and has a duct running forward through the cheek to open on the inside of the mouth at about the level of the upper molar teeth.

paroxysm 1 a sudden attack, such as a seizure, convulsion or spasm.
2 a sudden worsening of a disorder.

parthenogenesis the development of an unfertilized egg into an adult organism. Virgin birth. This occurs naturally in bees and ants and in some animal species development of an ovum can be induced chemically or by pricking with a fine glass fibre. The result is a clone of the mother cell identical in all respects. Only females can be produced by parthenogenesis, as no Y chromosome is present. If achieved, human parthenogenesis would make men biologically redundant. Very early human embryos derived only from ova have been produced experimentally by a parthenogenetic technique using chemicals that changed the concentration of ions in the ova.

partial reinforcement the process of intermittently re-strengthening of a CONDITIONED REFLEX by repetition of the association between the unconditioned and the conditioned stimuli.

parturition the process of giving birth.

passive euthanasia a form of euthanasia in which medical treatment that could keep a dying patient alive for a time is withdrawn.

passive immunity immunity, especially to specific infections, resulting from the acquisition of ANTIBODIES, either by injection or by transfer through the PLACENTA or ingestion in the breast milk. Sensitized T cells can also confer passive immunity.

passive smoking inhaling cigarette smoke exhaled by others. It has been shown that the rate of lung cancer in non-smokers rises significantly if they are regularly exposed to other people's cigarette smoke. At least 10 separate studies have shown an increase of up to 30 per cent in the risk of lung cancer among non-smokers living with smokers, compared with non-smokers living with non-smokers.

passive transport the movement of dissolved material through a biological membrane in the direction of fluid flow and without the expenditure of energy. Compare ACTIVE TRANSPORT.

pasteurization a method of destroying infective micro-organisms in milk and other liquid foods. The liquid is rapidly heated to about 78°C and maintained at that temperature for fifteen seconds. It is then rapidly cooled to below 10°C. (Louis Pasteur, 1822–95, French pioneer of bacteriology).

Pasteur, Louis the French chemist Louis Pasteur (1822–95), the founder of the science of microbiology, must be considered one of the greatest figures in biology and medicine. Pasteur was appointed professor of chemistry at the Lille Faculty of Science in 1854. Soon afterwards he was asked by the French government to investigate the reason for the souring of wine – a matter of huge economic importance. In a series of brilliant experiments he was able to show that if a substance in a flask was made sterile by boiling and air was prevented from reaching it, no fermentation or putrefaction would occur. If contact with the outside was allowed, these changes did occur. Clearly, something was entering from the outside. Later he was able to show that the fungus *Mycoderma aceti* was responsible for turning the wine to vinegar. Heating to 55° C solved the problem. The process, which became known as pasteurization, was then applied to beer and to milk, thereby saving the brewers a fortune and mankind much suffering from bovine tuberculosis.

On the basis of this work came the recognition that a great many diseases were caused by invisible 'germs' (micro-organisms). The great London surgeon Joseph Lister read about Pasteur's discovery and, following his principles, dictated practices that were to change the face of surgery. Lister adopted antisepsis using scrupulous cleanliness and a spray of carbolic acid, but this was soon followed by a method in which organisms were avoided altogether (asepsis).

Pasteur then showed that all organisms were derived from other organisms and that the idea of spontaneous generation was nonsense. Undeterred by a stroke in 1868, he went on to identify the bacillus responsible for the serious disease of anthrax. Then, inspired by the work, 80 years earlier, of Edward Jenner , showed that immunization was possible with a strain of anthrax whose virulence had been weakened by repeated passage through a series of animals. His researches provided a logical basis for hygiene and sanitation, and he showed how sterilization by heat or other means could prevent the spread of disease. In 1882 Pasteur established the cause of rabies and showed that it was transmitted by an organism too small to be seen with a microscope – a virus. Three years later he made an effective vaccine against rabies. In 1888, Pasteur, now world-famous, was appointed the first director of the new Pasteur Institute in Paris.

patella the knee cap. The patella is a large triangular SESAMOID bone lying on front of the knee joint within the tendon of the QUADRICEPS FEMORIS group of muscles.

patella disorders these include chondromalacia patellae and osgood-schlatter's disease.

patent 1 open or unobstructed.
2 a term still sometimes applied to proprietary medication, as in patent medicine. From Latin *patent*, open.

paternity tests tests designed to confirm or deny the claim that a man is the father of a particular child. When the man is not the father, this can be confirmed by normal blood-grouping tests in 97 per cent of cases. Blood-group and tissue typing (HLA) tests cannot prove that a particular man is the father, but they can offer strong supportive evidence. Genetic fingerprinting can prove paternity beyond any doubt.

pathogen any agent that causes disease, especially a micro-organism.

pathogenesis the mechanisms involved in the development of disease. Compare AETIOLOGY.

pathogenic able to cause disease.

pathognomonic of a symptom or physical sign that is so uniquely characteristic of a particular disease as to establish the diagnosis.

pathological pertaining to disease or to the study of disease (PATHOLOGY).

pathological gambling an addiction to the state of excitement experienced while gambling. There is progressive preoccupation with betting and a need to increase the size of wagers to achieve the desired mental effect. The syndrome includes lying to conceal losses, stealing and rationalising the theft as temporary borrowing. If gambling is prevented there is irritability, restlessness and even physical symptoms.

pathology the branch of medical science dealing with bodily disease processes, their causes, and their effects on body structure and function. Subspecialties in pathology include morbid anatomy, histopathology, haematology and clinical chemistry. Practitioners of forensic pathology apply all these disciplines to criminal investigation.

pathophysiology the discipline concerned with the effects of disease on body function.

-pathy *suffix denoting* perception, feeling or suffering.

patient-controlled analgesia a method of pain control in which the patient cooperates. An intravenous drip is set up and the patient has a small control unit with a button which, when pressed, inserts a small dose of a drug such as morphine, into the infusion fluid. Overdosage cannot occur. A disadvantage is the need to urinate at frequent intervals.

patrilineage descent through the male line.

patulous spreading apart or expanded.

PCO_2 the concentration of carbon dioxide in the blood. This is a sensitive indication of the efficiency or level of ventilation of the lungs. Hyperventilation causes a drop in PCO_2; underventilation causes a rise.

PCR *abbrev. for* POLYMERASE CHAIN REACTION.

peau d'orange an orange-skin-like dimpling of an area of the skin, affecting especially the breast and caused by LYMPHOEDEMA occurring in certain kinds of cancer.

pectinate having teeth like a comb.

pectoral muscles the group of muscles, consisting of the pectoralis major and minor on either side, covering the upper ribs on either side of the front of the chest. The pectoral muscles help to control the shoulder-blade (scapula) and, through it, to move the arm forward and down, and, acting directly on the arm, to pull it towards the body.

pederasty anal intercourse, especially with a boy.

pediculosis any kind of louse infestation.

pedigree a family tree, showing the members who have suffered from hereditary disorder, prepared for purposes of genetic diagnosis and research.

peduncle a stalk-like bundle of fibres, especially nerve fibres that connects different parts of the central nervous system.

pedunculated on a stalk. Compare with SESSILE.

peeping Tom a person who derives sexual pleasure and arousal from secretly watching others in a state of undress or engaged in sexual activity. A voyeur. The term derives from the legend of Tom the tailor of Coventry who was the only person to look at the naked Lady Godiva, and lived to regret it, being instantly struck blind. See also VOYEURISM.

Peking man early humans, of the species *Homo erectus* that lived in northern China some 500,000 years ago. A mass of skeletal remains of 44 people jumbled with thousands of broken and burned animal bones and many stone tools were found at Chou K'ou Tien, near Beijing, in the decade following 1927.

pelvic girdle the ring of bones at the lower end of the spine with which the legs articulate. Compare shoulder girdle. See also PELVIS.

pelvis **1** the basin-like bony girdle at the lower end of the spine with which the legs articulate. The pelvis consists of the SACRUM and COCCYX, behind, and the INNOMINATE bones on either side. Each innominate bone is made up of three bones – the pubis, the ilium and the ischium.
2 any funnel-shaped structure such as the pelvis of the kidney.

pendulous hanging loosely or suspended so as to be able to swing or sway.

penetrance the frequency with which a GENE manifests its effect. Failure to do so may result from the modifying effect of other genes or from environmental influences. A single hereditable dominant or recessive characteristic is either penetrant or not. Penetrance is measured as the proportion of individuals in a population with a particular genotype who show the corresponding PHENOTYPE.

penis the male organ of copulation containing the URETHRA through which urine and seminal fluid pass. The normally flaccid penis becomes enlarged and erect by virtue of three longitudinal cylindrical bodies of spongy tissue into which blood can flow under pressure under the influence of sexual excitement or other stimuli. One of these bodies, the corpus spongiosum, surrounds the urethra. The other two, the corpora cavernosa, lie side by side above the corpus spongiosum. Erection physiology is mediated partly by nitric oxide.

penis disorders these include balanitis, hypospadias, impotence, paraphimosis, penile warts, peyronie's disease, phimosis and pseudohermaphroditism.

penis envy the Freudian concept that all women have a repressed wish to have a penis and resent the fact that they are incomplete men. The idea is no longer taken seriously.

pentose a sugar with five carbon atoms in each molecule. The 'backbone' of DNA on each side of the helix consists of a chain of pentose sugars alternating with phosphate groups. The sugar in DNA is 2-deoxyribose, and in RNA is ribose. The NUCLEOTIDE chain is formed by linking the 5′ position of one pentose ring to the 3′ position on the next via a phosphate group. 5′ and 3′ are used to indicate the ends of a DNA fragment and the directions in which the 'backbones' run.

pepsin a digestive ENZYME whose precursor PEPSINOGEN is secreted by cells in the stomach lining. Pepsin breaks down protein to PEPTIDES. See also PEPTIDASE.

pepsinogen a biochemically inert substance produced by the cells of the stomach lining (gastric mucosa) that is converted to PEPSIN by the action of hydrochloric acid.

peptidase an enzyme that hydrolyses (see HYDROLYSIS) protein fragments (PEPTIDES), breaking them down to AMINO ACIDS.

peptide a chain of two or more AMINO ACIDS linked by peptide bonds between the amino and carboxyl groups of adjacent acids. Large peptides, containing many amino acids, are called polypeptides. Chains of linked polypeptides, are called PROTEINS. Peptides occur widely in the body. Many HORMONES are peptides.

peptide bond a covalent bond formed between amino acids during protein synthesis. The OH– on a carbon atom links with the H– on a nitrogen atom to form a water molecule which is given off as each peptide bond is formed. Amino acids linked by peptide bonds form dipeptides, tripeptides, polypeptides or PROTEINS.

peptide $YY_{3\text{-}36}$, PYY an intestinal hormone fragment that is produced during digestion and passes to the hypothalamus to signal satiety. When given by intravenous injection it reduces appetite and food intake for about 12 hours. Research has shown that obese people have much lower levels of $PYY_{3\text{-}36}$ than non-obese people.

peptidergic of a neuron that releases peptides.

peptidyl transferase the enzyme that catalyzes the formation of PEPTIDE BONDS during the synthesis of a polypeptide (translation).

peptones various protein derivatives obtained by acid or enzyme HYDROLYSIS of protein and used as nutrients or culture media.

percentile one of 100 equal parts in an ordered sequence of statistical data. Thus, for instance, the 93rd percentile is the value that equals 93 per cent of those in the series. If a value in a range of values falls in the 20th percentile, there are 19 lower and 80 higher. The method is widely used to record and check such parameters as the extent of body growth at a particular age.

perception the reception, selection and organization of sensory data. Perception is greatly influenced by previous experience and the stored data accumulated from such experience. The term is often also taken, in a metaphorical sense, to include the mental activity that follows the receipt of sense data.

perforation a hole through the full thickness of the wall of an organ or tissue made by disease, injury or deliberate surgical act.

perforin a molecule produced by cytotoxic T cells and natural killer cells which forms a pore in the membrane of the attacked cell resulting in LYSIS and cell death.

perfusion 1 the passage of blood or other fluids through the body.
2 the effectiveness with which a part, such as the brain, is supplied with blood.

peri- *prefix denoting* round about, surrounding.

pericardium the double-layered membranous sac that completely envelops the heart. The inner layer is attached to the heart and the outer layer to the DIAPHRAGM and the back of the breastbone (sternum). The two layers are separated by a thin film of lubricating fluid.

perichondrium the fibrous connective tissue that covers cartilage surfaces, apart from the bearing surfaces (articular cartilages) of joints.

pericranium the PERIOSTEUM covering the outer surface of the skull.

perikaryon 1 the PROTOPLASM surrounding the nucleus of a cell.
2 the cell body of a neuron containing the nucleus.

perilymph the fluid, similar to CEREBRO-SPINAL FLUID, that lies in the space between the membranous and bony labyrinths of the internal ear.

perimysium a sheath of connective tissue surrounding and separating bundles of muscle fibres.

perinatal pertaining to the period immediately before and after birth. For statistical purposes, the perinatal period is defined as the period from the 28th week of pregnancy to the end of the 1st week after birth. Perinatal mortality is the number of babies born dead together with the number dying during the first week after birth.

perinatology the study of the care of the pregnant woman, the developing fetus and the new-born baby, and especially of those cases in which risk is anticipated from conditions known to endanger the life or health of the fetus or mother.

perinephrium the connective and fatty tissue surrounding the kidney.

perineum that part of the floor of the PELVIS that lies between the tops of the thighs. In the male, the perineum lies between the anus and the scrotum. In the female, it includes the external genitalia.

perineurium a sheath of connective tissue surrounding and separating nerve fibre bundles.

period see MENSTRUAL PERIOD.

periodicity 1 recurring at intervals or in cycles.
2 of DNA, the number of base pairs in one complete turn of the double helix. The periodicity of DNA is about 10.

periodontium the layer of fibrous, supportive connective tissue between the root of the tooth and the tooth socket. For practical convenience, dentists extend this definition to include the CEMENTUM, the gum surrounding the neck of the tooth and the bone of the socket (alveolar bone).

periosteum the tissue that surrounds bone. Periosteum has an inner bone-forming (osteoblastic) layer, a middle fibrous layer and an outer layer containing many blood vessels and nerves.

periotic situated around the ear.

peripheral nervous system the entire complex of nerves that leave the confines of the brain and the spinal cord (the central nervous system) to supply the muscles, skeleton, organs and glands. The CRANIAL NERVES, the spinal nerves and the AUTONOMIC NERVOUS SYSTEM.

peristalsis a coordinated succession of contractions and relaxations of the muscular wall of a tubular structure, such as the OESOPHAGUS, small intestine or the URETER, producing a wave-like pattern whose effect is to move the contents along.

peritoneal cavity the potential space between the layers of the PERITONEUM.

peritoneum the double-layered, serum-secreting membrane that lines the inner wall of the ABDOMEN and covers, and to some extent supports, the abdominal organs. The fluid secreted by the peritoneum acts as a lubricant to allow free movement of organs such as the intestines. The peritoneum contains blood vessels, lymph vessels and nerves. From the Greek *peri*, round about, and *teinein*, to stretch.

peritubular capillaries blood capillary beds surrounding the renal tubules of nephrons.

permanent teeth the 32 teeth forming the second set that begins to appear (erupt) after the shedding of the primary or 'milk' teeth at about the age of 6. The permanent tooth complement of each jaw consists of four biting teeth (incisors) at the front, flanked by two eye teeth (canines), four premolars and six grinding teeth (molars). The 'wisdom teeth' are the pair of third molars that often do not erupt until well into adult life.

permeability constant a number that indicates the ratio between the concentration gradient across a membrane and the resulting flow of solution through the membrane. Permeability constant varies with the characteristics of the membrane and those of the solute.

peroneal pertaining to the outer side of the leg or to the FIBULA.

peroneus any one of the muscles that take origin from the FIBULA and are inserted into the bones of the foot. The peroneal muscles assist in walking and in everting the raised foot.

peroxisomes membrane-bound cell organelles containing oxidative enzymes concerned in the detoxification of various molecules and in the breakdown of fatty acids to acetyl-CoA.

persistent vegetative state a condition that has proved difficult to define because of uncertainties as to the real meaning of 'consciousness', 'awareness' and 'wakefulness'. Patients in a persistent vegetative state can breathe without mechanical assistance. Heart, kidney and intestinal functions are normal and the bladder and bowels empty automatically. At times they appear to be awake. They will respond to painful stimuli by opening their eyes, moving their limbs, breathing more quickly, and occasionally grimacing. The type and degree of brain damage indicates, however, that they cannot perform any of the higher neurological or mental functions known to be essential for any mental processes or appreciation of their situation.

personality

The term 'personality', as used in psychology, embraces a wide range of human qualities. It is the totality of all those mental and physical characteristics that contribute to uniqueness as a human being. It includes intellectual and educational attributes, the emotional disposition, the ethical standards and the behavioural tendencies, and incorporates both character and temperament.

Personality types

Many attempts have been made to classify personality into groups in accordance with behaviour traits, but these have not been particularly successful. This is mainly because the great majority of people fall into an indefinable group in which none of the extremes of personality type is particularly marked. There are, however, some who will be generally agreed to be of the outgoing extrovert type while others are of the inward-looking introvert type. Most people, however, cannot be realistically classified in this way at all.

In addition, a small proportion of people can be fairly categorized as having obsessional, hysterical or antisocial personalities. The obsessional person is overconscientious, meticulous, orderly and tidy, rigid in habit and hates disruption of the established patterns of behaviour. The hysterical person is demanding, histrionic, attention-seeking, manipulative, emotionally volatile and exaggerates consistently. The psychopath is conscienceless, overtly selfish, and behaves with no regard for the rights of others. These people are, in general, mentally more vulnerable than average.

If psychiatric disorder occurs, it commonly takes a form that seems to be determined by the personality (what has been called the 'prepsychotic personality'). Thus the introvert who becomes psychotic will nearly always become schizophrenic while the extrovert will develop a manic-depressive disorder. The obsessional may develop a severe obsessional neurosis, the hysteric a somatoform disorder (physical manifestations of mental upset) and the sociopath a psychopathic personality disorder.

The determinants of personality

Personality is the result of the interaction between inherited characteristics and the total personal environment, especially that operating in early childhood. Studies on twins have shown how strong genetic influences can be. Identical twins brought up apart, retain close similarities in personality. Even allowing for coincidence, some cases of separated identical twins demonstrate this dramatically. Hereditary factors largely determine the bodily characteristics (the somatotype), the pattern of physical and mental development, and, to some extent, the way in which the endocrine (hormonal) system operates. All of these can affect the personality.

Environmental factors have a more immediately obvious effect on the personality, and it is relatively easy to see how the qualities of the parents and other members of the family can mould and determine the personality of the growing child. Other important environmental factors include the wider social milieu, educational and cultural influences, life experience generally, nutritional standards, and major events such as serious illness. Among the most important family influences making for a healthy and stable personality are: freely expressed love and affection; unequivocally expressed standards of conduct; an unshaken sense of physical and emotional security; a consistent pattern of attitude and ethics by parents; ample and rich stimulation of all the senses; and freedom to express individual responses within the constraints of the rules.

The origins of the psychopathic personality disorder are often only too obvious. These unfortunate people have almost always had grossly deprived childhoods with little or no show of affection and often a pattern of early conditioning that could hardly fail to induce a criminal outlook. Material family circumstances seem almost irrelevant and psychopaths arise from every level of society.

Personality disorders

These must be distinguished from psychotic disease. There are four broad classes of personality disorder: schizoid personality disorder; obsessional personality disorder; hysterical personality disorder; and sociopathic or psychopathic personality disorder.

People suffering from a schizoid personality disorder may be severely disabled by being so shut in on themselves as to be socially isolated. However, they have full insight into their condition and are neither hallucinated nor are they deluded. Such people are concerned, even preoccupied, with the things of the mind and with material things, almost to the exclusion of interest in other human beings. Although often highly creative, artistic and intelligent, such people may be aloof and withdrawn, cut off from emotional contacts, easily offended and defensive.

People with obsessional personality disorder can also be seriously disabled. Such people may be unable to start work until they have arranged their desks in a particular way, may be unable to pass a colleague's telephone with a twisted cord without untwisting it; may become acutely uncomfortable and unable to listen, for instance, in a job interview, if the interviewer's security badge is at an angle; and may be outraged by the untidiness or disorderliness of others. A degree of obsessiveness can be an asset in many occupations and can add to efficiency, but an obsessional personality disorder often gives rise to conflict between spouses and close associates, and may cause serious trouble in conditions of stress when urgent action may be needed. Such a disorder commonly overlaps with the obsessive-compulsive neurosis in which anxiety is dealt with by the establishment of rituals. The chief distinction between this condition and an obsessional-compulsive neurosis is that, in the latter condition, the affected person is compelled to behave in a manner he or she recognizes to be irrational. Only about one third of the people who develop this neurosis have a previous obsessional personality.

People with a hysterical personality disorder express their feelings in exaggerated terms. They may be highly manipulative, and may use emotion and drama to achieve their aims. Anger is a prominent feature. Relationships tend to be superficial and usually brief as these people find it very difficult to form close attachments. They commonly complain of illness. Pain is always 'excruciating'; headaches are always 'migraine'; they do not vomit but are 'violently sick'. Some people with hysterical personality disorder simulate apparently serious illnesses by deliberate deception – putting sugar in a urine sample or damaging the skin to produce an unusual rash.

perspiration

Such 'factitious' illness is an attempt to arouse interest and concern. Extreme cases may amount to the Münchausen syndrome in which convincing but fabricated case histories are given, usually reproduced from medical textbooks, so that surgery is performed.

The sociopathic or psychopathic personality disorder manifests itself by repeated acts of criminal behaviour, usually dating back to childhood. The people concerned act as if they are quite unaware of the difference between accepted standards of right and wrong and appear indifferent to the effect their actions have on others. There is usually a rationalization of the conduct but this is superficial and seldom convincing. There is complete insight into the motives for conduct and no question of delusions or hallucinations. Psychopaths are fully aware of the possible or probable consequences of their actions and may be persuaded to amend behaviour, but only to avoid punishment.

perspiration sweating.

pes planus FLAT FOOT.

pessary 1 a device, often ring-shaped, that is placed in the vagina to support the womb or other pelvic organs.
2 a vehicle for medication that is placed in the vagina. Medicated pessaries are often made of cocoa butter which melts under body heat. They contain drugs to treat vaginal disorders, such as thrush (candidiasis) or trichomoniasis, or spermicides for contraceptive purposes.

petechiae tiny, flat red or purple spots in the skin or mucous membranes caused by bleeding from small blood vessels. Petechiae are a feature of the bleeding disorder PURPURA.

petrosal pertaining to, or near, the inner part of the temporal bone (the petrous portion) that surrounds the inner ear. From the Latin *petrosus*, rocky.

PET scanning positron emission tomography. This is a diagnostic imaging technique based on the detection of gamma rays produced by the annihilation of positively charged electrons (positrons) emitted by specially prepared radioactive substances that have been injected intravenously. Substances labelled with oxygen-15, fluorine-18, carbon-11 or nitrogen-13 are most commonly used. PET scanning provides uniquely valuable images of tissues showing local metabolic activity, especially in the brain, the rate of glucose and oxygen consumption at various sites, blood flow, neurotransmitter activity and the fate of drugs. Positron-emitting substances have a very short half-life and must be prepared on site in a cyclotron. This limits the application of the technique.

-pexy *combining form denoting* a fixing, securing or making fast.

pH an expression, widely used in medicine, of the acidity or alkalinity of a solution. pH is the logarithm to the base 10 of the concentration of free hydrogen ions in moles per litre, expressed as a positive number. The pH scale ranges from 0 to 14. Neutrality is 7. Figures below 7 indicate acidity, increasing towards zero; figures rising above 7 indicate increasing alkalinity. The pH of body fluids, in health, is accurately maintained between about 7.3–7.5. Below this range the condition of acidosis exists; above it, alkalosis. Both are dangerous.

-phage *suffix denoting* one that eats, as in MACROPHAGE.

phage a BACTERIOPHAGE.

phagocyte an AMOEBOID cell of the immune system that responds to contact with a foreign object, such as a bacterium, by surrounding, engulfing and digesting it. Phagocytes occur widely throughout the body wherever they are likely to be required. Some wander freely throughout the tissues. They include macrophages and neutrophil polymorphonuclear leukocytes ('polymorphs'). From the Greek *phago*, eating and *kutos*, a hollow or receptacle.

phagocytosis the envelopment and destruction of bacteria or other foreign bodies by PHAGOCYTES.

phagosome the vacuole, formed within a PHAGOCYTE, that surrounds material that has been taken up by cell membrane invagination. A phagosome may fuse with a lysosome to provide digestive enzyme.

phalanges the small bones of the fingers and toes. Fingers have three phalanges; the thumbs and big toes have two.

phalanx a finger or toe bone. Plural PHALANGES.

phallic pertaining to the PHALLUS.

phallus 1 the penis.
2 any object symbolizing the penis.

phantom limb a powerful sense that a limb which has been amputated is still present. The effect is due to nerve impulses arising in the cut nerves in the stump. These can only be interpreted as coming from the original limb.

pharmacogenomics the recognition of the fact and significance of human genetic variability in relation to drug action and its application to medical treatment. Genetic variability affects drug absorption, distribution, metabolism and excretion. As a result, many drugs which work very well in some people work poorly in others. At present, the best doctors can do is to prescribe a drug and then wait and see how well the patient responds. Fuller knowledge of pharmacogenomics will, in the future, allow more reliable treatment and avoid expensive waste.

pharmacology the science of DRUGS. Pharmacology is concerned with the origins, isolation, purification, chemical structure and synthesis, assay, effects, uses, side effects, relative effectiveness of drugs and the influence of genetic factors on drug action. It thus includes, among other disciplines, GENETICS, ORGANIC CHEMISTRY, PHARMACOKINETICS, THERAPEUTICS and TOXICOLOGY.

pharmacopoeia a book, known as a formulary, that lists and describes the characteristics of drugs used in medicine. The major pharmacopoeias, such as the *British Pharmacopoeia* (BP), the *Pharmaceutical Codex* and the *Extra Pharmacopoeia*, are large volumes dealing with all important drugs and offering a semiofficial guide to pharmacists, doctors and others as to their uses and disadvantages. A revised version of *The British National Formulary*, an 800-page paperback book, is published every six months by the British Medical Association and the Royal Pharmaceutical Society of Great Britain. It is also available on the Internet.

pharmacy 1 the process of preparing, compounding and dispensing drugs, usually to the prescription of a doctor.
2 a place where these activities are performed.

pharynx the common passage to the gullet (OESOPHAGUS) and the windpipe (TRACHEA) from the back of the mouth and the back of the nose. The pharynx is a muscular tube lined with MUCOUS MEMBRANE, and consists of the NASOPHARYNX, the OROPHARYNX and the LARYNGOPHARYNX.

-phasia *suffix denoting* a speech disorder, as in aphasia.

phase shift an alteration in the time relationship between the peaks (or troughs) of two different periodic functions. Thus, two circadian rhythms may suffer a phase shift as in jet lag.

phenocopy a PHENOTYPE or disorder caused by non-genetic factors that mimics, and may be mistaken for, a genetic disorder.

phenotype 1 the observable appearance of an organism which is the result of the interaction of its genetic constitution and its subsequent environmental experience.
2 any identifiable structural or functional feature of an organism. Compare GENOTYPE.

pheromone an odorous body secretion that affects the behaviour of other individuals of the same species, acting as a sex attractant or in other ways. Pheromones are important in many animal species but, until recently, were thought to be unimportant in humans. It has now been shown, however, that the timing of ovulation in women can be controlled by pheromones from the armpit. This is believed to be the explanation of the fact that women living together will frequently develop synchronized menstrual cycles.

Philadelphia chromosome an acquired chromosomal defect in which the long arm of chromosome 22 is deleted and attached (translocated) to another chromosome, usually number 9. Clones of cells with this defect cause chronic myeloid leukaemia.

-philia *suffix denoting* a tendency or attraction towards.

-philiac *suffix denoting* one that has a tendency toward.

phimosis tightness of the FORESKIN (prepuce) of such degree as to prevent retraction.

phleb-, phlebo- *combining form denoting* vein.

phlebotomy cutting into, or puncture of, a vein, usually for the purpose of removing blood.

-phobia *combining form denoting* abnormal fear of.

phobia an inappropriate, irrational or excessive fear of a particular object or situation, that interferes with normal life. Phobias may relate to many objects including reptiles or insects, open spaces or public places (agoraphobia), crowds, public speaking, performing or even eating in public and using public toilets. Exposure causes intense anxiety and sometimes a PANIC ATTACK. Treatment is by cognitive behaviour therapy.

phobophobia fear of developing a phobia. The term is in the dictionaries but, like many other alleged phobias, was probably invented by a lexicographer.

phocomelia a major, congenital limb defect featuring absence of all long bones so that the hands or feet are attached directly to the trunk and resemble flippers. Spontaneous cases of phocomelia are rare but the condition occurred in many children whose mothers were given thalidomide early in their pregnancy.

phonation the production of sounds in speech and singing.

phonetics the branch of linguistics concerned with the study of the speech sounds (phonemes) of a language and their classification and representation.

phono- *combining form denoting* sound or voice.

phosphatase an enzyme that removes phosphate groups from a molecule.

phosphates salts or esters of phosphoric acids or salts or esters containing a phosphorus atom. Most of the body phosphate is combined with calcium in the bones and teeth but an important fraction occurs in the blood and in all cells. Phosphates help to stabilize the pH of the blood and other tissue fluids and occur in the ADENOSINE TRIPHOSPHATE (ATP) which stores and releases energy in cells. They are also important groups in the structure of DNA and RNA NUCLEOTIDES.

phosphodiester bond the chemical linkages that join up the sugar, base and phosphate NUCLEOTIDES of DNA and RNA into polynucleotide strands. The subunits of the strand are triphosphate nucleosides, but when a number of these join up (polymerize) under the action of the enzyme DNA polymerase, two of the phosphates are cleaved off leaving only one phosphorous atom between each pair of adjacent sugar molecules. The two ester (diester) bonds in each linkage are Carbon–Oxygen–Phosphorus from the 5′-carbon on one sugar and Carbon–Oxygen–Phosphorus from the 3′-carbon on the next. The hydroxyl (–OH) on the 3′-carbon is also lost.

phospholipids a class of lipids containing a platform which may be glycerol or sphingosine, to which are attached one or more fatty acids and a phosphate group with an alcohol linked to it. Phosphoglycerides resemble triglycerides except that one of the fatty acids is replaced by the phosphate-alcohol group. The fatty acid elements are hydrophobic while the rest of the molecule is hydrophilic. The effect of this is that phospholipids readily form membranous structures in water and are abundant in all biological membranes.

phosphorylation the addition of a phosphate group to an organic molecule.

phot-, photo- *combining form denoting* light.

photic pertaining to light.

photomicrograph a photograph made through a microscope.

photophobia undue intolerance to light. This is a feature of certain eye disorders, especially corneal abrasions or ulcers, IRIDOCYCLITIS and congenital glaucoma. It also occurs in MIGRAINE and MENINGITIS.

photopsia visual sensations of light originating in the eyes or the nervous system and due to stimulation of the retinas or visual nerve pathways other than by light. Mechanical stimuli from traction on the retina by the VITREOUS BODY can cause photopsia. They are examples of entoptic phenomena. Also known as phosphenes.

photoreceptors specialized cells, such as the rods and cones of the RETINA, that originate nerve impulses when stimulated by light.

photosensitivity a state in which an abnormal reaction occurs on exposure to sunlight. The commonest reaction is a skin rash occurring as a combined effect of light and some substance that has been eaten or applied to the skin. Such substances are called photosensitizers and include various drugs, plant derivatives, dyes or other chemicals. Avoidance of either or both elements is important.

photosynthesis the vital process in nature by which plants are able to use the energy in sunlight to form organic compounds, such as sugars, from atmospheric carbon dioxide and water. This fixation of carbon is the ultimate process by which all living things are directly or indirectly nourished.

-phrenia *combining form denoting* the mind or the DIAPHRAGM.

phrenic pertaining to the mind or to the DIAPHRAGM.

phrenic nerves the two main nerves supplying the DIAPHRAGM and thus controlling breathing. Each phrenic nerve arises from the 3rd, 4th, and 5th cervical spinal nerves and passes down through the chest to supply one side of the diaphragm.

phrenology a theory, taken seriously for a time in the 18th century, that human characteristics were reflected in the relative growth of parts of the brain and that these could be detected by palpation of the skull bumps which, it was claimed, conformed to the shape of the brain.

phthisis wasting or consumption. An outmoded term for pulmonary TUBERCULOSIS.

phylogeny the evolutionary history ending in a species.

phylum the taxonomic group below kingdom and above class.

physical anthropology the study of the evolutionary biology of the species *Homo sapiens*. Its main purpose is to try to establish the process by which present-day man and his man-like ancestors (the hominids) evolved from early primates. Physical anthropology is an active science, constantly advancing as new fossil discoveries are made and advanced techniques are applied. Contemporary scientists use CARBON DATING methods to establish relative ages; biochemical studies to compare the physiology of man with other primates; detailed physical measurements of all parts of the body (anthropometry) to quantify comparisons; and genetic analysis especially the study of mitochondrial DNA – which is relatively unchanged through the generations – to help to determine human origins. Genetics have also been widely used in the studies of the more recent diversification of man. Various genetic markers such as the range of blood groups, tissue types and various gene mutations have proved valuable in tracing historic movement and intermarriage of human groups.

physical medicine a branch of medicine concerned with the treatment and rehabilitation of people disabled by injury or illness. Apart from specialist doctors, the discipline involves physiotherapists, occupational therapists and speech therapists.

physician 1 a person qualified and licensed to practise medicine.
2 a doctor specializing in a medical, as distinct from a surgical, speciality.

physician-assisted suicide see DOCTOR-ASSISTED SUICIDE.

physiology the study of the functioning of living organisms, especially the human organism. Physiology includes BIOCHEMISTRY and molecular cell biology, but these are such large disciplines that they are followed as separate specialities. Together with ANATOMY and PATHOLOGY, physiology is the basis of medical science.

physiotherapist a person engaged in the practice of PHYSIOTHERAPY.

physiotherapy the treatment discipline ancillary to medicine that uses physical methods such as active or passive exercises, gymnastics, weight-lifting, heat treatment, massage, ultrasound, short-wave diathermy and hydrotherapy. Physiotherapists aim to restore the maximum possible degree of function to any disabled part of the body and are also much concerned with patient motivation. See also PHYSICAL MEDICINE.

phyto- *combining form* meaning plant or vegetation.

phyto-oestrogens a range of naturally-occurring oestrogen-like substances derived from plants and present in the diet. Phyto-oestrogens have anticancer properties. Research has shown, for instance, that the risk of breast cancer is substantially lower in women with high levels of phyto-oestrogens, as measured by urinary secretion of these substances, than in those with a low intake. The main phyto-oestrogens are isoflavonoids (which are high in soya products) and lignans (which are high in whole grains, fruit, vegetables and berries). It has been found that the phyto-oestrogen genistein, found in soya, binds preferentially to the oestrogen receptors occurring mainly in the cardiovascular system rather than in those in the breast and uterus. This has encouraged the hope that phyto-oestrogens might be valuable in controlling atherosclerosis and osteoporosis without increasing the risk of cancer.

phytophilia a healthy dietary preference leading to an emphasis on vegetables and fruit rather than animal and dairy products.

Piaget, Jean the Swiss psychologist Jean Piaget (1896–1980), professor of child psychology at the University of Geneva, has been one of the most influential figures in the field of child intellectual, logical and perceptual development. Piaget's original intention was to unify biology and logic. To this end he began to study the development of thought in children so as to throw light on human thought processes generally. But he became so fascinated by what he found in his studies of children that he devoted the rest of his life to the subject. Piaget decided that mental growth was the result of the interaction between environmental influences and an innate and developing mental structure. Input of new information from the environment challenges the child's concept of the world (paradigm) so that this needs to be modified or expanded, and this process occurs repeatedly. Such a concept implies the existence of an innate form of logic. Intelligence is developed by a process of progressive refinement of the logical system as a result of experience.

Piaget defined four stages in development, for which he set specific age limits, and taught that each one had to be fully assimilated before the child could pass on to the next. After the first sensorimotor stage comes the preoperational stage, from two to seven years. In this stage the child struggles with the difference between fantasy and reality and, in particular, with the problem of causality. Many misinterpretations are made of the causal link between the child's own behaviour and major events in its life, such as a death or the divorce of parents. Magical thinking is common. From the ages of about seven to twelve the child gradually acquires the capacity for logical thought about objects, but it is not until the period from twelve onwards that logical manipulation of verbal propositions becomes possible.

Piaget was an exceptionally prolific writer and published many books and articles, some of which were written primarily as an aid to his own thought and inspiration. His most important works include

The Language and Thought of the Child (1923), *The Child's Concept of the World* (1926) and *The Child's Concept of Physical Reality* (1926).

pia mater the delicate, innermost layer of the MENINGES. The pia follows the convolutions of the brain, dipping into the SULCI.

pica a craving for, and the eating of, unsuitable material such as sand, earth, chalk or coal. Pica is commoner during pregnancy than at other times and may occur in cases of iron-deficiency anaemia. Also known as paroxia.

pidgin a language derived from parts of two or more other languages to allow communication between the speakers of other languages. Unlike creoles, pidgins are never the mother tongue of any society.

piebaldism 1 VITILIGO.
2 partial ALBINISM.
3 a rare hereditary disease in which there are patchy areas of skin with no pigment.

pigeon chest a deformity in which the chest is peaked forward, seen in people who have suffered from severe asthma from infancy. Also known as pectus carinatum.

pigeon toes a mainly cosmetic defect in which the leg or foot is rotated inwards.

pigmentation coloration of any part of the body, especially the skin. Normal pigmentation of skin, hair and eyes is occasioned by the presence of melanin – a brown or black pigment produced by cells called melanocytes. Abnormal pigmentation may occur from local or general loss of melanin. Albinism is caused by a general deficiency of melanin.

piles the common name for HAEMORRHOIDS.

pilo- *combining form denoting* hair.

piloerection hair 'standing on end' under the influence of the tiny pilomotor muscles attached to the hair follicles in the skin.

pilose hairy.

pilosebaceous pertaining to the hair follicles and their associated sebaceous glands.

pimple a PUSTULE or PAPULE commonly occurring in adolescents and young adults suffering from ACNE.

Pincus, Gregory the effect of the female sex hormones on egg production (ovulation) was known as early as the 1920s, but it was not until 1956 that the American physiologist and endocrinologist Gregory Pincus (1903–67), working in the Worcester Foundation for Experimental Biology, Boston, reported that ovulation could be temporarily stopped by means of progesterone and allied hormones. Working in collaboration with gynaecologists he developed an oestrogen–progestogen combination that could be taken by mouth and that was virtually one hundred per cent effective in preventing pregnancy. This was the contraceptive pill, approved by the American Food and Drug Administration in 1960 and first put on the market in Britain in 1962. Since then it has been used by many millions of women throughout the world and has had a considerable effect on attitudes to sexual behaviour.

pineal gland a tiny, cone-shaped structure within the brain, whose sole function appears to be the secretion of the hormone melatonin. The amount of hormone secreted varies over a 24-hour cycle, being greatest at night. Control over this secretion is possibly exerted through nerve pathways from the retina in the eye; a high light level seems to inhibit secretion. The exact function of melatonin is not understood, but it may help to synchronize circadian (24-hour) or other biorhythms. The pineal gland is situated deep within the brain, just below the back part of the corpus callosum (the band of nerve fibres that connects the two halves of the cerebrum).

pinna the visible external ear. The pinna consists of a skin flap on a skeleton of cartilage. It has comparatively little effect on the acuity of hearing. Also known as the auricle.

pinocytosis the process in which cells engulf fluid to form tiny clear spherical containers (vacuoles) which then move through the cell cytoplasm, sometimes acting as scavenging vehicles to be discarded through another part of the cell membrane. Extracellular fluid with dissolved molecules may be moved intracellularly by pinocytosis.

pisiform 1 resembling a pea.
2 the pea-shaped bone of the wrist near the lower end of the ULNA.

pitting oedema OEDEMA that allows visible but temporary indenting of the skin by finger pressure.

pituitary dwarfism stunted growth due to deficiency of pituitary growth hormone during the early years of life. This is now seldom allowed to occur as treatment with growth hormone is readily available.

pituitary gland the central controlling gland in the ENDOCRINE system. The pituitary is a pea-sized gland that hangs by a stalk from the underside of the brain and rests in a central bony hollow on the floor of the skull. The pituitary stalk emerges immediately under the HYPOTHALAMUS and there are numerous connections, both nervous and hormonal, between the hypothalamus and the gland. The pituitary releases many hormones that regulate and control the activities of other endocrine glands as well as many body processes. Hormonal feedback information from the various endocrine glands to the hypothalamus and the pituitary ensures that the pituitary is able to perform its control function effectively. See also ADRENOCORTICOTROPIC HORMONE, ANTIDIURETIC HORMONE, FOLLICLE-STIMULATING HORMONE, LUTEINIZING HORMONE, MELANOCYTE-STIMULATING HORMONE, OXYTOCIN, PROLACTIN, and THYROID-STIMULATING HORMONE,

placebo 1 a pharmacologically inactive substance made up in a form apparently identical to an active drug that is under trial. Both the placebo and the active drug are given, but the subjects are unaware which is which. This is done for the purpose of eliminating effects due to purely psychological causes.
2 a harmless preparation prescribed to satisfy a patient who does not require active medication. From the Latin *placere*, to please. See also PLACEBO EFFECT.

placebo effect the often significant, but usually temporary, alteration in a patient's condition, following the exhibition of a drug or other form of treatment, which is due to the patient's expectations or to other unexplained psychological effects, rather than to any direct physiological or pharmacological action of the drug or treatment. The placebo effect has done much to foster the reputation of many valueless forms of treatment.

placenta the part of the early developing EMBRYO, that differentiates to form an organ attached to the lining of the womb and provides a functional linkage between the blood supplies of the mother and the fetus. This allows for the passage of oxygen and nutrients from the mother to the fetus. The placenta is connected to the fetus by the UMBILICAL CORD and is discharged from the womb after the birth of the baby. Also known as afterbirth.

placental lactogen a hormone produced by the placenta with effects similar to those of pituitary prolactin and somatotropin (growth hormone).

placoid plate-like.

plagiocephaly a skull deformity in which the major dimension is on a diagonal because of premature closure of the SUTURES on one side.

planes of the body imaginary, two-dimensional surfaces used for anatomical description. With the body supposed to be standing upright, the sagittal plane is a vertical plane that cuts the body into right and left halves; the coronal plane is a vertical plane that separates the front half of the body from the back; the transverse plane cuts the body horizontally at any level.

plantar pertaining to the sole of the foot. Compare PALMAR.

plasma 1 the fluid in which the blood cells are suspended.
2 blood from which all cells have been removed. Plasma contains proteins, electrolytes and various nutrients and is capable of clotting.

plasma cell one of a number of large oval cells, cloned from a selected B LYMPHOCYTE, that synthesize large quantities of the required IMMUNOGLOBULIN (antibody).

plasma membrane the membrane that defines the boundary of a cell.

plasma proteins the proteins present in blood plasma. They include albumin, the range of IMMUNOGLOBULINS and prothrombin and fibrinogen necessary for blood clotting. Plasma proteins help to maintain the

OSMOTIC PRESSURE of the blood and hence the blood volume.

plasmid a ring-shaped, double-stranded, piece of DNA in bacterial cells that contains genes extra to those in the chromosome. Plasmid genes code for characteristics such as toxin production and the factors that cause antibiotic resistance. Plasmids are convenient vehicles for the introduction of new genes into organisms in recombinant DNA technology (genetic engineering).

plasmin a protein-splitting enzyme (proteinase) in the blood that dissolves FIBRIN clots.

plasminogen the precursor to PLASMIN.

plasminogen activation system the cascade that regulates the conversion of plasminogen to the proteinase plasmin. It involves a urokinase-type plasminogen activator, a tissue-type activator and at least two inhibitors.

-plastic *combining form denoting* forming, developing or growing.

plasticity the ability of nervous system to be functionally modified as a result of repetitive activation. Thus the formation of functional links between the retina and the visual cortex in early infancy require the exercise of the visual function. If for any reason one eye is not used during the first six or seven years of life (the period of plasticity) that eye will remain effectively blind.

platelet a fragment of the CYTOPLASM of a MEGAKARYOCYTE 2–4 mm in diameter. Each megakaryocyte produces 1000–3000 platelets, which are present in large numbers in the blood – 50,000–300,000 per cu. mm. Platelets survive for about 10 days and play an essential part in blood clotting. Platelet plasma membranes contain a range of glycoproteins by means of which they bind to different materials including collagen, fibrinogen and von Willebrand factor. Platelets are by no means the passive tissue fragments they were formerly thought to be. They carry many granules and a canalicular system by which the granules are released. They also have a dense tubular membrane system in which prostaglandins and thromboxanes are synthesized. Platelet granules contain heparin-neutralizing factor, von Willebrand factor, smooth muscle growth factor and fibrinogen. Deficiency of platelets is known as thrombocytopenia.

platelet activating factor a factor which, like PROSTAGLANDINS and LEUKOTRIENES, is released from cell membranes. It is a potent cause of INFLAMMATION and an activator of several cell types of the immune system. It is rapidly destroyed and its effect is local.

platelet glycoprotein receptor one of a range of receptors on blood platelets that bind glycoproteins. Platelet glycoprotein receptor IIb/IIIa binds fibrinogen, thereby promoting platelet aggregation and the formation of a blood clot within a blood vessel. This receptor can be blocked by various substances such as HIRUDIN or a number of synthetic drugs such as tirofiban.

platy- *combining form denoting* flat or broad.

platysma the broad, flat muscle lying immediately under the skin of the neck, from the shoulders to the point of the chin. The action of the platysma is to tighten the skin of the neck, pull down the corners of the mouth and lower the jaw.

pleasure any enjoyable or agreeable emotion or sensation, to the pursuit of which most people, who are free to do so, devote their lives.

pleasure principle the tendency to seek immediate gratification of instinctual desires and to avoid pain. In the Freudian model, this primitive id reaction is gradually modified by the reality principle, a more mature ego function. See also FREUDIAN THEORY.

Pleistocene epoch the space of time from about 1.7 million years ago until about 10,000 years ago. During this epoch repeated ice ages occurred, possibly as many as 30, with four or five expansion of kilometre-thick polar ice caps as far south as what is now London, Berlin and New York. Many mammals existed during this period that are now extinct. Three species of humans, *Homo erectus*, *Homo habilis* and *Homo sapiens*, existed during the Pleistocene epoch.

pleo- *combining form denoting* excessive or multiple.

pleomorphism taking different physical forms. Also known as polymorphism.

plethoric having a ruddy complexion from widening of blood vessels under the skin or, rarely, from POLYCYTHAEMIA.

pleura the thin, double-layered membrane that separated the lungs from the inside of the chest wall. The inner layer is attached to the lung and the outer to the inside of the chest cavity. A film of fluid between the two layers provides lubrication to allow smooth movement during breathing.

pleural disorders these include pleural effusion, pleurisy and pneumothorax.

plexus any interlacing network, as of nerves, blood or lymph vessels. The solar plexus is a network of AUTONOMIC nerve fibres lying on the abdominal AORTA.

-ploid *combining form denoting* some multiple of a chromosome set, as in haploid, diploid, triploid.

ploidy the number of copies of the set of chromosomes in a cell. A normal diploid cell with two copies has a ploidy of two; a haploid cell with one copy, such as an ovum or a spermatozoon, has a ploidy of one.

pluripotentiality the ability of stem cells to differentiate into almost all cells that arise from the three germ layers but not to form placenta or supporting structures. See also UNIPOTENTIALITY.

PMS *abbrev. for* PREMENSTRUAL SYNDROME.

PMT *abbrev. for* premenstrual tension. See PREMENSTRUAL SYNDROME.

pneumo- *prefix denoting* the lung.

pneumonia inflammation of the lower air passages (bronchioles) and air sacs (alveoli) of the lungs due to contact with irritant or toxic material or to infection with any of a wide spectrum of micro-organisms, including viruses, bacteria, fungi and microscopic parasites. Pneumonia is commonest at the extremes of life and is a common terminal event in the elderly.

PO_2 the concentration of OXYGEN in the blood. This commonly measured parameter is an important indicator of the efficiency with which oxygen is transferred from atmosphere to blood. When PO_2 drops, respiration is automatically stimulated.

pod-, podo- *combining form denoting* the foot.

podiatry chiropody. Currently the usage is mainly American.

poikilo- *combining form denoting* varied or variegated.

point mutation the replacement of one NUCLEOTIDE with another. This need not necessarily cause any change in the protein produced by the affected gene because 18 of the 20 amino acids have more than one coding triplet of base pairs (codon). Glycine, for instance, is coded for by GGA, GGU, GGG and GGC. This redundancy feature of the genetic code arises because the four bases, taken three at a time, allow 64 triplets to code the 20 amino acids and the three stop codons. It is called degeneracy.

poison any substance capable, in small amounts, of damaging the structure or function of living organisms or of causing their death. The virulence of a poison is assessed by the smallness of the dose required to produce its effect and by the severity of the effect. Many of the most poisonous substances act by interfering with fundamental cell enzyme systems. Bacterial toxins are amongst the most poisonous substances known.

pokeweed mitogen an agent, useful for promoting MITOSIS especially in B lymphocytes, derived from the North American plant, *Phytolacca americana.*

polar of a molecule or chemical group whose electric charges are separated so that one end is positive and one negative (forming a dipole). Cell plasma membranes are made of a double layer of phosopholipid molecules each containing a polar head group with a strong affinity for water (hydrophilic) and a non-polar hydrocarbon tail that avoids water (hydrophobic). The polar head groups in both layers are oriented outwards in the membrane so as to form both free surfaces. So, in this context, the term 'polar' is used to refer to a hydrophilic chemical group, and 'non-polar' refers to a hydrophobic group.

polar covalent bond a chemical bond in which electrons are shared unequally between the two linked atoms. The effect is that one end of the resulting molecule has a slight negative charge while the other end is slightly positive.

poly- *combining form denoting* many, much or excessive.

polyandry marriage to more than one man, simultaneously.

polycoria more than one pupillary opening in a single iris.

polycythaemia an abnormal increase in the number of red blood cells as a result of increased red cell production by the bone marrow. This occurs naturally in people living at high altitudes (secondary polycythaemia) but may occur for no apparent reason (polycythaemia vera) causing flushed skin, headaches, high blood pressure (HYPERTENSION) and blurred vision. STROKE is a common complication.

polydactyly the possession of more than the normal number of fingers or toes.

polydipsia excessive thirst leading to excessive fluid intake. This is a feature of untreated severe DIABETES MELLITUS and of DIABETES INSIPIDUS.

polygamy marriage to more than one spouse, simultaneously.

polygyny marriage to more than one woman, simultaneously.

polymastia having more than two breasts.

polymer a chain molecule made up of repetitions of smaller chemical units or molecules called monomers. Polysaccharides, for instance, are long chains made up of repeated units of simpler monosaccharide sugars. Proteins are polymers of AMINO ACIDS. Polymerization is the process of causing many similar or identical small chemical groups to link up to form a long chain. From Greek, *poly*, many and *meros*, a part.

polymerase any enzyme that promotes the linkage of a number of similar or identical chemical subunits into repetitive long-chain molecules (polymers), especially of NUCLEOTIDES to form DNA or RNA. Derivation as in POLYMER with the -ase suffix denoting an enzyme.

polymerase chain reaction an important technique for rapidly producing large numbers of copies of any required sequence of DNA. DNA is separated by heat into its two strands, small molecules called primers are attached to the sequences at either end of the target sequence, and an enzyme, DNA polymerase, is used to build a new strand of the section between the primers. This becomes a template for the production of further strands and in twenty cycles a million copies are made. The polymerase chain reaction is one of the most powerful techniques currently in use in biological science. The American biochemist inventor of the process, Karry B. Mullis, was awarded the Nobel Prize for Chemistry in 1993.

polymerization the formation of POLYMERS from monomers.

polymorph the common abbreviated term for polymorphonuclear (many-shaped nucleus) neutrophil leukocyte, the small phagocytic white cell of the immune system present in enormous numbers in the blood and the tissues. Polymorphs are amoeboid and highly motile scavengers of foreign material especially antigen-bearing bacteria and other antigenic material to which antibodies have been bound. See also POLYMORPHIC.

polymorphism 1 occurring in many different shapes.

2 in genetics, the existence of different ALLELES of the same gene in different genomes. See also RESTRICTION FRAGMENT LENGTH POLYMORPHISM.

polymorphic occurring in a variety of shapes, as in the case of the nucleus of the neutrophil polymorphonuclear leukocyte (POLYMORPH).

polymorphonuclear having a lobed nucleus of varying lobe number or shape. The term is used of three classes of LEUKOCYTES, neutrophil, basophil and eosinophil.

polyopia the perception of multiple images of a single object.

polypeptide a molecule consisting of a chain of AMINO ACIDS linked together by peptide bonds. A POLYMER of amino acids that may form part of a protein molecule. Polypeptides link together to form proteins.

polyphagia the state of having an excessive or pathological desire to eat.

polypharmacy a mildly facetious term for the generally disapproved practice of prescribing several different drugs to one person at the same time. Polypharmacy increases the risk of unwanted side effects and of dangerous interactions between different drugs.

polyploid having more than twice the normal HAPLOID number of chromosomes. See also DIPLOID.

polyprotein a large gene product that is split into two or more independent proteins.

polyribosome a cluster or string of ribosomes held together by a molecule of messenger RNA.

polysaccharide a POLYMER of linked monosaccharide molecules. Thus, glycogen is a polysaccharide of glucose units. Like proteins, polysaccharides may have molecular weights of several million. The polymer is often branched. They include glycogen, starch and cellulose.

polyubiquitination the process of tagging proteins destined for degradation by attaching a chain of UBIQUITIN molecules to them. Such tagged proteins are then taken up by PROTEASOMES and enzymatically degraded to peptides.

polyunsaturated pertaining to long chain carbon compounds, especially fatty acids, that contain more than one carbon to carbon double bonds (unsaturated bonds).

polyunsaturated fatty acid a fatty acid with more than one double bond linking two carbon atoms.

polyuria the formation of abnormally large quantities of urine. See also POLYDIPSIA.

pons 1 any anatomical structure joining two parts or bridging between them.
2 the middle part of the BRAINSTEM, lying below the cerebral peduncles of the midbrain and above the MEDULLA OBLONGATA. From the Latin *pons*, a bridge.

popliteal pertaining to the hollow surface behind the knee joint.

popliteus muscle a short diagonally placed muscle running from the outer side of the lower end of the thigh bone (femur) to the back of the upper part of the main lower leg bone (TIBIA). Its action is to rotate the femur on the tibia, or vice versa.

pore a tiny opening, especially opening in the skin through which sweat or sebaceous secretion (SEBUM) pass to the surface. Most of the sebaceous pores are also hair follicles.

portal pertaining to an entrance or gateway, especially to the porta hepatis, the fissure under the liver at which the PORTAL VEIN, the hepatic artery and the hepatic bile ducts pass through.

portal vein the large vein that carries blood from the intestines, the STOMACH, the lower end of the OESOPHAGUS and the SPLEEN into the liver. After a meal the portal vein contains large quantities of digested nutrients.

port-wine stain a flat, permanent, purple-red birthmark caused by a benign tumour of small skin blood vessels. A capillary haemangioma. Port-wine stains can be treated by skin grafting or with laser burns.

position effect the effect on the expression of a gene resulting from its translocation to a different part of the genome. A gene may, for instance, be inactivated if moved to a region that is permanently in a highly condensed condition (heterochromatin).

positive feedback the characteristic of any system with an output proportional to its input in which a portion of the output is fed back to the input in such a phase as to increase the input. The effect of this is rapidly, and sometimes dangerously, to increase the output. Compare NEGATIVE FEEDBACK.

positivism a school of philosophy that rejects value judgements, metaphysics and theology and holds that the only path to reliable knowledge is that of scientific observation and experiment.

post- *prefix denoting* after.

postcoital contraception the use of any measure to prevent pregnancy after sexual intercourse has occurred. Such measures are not strictly contraceptive as they are likely to act after conception has occurred. They include taking two high-dose contraceptive pills as soon as possible and then 12 hours later; the taking of a single dose of mifepristone; and the insertion of a copper-releasing IUCD. Mifepristone is now widely licensed outside China for postcoital contraception. Also known as emergency contraception or the 'morning after pill'.

posterior pertaining to the back of the body or a part. Dorsal. Compare ANTERIOR.

postganglionic pertaining to the nerve fibres whose cell bodies are in the ganglia of the AUTONOMIC NERVOUS SYSTEM, and to any synaptic connection with these fibres. The ganglia are collections of the bodies of neurons whose axons conduct impulses peripherally. Compare PREGANGLIONIC.

posthypnotic suggestion a command to a hypnotized person to perform a specified action after restoration from the hypnotic state.

postmaturity the state in which a pregnancy lasts longer than two weeks beyond the normal 40 weeks from the first day of the last menstrual period.

postmortem examination see AUTOPSY.

postnatal occurring subsequent to birth.

postoperative after surgery.

postpartum following childbirth.

postprandial after a meal.

post-traumatic stress disorder an anxiety disorder caused by the major personal stress of a serious or frightening event such as injury, assault, rape or exposure to warfare or a natural or transportational disaster. The reaction may be immediate or delayed for months. There are nightmares, insomnia, 'flash-backs' in which the causal event is vividly relived, a sense of isolation, guilt, irritability and loss of concentration. Emotions may be deadened or depression may develop. Most cases settle in time, but support and skilled counselling may be needed. Some persist for a lifetime.

postural hypotension a drop in the blood pressure caused by standing or by rising suddenly. The brain is temporarily deprived of a full blood supply. This is a common cause of fainting.

posture the relationship of different parts of the body to each other and to the vertical. In youth, posture is fully under voluntary control. Faulty posture tends to become permanent and may affect health as well as appearance.

potassium an important body mineral present in carefully controlled concentration. Potassium is necessary for normal heart rhythm, for the regulation of the body's water balance and for the conduction of nerve impulses and the contraction of muscles. Many diuretic drugs result in a loss of potassium from the body and this can be dangerous. Supplementary potassium is often included in the formulation of these preparations.

potato nose a popular term for the condition of RHINOPHYMA in which overgrowth of sebaceous tissue and blood vessels causes a bulbous deformity. A feature of some cases of ACNE ROSACEA.

potency 1 the ability of a man to obtain an erection and so perform sexual intercourse.
2 the strength of a drug based on its effectiveness to cause change.
3 the claimed increase in the power of homeopathic remedies (see HOMEOPATHY) with increasing dilution and shaking.

pothead a slang term for a habitual MARIJUANA smoker.

pouch of Douglas the space, lined with PERITONEUM, between the womb and the rectum. (James Douglas, 1675–1742, Scottish anatomist).

pragmatism 1 action determined by the need to respond to immediate necessity or to achieve a particular practical result, rather than by established policy or dogma.
2 the philosophic principle that the truth and meaning of an idea is entirely relative to its practical outcome.

prandial referring to a meal, or to the effects of a meal. Prandial insulin is insulin given in the attempt to mimic the response of endogenous insulin to food intake.

praxis 1 action.
2 the ability to perform skilled actions.

pre- *prefix denoting* before, preceding or prior to.

precocious puberty the onset of physical sexual maturity at an abnormally early age, arbitrarily set at prior to 6 years for girls and 7 years for boys. If there are signs of unusually rapid development of the secondary sexual characteristics or growth before 8 in girls or 9 in boys, precocious puberty should be suspected.

prediabetes a syndrome in severely obese children featuring impaired glucose tolerance, insulin resistance, fat accumulation within muscle cells and in the abdomen, and a high risk of diabetic complications from Type II diabetes.

predisposition a special susceptibility to a disease or disorder, as by the action of direct or indirect genetic or environmental factors.

prefrontal pertaining to the front part of the frontal lobe of the brain.

preganglionic pertaining to the nerves of the spinal cord that connect to the chains of ganglia of the AUTONOMIC NERVOUS SYSTEM lying alongside the spine.

pregnancy the state of a woman during the period from fertilization of an ovum (conception) to the birth of a baby or termination by ABORTION.

pregnancy disorders these include abortion, antepartum haemorrhage, miscarriage, oligohydramnios, polyhydramnios, pre-eclampsia, prematurity and rhesus incompatibility.

pregnancy tests tests on urine or blood for the presence of human chorionic gonadotrophin, a hormone produced by the PLACENTA and thus occurring only during pregnancy. Urine pregnancy tests are about 97 per cent accurate if positive and about 80 per cent accurate if negative.

preimplantation genetic diagnosis the use of genetic analysis in the course of vitro fertilization to ensure that a baby does not possess a known genetic defect of either parent. After analysis of the embryos formed, only those free of defect are implanted in the mother's womb.

premature ejaculation the occurrence of a male orgasm at such an early stage in sexual intercourse as to deprive both partners of satisfaction. Premature ejaculation is usually due to sexual inexperience or over-excitement.

prematurity birth of a baby before the 37th week of pregnancy gestation or of a baby weighing less than 2500 g, regardless of the gestation period.

premenstrual dysphoria disorder a syndrome affecting some women during the period from about a week before the start of menstruation to about three days after and featuring marked swings of mood, depression and self-critical thoughts, anger, irritability, crying, fatigue, increased appetite and sometimes a craving for high-carbohydrate foods. The disorder is often severe enough to disrupt normal social activity.

premenstrual syndrome a group of physical and emotional symptoms that may affect women during the week or two before the start of each menstrual period. The cause of the syndrome remains unclear. It features irritability, depression, fatigue, tension, headache, breast tenderness, a sense of abdominal fullness and pain, fluid retention and backache. Various treatments may have to be tried before relief is obtained.

premenstrual tension see PREMENSTRUAL SYNDROME.

premolars the four pairs of permanent grinding teeth situated on either side of each jaw between the CANINES and the MOLARS.

prenatal existing or occurring before birth.

prepuce see FORESKIN.

presbyacusis progressive loss of hearing for the higher frequencies that occurs with age. Presbyacusis is a sensorineural type of deafness probably caused by several factors including inner ear trauma from exposure to noise. From the Greek *presbys*, an old man, and *akousis*, hearing.

presbyopia progressive loss of focusing power of the eyes associated with loss of elasticity of the internal (crystalline) lenses. The condition is closely age-related and the effect is to make close work, such as reading, increasingly difficult. The term derives from the Greek *presbys*, an old man, and *ops*, an eye.

pressure points places where major arteries lie near the surface and over bones. Direct pressure applied to these points can limit the blood flow and control severe bleeding as a first aid measure in cases of injury.

prevalence the number of people suffering from a particular disease at any one time in a defined population. Prevalence is usually expressed as a rate per 100,000 of the population. Compare INCIDENCE.

priapism persistent, usually painful erection of the corpora cavernosa of the penis without sexual interest. Priapism results from the failure of blood to drain from the penis and may lead to permanent damage from blood clotting. Urgent treatment is needed to withdraw the blood through a wide-bore needle or to obtain detumescence by other means. From the Greek *Priapos*, the god of procreation.

primary 1 occurring as an initial, rather than as a secondary, event or complication. 2 originating within the affected organ or tissue, rather than having spread from another source. 3 the first of several diseases to affect a part. 4 of unknown cause.

primary immune response the weak initial reaction caused by the first encounter of a 'naive' lymphocyte with a particular antigen. Antibodies are not detectable for several days then rise to a (low) peak concentration and then fall again. If now there is a second challenge with the same antigen, there is a more rapid and much higher rise in the level of antibodies. This is the secondary immune response.

primary teeth the first set of teeth to appear, usually beginning to erupt around the age of 6 months. There are 20 primary teeth consisting of 8 incisors, 4 canines and 8 molars. Also known as deciduous or milk teeth. Compare PERMANENT TEETH.

primates the order of mammals that includes some 200 species of monkeys and apes including capuchin monkeys, baboons, gorillas, orangutans, chimpanzees and man. The order contains over 50 genera. Primates enjoy single binocular vision and depth perception and a correspondingly highly developed visual cortex in the brain. The olfactory part of the brain is relatively less well developed than in lower animals. The thumb is prehensile – adapted for grasping – by being able to be moved across the other fingers. Claws have, in most cases, been modified to fingernails. Primates have collar bones (clavicles). Female primates have a regular menstrual cycle and, in many cases, will accept males most readily around the time of ovulation. Males will copulate at any time. The penis is pendant, except when erect. Most primates are vegetarian, but those closest to man are omnivorous. They all enjoy a highly social life.

primer 1 a short RNA sequence paired with one DNA strand that provides a free 3′–OH end on which synthesis of a deoxyribonycleotide chain can start. 2 the short nucleotide sequence used to start the POLYMERASE CHAIN REACTION (PCR) process.

primigravida a woman pregnant for the first time.

primum non nocere a dictum, universally-respected among doctors, to the effect that one's first concern should be to do no harm to the patient.

prion protein a protease-resistant sialoglycoprotein that is a normal constituent of the brain. Abnormal forms of the protein are now generally accepted as the causal agents in Creutzfeldt-Jakob disease (CJD) and bovine spongiform encephalopathy (BSE). The protein was isolated by Stanley Prusiner in 1982, the term prion (an abbreviation of 'proteinaceous infectious particle') being proposed by Prusiner to make the point that it was not a virus. Prion protein (PrP) is found in high concentration in brains affected with spongiform encephalopathy, and forms amyloid deposits in these brains. This structurally simple, seemingly infectious agent of simpler constitution than any virus, is capable of causing a severe and invariably fatal disease of the nervous system. Prions resist sterilization by normal methods and have been spread on surgical instruments and in donated human growth hormone. Prusiner was awarded the Nobel Prize in 1998 for his work on prions. See also PRION PROTEIN DISEASE.

prion protein disease one of a number of transmissible diseases, including Creutzfeldt-Jakob disease, Gerstmann-Straussler syndrome and kuru, all of which feature a spongiform encephalopathy. They are all associated with an abnormal form of a normal cell protein called prion protein. They feature plaques of amyloid in the brain tissue, and the prion protein is the main constituent of these plaques.

pro- *prefix denoting* forward, first, to, towards the front or preceding. It may or may not be a coincidence that hundreds of drug names, especially brand names, have this prefix. More than 50 of the names of those currently being commonly prescribed, start with 'pro-' and the practice appears to be spreading to the naming of products outside pharmacology.

probability see P-VALUE.

proband the presenting patient of a group with an identical or similar disorder, especially the member of a family first found to have an inheritable disorder. Also known as propositus.

proctology the branch of medicine concerned with the disorders of the lower (sigmoid) colon, the rectum and the anus.

proenzyme a protein that can give rise to an enzyme.

progeria premature ageing. There are two types: Hutchinson-Gilford syndrome and Werner's syndrome. In the former, a child of 10 may show all the characteristics of old age – baldness, grey hair, wrinkled skin, loss of body fat and degenerative diseases of the arteries. In the latter, the disease starts in adult life and runs a rapid course over about 10 years. Both types are now believed to result from single spontaneous gene mutations. In HG syndrome the mutation is in the gene coding for lamin A, an important structural protein. In Werner's syndrome the mutation is in a helicase – a gene that unwinds double-strand DNA into two single strands. From the Greek *pro*, before and *geras*, old age.

progesterone the hormone secreted by the CORPUS LUTEUM of the ovary and by the PLACENTA. Progesterone acts during the menstrual cycle to predispose the lining of the womb (endometrium) to receive and retain the fertilized ovum. During pregnancy, progesterone from the placenta ensures the continued health and growth of the womb and promotes the growth of the milk-secreting cells of the breasts. Progesterone-like substances (progestogens) are widely used in medicine and are common constituents of oral contraceptives. The hormone is used to treat menstrual symptoms and infertility and as an adjunct to oestrogen in post-menopausal hormone replacement therapy (HRT).

prognathism abnormal protrusion of either jaw, especially the lower.

prognosis an informed medical guess as to the probable course and outcome of a disease. Prognosis is based on a knowledge of the natural history of the disease and of any special factors in the case under consideration.

prohormones molecules that are split by enzymes to form hormones. Most are proteins of at least moderate size. Peptide hormones cannot be directly synthesized because cells cannot produce proteins below a particular size.

prohormone convertases enzymes that cleave prohormones to yield active hormones.

prokaryote, procaryote a class of primitive single-cell living organisms, containing the bacteria and the blue-green algae, and so called because the members do not possess a discrete nucleus with a nuclear membrane. The genome is merely dispersed throughout the cell. All nucleated cells and organisms are said to be eukaryotic or eucaryotic.

prolactin one of the PITUITARY GLAND hormones. Prolactin stimulates the development and growth of the breasts (mammary glands) and helps to start and maintain milk production at the end of pregnancy.

prolactin-inhibiting hormone the hormone dopamine.

prolapse the downward displacement, or movement to an abnormal position, of a body part or tissue. Common examples are prolapse of the uterus (PROCIDENTIA), prolapse of the RECTUM and prolapse of the pulpy centre of an intervertebral disc.

proliferation multiplication. The process of increasing in number by reproduction.

proliferative phase the part of the menstrual cycle between the end of menstruation and the next ovulation. During this stage the womb lining (endometrium) thickens and becomes suitable for implantation of the fertilized ovum.

promiscuity a loose term for sexual promiscuity – a common pattern of behaviour in which sex is valued for its variety and immediate gratification rather than as one of the important bases for a long-term relationship. Some psychologists equate promiscuity with social immaturity; others hold it to be unrelated. Promiscuity has always carried penalties of some kind, often in the form of sexually transmitted diseases such as gonorrhoea, herpes and Chlamydial infections. In an AIDS context, promiscuity has become much more significant and dangerous.

promoter in genetics, a nucleotide sequence at the start of a gene to which RNA polymerase must bind before the process of transcription can start. The promoter determined which of the two strands of DNA is transcribed into RNA. It is located UPSTREAM of the gene it regulates.

pronation the act of turning to a face down (prone) position, or of rotating the horizontal forearm so that the palm of the hand faces the ground. The opposite movements are called supination.

prone lying with the front of the body downward. Face downward. Compare SUPINE.

proofreading in genetics, the correction, performed by DNA polymerase, of mistakes in the incorporation of nucleotides in a sequence, the corrections being made after individual units have been added to the chain.

proopiomelanocortin (POMC) the pro-hormone from which ACTH, beta-lipotropin, and beta-ENDORPHIN are produced by cleavage. POMC is found most abundantly in the pituitary and hypothalamus, but also occurs in the sex glands and elsewhere.

propaganda a term derived from an institution of the Roman Catholic Church concerned with missionary work. It is now taken to mean any system of sustained and organized dissemination of information, whether true or false, for the purposes of modifying popular opinion in favour of a government or movement or against an enemy. Once seen for what it is, propaganda is often counter-productive.

prophase the first stage in cell division by MITOSIS and MEIOSIS, during which CHROMATIN coils up to form chromosomes.

prophylactic and any act, procedure, drug or equipment used to guard against or prevent an unwanted outcome, such as a disease.

proprioception awareness of the position in space, and of the relation to the rest of the body, of any body part. Proprioceptive information is essential to the normal functioning of the body's mechanical control system and is normally acquired unconsciously from sense receptors in the muscles, joints, tendons and the balance organ of the inner ear.

proptosis abnormal protrusion of the eyeball. Also known as EXOPHTHALMOS. Proptosis is cause by any process that increases the bulk of the soft tissues in the ORBIT behind the eyeball.

prosencephalon the forebrain.

prostacyclin a short-acting hormone produced by the lining of blood vessels (ENDOTHELIUM) and by the lungs, that limits the aggregation of PLATELETS and is probably of major importance in preventing THROMBOSIS. Prostacyclin has a half-life of only 2–3 minutes.

prostaglandins a group of unsaturated fatty acid mediators occurring throughout the tissues and body fluids. They are generated from cell membrane ARACHIDONIC ACID by the action of phospholipase A_2 and function as hormones. They have many different actions. They cause constriction or widening of arteries, they stimulate pain nerve endings, they promote or inhibit aggregation of blood PLATELETS and hence influence blood clotting, they induce abortion, reduce stomach acid secretion and relieve asthma. They can both stimulate and inhibit immune responses. Some painkilling drugs, such as aspirin, act by preventing the release of prostaglandins from injured tissue.

prostate see PROSTATE GLAND.

prostate gland a solid, chestnut-like organ situated under the bladder surrounding the first part of the urine tube (urethra) in the male. The prostate gland secretes part of the seminal fluid.

prostate gland disorders these include benign prostatic enlargement (hyperplasia), symptoms resulting from enlargement of the gland – urgency to urinate, undue frequency of urination, a weak urinary stream and burning pain on urination – (prostatism), prostate cancer, and inflammation (prostatitis).

prostate-specific antigen (PSA) an enzyme produced by the epithelial cells of the prostate gland, whether healthy or malignant, to liquefy the seminal fluid. Small quantities of PSA enter the bloodstream and the levels can be measured. Raised levels imply an increase in the bulk of prostate tissue and can thus be used as a marker for prostatic hyperplasia or prostate cancer. PSA levels below 4 nanograms per decilitre are, in general, considered normal, but do not preclude cancer. Levels above 10 ng/dL suggest a 70 per cent risk of prostate cancer. The test is made more sensitive for men with levels below 10 ng/dL by noting the ratio of free PSA to PSA complexed with antichymotrypsin. If free PSA is 25 per cent or more of total PSA and the gland feels normal, biopsy is considered unnecessary. A free PSA of 15 per cent or less suggests cancer. The opinion is growing among experts that the PSA is a less reliable test for prostate cancer than was formerly thought.

prosthesis any artificial replacement for a part of the body. Prostheses may be functional or purely cosmetic and may be permanently installed internally or worn externally. The range of prosthetic devices is wide – from artificial eyes and legs to heart valves and testicles.

prosthetic group a non-peptide organic molecule or a metallic ion that binds to a protein and is necessary for its particular function. The haem in haemoglobin is a prosthetic group.

protanopia partial colour blindness with defective perception of red.

protease one of a range of protein-splitting enzymes. One focus of current interest in proteases is in their role in breaking down tissue barriers in the spread of cancer. High concentrations of the activator of one of these proteases has been found to be associated with a poor outlook in cancers of the colon and rectum.

proteasome a large, cylindrical protein complex of several sub-units, present in the cytoplasm and nucleus of all cells and an essential component in cell metabolism. The function of the proteasome is to act as a kind of shredder, degrading unwanted proteins that have been tagged for destruction with UBIQUITIN chains. It strips proteins of their ubiquitin, unfolds them and catalyzes them to peptides. Proteasomes have aroused much interest as therapeutic targets in cancer.

proteins

The whole structure of the body is based on protein. The bones are made of protein impregnated with minerals. The muscles are almost pure protein, as are the tendons and ligaments. The skin and its appendages – the hair and the nails – are mainly composed of protein. All the connective tissue of the body, the structural material that forms the scaffolding of the organs, is made of protein. The blood contains a considerable quantity of protein. Some of this is nutritional, some maintains the necessary viscosity and osmotic pressure, some is needed for blood clotting and some constitutes the large range of antibodies that provide us with protection against infection. The blood also contains large numbers of enzymes, all of which are proteins. Every cell in the body contains thousands of different enzymes and other proteins. About half the dried weight of the average cells consists of protein. This central importance is reflected in the derivation of the term: the Greek word *protos* means 'first' or 'earliest'.

Proteins are large, complex molecules made up of chains of varying numbers of 20 different amino acids linked together in different combinations. Chemically, the amino acids are fairly simple. Each one contains two functional groups; an amino group (a nitrogen atom linked to two hydrogen atoms $-NH_2$) and a carboxylic acid group (a carbon atom linked to an oxygen atom and a hydroxide group $-COOH$). Both of these groups are connected to the same central carbon atom. The amino acids differ from each other by virtue of another group – the R group – also connected to the same central carbon atom.

Protein molecules are built up by virtue of the ease with which the carboxyl group of one amino acid becomes attached to the amino group of another. When this happens a molecule of water (H_2O) is eliminated. This – one of the most important chemical linkages in all science – is called a peptide bond, and it allows any amino acid to link with any other. Two or three amino acids linked together by peptide bonds constitute a dipeptide and a tripeptide respectively. When many are linked we have a polypeptide. When a polypeptide forms it has a free amino group at one end and a free carboxyl group at the other. One or more polypeptide chains constitutes a protein molecule. The difference in the two ends of the polypeptide chain allows the chain to be incorporated in the protein molecule in a particular direction.

Protein molecules also have three-dimensional structures that determine their biological properties. This arises because the long polypeptide chain becomes spontaneously folded or twisted in a particular way as a result of interaction between amino acids that may be quite distant from each other along the chain. Often the chain twists into a regular helix. Helical molecules may, in turn, twist into a tertiary, or even quaternary configuration, mainly under the influence of the R groups.

Structural proteins, such as collagen, of which the skeleton and the connective tissue of the body are made, and the keratin of the hair and nails, need be designed only for strength and stability.

These are fibrous proteins, consisting of bundles of helical molecules. The keratin of hair and the myosin of muscles is a regularly repeating single helix. The strong collagen of bones and tendons is a triple helix. Proteins with a more active function, such as the enzymes and the antibodies, require a specific, three-dimensional shape to perform their actions. The remarkable specificity of enzymes is the result of the unique three-dimensional configuration of each – allowing the right enzyme to slot into the right place to promote a chemical reaction.

Disruption of the three-dimensional shape of a protein molecule occurs when it is heated or acted on by strong reagents. This is called 'denaturation'. A denatured protein can never recover its original shape or properties. So the transparency of egg white or of the crystalline lens of the eye cannot be restored if the former has been boiled or the latter has become cataractous.

See also ABC TRANSPORTER PROTEINS, ACTIN BINDING PROTEINS, ACUTE PHASE PROTEINS, ALPHAFETOPROTEIN, AMYLOID, AMYLOID PRECURSOR PROTEIN, ANNEXINS, ANTIBODY, CHAPERONES, CHEMOKINES, COLLAGEN, COMPLEMENT, CONFORMATIONAL EPITOPE, CORE OCTAMER, CYTOKINES, DENATURATION, ELASTINS, FIBROBLAST, GLUTEN, GLYCOPROTEINS, G PROTEINS, GRANINS, HEAT-SHOCK PROTEINS, IMMUNOGLOBULINS, INTERMEDIATE FILAMENT PROTEINS, KERATIN, LECTINS, LIGAMENTS, LIPOPROTEINS, MEMBRANE-PROTEIN ION CHANNELS, MOTOR PROTEINS, MUSCLE, PLASMA PROTEINS, POLYMER, PRION PROTEINS, PROTEASE, PROTEIN BINDING, PROTEIN FOLDING, PROTEIN SYNTHESIS, REPRESSOR PROTEIN, RNA, SINGLE STRAND BINDING PROTEINS, TAU PROTEINS, TERTIARY STRUCTURE AND UBIQUITIN.

protein binding the attachment of proteins to other molecules. Proteins are so structured that they are able to bind to a wide range of molecules with a high degree of specificity. Molecules they bind to are called ligands. An example is the binding of a protein to a particular sequence of base pairs in DNA. Such binding can turn on or off the expression of nearby genes.

protein C a plasma protein that inhibits clotting.

protein folding the process by which proteins acquire their normal, energetically-favourable, three-dimensional form. The folding has long been believed to be specified by the amino acid sequence, but recent research suggests that proteins may fold into abnormal shapes by mechanisms not yet determined. Abnormal folding results in loss of normal protein function but also in proteolytic degradation and the accumulation of fragments that form insoluble plaques in the various organs including the brain and the liver. This is believed to be the way in which the characteristic plaques and fibrillary tangles of ALZHEIMER'S DISEASE occur.

protein S a natural anticoagulant the deficiency of which causes an increased risk of blood clotting within the vessels (thrombosis) and embolism. Protein S deficiency may be of genetic origin or may have other causes. It has been suggested that women with low free protein S may be at increased risk of thromboembolism if they are using oral contraceptives.

protein synthesis the construction of protein molecules from AMINO ACIDS. This occurs in the cell CYTOPLASM on the basis of the GENETIC CODE in the DNA. Sections of DNA that code for the particular protein are first transcribed to MESSENGER RNA and this passes out of the cell nucleus to the cytoplasm. There, one or more ribosomes attach themselves to one end of the mRNA molecule and move along it to effect transcription, using, in the process, the sequence of RNA bases to indicate which amino acids should be selected from the cell pool and in what order. In this way the correct amino acids are linked together to form polypeptides and these are then joined to form the particular protein.

proteoglycans bimolecule complexes of proteins and GLYCOSAMINOGLYCANS with a high proportion of carbohydrate (up to 96 per cent) to protein. They are structural components of connective tissues and lubricants, help in the adhesion of cells to the extracellular matrix, and bind factors that promote cell proliferation.

proteolytic able to split protein molecules into polypeptides and amino acids, as of enzymes such as PEPSIN and CHYMOTRYPSIN. Proteolysis occurs by HYDROLYSIS of peptide bonds.

proteome the totality of all the proteins in an organism such as the human body.

proteomics the study of the proteome – the proteins expressed by the approximately 22,000 genes in the GENOME or by a cell. The form and quantity of the proteins produced by a cell cannot be fully predicted from DNA or RNA analysis alone. This is because of the controls and the many modifications that can occur in the stages between transcription and protein formation. Thus the totality of the genes can result in at least several hundred thousand different proteins. Proteomics includes the study of the factors that cause this multiplication. The discipline is being applied effectively to cancer studies.

prothrombin a soluble protein in the blood that is converted to the insoluble form thrombin, under the action of the enzyme prothrombinase, at the end of the cascade of events involved in blood clotting. Thrombin is the main ingredient of the blood clot.

proto- *combining form denoting* first, earliest or primitive.

proton the positively-charged nucleus of a hydrogen atom. A hydrogen ion and the basis of acids.

proton pumps systems of enzymes that transport hydrogen ions across membranes, using energy in the process. Proton pumps that produce hydrochloric acid in the stomach derive their energy from hydrolysis of ATP.

proto-oncogene any gene capable of becoming a cancer-producing gene (an oncogene). Proto-oncogenes have important functions in the normal cell, but, by mutation or by the acquisition of genetic control elements from oncoviruses they can lose their normal regulatory functions and lead to uncontrolled multiplication.

protoplasm the whole of the internal substance of the living cell, consisting of the material surrounding the nucleus (the CYTOPLASM) and the nuclear material (nucleoplasm).

protoporphyrin a porphyrin occurring in the course of the synthesis of HAEMOGLOBIN. Excess of this porphyrin, protoporphyria, causes intense itching, swelling and redness of the skin on exposure to sunlight and leads to a strikingly weatherbeaten appearance.

protozoa primitive, single-celled, microscopic animals able to move by amoeboid action or by means of CILIA or whip-like appendages (flagella). Many protozoa are parasitic on humans and are of medical importance. These include the organisms that cause amoebiasis, balantidiasis, cryptosporidiosis, giardiasis, isosporidiosis, leishmaniasis, malaria, sleeping sickness, toxoplasmosis and trichomoniasis.

provitamin a substance that is converted in the body to a VITAMIN.

proximal pertaining to any point on the body nearer to, or nearest to, the centre. The upper arm is proximal to the hand and is the proximal part of the arm. Compare DISTAL.

proximal tubule the first part of the tubular system of a nephron after the Bowman's capsule. It has convoluted and straight segments and is followed by the distal tubule.

pruritus itching. The term is often linked with a word that indicates the site, as in pruritus ani or PRURITUS VULVAE.

PSA see PROSTATE-SPECIFIC ANTIGEN.

pseud-, pseudo- *prefix denoting* false.

pseudocyesis the occurrence of the signs and symptoms of pregnancy when no pregnancy exists. This may sometimes be a manifestation of an overwhelming desire for conception in an infertile woman.

pseudogenes the stable and inactive, long-term consequences of earlier mutations that occurred during the process of evolution. Pseudogenes might be considered as analogues of fossils in geology.

pseudohermaphroditism a congenital abnormality of the GENITALIA in which they resemble those of the opposite sex. The testes and ovaries (gonads) are, however, those of the genetically correct sex.

psilocybin a powerfully hallucinogenic phosphorylated tryptamine present in the fungi contaminating some types of mushrooms, especially in *Psilocybe mexicana.* It is a powerful hallucinogenic drug with properties similar to those of LSD.

psoas muscle a two-part muscle running from the front of the lower spine to the margin of the pelvis (psoas minor) and to the front of the top of the thigh bone (femur) (psoas major). Its action is to raise the thigh forwards.

psych- *prefix denoting* mind, mental, psyche.

psyche the mind, as opposed to, or in contradistinction to, the body.

psychiatry the branch of medicine concerned with the management of mental illness and emotional and behavioural problems. Compare PSYCHOLOGY.

psychoanalysis **1** a purported treatment for psychiatric disorders in which the patient is encouraged to reminisce freely about his or her past life while the analyst silently interprets these free associations in the light of FREUDIAN THEORY. Success is said to be unlikely unless the subject falls in love with the analyst (transference). Classical psychoanalysis involves sessions of about an hour, up to six times a week, for several years. **2** a dogmatic theory of human behaviour.

psychodrama a technique in PSYCHOTHERAPY in which the subject acts out relevant incidents or adopts particular roles, so allowing the expression of troublesome emotions or the contemplation of deep conflicts.

psychogenic of mental rather than of physical origin. The term is usually applied to symptoms or disorders thought to be due to problems of social or personal adjustment rather than to organic disease.

psychology

A central concern of psychology is the systematic study of human behaviour. Psychologists try to understand, describe, predict, and sometimes change, human behaviour. They are concerned with behaviour at all stages in life, from birth, through the developmental period and maturity, to old age, and with the behaviour of people in all kinds of contexts, occupations and groupings. Psychology is a very large subject with few boundaries and has many branches.

A distinction must be made between the schools of psychology and the branches. In general, the schools are sub-divisions based on the different theoretical systems to which their adherents subscribe and upon which their practice is claimed to be based. The branches, on the other hand, are functional or professional sub-specialties based on practical considerations. Many practical psychologists profess not to belong to any school. Some take an eclectic position.

The many branches of psychology are designated by a qualifier. These include: abnormal, analytic, applied, clinical, comparative, developmental, educational, experimental, geriatric, industrial, infant, physiological and social. The list is not exhaustive.

Clinical psychology

This should not be confused with psychiatry which is a medical speciality, practised by doctors concerned with the treatment of mental disorders of all kinds. Clinical psychology is a discipline ancillary to medicine and concerned, broadly, with the diagnosis and non-medical treatment of emotional disorders and aberrant or maladaptive behaviour. It may also be concerned with the treatment of interpersonal problems, marital difficulties, sexual problems and drug abuse.

Clinical psychologists may work in hospitals, in which case their activities are largely diagnostic, or they may work privately or in clinics where they often provide psychotherapy of one kind or another. Psychotherapists usually belong to one of the many schools of psychology – Adlerian, behavioural, client-centred, cognitive-behavioural, encounter, family, Freudian, gestalt, group-therapeutic, holistic, Jungian, psychoanalytic, transactional analytic and so on. Clinical psychologists in hospital are often concerned with the assessment of mental retardation, whether present from birth or the result of later brain damage, and with the progress of functional recovery from brain injury.

Industrial psychology

This is the application of psychology to the problems of industry, management, business and the workplace. It is a large subject, encompassing such matters as personnel selection and placement, training, organizational planning, job specifications, work incentive, career planning, motivation and morale, human relations, communication, working conditions, ergonomics, advertising and consumer psychology. Inevitably, industrial psychologists experience conflicting aims, even divided loyalties. Employers may be interested primarily in productivity and profit while employees may be more concerned with working conditions and individual rights. Ideally, industrial psychology should be concerned with both and should operate on the assumption that these aims are not inherently antagonistic.

Educational psychology

This branch of applied psychology is concerned with the practical problems of teaching and learning and is directed both to the production of improved teaching methods and to the assistance of those with learning difficulties. Educational psychologists work in association with teachers and children and are also concerned with counselling older students about courses and personal educational problems.

Although the mechanisms of learning are a central concern of research psychology and much that has been discovered can be applied to facilitate effective study, educational psychologists are more concerned with the human elements in education. They are concerned with the effects on learning of such factors as ability, intelligence, aptitude, personality traits, classroom and study environmental conditions, social status and emotional development.

Educational psychologists may subscribe to any of the schools of psychology. Behaviourist psychologists believe in progressing by small stages with the application of strong positive reinforcement. Developmental psychologists often follow the theories of the Swiss psychologist Jean Piaget (1896–1980), applying his ideas of the way in which children develop in stages by constructing their own concepts of reality through the solution of problems encountered in real life. Freudian psychologists concentrate on the role of the emotions, the exploration of feelings and the attempt to ensure the satisfaction of basic emotional needs. Others take a humanistic standpoint and base their work on the assumption that children should be allowed to direct their own learning in a loving, accepting environment.

Psychometrics

This term literally means 'measuring the mind', but, in practice, psychometrics is concerned with the somewhat more limited goal of trying to find ways of reliably quantifying the various forms of mental ability. It is concerned with assessing intelligence, aptitudes, talents, skills of all kinds including the verbal, potential in various fields, personality traits and potential for psychological breakdown.

Critical psychometric workers are constantly aware of the need to ensure that tests really do measure what they purport to measure, that their results are reliable and valid over the long term and that they can be properly related to statistical norms in large groups. The methods used include interviews, direct observation, formal written tests, manual skill tests, matrix tests, vocational tests, personality inventories, situational testing and such 'projective' methods as the Rorschach (ink blot) test.

Psychometrics have come in for a good deal of criticism partly on the grounds of political correctness. One of the most virulent critics was William H. Whyte Jr. who, in his best-selling book *The Organization Man* (1956) scathingly denounced the use of personality tests in personnel selection and amusingly instructed his readers how to cheat and obtain a high score. Since that time, psychometrics has advanced and become more scientific, but it is still viewed with suspicion and scepticism by many.

psychometry the measurement of psychological functions, including correlative ability, memory, aptitudes, concentration and response to logical puzzles. Intelligence has never been adequately defined and so there are no tests for pure intelligence.

psychoneuroimmunology the discipline concerned with the effect of the emotions on the immune system and hence on the development of disease. Psychoneuroimmunology is not predicated on the proposition that the mind and the body are discrete entities that interact via the hypothalamic-pituitary-adrenal axis; it is based on the growing recognition that mental and physical events are so inextricably interrelated that nothing of importance can happen to one without affecting the other.

psychopath a person whose behaviour suggests indifference to the rights and feelings of others. A person in whom prior experience has induced an antisocial personality disorder.

psychopathology the study of the nature of abnormal mental processes and their effects on behaviour.

psychosexual disorders a range of conditions affecting the mental and emotional attitudes to sexuality and which may interfere with normal sexual responses. They include impotence, premature ejaculation, non-organic dyspareunia, transsexualism, and various sexual deviations. Many of these conditions respond well to expert treatment.

psychosis one of a group of mental disorders that includes schizophrenia, major affective disorders, major PARANOID states and organic mental disorders. Psychotic disorders manifest some of the following: DELUSIONS, HALLUCINATIONS, severe thought disturbances, abnormal alteration of mood, poverty of thought and grossly abnormal behaviour. Many cases of psychotic illness respond well to antipsychotic drugs in the sense that these drugs, while they are being taken, often induce a state of docility, acquiescence, apparent mental normality and conformity with social norms readily acceptable to medical staff and relatives.

psychosomatic **1** pertaining to the relationship between the mind and the body.

2 pertaining to the apparent effect of mental and emotional factors in contributing to physical disorders. These definitions imply the possibly untenable assumptions enshrined in the long-held view (Cartesian dualism) that the mind and the body are distinct, separable entities.

psychotherapy any purely psychological method of treatment for mental or emotional disorders. There are many schools of psychotherapy but results appear to depend on the personal qualities, experience and worldly wisdom of the therapist rather than on the theoretical basis of the method. Currently the most fashionable, and seeming successful, school is that of cognitive behaviour therapy.

pterygoid 1 wing-shaped. 2 pertaining to two processes attached like wings to the body of the SPHENOID bone in the skull.

ptosis drooping of the upper eyelid. Ptosis may be a congenital weakness of the lid elevating muscle or may result from later injury or disease.

ptyalin the salivary enzyme MALTASE.

ptyalism excessive salivation from any cause, such as mercury or organophosphorous poisoning, mouth irritation, OESOPHAGITIS or PEPTIC ULCER.

puberty

Puberty is the period of physical development during which the bodily changes that characterize the sexually capable adult occur. When puberty is accomplished, the individual is capable of reproduction. Adolescence, on the other hand, is the sometimes prolonged period of transition from childhood to adulthood. Puberty occurs during adolescence, but adolescence involves mental and emotional changes in addition to the those concerned with the reproductive function. Prior to puberty, children grow in body and mind but the changes occurring are quantitative only. At puberty, qualitative changes begin that fully differentiate relatively asexual girls and boys into women and men.

At around the age of 10 in girls and 12 in boys there is a striking growth spurt. The rate of growth increases to some 9 cm per year in girls and over 10 cm per year in boys. This prepubertal growth spurt affects different parts of the body at different times so that, for a time, the body may appear disproportioned. Growth acceleration affects first the feet, then the legs, then the trunk, and finally the face, especially the lower jaw. Because puberty occurs about two years later in boys than in girls, boys have a longer period of preadolescent growth spurt and tend to gain a significant advantage in height over girls.

Puberty in girls

Since the middle of the 19th century, puberty in Western European and American girls has been occurring progressively earlier. The age of the onset has been getting lower at a rate of four to six months with every passing decade. This phenomenon is thought to be due to improvements in nutrition. Puberty in girls starts at any age between

10 and 13 years and takes three or four years. Because of this variability, some girls may have reached physical sexual maturity while others of the same age may still be of a childlike physique. A similar phenomenon occurs with boys and, in both cases, can be a cause of distress. The effect is, however, temporary and by the age of 16 or 17 almost all the late starters will have caught up. Late onset of puberty has no significance for future sexuality.

The period of puberty features considerable body growth, changes in body proportions and major changes in the sexual organs. The first sign is usually breast budding but may be the appearance of pubic and underarm hair. Breast buds may appear as early as age 8 but occurs, on average, at about 11. From this point it is usual for about a year to pass before menstruation starts. By the time the girl menstruates the other changes of puberty are well established. Breast growth is often rapid, and, for a time, one side may grow more rapidly than the other. It is rare, however, for the breasts to remain of different sizes.

Bony changes in the pelvis cause it to widen relative to the rest of the skeleton and the effect of this is emphasized by new deposits of fat laid down around the hips. The general contours of the female body are considerably influenced by these specific fat deposits, which also occur under the skin of the breasts and buttocks. In girls, puberty is deemed complete when menstruation is occurring at regular, predictable intervals. This implies ovulation and the possibility of conception. The onset of regular menstruation can, however, be affected by various factors such as obesity, which tends to bring on the periods earlier, and malnutrition and excessive athletic activity, which tend to delay onset.

Puberty in boys

All the physical changes that occur during male puberty are caused by the male sex hormone testosterone. This is an anabolic steroid produced by cells in the testicles lying between the sperm-producing tubules (the interstitial cells). Until shortly before puberty the testicles contain only numerous solid cords of pale cells, and there is no sign of sperm production. At puberty, the cells at the centre of these cords disappear so that the cords become tubes. These are the seminiferous tubules in which the sperm develop. Once this process is complete, early in puberty, sperm production is active and rapid. Very large numbers of sperm are produced – between 300 and 600 per gram of testicle every second.

Testosterone is a powerful hormone and has many effects. It enlarges the testicles, scrotum and penis; it causes the seminiferous tubules to begin to produce sperm; it enables the penis to erect fully; it causes the sperm-carrying ducts and the semen-storage sacs (the seminal vesicles) to enlarge and mature; it causes the prostate gland to enlarge and begin to secrete fluid that makes up part of the seminal fluid; it causes the voice box (larynx) to enlarge and the voice to drop in pitch as a result; it promotes the growth of pubic and underarm hair and of the beard and may cause hair to grow on the chest and abdomen; and it accelerates general body growth and muscular development.

These changes usually occur roughly in the order given. Puberty may start at any age from about 10 to 15 and usually takes about two and a half years, so it may not be complete until after 17. As a result some boys of about 13 may be completely sexually mature while friends of the same age may still be sexually infantile. Most boys have reached about 80 per cent of their adult height before puberty, but there is nearly always a large growth spurt during puberty.

Psychological aspects of puberty

The main psychological consequence of puberty is the initiation and growth of sexual interest and sexual drive. Both boys and girls quickly become aware of the physical differences between the sexes and this awareness often arouses intense interest in the opposite sex. In many cases this becomes the chief preoccupation. There is often a new and uncharacteristic concern over personal appearance and clothes and increasing anxiety to conform to current fashions in dress.

Adolescence is a difficult time for many. Growing knowledge and intellectual powers often bring with them a kind of arrogance and a contempt for the opinions of older and more experienced people. Adult mores are often rejected and there may be extreme impatience with imposed rules and regulations. There is a growing recognition of the importance of success, and failure to achieve this in an adult-oriented context may lead to attempts to succeed in the eyes of a peer group. This commonly leads to delinquency, but this if usually a temporary phase.

Many adolescents never reach Piaget's fourth stage of fully adult operational thinking in which abstract reasoning predominates. In many, thinking continues to be concerned with the particular rather than with the general, the concrete rather than the conceptual. Thinking often remains egocentric and some have difficulty in appreciating the relationship between present action and future consequence.

Abnormal puberty

The whole process of puberty is initiated by the production of gonadotrophin releasing hormone by the hypothalamus of the brain. Why this occurs is not clear. This hormone acts on the pituitary gland causing it to begin to secrete a new group of hormones, the gonadotrophins around the age of 10–14 years. The term means 'sex gland stimulators'. Gonadotrophins cause the ovaries to secrete oestrogens and the testicles to produce testosterone. Very rarely, abnormal changes in the hypothalamus, such as a brain tumour may start this sequence going at a much earlier age than normal. As a result, puberty may occur at almost any earlier age, even in infancy. Because the adrenal glands also secrete male sex hormones, a tumour of an adrenal can have a similar effect in a small boy. Cases of extreme precocious sexual development, in which full sexual maturity, with the power to reproduce, has been reached at the age of five years in both boys and girls. The youngest known mother gave birth to a healthy baby four months before her sixth birthday.

pubes the pubic hair or the PUBIC region of the body.

pubic pertaining to the pubic bone (os pubis) at the front of the pelvis or to the region at the central point at the front of the lowest part of the abdomen at which the inner aspects of the two pubic bones normally lie in close apposition.

pubic lice see LICE.

pubic symphysis the mid-line cartilaginous joint between the inner surfaces of the pubic bones.

pudenda the external genitalia. From the Latin *pudere*, to be ashamed.

puerperium the period after childbirth when the womb (uterus) and VAGINA are returning to their normal state.

pulmonary pertaining to the lungs.

pulmonary function tests a range of tests of the efficiency of the lungs and of diagnostic procedures to detect lung disease. They include tests of chest expansion, air lung volume, the maximum volume of air that can be expired (vital capacity), the peak air flow rate achievable and tests of blood concentrations of oxygen and carbon dioxide.

pulp 1 the soft tissue in the middle of each tooth that contains blood vessels and nerves. 2 the soft tissue, the NUCLEUS PULPOSUS, in the centre of each INTERVERTEBRAL DISC.

pulse the rhythmic expansion of an artery from the force of the heart beat. In health, the pulse is regular, moderately full and at a rate of between about 50 and 80 beats per minute.

pulse pressure the difference between the diastolic and the systolic blood pressure.

punch-drunk pertaining to the state of brain damage resulting from repeated blows to the head as are sustained in boxing. A punch-drunk person has slow mental processes, slurred speech and impaired concentration. Recovery is unlikely.

punctuation codon a CODON that marks the start or the end of a gene.

pupil the circular opening in the centre of the iris of the eye. The pupil becomes smaller (constricts) in bright light and widens in dim light under the action, respectively of its circular and radial muscle fibres.

purines a group of nitrogen-containing compounds that includes adenine and guanine, the bases whose sequence forms the genetic code. An excess of purines can cause gout.

Purkinje cells nerve cells with very large, pear-shaped bodies and a profuse collection of dendrites that are found in the middle layer of the surface zone (cortex) of the CEREBELLUM. (Johannes Evangelista von Purkinje, 1787–1869, Czech polymath and physiologist).

Purkinje fibres the large specialized heart muscle fibres that distribute electrical impulses from the lower ends of the right and left conducting bundles to the musculature of the ventricles thereby coordinating the heart beat. (Johannes Evangelista von Purkinje, 1787–1869, Czech polymath and physiologist).

purulent pertaining to PUS.

pus a yellowish or green viscous fluid consisting of dead white blood cells, bacteria, partly destroyed tissue and protein. Pus is formed at the site of bacterial infection but may occur in sterile situations as a result of inflammation from other causes.

putamen the outer shell of the lentiform nucleus of the BASAL GANGLIA of the brain.

p-value the probability, expressed as a number, that a particular effect or association is real or that a given statement or hypothesis is true. If a trial has n possible outcomes and m of these are the desired outcome, then the probability (p) of obtaining the desired outcome is m/n. If all the outcomes are as desired then $m = n$ and $p = 1$. If none are as desired $p = 0$. So p is a measure that will always range from 0 to 1.

PWA *abbrev. for* person with AIDS.

py-, pyo- *combining form denoting* PUS.

pyel-, pyelo- *combining form denoting* the pelvis of the kidney.

pyle- *combining form denoting* the PORTAL VEIN.

pylorus the narrowed outlet of the stomach where it opens into the DUODENUM. At the pylorus, the muscular coats of the stomach wall are thickened to form a strong muscle ring (a SPHINCTER) capable of closing and opening to control the movement of food.

pyr-, pyro- *combining form denoting* fire, heat or fever.

pyramidal tract the great inverted pyramid of motor nerve fibres descending from the motor cortex of the cerebrum through the internal capsule and down into the brainstem where the fibre bundles on each side cross to the other side. This is why a stroke on the right side causes paralysis on the left side of the body.

pyrexia fever.

pyrimidine a nitrogenous base compound. Two pyrimidines, cytosine and thymine, are the DNA bases which, with two PURINES, form the genetic code. A third pyrimidine, uracil, takes the place of thymine in RNA.

pyrogen any substance that causes fever. Endogenous pyrogens are proteins, such as interleukin-1, released by white blood cells in response to bacterial or viral infections. These act on the temperature-regulating centre in the brain, effectively resetting the thermostat at a higher level and causing the muscles to contract repeatedly and rapidly (shivering) so as to raise body temperature.

pyromania a compulsion to start fires.

Qq

QRS complex the part of the electrocardiograph tracing corresponding to the contractions of the main chambers of the heart (the ventricles). The Q wave is a short downwards deflection, the R wave a conspicuous upwards stroke and the S wave a return to below the level of the base-line.

quadri- *combining form denoting* four.

quadriceps muscles the bulky muscle group on the front of the thigh, consisting of four muscles arising from the thigh bone (femur), and from the front of the pelvis. These muscles end in a stout tendon which incorporates the kneecap (patella) and is attached to a bony ridge on the upper end of the front of the main bone of the lower leg (tibia). The quadriceps group powerfully straighten the knee.

quadrigemina pertaining to the four bodies, the corpora quadrigemina, or colliculi, on the roof of the fourth ventricle of the midbrain. The upper pair of quadrigeminal bodies are concerned with visual tracking and the lower pair are associated with hearing.

quadriplegia paralysis of the muscles of both arms, both legs and of the trunk. Quadriplegia results from severe spinal cord damage in the neck, usually as a result of a fracture-dislocation, but sometimes as a result of neurological disease.

qualm a sudden feeling of nausea or faintness.

queer a slang term referring to male homosexuality.

quickening perception by the mother-to-be of movements of the fetus in the womb. In first pregnancies this is usually noticed around the 20th week, but, with experience, quickening may be recognized as early as the 16th week.

quintuplets five babies born in a single gestation. Quins were once rare but since the introduction of fertility drugs that promote multiple ovulation they have become more common.

Rr

race a loaded term that, although widely used, can scarcely be employed without risk of causing offence to someone. This is because the word is so often used in common language to express chauvinistic prejudices. Scientific considerations of race, however, have nothing to do with racism. In the case of the species *Homo sapiens*, races are not easily determined because of the extreme variability of characteristics in any particular population group. Today, as a result of generations of high population mobility and the resultant 'gene flow', it is no longer possible to identify races on purely geographic grounds. Even the arbitrary and readily visible features commonly used in the past – skin colour, facial configuration, hair patterns, head shape, body shape and form – are no longer reliable as criteria. Genes that transmit such characteristics do not do so in clusters, so different members of a group will tend to show only some of the features commonly ascribed to a particular race. There may be more genetic variation within a given race than between it and another race.

The increasing trends to intermarriage between groups formerly attributed to different races has increased the diversification to the point where it is now difficult to sustain, in any realistic sense, the former concepts of race. Today, many scientists reject the idea of race, or use it as a convenient expression for the variations in anatomical features found in different indigenous populations occupying broad geographic areas.

George Bernard Shaw, somewhat daringly for the time, expressed the hope that racial problems would be solved when the whole of mankind was reduced to 'one charming coffee colour'. This is a hope that might usefully be expressed again today, perhaps with a greater likelihood of fulfilment.

racemose having a structure of clustered parts, especially of a gland. From the Latin *racemosus*, clustered, as in *racemus*, a bunch of grapes.

rachi- *combining form denoting* spine or spinal cord. From Greek *rachis*, the spine.

rad a unit of dosage of absorbed ionizing radiation. The rad is equal to 0.01 joule of absorbed energy per kilogram of the exposed tissue or other material.

radial 1 branching out, like rays, from a point.

2 pertaining to the RADIUS bone or its associated artery or nerve.

radial nerve one of the main nerves of the arm and hand. The radial nerve is a mixed motor and sensory nerve. It supplies the forearm muscles that straighten the flexed wrist and conveys sensation from the back of the forearm and hand.

radiation the emission and almost instantaneous propagation of electromagnetic waves ranging in wavelength from thousands of metres (radio waves) to millionths of millionths of millimetres (gamma rays). Radiation of long wavelength may cause body atoms and molecules to vibrate but does not, so far as is known, significantly damage them (non-ionizing radiation). Very short wavelength radiation, such as X-rays and gamma rays (ionizing radiation), however, can knock out linking electrons from molecules, causing them to separate into smaller charged bodies or chemical groups called ions, or free radicals. Ionizing radiation can damage any body molecules, including DNA, and this may kill cells or alter their genetic structure. Such mutations in surviving cells may lead to cancer. At the same time, rapidly dividing cancer cells are more susceptible to the effects of ionizing radiation than normal cells. This is the basis of radiotherapy.

radic-, radicul-, radiculo- *combining form denoting* root.

radicle any small structure resembling a root. From the Latin *radicula*, diminutive of radix, a root.

radio- *combining form denoting* radiation.

radioactive marker a compound containing a radio-isotope atom whose movement through a chemical reaction can be monitored by virtue of the radiation emitted.

radioactivity 1 spontaneous emission of RADIATION.
2 the radiation emitted by unstable atomic nuclei or in the course of a nuclear reaction. Radioactivity includes alpha particles (helium nuclei), beta particles (high speed electrons), neutrons and gamma rays.

radiobiology the study of the effects of radiation, especially ionizing radiation, on living organisms.

radioimmunoassay any method of measuring the extent of linkage between ANTIGEN and ANTIBODY in which one or other of these is labelled with a radioactive substance (radionuclide). Measurement of radiation can be remarkably precise.

radiolabelling a method of tagging and following the movement of a molecule by incorporating it into a radioactive atom. In genetics, radiolabelling is commonly used for the purposes of HYBRIDIZATION PROBING.

radiologist a doctor who specializes in medical imaging and who is skilled in the interpretation of X-ray, CT scan, MRI, PET scan and radionuclide scanning films.

radiology the medical specialty concerned with the use of RADIATION for diagnosis and treatment.

radius one of the two forearm bones, the other being the ULNA. The radius is on the thumb side and lies parallel to the ulna when the palm is facing forward. When the hand is rotated so that the thumb turns inward, the upper end of the radius rotates on a boss on the lower end of the upper arm bone (HUMERUS) but the lower end crosses over the ulna.

ragged red fibres a feature of several MITOCHONDRIAL DNA disorders in which a high proportion of the muscle fibres contain structurally and functionally abnormal mitochondria. When stained and examined under the microscope these fibres appear red and ragged. The finding is commonly associated with slowly progressive weakness of the muscles of the limbs, defects of ocular movements, abnormally rapid tiredness on exertion, and a marked rise in the levels of lactic acid in the blood on exertion or even at rest.

ramose many-branched.

ramus a branch or subdivision arising from the division (bifurcation) of a blood or lymphatic vessel or a nerve.

raphe a seam. The line or ridge of union between two mirror-image or symmetrical body parts.

rash any inflammatory skin eruption of reasonable extent and of whatever cause.

ras proto-oncogenes genes in humans and many other animals that influence cell differentiation. Mutations of ras genes may contribute to one third of all cases of cancer.

rate-limiting of a delay at any stage in a metabolic pathway that restricts the speed of the whole process. The delay point might be an enzyme that is more easily saturated with substrate than any other.

rationalism a general term for the group of philosophic schools that reject received or authoritarian wisdom and dogmatic religion and hold that knowledge is to be obtained only from observation and the application of logic to data so derived. Rationalism does not necessarily exclude religious beliefs, but tends to do so.

RDA *abbrev. for* recommended daily allowance. The daily intake of a particular nutrient considered adequate to maintain health. RDAs are decided upon by various official nutritional committees and tend to be diverse in detail.

re- *prefix denoting* again or back or backward.

reaction time the interval between the application of a stimulus and the first sign of a response.

reading frame an imaginary window through which base pairs can be inspected three at a time, and which provides a way in which a nucleotide sequence can be viewed as three different sets of CODONs only one of which is correct.

reason the faculty by which new information is derived from old, judgement exercised and argument pursued.

receding chin congenital underdevelopment of the lower jaw bone (mandible) of cosmetic importance only. Various plastic operations are possible. Also known as retrognathism.

receptor 1 any structure on or penetrating a plasma cell membrane or other membrane, capable of binding a specific external substance, such as a HORMONE, CYTOKINE, STEROID or NEUROTRANSMITTER, and, as a result, effecting a response within the cell. Plasma membrane receptors commonly respond by releasing a 'second messenger' within the cell.
2 a sensory nerve ending capable of receiving stimuli of various kinds and responding by the production of nerve impulses. The receptor has gradually come to be recognized as one of the fundamentally important entities in physiology, pathology, pharmacology and medical science generally.

recessive pertaining to an alternative form of a gene (ALLELE) that produces an effect only when carried by both members of the pair of homologous chromosomes (only when HOMOZYGOUS). People with HETEROZYGOUS alleles for a condition are called carriers. A recessive gene has no effect in the presence of a DOMINANT allele either because of its inactivity or because of the absence of a product.

recessive lethal a gene that has a lethal effect when present in both loci (homozygous).

reciprocal innervation the reduction in nerve impulses to the antagonists of muscles that are contracting to cause movement.

recognition the process of binding of an antigen to a specific receptor on a cell of the immune system.

recombinant pertaining to an organism, chromosome or segment of DNA produced by genetic material from more than one source.

recombinant DNA DNA produced by the artificial linkage, in the laboratory or factory, of DNA from different sources. See also GENETIC ENGINEERING.

recombinant protein a protein synthesized in a cell from DNA into which the gene sequence that codes for the protein has been artificially introduced. Recombinant proteins can be produced in large quantities by GENETIC ENGINEERING.

recombination the formation in offspring of a combination of two or more genes that differs from the arrangement of these genes in either parent. This is the result of the exchange of segments of DNA during the germ cell divisions that resulted in the formation of paternal sperm and maternal ova.

recombination repair the process of making good a gap in one strand of DNA by plugging it with a homologous strand taken from another length of double helix.

recon the smallest genetic unit capable of recombination.

recreational drugs a dubious term that trivializes the dangers and serious social implications of the use of drugs such as cocaine, amphetamine (amfetamine), various hallucinogenic drugs and marijuana.

recruitment 1 activation of an increasing number of responsive cells as the size of the stimulus increases.
2 an unpleasant blasting sensation experienced by people with sensorineural deafness when exposed to loud noises.

rectal pertaining to the RECTUM.

rectum the 12.5 cm long, very distensible terminal segment of the large intestine, situated immediately above the anal canal. In spite of its name (*rectus* is Latin for straight), the rectum is curved and follows the hollow of the SACRUM. Its lining is smooth and the whole of the inside is accessible to the examining finger. Movement of bowel contents into the rectum causes the desire to defaecate.

rectus any of several straight muscles, such as the central vertical muscle on either side of the midline of the abdomen (the rectus abdominis), the rectus femoris on the front of the thigh or the four rectus muscles that move the eyeball. From the Latin *rectus*, straight.

recumbent lying down or reclining.

recurrent laryngeal nerves branches of the vagus, the 10th pair of cranial nerves. The recurrent laryngeal nerves leave the main trunk low in the neck, especially on the left side, and run up again to supply the muscles of larynx concerned with phonation. One of these nerves is commonly involved in neck cancer, the first sign of which may be severe loss of voice from paralysis of one vocal cord.

red blood cell see ERYTHROCYTE.

reduction the restoration of a displaced or broken part of the body to its proper position or alignment by manipulation or other surgical procedure. Reduction of bone fractures involves energetic pulling (traction) under anaesthesia to correct overlap, and local moulding pressure to realign the bone.

reduction division see MEIOSIS.

re-epithelialization the usually final healing stage of a wound in which the surface layer (EPITHELIUM) regenerates from the edges to cover the wound site.

referred pain pain felt in a place other than the site of the causal disorder. Stimulation of a sensory nerve at any point, including its root ganglion (as in shingles), always causes a sensation in the area of the peripheral distribution of the nerve. Referred pain also arises because nerves running to a particular segment of the spinal cord may come from widely separated points, and impulses coming from one of these areas may be interpreted as coming from the other. Stimulation from an inflamed gall bladder, for instance, causes pain in the right shoulder, the pain of angina pectoris is felt in the arms and neck, and pain from a URETER is often felt in the testicle.

reflex 1 an automatic, involuntary and predictable response to a stimulus applied to the body or arising within it.
2 the point of light reflected from a curved smooth surface, such as the CORNEA.

reflex bladder a urinary bladder no longer under voluntary control that empties automatically from time to time under the control of a local spinal reflex. This is a common sequel to severe spinal cord injury or disease.

reflux movement of fluid or semifluid material in a direction opposite to the normal. Regurgitation. Examples are reflux of acid material from the STOMACH into the OESOPHAGUS, of urine from the bladder up the URETERS to the kidneys or of the abnormal movement of blood back through a leaking (incompetent) valve in the heart.

refraction 1 the bending of light rays that occurs when they pass obliquely from a transparent medium of one density to one of another density.
2 the assessment of the optical errors of the eyes so that appropriate correcting spectacles can be prescribed.

refractory period the period immediately following the passage of a nerve impulse or the contraction of a muscle fibre during which a stimulus, normally capable of promoting a response, has no effect.

regimen any system or course of treatment, especially one involving special diet or exercise.

regression 1 a psychoanalytic term implying a return to childish or a more primitive form of behaviour or thought, as from a genital to an oral stage.
2 a psychological term denoting a temporary falling back to a less mature form of thinking in the process of learning how to manage new complexity. Cognitive psychologists view such regression as a normal part of mental development.
3 a statistical term defining the relationship two variables such that a change in one (the

independent variable) is always associated with a change in the average value of the other (the dependent variable).

regulator gene a gene that codes for RNA or for a protein whose function is to controls the expression of one or more other genes.

regurgitation see REFLUX.

rehabilitation 1 restoration of the physically, mentally or socially disabled to a normally functional life.
2 the specialty of physical medicine which is concerned with such restoration.

reimplantation the reattachment of an amputated body part or the restoration of a dislodged tooth to its socket.

reinforcement a term used in learning theory and in behaviour therapy that refers to the strengthening of a tendency to respond to particular stimuli in particular ways. In classical conditioning, the occurrence or deliberate introduction of an unconditioned stimulus along with a conditioned stimulus; in operant conditioning, a reinforcer is a stimulus, such as a reward, that strengthens a desired response.

relative refractory period the period during which a neuron which has recently produced an action potential cannot respond to a normal stimulus but will do so if a stronger than usual stimulus is applied.

releasing hormones hormones secreted by neurons in the hypothalamus which control the production of hormones by the PITUITARY GLAND.

remediable capable of being remedied.

remedial curative. Pertaining to, or providing, a remedy.

remission a marked reduction in the severity of the symptoms or signs of a disease, or its temporary disappearance.

REM sleep rapid eye movement sleep. This occurs during about 20 per cent of the sleeping time and features constant movement of the eyeballs, twitching of the muscles and erection of the penis in men. REM sleep is deep and is associated with dreaming. It is necessary for health.

renal pertaining to the kidneys.

renal disorders see KIDNEY DISORDERS.

renal dwarfism failure of body growth as a result of persistent kidney disease in childhood.

renal pelvis the conical cavity lying on the inner side of the kidney into which all urine secreted by the kidney runs. The renal pelvis is connected directly to the ureter.

reniform kidney-shaped.

renin an enzyme produced by the kidney in conditions of abnormally low blood pressure. Renin catalyses the release of angiotensin I from a blood globulin angiotensinogen, and this, in turn, is converted to angiotensin II by a converting enzyme found in the lung. Angiotensin II causes the adrenal glands to secrete the hormone aldosterone which acts on the kidneys to reduce the loss of sodium in the urine. The increased blood sodium raises the blood pressure.

repetitive strain injury a disorder of motor function caused by any often-repeated activity that is persisted in beyond a particular threshold, especially if the activity involves an inherently awkward or uncomfortable position of the body. RSI particularly affects musicians, keyboard operators, cleaners, packers and machine operators. There is acute pain and cramp-like stiffness, and sometimes total inability to continue in the associated occupation.

repolarization restoration of the resting polarized state in a muscle or nerve fibre. Polarization implies a balanced electrical charge on either side of the fibre membrane, being, in the resting state, negative on the inside and positive on the outside. In depolarization the charges are locally reversed.

repressed of a gene that is inhibited from transcription by the bonding to it of a REPRESSOR PROTEIN.

repression 1 inhibition of transcription at a particular site on DNA or MESSENGER RNA by the binding of REPRESSOR PROTEIN to the site.
2 the prevention of the synthesis of certain enzymes by bacterial products.

repressor gene a gene that codes for a repressor protein.

repressor protein a protein that binds to an operator gene on DNA to inhibit its transcription into messenger RNA, or that binds to RNA to prevent its translation.

reproduction any process by which an organism gives rise to a new individual. Most biological reproduction is cellular and asexual and occurs by chromosomal duplication followed by elongation and splitting of the cell into two individual cells identical to the parent. Sexual reproduction is more complex and involves the production of specialized body cells called gametes which have experienced two stages of shuffling and redistribution of chromosomal segments and a reduction to half the full number of CHROMOSOMES (haploid). In the fusion of the male and female gametes, sperm and egg respectively (fertilization), the full complement of chromosomes is made up. The potential new individual now has a GENOME different from that of either parent and will differ in many respects. A fertilized ovum divides rapidly and repeatedly, but the reproduced cells do not usually separate, but continue to duplicate and specialize until a new individual is formed. Sometimes, after the first or second division, the reproduced cells separate to form genetically identical siblings.

reproductive system

Reproduction is not necessarily sexual; indeed the great majority of living things, such as the cells of our bodies and most micro-organisms, reproduce asexually. But all highly organized animals and plants reproduce themselves sexually and possess specialized organs for the purpose. Asexual reproduction implies that the daughters are identical to the mother, because no new genetic information is involved. Sexual reproduction, on the other hand, is the basis for the remarkable diversity of individuals within a species. In humans, sexual reproduction also has major emotional and social elements relating to sexual attraction, mating, family formation and the protection of the young.

The essential feature of human reproduction is that each new individual begins life as a single cell (the maternal egg or ovum) that has been penetrated by the paternal spermatozoon. Both the unfertilized ovum and the sperm contain half the normal number of chromosomes, so the genetic characteristics of the person-to-be come half from the mother and half from the father. Fertilization and the subsequent nurturing of the growing embryo and fetus occur within the body of the mother; sperm production and emission are all that are required of the male. These facts imply considerable differences between those parts of the male and female anatomy concerned with reproduction. In other respects, although there are many quantitative differences, male and female anatomy are essentially the same.

The female requires structures for egg production (ovaries), ducts in which the eggs can be fertilized (the fallopian tubes), a structure for the accommodation of the growing embryo and fetus (the womb or uterus) and a suitable duct, accessible from the outside, that allows sperm to be deposited near the site of the egg (the vagina) and provides an exit route for the new individual. Because the developing offspring must be nourished within the body of the mother until it is mature enough to exist independently, the womb must be capable of extending considerably, of providing large quantities of all the substances needed for growth, and of actively expelling the mature fetus at the appropriate time.

Female reproductive organs

The externally visible parts of the female genitalia (the vulva) consist of the large and small lips (the labia majora and minora) and the clitoris. These are not essential for reproduction. The large lips are usually in contact so that the entrance to the vagina (the introitus) is covered. Just within them are the thin skin folds constituting the labia minora. At the front these join to form a kind of hood for the clitoris. Both the clitoris and the labia minora contain many blood vessels which, under the influence of sexual excitement, become widened so that the parts become engorged, swollen and hot. The same stimulus causes secretion of clear mucus from the two Bartholin's glands, which lie in the rear parts of the labia majora. This mucus lubricates the entrance to the vagina and facilitates sexual intercourse. Between the clitoris, at the front, and the vaginal opening near the back of the vulva, lies the small pouting opening of the urethra – the short tube that carries urine from the bladder to the exterior.

The clitoris is a miniature penis containing spongy tissue and is capable, like its male counterpart, of becoming engorged with blood and erect. During sexual intercourse, movement from the vagina causes the hood of the labia minora to massage gently the sensitive bulb of the clitoris. The resulting pleasure (given an acceptable partner), and the anticipation of orgasm, may prompt the female to allow the process to continue until male ejaculation has occurred.

The vagina is a remarkably extensible tube of muscle and fibrous tissue, about 10 cm long. It contains no glands, but fluid can pass though its lining from the underlying tissue. It runs upwards and slightly backwards and is normally passively closed. Tight closure may occur voluntarily or from fear by contraction of the muscles of the floor of the pelvis. Many bacteria, known as lactobacilli, inhabit the healthy vagina. These act on a carbohydrate stored in the surface cells to produce lactic acid. Vaginal acidity is important in discouraging growth of undesirable organisms such as the thrush fungus.

The uterus is a small, hollow, pear-shaped, thick-walled, muscular organ, lying inclined forward almost at right angles to the vagina. The neck of the uterus, or cervix, protrudes into the upper part of the vagina. The cavity of the uterus is about eight cm long, and the lining varies in thickness, from 1 to 5mm depending on the stage of the menstrual cycle. From the right and left sides of the upper part of the body of the uterus narrow muscular tubes, the fallopian or uterine tubes, run outwards to the ovaries. The outer end of each of these tubes is open and bears a large number of finger-like processes (fimbriae) which almost cover the surface of the corresponding ovary. These processes ensure that eggs released from the ovaries are carried into the fallopian tubes, where fertilization occurs. Each ovary is an almond-shaped body, 3–4 cm long, hanging from the back of a membrane that descends from the fallopian tube on each side. The eggs develop in cell collections, known as Graafian follicles, on the surface of the ovaries.

The menstrual cycle

This is the result of the production of a sequence of sex hormones from the ovaries. These act on the lining of the uterus (the endometrium) causing it to thicken and become suitable for the reception of a fertilized egg. If pregnancy occurs the cycle proceeds no further, but if it does not, the thickened lining is cast off over the course of three or four days. This is called menstruation. The whole cycle takes about 28 days and is timed around the event of ovulation. In the absence of pregnancy, menstruation normally occurs about 14 days after ovulation.

The thickening of the womb lining is caused by oestrogen hormones, the output of which begins to rise at the end of menstruation. During this time a follicle is maturing in an ovary as a result of follicle-stimulating hormone from the pituitary gland. The follicle cells produce the oestrogen. After the egg is released a mass of large yellowish cells (the corpus luteum) forms in the empty follicle. These cells secrete increasing quantities of another hormone, progesterone. This causes the womb lining to thicken further by fluid retention. If pregnancy occurs the corpus luteum persists and continues to maintain the womb lining. If it does not, the corpus luteum shrinks and the output of both oestrogen and progesterone drops. As a result, the womb lining is shed as a menstrual period.

Male reproductive system

The pair of testes, or testicles, are the site of sperm production. They are accommodated in the hanging skin bag, the scrotum where

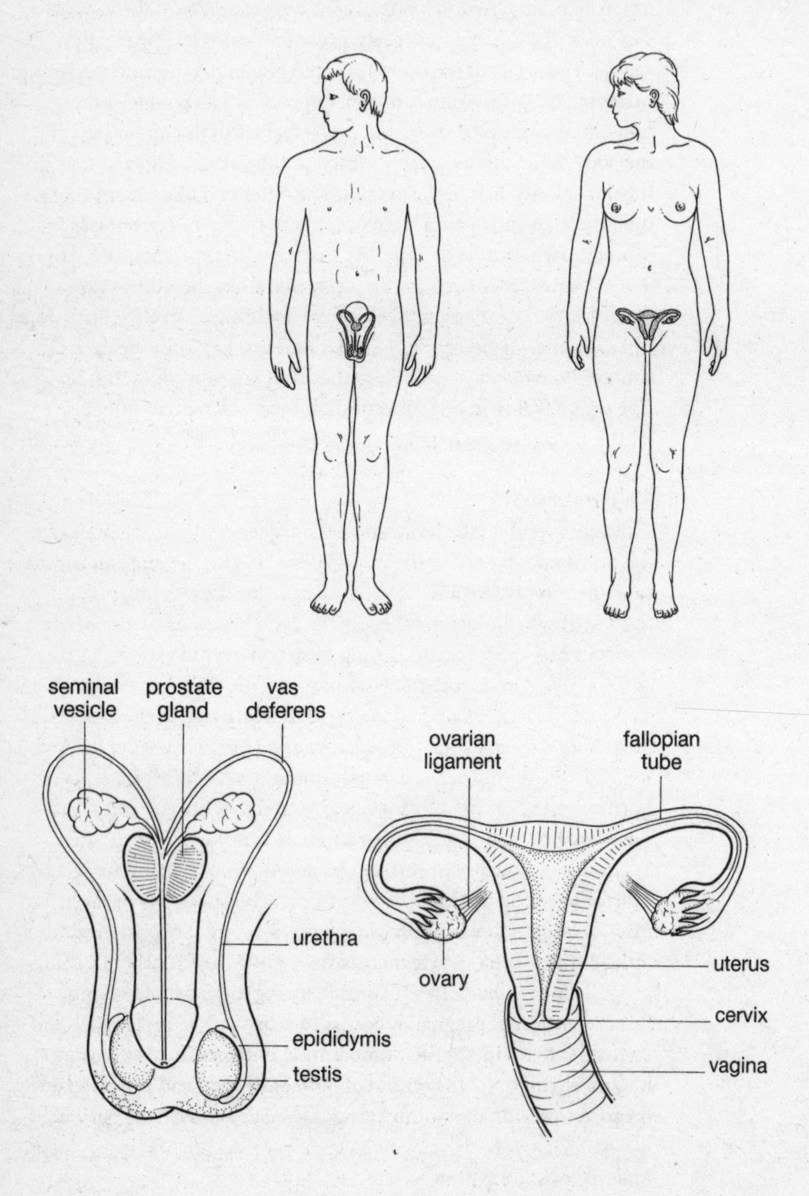

Fig. 14 **Reproductive system**

they are exposed to a temperature, somewhat lower than internal body temperature, necessary for sperm production. Each testis is divided into 300–400 tiny lobes and each lobe contains one to three coiled tubes in the walls of which the sperm develop. The total length of these seminiferous tubules is about 500 metres and they join up into a single, highly coiled tube that forms a slug-shaped mass, on the outside of the testis, called the epididymis. This is several metres long and it takes several days for sperm to traverse it. This time is necessary for their maturation.

Each epididymis ends in a vas deferens – the tube that carries the sperm up to the level of the outlet tube, the urethra. The spermatozoa do not swim up the vasa deferentia; they are forced up from below by pressure of new sperm production, and accumulate near the urine outlet tube (urethra). The seminal vesicles are two small, elongated sacs lying on either side of the prostate gland which surrounds the start of the urethra immediately under the urinary bladder. The seminal vesicles do not store spermatozoa; they produce a fluid that contributes to the volume of the ejaculate. The prostate secretes about one third of the fluid content of the seminal fluid and ducts pass through it from the seminal vesicles to enter the urethra that runs along the inside of the penis.

The main bulk of the penis consists of three longitudinal columns of spongy erectile tissue – the two corpora cavernosa that lie side by side along the upper aspect of the organ, and the corpus spongiosum lying centrally behind and encloses the urethra. The front end of the corpus spongiosum expands into an acorn-shaped swelling called the glans penis and this is enclosed in a sheath of skin, the foreskin or prepuce. Sexual interest causes the arteries supplying the erectile tissue to widen so that a quantity of blood flows under pressure into the three columns. As a result they expand considerably and the penis becomes stiffened, enlarged and erect. This causes compression of the veins that drain blood from the columns and the erection is maintained until sexual interest diminishes.

Penile erection is readily induced, especially in the young, by anticipation of sexual activity, sexual thoughts and fantasizing, erotic literature or art, physical contact of a suggestive nature or genital stimulation. The same stimuli cause secretion of clear mucus by glands near the base of the penis and this appears at the urethral opening at the tip of the penis. This lubrication of the rigid organ permits relatively easy insertion into the vagina of a cooperative partner.

Sexual intercourse and fertilization

Rhythmical massage of the sensitive glans penis by the walls of the vagina produced a strongly pleasurable sensation. This, together with the psychic stimulus of close human intimacy, soon produces the male orgasm. In most cases this occurs within four or five minutes of penetration – often much sooner. First the fluid contents of the seminal vesicles and the concentration of spermatozoa at the top of the vasa deferentia are squeezed into the urethra by contraction of the muscles in the vesicle walls and the prostate. Then the muscles on the floor of the pelvis and around the base of the penis contract strongly, rhythmically and repeatedly, driving the seminal fluid out of the penis (ejaculation). The male orgasm is associated with a pleasurable release of tension.

Within a few minutes of ejaculation, some sperm have passed into the canal of the cervix. The state of the cervical mucus greatly influences the ability of sperm to pass. The mode of sperm transport remains uncertain. Because of their very small size relative to the length of their journey, it is unlikely that tail-lashing propulsive effort would suffice. It seems probable that they are also carried by fluid currents set up in the uterus by the action of hair cells (cilia) in the lining. Only one sperm in a hundred reaches the inside of the womb and for 30–45 hours they remain capable of penetrating an egg. For pregnancy to occur, eggs must be fertilized within the fallopian tube and this must occur between six and twenty-four hours after release from the ovary.

The relatively few sperm that reach the ovum surround it, butting with their heads and releasing an enzyme, hyaluronidase, that softens and thins the outer membrane of the egg. Normally, only one sperm penetrates and as soon as this has happened a physical barrier to further sperm penetration forms as a result of a change in the electrical charge on the egg. If more than one sperm gains admission the number of chromosomes exceeds the normal and this leads to gross abnormality in the embryo and usually its early death.

residual volume the volume of air remaining in the lungs after maximal expiration.

respiration 1 breathing.
2 the whole process by which oxygen is transferred from the atmosphere to the body cells and carbon dioxide is moved from the cells to the atmosphere. Respiration is vital to life and cessation for more than a few minutes is fatal.

respirator 1 any mechanical device used to maintain the breathing and the supply of air or oxygen to the lungs. Most modern respirators are of the intermittent positive pressure type.
2 a filtering device that covers the face and removes toxic elements from the inspired air.

respiratory arrest cessation of breathing.

respiratory rate the number of breaths per minute.

respiratory system

A person can survive for weeks without food and for days without water but even a few minutes without oxygen is likely to be fatal. Oxidation – combination with oxygen – is probably the most important of all chemical reactions, and without it, life is inconceivable. Oxidation is the ultimate source of energy for nearly all living organisms and is the reaction involved in the natural decomposition of organic matter. Oxidation can be rapid, as in burning, when energy is rapidly released, or slow as in the reactions in the cells of the human body or in decomposition. Slow organic oxidation reactions are usually brought about by the catalytic action of enzymes. The oxidation of body fuels – glucose, fatty acids and sometimes amino acids – ends in the production of carbon dioxide and water.

The necessity for a constant supply of oxygen, literally vital for survival, predicates a body system in which oxygen can be taken from the atmosphere and rapidly distributed to every cell in the body. This requirement is effected by close cooperation between the circulatory system and the respiratory system. The air we breathe consists of about 20 per cent oxygen and 78 per cent nitrogen, with traces of rare gases and about 0.03 per cent of carbon dioxide. Nitrogen is an inert gas that is largely ignored by the body and becomes important only when the body is exposed to such high pressures that the gas dissolves in the body fluids in large quantity.

Respiration

Air is drawn into the lungs by the continuous, and largely unconscious, action of the muscles of respiration under the control of the respiratory centre in the brainstem. These muscles are, however, voluntary, and, unlike the heart, have no inherent tendency to contract. They do so only when stimulated by nerve impulses from a controlled source in the respiratory centre. If the respiratory centre is destroyed by disease or injury breathing stops and only artificial respiration by mechanical means can preserve life.

When the ribs are pulled upwards by the muscles lying between them (the intercostal muscles), in a manner resembling venetian blinds, they also rotate outwards. Because of their curved shape, this movement widens the chest and increases its volume. At the same time, the muscles of the upwardly-domed diaphragm contract, pulling down its central fibrous portion and causing the whole structure to flatten. This, too, adds to the internal volume of the chest. The reduction in air pressure in the chest immediately results in air being pushed into it by the higher atmospheric pressure.

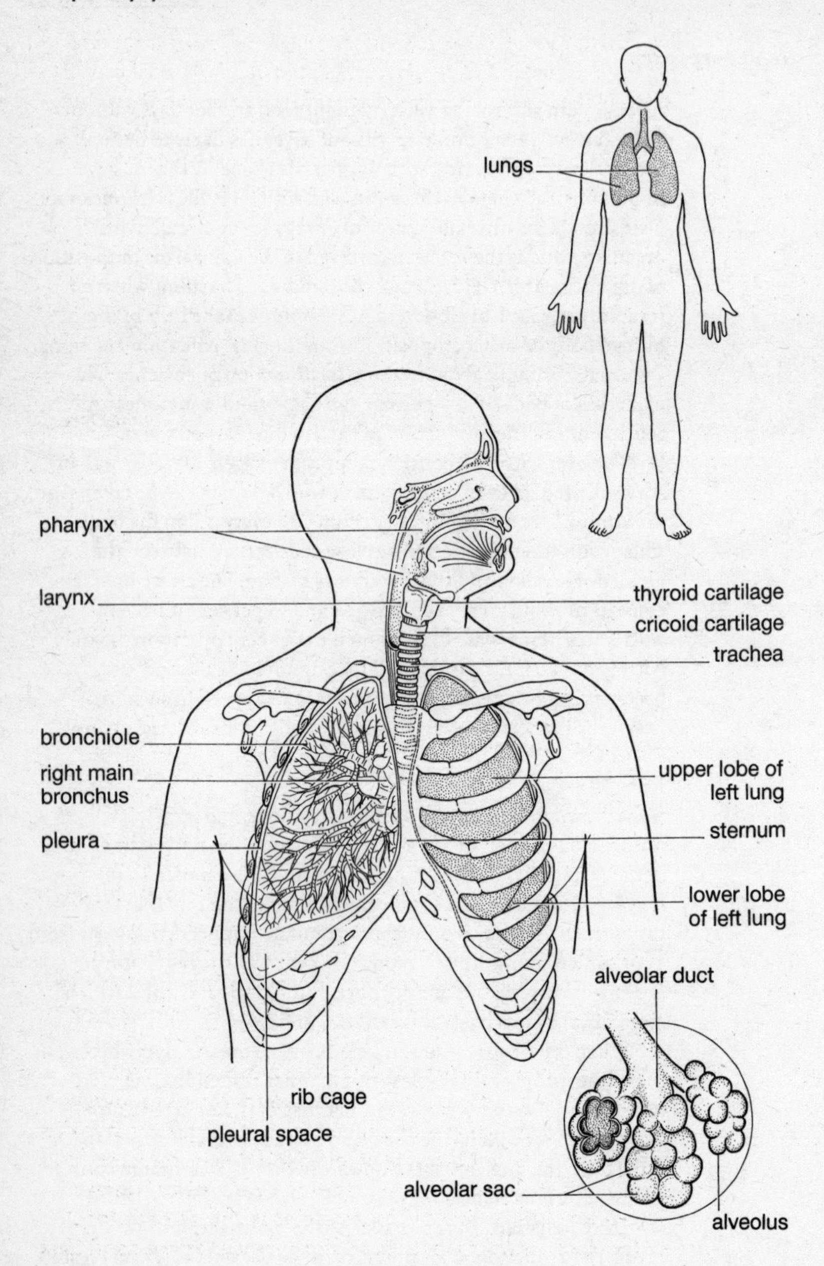

Fig. 15 **Respiratory system**

This air passes through the nose or mouth, down the throat, through the opening into the voice box (larynx), between the vocal cords in the larynx, down the windpipe (trachea) and the branching air tubes (bronchi), down the smaller air tubes (bronchioles) to end up in the 300 million or so tiny air sacs of the lungs (the alveoli). When inspiration is complete, the rib cage falls, the diaphragm returns to its domed shape and the lungs deflate by their natural elasticity. This whole cycle is repeated at a rate which varies, with the needs of the body or the state of the emotions, from about 12 per minute at rest rising to as high as 180 per minute during strenuous exercise or extreme anxiety.

Gas exchange between the atmosphere and the blood occurs in the alveoli of the lungs. Each of these tiny air sacs is surrounded by a network of blood capillaries which are part of the blood circulation system of the lungs. The walls of the alveoli and of the surrounding capillaries are both so thin that the blood, spread out in the fine capillaries, is brought into intimate relationship with the air in the alveoli.

Oxygen transport

Each millilitre (cubic centimetre) of blood contains about five billion oxygen carriers. These are the red blood cells (erythrocytes), each about 7 thousandths of a millimetre in diameter. Erythrocytes are tiny, disc-shaped, non-nucleated cells, concave on both sides to increase their surface area and each filled with about 300 million molecules of the iron-containing protein haemoglobin. Globins are proteins and each haemoglobin molecule consists of four such protein chains folded together, each being wrapped round a complex of iron and protoporphyrin called a haem. The iron atoms in the haem can readily form loose bonds with oxygen, each atom linking to one molecule of oxygen (O_2). This is the secret of oxygen transportation. Whenever haemoglobin finds itself in an environment of high oxygen tension (as in the alveoli) the iron atoms link up with oxygen molecules. In an environment of low oxygen tension, the haemoglobin gives up oxygen.

The demand for oxygen in the tissues is high and oxygen is rapidly taken up. At the end of the metabolic processes the oxygen comes into combination with carbon to make CO_2 and with hydrogen to make water (H_2O). The result of this is that the environment through which the blood in the tissue capillaries passes is low in oxygen. Any oxygen linked to haemoglobin is thus given up to it. All the blood returning to the heart and pumped to the lungs is thus low in oxygen. As it passes through the capillaries surrounding the alveoli it enters an environment of high oxygen tension and is immediately, and automatically, replenished, every haemoglobin molecule picking up four atoms of oxygen. Haemoglobin linked with oxygen is a bright red colour; haemoglobin without oxygen is a dull purplish colour. These colour differences readily distinguish arterial blood, which has just come from the lungs, from venous blood, which has just come from the tissues.

Carbon dioxide is carried mainly in the form of bicarbonate dissolved in the blood serum and in the red blood cells, and leaves the blood in the alveoli simply by diffusing through the alveolar walls into an atmosphere of lower carbon dioxide tension. So the expired air always contain substantially more CO_2 than the inspired air.

Control of respiration

The maintenance of an adequate supply of oxygen to the body is so important that without a feed-back control mechanism to correct oxygen lack, we would be unlikely to survive. Danger arises when there is either a drop in the level of oxygen, or a rise in the level of carbon dioxide, in the blood. Both parameters are monitored and automatically regulated. The main artery of the body (the aorta) and two of its branches to the head (the carotid arteries) carry in their walls collections of monitoring cells which can detect changes in the amounts of oxygen carried by the blood. When a drop is detected, messages are sent to the respiratory centre in the brainstem to cause it to increase the rate of respiration by sending impulses more rapidly to the respiratory muscles.

Any rise in the carbon dioxide level in the tissues is at once reflected in a rise in the amount of carbon dioxide carried by the blood. This blood, passes through all parts of the body, including the respiratory centre. As it does so it is monitored and if a rise in CO_2 is detected, the centre immediately increases the rate of respiration. A high rate of respiration often occurs as a response to acute anxiety. This is normally helpful as an appropriate response to physical emergency or danger. But if the anxiety is inappropriate and exertion is not needed, overbreathing (hyperventilation) merely results in the abnormal loss of carbon dioxide, an increase in the alkalinity of the blood leading to a drop in calcium and possibly spasms of the muscles (tetany).

Respiration and exercise

Strenuous exercise demands a greatly increased supply of oxygen and a greatly increased rate of disposal of carbon dioxide. During exercise, the ventilation of the lungs may increase by as much as twenty times, but this, in itself may not be sufficient to supply the extra oxygen needed by the muscles. During sedentary breathing many of the capillaries surrounding the alveoli are closed simply because the blood pressure in the lungs, caused by the contraction of the right ventricle, is insufficient to open them. During exercise, however, the heart rate and force increases and the pressure in the pulmonary circulation rises. This causes more of the capillaries to open so that there is a greater surface area available for the passage of oxygen from the alveoli into the blood and for the passage of carbon dioxide in the opposite direction.

The increased efficiency of gas exchange during exercise prevents any marked change occurring in the levels of oxygen and carbon dioxide in the blood. This poses the question: What is the stimulus for the greatly increased respiratory rate during exercise? Surprisingly, in spite of much research, this remains unexplained. It seems probable, although unproved, that the muscles and joints send stimulating messages to the respiratory centre. Possibly, the abrupt increase in ventilation during exercise is the result of a conditioned reflex.

respiratory tract infections these include bronchiolitis, bronchitis, common cold, croup, laryngitis, pharyngitis, pneumonia, sinusitis and tonsillitis.

resting membrane potential the voltage difference between the inside and the outside of a cell when no stimulus is applied.

restriction enzymes enzymes that break DNA at specific sites. These enzymes are extensively used in research and in GENETIC ENGINEERING. Also known as restriction endonucleases.

restriction fragment length polymorphism (RflP) variations within a species in the lengths of fragments of DNA caused by RESTRICTION ENZYMES. The variations are caused by mutations that either abolish the normal sites of breakage or create new ones, characteristic of the mutations. RflP analysis may allow genetic abnormalities to be detected, often before birth, even if the location of the mutated gene or genes is unknown.

restriction map a diagram of the sites on DNA that are cut by different RESTRICTION ENZYMES.

resuscitation 1 restoration of a stable physiological condition to a person whose heart action, blood pressure or body oxygenation have dropped to critical levels. 2 active measures to treat shock.

retardation a state of backwardness or delayed development, especially of the intellectual functions. Learning difficulty or mental deficiency.

rete a mesh or network, as of blood vessels. From the Latin *rete*, a net.

rete testis a network of cords in the testicle that canalize to become the SEMINIFEROUS TUBULES in which the spermatozoa develop.

reticular net-like.

reticular formation a network of islets of grey matter, consisting of large and small nerve cells and their connections, scattered throughout the brainstem and extending into the THALAMUS and HYPOTHALAMUS. The formation receives information from many other parts of the brain and is concerned with alertness and direction of attention to external events, as well as sleep. It has a major effect on the sensory and motor systems.

reticulo- *combining form denoting* a net or a net-like structure.

reticulocytes immature red blood cells that contains a network that can be stained blue with basic dyes. Reticulocytes appear in the circulation at times of increased red cell formation.

reticulocytosis a greater than normal proportion of RETICULOCYTES in the blood. This indicates an increased rate of red cell formation, as during the treatment of anaemia or following haemorrhage.

reticuloendothelial system an obsolescent term for the widespread system of protective MACROPHAGE (phagocyte) cells and endothelial cells found in the bone marrow, liver, spleen and elsewhere. The cells of the reticuloendothelial system include HISTIOCYTES, MONOCYTES, the KUPFFER CELLS of the liver and lung macrophages.

reticulum any netlike structure of the body.

retina the complex membranous network of nerve cells, fibres and photoreceptors that lines the inside of the back of the eye and converts optical images formed by the lens system of the eye into nerve impulses. The retina contains colour-blind but very sensitive rods and colour-sensitive cones and a computing system that refines the signals produced by these. The impulses leave each retina by way of about one million nerve fibres that form the optic nerve.

retinaculum a fibrous band, strap or ligament that holds another part in place. The flexor retinaculum prevents the flexor tendons from springing away from the front of the wrist when it is bent.

retinal the aldehyde found in visual pigments, such as visual purple. Also known as retinaldehyde.

retinal disorders these include central serous retinopathy, macular degeneration, retinal detachment, retinal haemorrhage, retinitis pigmentosa, retinoblastoma, retinopathy, toxocariasis and toxoplasmosis.

retino- *combining form denoting* RETINA.

retinoids a class of compounds containing 20 carbon atoms and related to RETINAL and vitamin A. They include retinol, tretinoin, isotretinoin, etretinate, acitretin and arotinoid. Retinoids bind to cell nuclear retinoid receptors and the latter bind to DNA. The effect of the retinoids is to promote transcription of DNA in much the same way as do steroids and thyroid hormone. They have a wide range of biological actions and have been found useful in acne, certain skin cancers, psoriasis, skin ageing and other skin disorders. Retinoids taken in the first three weeks of pregnancy are liable to cause abortion or serious fetal defects. For this reason the use of systemic retinoids in women is strictly controlled.

retinol the common form of vitamin A found in animal and fish livers and other foods of animal origin. It is easily converted in the body into RETINAL.

retinopathy any non-inflammatory disease of the RETINA.

retirement the period, demonstrably dangerous to health and longevity, after the permanent cessation of work.

RET oncogene a proto-oncogene on chromosome 10 that codes for one of the tyrosine kinase receptors. Several mutations are known and all of them are associated with thyroid tumours. A number of children living in the Chernobyl region at the time of the power station disaster have been found to have rearrangements of the RET oncogene. Thyroid cancers, normally very rare in children are now occurring much more frequently than usual.

retro- *combining form denoting* back, backward or behind.

retrograde a going backwards or a reversion of the usual sequence.

retrograde amnesia loss of memory for a period before the time of a head injury. In general, the more severe the injury and the longer the period of loss of memory after the injury, the longer will be the retrograde amnesia.

retrograde ejaculation the passage of seminal fluid into the bladder during the male orgasm. This is due to failure of closure of the internal urethral sphincter as a result of injury during prostatectomy, damage to the sympathetic nerves in the region during local surgery, spinal injury or diabetic neuropathy. Infertility may result but various measures may be used to overcome this.

retroperitoneal behind the peritoneal membrane of the abdomen.

retrosternal behind the breastbone.

retroversion a turning or tilting backward or the state of being turned or tilted back.

retroverted uterus a UTERUS that lies in line with the long axis of the vagina or at an angle that slopes back from this axis. Normally, the uterus is inclined forward at about a right angle to the vagina. Retroversion is the condition of about 20 per cent of women and is not, in itself, considered to be in any way harmful.

retrovirus a virus with a GENOME consisting of a single strand of RNA from which DUPLEX DNA is synthesized under the catalytic influence of an enzyme called reverse transcriptase. This is the reverse of the much more common DNA to RNA process. The AIDS virus HIV is a retrovirus.

reverse transcriptase an enzyme that allows certain viruses, notable HIV, to synthesize double strand DNA from a single strand of RNA so that it can be incorporated into the genome of the host cell. Such viruses are called retroviruses.

reward deficiency syndrome a name for a relative failure of the dopaminergic system which plays a major part in brain-reward mechanisms. The syndrome, which has been linked to dysfunction of the D_2 dopamine receptors, includes various conditions, such as drug and alcohol abuse, smoking, obesity, pathological gambling and attention deficit hyperactivity disorder, in which the subject seems to be unusually concerned to achieve reward. The D_2 dopamine receptor gene is on chromosome 11 and has multiple allelic forms. Variants have been correlated with these and other reward-seeking behaviours.

rhabdo- *combining form denoting* rod-shaped or striped.

rhabdomyo- *combining form denoting* striped muscle.

rheology the study of the deformation and flow of matter in tubes and elsewhere. Rheology has become important in studies of blood flow in vessels.

rhesus factor Rh factor. A group of antigens occurring on red blood cells in a proportion of people, the most important of which was first found in a rhesus monkey. The gene locus for the Rh factor is on one end of the long arm of chromosome 1. See RHESUS FACTOR DISEASE.

rhesus factor disease a severe blood disorder caused by incompatibility between the blood group of the fetus and that of the mother. The rhesus factor is acquired by dominant inheritance and is present in 85 per cent of the population. The danger arises if the mother is rhesus negative and the father rhesus positive and passes this on to the fetus. When the fetus is rhesus positive, its red blood cells act as ANTIGENS causing the mother to produce ANTIBODIES against them. The danger is minimal in the first pregnancy but increases thereafter as antibody levels rise. In the most severe cases, the fetus may die in the womb. If born alive, the baby is deeply jaundiced with an enlarged liver and spleen and a low haemoglobin level. Treatment is by an exchange transfusion, via the UMBILICAL CORD, immediately after birth or even while still in the womb. Rhesus negative women can be prevented from developing antibodies by being given an anti-D GAMMAGLOBULIN.

rheumatism a common term for pain and stiffness in joints and muscles as well as for major disorders such as rheumatoid arthritis, osteoarthritis and polymyalgia rheumatica.

rheumatology the medical specialty concerned with the causes, pathology, diagnosis, and treatment of diseases affecting the joints, muscles, and connective tissue.

rhin-, rhino- *combining form denoting* nose.

rhinencephalon the part of the brain concerned with smell.

rhinogenous originating in the nose.

rhinology the medical and surgical specialty concerned with the structure, functions and diseases of the nose.

Rh negative lacking the rhesus factor on the red blood cells.

rhodopsin the retinal rod photoreceptor pigment. Also known as visual purple.

rhythm method an unreliable method of contraception based on an attempt to predict, and avoid, the part of the menstrual cycle on which OVULATION is most likely to occur.

riboflavin vitamin B_2.

ribonuclease an enzymes that promotes the HYDROLYSIS of ribonucleic acid (RNA).

ribonucleic acid see RNA.

ribose a pentose sugar that is a component of nucleic acids.

ribosomal RNA (rRNA) ribonucleic acid that is a permanent structural feature of RIBOSOMES.

ribosome a spherical cell ORGANELLE made of RNA and protein which is the site of protein synthesis in the cell by linking amino acids into chains. Ribosomes may be free or may be attached to the endoplasmic reticulum. During translation, ribosomes attach to MESSENGER RNA molecules and travel along them, synthesizing polypeptides as they go.

ribosome binding site the base sequence on a MESSENGER RNA molecule to which a RIBOSOME attaches.

riboswitches the mechanism that regulates gene expression (the turning on and off of genes) by means of a set of molecular switches. The structure of RNA can readily be altered by the binding on of small molecules or oligonucleotides.

ribozyme one of a unique class of RNA molecules that can act as cleaving enzymes in addition to storing genetic information. This is a notable exception to the general rule that all enzymes are proteins. Ribozymes form complementary base pairs in the normal manner but can cleave segments of nascent RNA during the splicing process of the formation of mature RNA transcripts of DNA. Ribozymes can be used in various ways as treatment modalities.

ribs the flat, curved bones that form a protective cage for the chest organs and provide the means of varying the volume of the chest so as to effect respiration. There are 12 pairs of ribs, and each pair articulates with a vertebra in the spine, at the back. The upper 10 pairs are joined to cartilages connected to the breastbone at the front, but the 11th and 12th pairs are free at the front (floating ribs).

rictus a facial contortion or grimace of pain.

rigidity sustained muscle tension causing the affected part of the body to become stiff and inflexible. Rigidity may be due to muscle injury, neurological disease such as Parkinson's disease, underlying inflammation as in peritonitis, or arthritis in an adjacent joint.

rigor a violent attack of shivering causing a rapid rise in body temperature.

rigor mortis the stiffening of muscles which occurs after death as a result of the loss of ATP so that the cross bridges cannot disconnect from actin. Rigor usually starts about 3 or 4 hours after death and is usually complete in about 12 hours. It may start much earlier, sometimes almost immediately, if the subject was engaged in strenuous activity or was fevered or suffering convulsions before death. It passes off as enzymes break down and soften the muscles over the course of the next 2 or 3 days.

ringing in the ears see TINNITUS.

risus sardonicus a characteristic facial expression, as of a sardonic grin, caused by spasm of the muscles of the forehead and the corners of the mouth in acute TETANUS.

RNA

RNA is an abbreviation for ribonucleic acid. This molecule, with DNA and MITOCHONDRIAL DNA, carries coded instructions for the synthesis of specific proteins from AMINO ACIDS.

Proteins are not made directly on the DNA molecule itself but in bodies known as *ribosomes*. These lie within the cytoplasm, the part of the cell outside the nucleus. First, a short length of double helix separates longitudinally to form a loop, exposing the sequence of single bases that constitute the gene. Base triplets different from those coding for amino acids indicate where the gene starts and where it ends. A new, short, complementary strand is now made on the exposed bases. This strand is called RNA and it forms at a rate of about 50 bases per second. In RNA production the accuracy of copying is poor, about one mistake occurring every 100,000 bases. But many copies are made and the occasional error doesn't matter. RNA uses three of the same bases as DNA, but thymine is not used. Another base called uracil is substituted for thymine, and this links with the DNA adenine.

The RNA chains formed in this way are called messenger RNA (mRNA) because they carry the code of the gene out through pores in the nucleus of the cell to the ribosomes in the cytoplasm. Before the codes are used to sort the amino acids, each mRNA length is edited to get rid of unneeded sections and the remaining coding sequences are spliced together. It is this edited version of the mRNA that is read by the ribosomes.

The cell fluid contains millions of amino acids, most of which are derived from food but some of which are synthesized in the body. Before the ribosomes can join these together in the right order to form new protein molecules they have to be brought to the ribosomes in the right sequence. This is done by yet another kind of RNA, transfer RNA (tRNA), which is replicated off from the edited mRNA. Transfer RNA moves around in the cell fluid picking up the twenty different amino acids and carrying them to the ribosome site. There, the mRNA, the tRNA and the ribosome all work together to form the chain of protein. The ribosome is a tiny protein body that moves along the strand of mRNA checking the sequence of bases, selecting amino acids from the tRNA in the right order, and linking them together to form proteins.

RNA interference RNAi a pathway that blocks gene expression by degrading RNA. RNA interference occurs naturally against viruses and to control gene activity and has attracted much attention for its potential as a therapeutic process. It involves an attack on MESSENGER RNA using small RNA molecule fragments (siRNA) that matches part of the sequence of the target gene. In the search for effective artificial RNA interference methods, arbitrary short sequences have often been found to fail because they are 'decoyed' by other genes in the genome. MicroRNAs have been shown to be important in the development of flower structure and may also be important in animal development. RNAi has been used in genetic research in determining the function of genes.

RNA polymerase an enzyme that catalyses the joining of appropriate NUCLEOTIDES to form a molecule of RNA, using DNA as a template.

RNA replicase an enzyme that catalyses the synthesis of RNA, using RNA as a template. The enzyme is used by RNA viruses for their replication within a cell.

Rnase RIBONUCLEASE.

RNA transcript an RNA complementary copy of a gene. The RNA transcript is always longer than the gene because the RNA polymerase also transcribes a leader segment prior to the gene code and a trailer segment after it.

robotics the branch of technology concerned with the development of machines capable of performing complex tasks of a kind normally limited to humans. Robotic machines of limited function controlled by computer have now become commonplace in the manufacturing industries, but the expected development of anthropomorphic or humanoid robots, in the manner predicted by the writer Karel Capek in his 1921 novel Rossum's Universal Robots, has not been fulfilled in any but a trivial sense. Robotics is now impinging on surgery and is likely to be important in the future.

role-playing 1 the acting out of a pattern of behaviour considered appropriate to one's social, educational or professional position or to one's current health status.
2 the adoption of a role foreign to one's normal situation.

root canal the pulp cavity in the root of a tooth.

Rorschach test see INK BLOT TEST. (Hermann Rorschach, 1884–1922, German-born Swiss psychiatrist).

rotator cuff the tendinous structure around the shoulder joint consisting of the tendons of four adjacent muscles blended with the capsule of the joint. Tearing or degeneration of any of these fibres may cause the common, painful and disabling rotator cuff syndrome in which there may be inability to raise the arm in a particular direction. Surgical repair may be necessary.

roughage dietary fibre, consisting of polysaccharides such as celluloses, pectins and gums for which no digestive enzymes are present in the intestinal canal. Roughage is effective in treating CONSTIPATION, DIVERTICULITIS and the IRRITABLE BOWEL SYNDROME, and may reduce the probability of developing cancer of the colon.

round window the membranous, elastic opening in the outer wall of the inner ear that allows free movement of the fluid within the COCHLEA of the inner ear when sound vibrations are conveyed to it by the AUDITORY OSSICLES.

RSI *abbrev. for* repetitive strain injury.

rugae folds, wrinkles or creases, as in the skin of the scrotum.

rumination voluntary regurgitation of food from the stomach which is then again chewed and swallowed. Rumination sometimes occurs in mentally disturbed people.

rump the buttocks.

rupture a popular term for an abdominal hernia.

Ss

Sabin vaccine an effective oral vaccine used to immunize against poliomyelitis. This vaccine contains live attenuated viruses that spread by the fecal-oral route in the manner of the original disease, thus effectively disseminating the protection. It was produced in 1955 and has been highly successful. (Albert Bruce Sabin, 1906–93, Russian-born American bacteriologist).

sac any bag-like organ or body structure.

saccades rapid intermittent eye movements made as the attention switches from one point to another.

saccharine a sweetening agent with no caloric or nutritional value, that is excreted unchanged in the urine. It is said to be 550 times sweeter than sugar.

sacral pertaining to the SACRUM.

sacralization congenital fusion of the lowest lumbar vertebra to the top of the SACRUM; a harmless condition.

sacro- *combining form denoting* SACRUM.

sacroiliac joints the firm ligamentous junctions between the sides of the SACRUM and the two outer bones of the pelvis (iliac bones). Normally the sacro-iliac joints are semi-rigid, but in late pregnancy they relax a little to allow easier childbirth. See also RELAXIN.

sacrum the large, triangular, wedge-like bone that forms the centre of the back of the PELVIS and the lower part of the vertebral column. The sacrum consists of five fused, broad vertebrae and terminates in the tail-like COCCYX.

sadism a form of deviant sexuality in which pleasure and sexual arousal are derived from the infliction, or contemplation, of another's pain. From the name of the Marquis de Sade (1740–1814), a French writer of pornographic pseudophilosophy. Compare MASOCHISM.

sadomasochism a sexual deviation in which arousal is achieved by inflicting pain (SADISM) or by experiencing pain or abuse of various kinds (MASOCHISM).

SADS *abbrev. for* 1 seasonal affective disorder syndrome in which the mood changes according to the season of the year and which is treated by exposure to bright light. 2 sudden adult death syndrome: the sudden death of an apparently healthy adult, for which no cause can be found at postmortem.

safe period the part of the menstrual cycle in which coitus cannot result in fertilization because of the absence of an ovum. As a basis for contraception the term is far from appropriate.

safe sex measures taken to try to minimize the risk of sexually transmitted disease, especially AIDS. These include the avoidance of promiscuity, fidelity to the partner, the use of condoms and non-penetrative sexual activity.

sagittal 1 pertaining to the SUTURE that joins the two parietal bones of the skull. 2 pertaining to the SAGITTAL PLANE. From the Latin *sagitta*, an arrow (referring to the vertical feather of the flight).

sagittal plane the front-to-back longitudinal vertical plane that divides the upright body into right and left halves.

saline a solution of salt (sodium chloride) in water. Normal saline is a solution with the same concentration of salt as body fluids and is suitable for infusion into a vein. Also known as physiological saline.

saliva a slightly alkaline, watery fluid secreted into the mouth by the SALIVARY GLANDS. Saliva contains the digestive enzyme amylase capable of breaking down starch to simpler sugars. Saliva keeps the mouth moist, dissolves taste particles in food so that they can stimulate the taste buds on the tongue and lubricates food during mastication to assist in swallowing.

salivary glands three pairs of glands that open into the mouth to provide a cleaning, lubricating and digestive fluid. The largest pair, the parotid glands, lie in the cheek in front of the ear. The other two pairs, the sublingual and the submandibular glands are in the floor of the mouth.

salivary gland disorders these include mixed parotid tumour, mumps and Sjögren's syndrome.

salivation the production of saliva. Excessive salivation is a feature of mouth ulcers and other causes of mouth irritation, PARKINSON'S DISEASE, nerve gas poisoning, organophosphorus insecticide poisoning, mercury poisoning, RABIES and overactivity of the parasympathetic nervous system.

Salk vaccine a killed virus anti-POLIOMYELITIS vaccine. (Jonas Salk, American microbiologist, 1914–95).

Salmonella a genus of bacteria containing over 2000 strains, no longer considered to be separate species. Some have species-specific infectivity. About half of the strains are known to cause food poisoning in humans. Salmonella organisms also cause typhoid and paratyphoid fevers. Common contaminants of food include *Salmonella typhimurium, S. hadar, S. enteritidis* and *S. virchow*. (Daniel E. Salmon, American pathologist, 1850–1914).

salping-, salpingo- *combining form denoting* the FALLOPIAN TUBE or the EUSTACHIAN TUBE.

salpinges the Fallopian tubes. The singular is salpinx.

salt 1 any substance that dissociates in solution into ions of opposite charge. 2 common salt, sodium chloride (NaCl).

saltatory conduction conduction of a nerve impulse along an axon in a process of leaps from one node of Ranvier to the next.

sane of sound mind. This is a legal rather than a medical term, as in the case of 'insanity'. The lawyers are forced by their trade to assume that a person must either be sane or insane; the doctors are not so sure.

Sanger, Frederick the English biochemist Frederick Sanger was born in Rendcomb, Gloucestershire in 1918. The son of a medical doctor he was educated at Bryanston School, Blandford Forum, Dorset and at St John's College, Cambridge University, taking his degree in 1939 and his PhD in 1943. He was a Quaker and was allowed exemption from war service. During the whole of his scientific career he never moved away from Cambridge. In 1951 he joined the Medical Research Council as a scientist and in 1961 he was appointed Head of the Protein Chemistry Division of the MRC's Molecular Biology Laboratory.

Sanger was remarkable in that he won the Nobel Prize twice for two important biochemical advances. The 1958 prize was for working out the complete chemical structure of the insulin molecule – a sequence of 51 amino acids in two chains. The second prize, in 1980, shared with the American scientists, Paul Berg and Walter Gilbert, was for determining the entire base structure of a strand of DNA from a virus.

Sanger's work on insulin was of great biological and medical importance because it encouraged other scientists to study and analyse proteins, it showed that insulin from different animals had slight amino acid differences from human insulin, and it made possible the artificial synthesis of human insulin. His work on nucleotide sequencing was even more scientifically significant. He used new techniques involving the use of enzymes (restriction enzymes) that cut nucleic acid chains into small lengths, a few hundred nucleotides long, at specific sequences of a few nucleotides. These fragments were then separated by gel electrophoresis. Radioactive labelling was also used. Sanger and his colleagues later performed the remarkable feat of sequencing the roughly 17,000 nucleotides in mitochondrial DNA.

sanguine of a ruddy complexion.

sanguineous pertaining to blood.

sanitary pertaining to health and hygiene.

sanitary protection pads or tampons used to avoid blood staining of clothing during the menstrual period.

sanitation measures concerned with the protection or promotion of public health, especially those relating to the disposal of sewage.

sapon- *combining form denoting* soap.

saponaceous having soap-like properties.

saponification hydrolysis of a fat by an alkali to form a soap and an alcohol, usually glycerol.

Sapphic pertaining to female homosexuality. From the Greek female poet Sappho who lived on the island of Lesbos.

sapr-, sapro- *prefix denoting* decay or putrefaction.

saprogenic causing, or resulting from, putrefaction.

sarc-, sarco- *prefix denoting* flesh or striped muscle.

sarcolemma a delicate membrane surrounding a striped muscle fibre.

sarcoma one of the two general types of cancer, the other being carcinoma. Sarcomas are malignant tumours of connective tissue such as bone, muscle, cartilage, fibrous tissue and blood vessels. Sarcomas are named after the parent tissue and include osteosarcoma, myosarcoma, chondrosarcoma, fibrosarcoma and angiosarcoma. Kaposi's sarcoma is a tumour of blood vessels.

sarcomere the structural unit of a striped muscle fibre (myofibril) consisting of thick and thin filaments.

sarcoplasm the cytoplasm of muscle cells.

sarcoplasmic reticulum the endoplasmic reticulum of a muscle cell which stores and releases calcium ions.

Sarcoptes scabei the mite that causes scabies.

sarin a military and terrorist nerve gas agent that causes breathing difficulty, cyanosis, running nose, salivation, profuse sweating and vomiting, constricted pupils, major tonic-clonic convulsions, coma and death.

sartorius a long, narrow, flat, strap-like muscle that crosses the front of the thigh obliquely from the hip to the inner side of the top of the main lower leg bone (tibia). From the Latin *sartor*, a tailor – a reference to the historic cross-legged sitting posture of the tailor, which the muscle assists in adopting.

satyriasis an almost uncontrollable male craving for sexual intercourse. The male equivalent of NYMPHOMANIA.

saturated fats fats containing FATTY ACIDS in which the carbon atoms are linked by single covalent bonds.

scab a skin crust formed when serum leaking from a damaged area mixes with pus and dead skin and then clots.

scala tympani a fluid-filled passage in the cochlea running on the outside of the cochlear duct from the tip of the basilar membrane at the HELICOTREMA to the round window in the middle ear where it communicates with the scala vestibuli.

scala vestibuli a fluid-filled passage in the cochlea running on the outside of the cochlear duct from the oval window in the middle ear to the HELICOTREMA where it communicates with the scala tympani.

scald a burn caused by hot liquid or steam.

scaling removal of dental calculus from the teeth to prevent or treat PERIODONTAL DISEASE. The hard calculus is levered or scraped off with a sharp-pointed steel scaler and the teeth are polished with an abrasive.

scalp the soft tissue layers covering the bone of the vault of the skull and consisting of a thin sheet of muscle, the epicranius, a layer of connective tissue richly supplied with blood vessels and the skin.

scanning optical microscopy a method of light microscopy in which the object, often a block of wet or dry tissue, is illuminated with a scanned laser beam in such a way as to bring a narrow plane into focus while leaving the remainder out of focus. Serial observations, with photographs, of different planes can be made so that cutting sections is unnecessary. A three-dimensional image can be built up. The sites of binding of fluorescence-labelled antibodies can be determined with great accuracy.

scanning techniques these include CT scanning, digital radiography, MRI, PET scanning, radionuclide scanning, scanning optical microscopy and ultrasound scanning.

scaphoid bone one of the eight bones of the wrist. The scaphoid is roughly boat-shaped and is also known as the navicular bone. Because of its shape it is more readily fractured than the other carpal bones. From Greek *skaphoidis*, a boat, and Latin *navis*, a boat.

scapula the shoulder blade. A flat, triangular bone with a prominent, near-horizontal raised spine, lying over the upper ribs of the back. At its upper and outer angle the scapula bears a shallow hollow with which the rounded head of the upper arm bone (the humerus) articulates. The spine ends in a bony process, the coracoid process, the end of which connects with the outer end of the collar bone (clavicle).

scar tissue fibrous COLLAGEN formed by the body to repair any wound, whether on the skin or internally. If the edges of the wound are brought close together the amount of scar tissue formed is minimal.

Scheele, Karl Wilhelm in 1771 the Swedish apothecary and chemist Karl Wilhelm Scheele (1742–86), while heating mercuric oxide, discovered a new gas with remarkable properties. It was colourless and odourless and could keep small animals alive and frisky. A glowing wood splint plunged into it would burst into flame. Scheele called the gas 'fire air' and wrote a book about it in which he described his experiments. Unfortunately, his dilatory publisher did not get the book out until 1777 and by that time the English scientist Joseph Priestley (1733–1804) had reported his own similar experiments and had taken the credit – still generally acknowledged – as the discoverer of oxygen.

Scheele was a remarkable man, the extent of whose chemical discoveries is almost unrivalled in science and is matched only by his inexplicable obscurity. In a life devoted exclusively to the pursuit of science – and probably shortened by his habit of tasting all the new substances he came across – he isolated the elements barium, chlorine, manganese, molybdenum, nitrogen and oxygen, and produced scores of important new compounds. His contributions to chemistry probably exceeded those of any other scientist of his time.

Neither Scheele nor Priestley fully grasped the role of oxygen in combustion and respiration. In 1774 Priestley visited the French scientist Antoine-Laurent Lavoisier (1743–94) and told him about his experiments. Lavoisier immediately repeated these and soon saw the importance of the newly discovered gas and its relation to air. He gave oxygen its current name (mistakenly derived from Greek roots meaning 'acid maker') and showed that air contained two main gases, one which supported combustion and one which did not. He studied the heat produced by animals breathing oxygen and demonstrated the relationship of respiration to combustion. Lavoisier, anxious to be known as the discoverer of an element, did not acknowledge the help of Priestley – whom he regarded as an amateur. Scheele, the real hero, was ignored and forgotten.

-schisis *combining form denoting* a splitting or cleaving.

schiz-, schizo- *prefix denoting* split or cleft, cleavage or fission.

schizogenesis reproduction by simple division into two or more individuals (fission).

schizogony multiple fission of a unicellular organism so that many daughter organisms are formed. This occurs in the life cycle of malarial parasites.

schizoid personality a term describing people who are withdrawn, solitary, socially isolated, often appearing cold and aloof and sometimes eccentric. About 10 per cent of people of this personality type develop overt SCHIZOPHRENIA.

schizont a stage in the life-cycle of a sporozoan parasite, especially a malarial parasite. A schizont reproduces by SCHIZOGONY producing multiple trophozoites or merozoites.

schizophrenia the commonest major psychiatric disorder affecting about 1 per cent of Western populations and usually appearing in adolescence or early adult life. Schizophrenia is not a disease in the normal medical sense and the diagnosis is based entirely on behaviour and on the statements of the affected person. Schizophrenics have DELUSIONS, HALLUCINATIONS, disordered thinking and loss of contact with reality. They indulge in non-logical free associations, appear to confuse literal and metaphorical meaning and use invented words. The cause has not been established but life case histories suggest that some schizophrenics have adopted an alternative reality as an escape from an intolerable life situation. There also appears to be some genetic basis, but this does not exclude an important environmental causation. A mutation in the serotonin receptor gene has been proposed as a cause. Schizophrenics can usually be made to conform to conventional social mores by treatment with antipsychotic drugs. Psychoanalysis has no value in the treatment of schizophrenia.

Schwann cell a NEURILEMMA cell of a nerve fibre that produces the myelin sheath. (Theodor Schwann, 1810–82, German anatomist).

Schwann, Theodor the German physiologist Theodor Schwann (1810–82) was born in Neuss, Rhenish Prussia, and trained as a doctor, graduating in 1834. He then worked in the Berlin Museum of Anatomy as an assistant to German physiologist Johannes Peter Müller and, almost immediately made an important discovery. It was at that time assumed that the hydrochloric acid in the stomach was alone responsible for breaking down food into its simpler components. But in 1834 Schwann demonstrated that, in the laboratory, acid alone acted very slowly on food. But if extracts from the glandular lining of the stomach were added to the acid, the speed of breakdown was greatly accelerated. Two years later Schwann succeeded in isolating the active ingredient from the lining tissue and called it pepsin – a term derived from the Greek word for digestion.

Schwann carried out experiments to show that the then current theory of spontaneous generation of living things, such as maggots, was nonsense. He also introduced and explained the term metabolism for the range of chemical syntheses and breakdowns that occur in living organisms. But his greatest contribution to biology was his assertion that all living tissues were made up of cells, and that each cell had essential constituents such as a nucleus and an enclosing membrane. He demonstrated that both animals and plants were constructed of cells and acellular connective tissue, and that eggs were cells that repeatedly divided to form large masses of new cells. The German botanist Matthias Schleiden had also proposed a cellular theory, but it was Schwann who made the stronger and more convincing case. The English microscopist Robert Hooke had described what he called cells in plant tissue, but what he was describing were the empty spaces in cellulose that had been occupied by cells. Hooke, however, must have the credit for using the term cell in a biological context.

After the publication of Schwann's book *Microscopic Studies on the Similarity in the Structure of Animals and Plants* (1839) there were no further arguments about the cellular nature of living things.

sciatic nerve the main nerve of the leg and the largest nerve in the body. The sciatic nerve is a mixed MOTOR and sensory nerve. It arises from the lower end of the spinal cord and runs down through the buttock to the back of the thigh. It supplies the muscles of the hip, many of the thigh muscles, all the muscles of the lower leg and foot, and most of the skin of the leg.

scintigraphy radionuclide scanning.

scirrhous hard and fibrous, especially of malignant tumours containing dense fibrous tissue.

scler-, sclero- *combining form denoting* hard or the SCLERA.

sclera the white of the eye. The tough outer coating of dense, interwoven collagen fibrils visible through the transparent overlying CONJUNCTIVA.

sclerosed hardened. Affected with SCLEROSIS.

sclerosis hardening of tissues usually from deposition of fibrous tissue, following persistent INFLAMMATION.

sclerotic 1 the white outer coat (sclera) of the eye.
2 pertaining to the SCLERA.

scolio- *combining form denoting* twisted.

scoliosis a spinal deformity in which the column is bent to one side usually in the chest or lower back regions. This may cause crowding of the ribs on one side.

scoto- *combining form denoting* darkness.

scotoma a blind spot or area in the field of vision. This may be caused by glaucoma, multiple sclerosis, migraine, retinal disorders or a brain tumour.

scotopia ability to see in poor light. Vision adapted to night-time conditions.

screening the routine examination of numbers of apparently healthy people to identify those with a particular disease at an early stage.

scrotum the skin and muscle sac containing the testicles and the start of the SPERMATIC CORD. The wrinkled appearance of the scrotal skin is due to the thin layer of DARTOS MUSCLE under the skin.

scurf an exaggerated loss (desquamation) of the surface layers of the EPIDERMIS of the skin, especially from the scalp. Dandruff.

scurvy a deficiency disease caused by an inadequate intake of vitamin C (ascorbic acid). This vitamin C is needed for the formation of stable COLLAGEN; deficiency leads to weakness of small blood vessels and poor healing of wounds, and spontaneous bleeding occurs into gums, skin, joints and muscles. Treatment with large doses of ascorbic acid is rapidly effective.

seasonal affective disorder syndrome see SADS.

seb-, sebo- *combining form denoting* fat, SEBUM.

sebaceous glands tiny skin glands that secrete an oily lubricating substance, called SEBUM, either into hair follicles or directly on to the surface of the skin.

sebum the secretion of the SEBACEOUS GLANDS. Sebum is chemically complex and consists mainly of triacyl glycerols, wax esters and squalene. Blackheads (comedones) are plugs of sebum, the darkened tips being due to oxidation.

secretin a hormone secreted in the DUODENUM that prompts the production of PANCREATIC JUICE.

secretion the synthesis and release of chemical substances by cells or glands. Substances secreted include enzymes, hormones, lubricants, surfactants and neurotransmitters. Internal secretion is secretion into the bloodstream. External secretion may be into the intestinal canal or other organs or on to the skin. Compare EXCRETION.

secretor status the ability of an individual to secrete the water-soluble form of the ABO blood group ANTIGENS into body fluids. There is a genetically determined inability to secrete these antigens and those so affected are unduly susceptible to various bacterial and superficial fungal infections.

secretory phase the stage of the menstrual cycle immediately following ovulation, during which the womb lining is at full thickness and its mucus glands are actively secreting.

sectioning 1 an informal term, or euphemism, used to describe the implementation of a section of the UK Mental Health Act so that a person suffering from a psychiatric disorder can be detained.
2 cutting a thin slice of tissue for microscopic examination.

security object any object, such as a teddy bear, a baby blanket or a former night garment, that brings comfort and a sense of security to a young child.

sedation the use of a mild drug to calm, alleviate anxiety and promote sleep.

segmentation repetitive bowel contractions affecting short fixed lengths but without shift of position as in peristalsis. Segmentation functions to mix and homogenize the bowel contents but is not concerned with their transport.

seizure an episode in which uncoordinated electrical activity in the brain causes sudden muscle contraction, either local (partial seizure) or widespread (generalized seizure). Recurrent seizures are called epilepsy. Also known as a fit.

selenium a trace element recently found to be an essential component of the enzyme deiodinase which catalyses the production of the hormone triiodothyronine (T_3) from thyroxine (T_4) in the thyroid gland. Selenium deficiency prevents the formation of T_3.

self-image a person's conception of his or her own appearance, personality and capabilities.

self-mutilation acts of destruction of parts of one's own body, such as amputation of fingers, limbs or genitals, gouging out of eyes, and so on.

semantics the study of meaning, of the effectiveness with which thought is translated into language, and of the relationship between words and symbols and meaning.

semeiography a description of the signs and symptoms of a disease. Also semiography.

semen see SEMINAL FLUID.

semi- *prefix denoting* half or partial.

semicircular canals the three tubular structures of the labyrinth of the inner ear, which lie in three different planes and which contain fluid. Movement of the head leads to differential stimulation of nerve endings within the canals and provides the brain with information about orientation.

semicoma a partially comatose state in which the patient appears to be unconscious but can be roused by painful stimuli and usually makes purposeful movements.

semiconscious half-conscious.

semilunar or **semilunate** shaped like a half moon or crescent.

semilunar bone the LUNATE bone in the wrist.

semilunar valves the crescent-shaped valves, each having three cusps, situated one at the beginning of the AORTA and the other at the origin of the PULMONARY ARTERY. These valves allow blood to flow only in a direction away from the heart.

semilunate semilunar.

seminal fluid a creamy, greyish-yellow, sticky fluid that is forced out of the penis during the ejaculation that accompanies the sexual orgasm. Seminal fluid is secreted by the PROSTATE GLAND, the SEMINAL VESICLES, the lining of the sperm tubes and some small associated glands. It contains the male gametes, the SPERMATOZOA. Also known as sperm.

seminal vesicles two small, elongated, sac-like containers situated on the PROSTATE GLAND at the point at which the VAS DEFERENS passes through to join the URETHRA. The seminal vesicles store part of the SEMINAL FLUID until it is discharged by ejaculation. Their secretion contributes most of the volume of the ejaculate but the vesicles do not store spermatozoa.

seminiferous tubules the many, long, coiled-up, fine tubes, hundreds of which are present in each testicle and are said to total in length over 400 m. The spermatozoa develop from cells in the walls of these tubules.

semiotics the study of signs, including words, symbols, gestures and body language, and of their cardinal role in conveying information. Semiotic studies suggest that meaning, although it may often seem self-evident, is always the result of social conventions. Cultures can be analyzed in terms of a series of sign systems. One difficulty, perhaps responsible for a certain vagueness in discussion of the subject, is that the experts have never been able to reach full agreement on the exact definition of the central terms 'sign', 'symbol' and 'signal'.

semipermeable able to allow the passage of molecules below a certain size and to retain those above this size. Much of PHYSIOLOGY depends on the semipermeability of body membranes.

semipermeable membrane any membrane permeable to water and other liquids but which does not allow the passage of dissolved molecules larger than a certain size. The association of semipermeable membranes with solutions allows selective movement of substances into and out of various body compartments and cells and subserves the principle of OSMOTIC PRESSURE.

senescent ageing.

senile pertaining to old age. From the Latin *senilis*, old. The term ought not to imply physical or mental deterioration, but often does.

senility old age, usually with the connotation of mental or physical deterioration. From the Latin *senilis*, old (which had no negative significance).

senopia an acquired form of MYOPIA, occurring in the elderly, that may allow reading without glasses. Senopia is due to an increase in the refractive index of the crystalline lenses in incipient CATARACT. Also known as index myopia.

sensate perceived by the senses.

sensate focus technique a method of managing male impotence and female lack of sexual interest, essentially by prohibiting coitus and encouraging sensual massage, performed in stages, until, willy-nilly, nature has its way. The formal prohibition relieves anxiety which is at the root of most cases of impotence.

sensation the conscious experience produced by the stimulation of any sense organ such as the eye, ear, nose, tongue, skin, or any internal sensory receptor.

sensitization the preliminary exposure of a person to an ALLERGEN that leads to ANTIBODY production by the immune system and, on subsequent exposure, to an ALLERGIC or hypersensitivity reaction. Immunoglobulin Type E (IgE) is the main type of antibody involved.

sensory cortex an area on the outer layer (cortex) of the CEREBRUM through the stimulation of which consciousness of sensation is mediated. The sensory cortex for pressure, pain, taste and temperature lies immediately behind the motor cortex on the side of each cerebral hemisphere, about half way back. The cortex for visual sensation is at the back, and that for sound is in the temporal lobes on either side.

sensory deprivation the effecting of a major reduction in incoming sensory information. Sensory deprivation is damaging because the body depends for its normal functioning on constant stimulation. Sensory deprivation early in life is the most damaging of all and can lead to severe retardation and permanent malfunctioning of the deprived modality.

sensorimotor cortex those parts of the outer layer of the brain concerned with the control of skeletal muscles.

separation anxiety excessive and inappropriate levels of anxiety experienced by children during separation, or the threat of separation, from a parent or from a person in loco parentis.

sepsis the condition associated with the presence in the body tissues or the blood of micro-organisms that cause infection or of the toxins produced by such organisms. Sepsis varies in severity from a purely local problem to an overwhelming and fatal bacterial intoxication. Sepsis has been defined as the systemic inflammatory response to infection based on the clinical criteria of a temperature over 38°C, a heart rate of over 90 beats per minute, a respiratory rate of over 20 per minute and a white blood cell count increase of more than 12,000 or with more than 10 per cent immature neutrophil polymorphs. Severe sepsis is defined as sepsis associated with organ dysfunction. Severe sepsis has a mortality of up to 50 per cent.

septal defect an abnormal opening in the central wall of the heart providing a communication from the right side to the left or vice versa. This is a developmental congenital defect and its significance varies from trivial to grave depending on its position and size. Popularly known as 'hole in the heart'.

septate having a partition (SEPTUM) or several partitions (septa).

septic pertaining to SEPSIS.

septicaemia the presence in the circulating blood of large numbers of disease-producing organisms. Septicaemia causes high fever, shivering, headache and rapid breathing and may progress to delirium, coma and death. Treatment is with antibiotics and sometimes transfusion. Also known as blood poisoning. See also SEPTIC SHOCK.

septic shock a dangerous condition caused by damage to immune cells, the endothelium of blood vessels and the malfunction of most of the organs of the body by the uncontrolled production of potent CYTOKINES mainly by bacterial pathogens. These include tumour necrosis factor and interleukin-1 which are produced as a response to the toxins of bacteria in the course of SEPTICAEMIA. Small blood vessels become leaky and so much fluid is lost into the tissue spaces that the normal circulation cannot be maintained and the blood pressure drops. This is called shock and it is similar in effect to shock from other causes such as severe blood loss or severe burns. Urgent treatment with antibiotics, transfusion and organ support is necessary to save life.

septum any thin dividing wall or partition within or between parts of the body.

sequela a sequel. Any condition or state that follows a disease, disorder, or injury, especially one that is a consequence of it. A complication. The term is most often used in the plural form – sequelae.

serine a non-essential amino acid found as a component of most proteins. It is a precursor of choline, glucine, cysteine and pyruvate. Serine is present in most diets but most of the body serine is synthesized.

sero- *prefix denoting* SERUM or serotherapy.

serology the branch of laboratory medicine concerned with the investigation of blood SERUM with special reference to its antibody (immunoglobulin) content. Detection of antibodies and ANTIGENS is of considerable medical importance especially in diagnosis.

serotonin 5-hydroxytryptamine (5-HT). A NEUROTRANSMITTER and HORMONE found in many tissues, especially the brain, the intestinal lining and the blood platelets. Serotonin is concerned in controlling mood and levels of consciousness. Its action is disturbed by some hallucinogenic drugs and imitated by others. It constricts small blood vessels, cuts down acid secretion by the stomach and contracts the muscles in the wall of the intestine.

serotype 1 a subgroup of a genus of microorganisms identifiable by the ANTIGENS carried by the members.
2 a category into which material is placed, based on its serological activity, especially in terms of the antigens it carries or the antibodies it produces in the body.

Sertoli cells elongated cells in the walls of the seminiferous tubules that closely surround ('nurse') the spermatids, encouraging spermatogenesis. Together, they provide a blood-testis barrier and secrete an androgen-binding protein. (Enrico Sertoli, Italian histologist, 1842–1910).

serum the clear, straw-coloured fluid that separates from blood when it is allowed to clot and then to stand. Serum is blood less the red cells and the proteins which form the clot. It contains many substances in solution including sodium, potassium, calcium, magnesium, chloride, bicarbonate, phosphate, albumin, globulins, amino acids, carbohydrates, vitamins, hormones, urea, creatinine, uric acid and bilirubin.

serum albumin one of the soluble protein fractions of blood serum. Albumin is important in maintaining the OSMOTIC PRESSURE of the blood.

serum globulin one of the soluble protein fractions of blood serum. Most of the serum globulins are ANTIBODIES (immunoglobulins).

sesamoid bone a bone lying within a tendon to assist in its mechanical action and to bear pressure. The most conspicuous sesamoid bone is the knee cap (patella).

sessile having no stalk, flat and wide-based. Compare PEDUNCULATED.

sex 1 gender, as genetically determined.
2 the condition of being male or female.
3 the urge or instinct manifesting itself in behaviour directed towards copulation.
4 the genitalia.
5 a popular term for COITUS.

sex cell a GAMETE.

sex chromosomes the pair of chromosomes that determines gender. The other 22 pairs of chromosomes are known as autosomes. Women have two sex chromosomes of similar appearance called X chromosomes (XX). Men have one X chromosome and another, much smaller chromosome, called a Y (XY). Sperm have half the normal complement of chromosomes (haploid) and contain either an X or a Y. If an X chromosome effects fertilization the result will be a female, if a Y, a male.

sex gland a testis or ovary. A gonad.

sex hormones HORMONES that bring about the development of bodily sexual characteristics and regulate sperm and egg production and the menstrual cycle. There are steroids of three main types – androgens (male), oestrogens (female) and PROGESTERONES that prepare women for, and maintain, pregnancy.

sex-linkage inheritance in which the gene for the condition is carried on one of the X or Y sex chromosomes. Recessive sex-linked conditions almost always affect males, are carried on the male X chromosome and thus cannot be transmitted directly from father to son. The best-known sex-linked recessive condition is haemophilia. Y-linked conditions are rare; the Y chromosome is very small and carries few genes.

sex-linked see SEX-LINKAGE.

sexology the study of sexual behaviour, especially in humans.

sex selection the determination of the sex of a future individual before conception by the separation of sperm bearing Y chromosomes (male) from those bearing X chromosomes (female). Such techniques are still experimental and uncertain and raise major ethical issues.

sex therapy specialized methods of treatment for problems such as erectile dysfunction (impotence), premature ejaculation, anorgasmia (failure to achieve orgasm), vaginismus, sexual phobias and painful intercourse (dyspareunia). See also SENSATE FOCUS TECHNIQUE.

sextuplets six SIBLINGS delivered at one birth.

sexual pertaining to sex, sexuality, the sexes, reproduction, erotic desires or activity, the sex organs or the union of male and female GAMETES.

sexual abuse subjection of any person, but especially a minor, to sexual activity likely to cause physical or psychological harm. See also CHILD ABUSE, RAPE.

sexual deviation an arbitrary term whose meaning varies with the attitudes and views of the user. Few, however, would exclude from its definition SADISM, SADOMASOCHISM, PAEDOPHILIA, bestiality, sexual exhibitionism, sexual fetishism, voyeurism, telephone scatologia, frotteurism and coprophilia.

sexual disorders these include anorgasmia, impotence, kraurosis vulvae, Peyronie's disease, premature ejaculation, sexual deviation, sexually transmitted diseases and vaginismus.

sexual intercourse 1 the totality of the physical and mental interplay between humans in which the explicit or implicit goal is bodily union and, ideally, the expression of love and affection.

2 COITUS.

sexuality 1 the structural differences between male and female.

2 a person's sexual attitudes, drive, interest or activity.

3 all the emotions, sensations, behaviour patterns and drives connected with reproduction and with the use of the sex organs.

4 (Freudian) all drives connected with bodily satisfaction. See also HETEROSEXUALITY, HOMOSEXUALITY and BISEXUALITY.

sexually transmitted infections, STI, STD these include acquired immune deficiency syndrome (AIDS), amoebiasis, candidiasis, chancroid, chlamydial infections, crab lice, *Gardnerella vaginalis* infection, genital herpes, genital warts, genital thrush, giardiasis, gonorrhoea, hepatitis a, b and c, granuloma inguinale, lymphogranuloma venereum, non-specific urethritis, scabies, syphilis, trichomoniasis and yaws.

shaking palsy Parkinson's disease.

shame a distressing emotion involving a strong sense of having transgressed against a social or moral code. Shame is always relative to current mores or to the upbringing of the person concerned.

sheath 1 an enveloping structure or part, usually tubular.

2 a condom.

shin the front part of the leg below the knee and above the ankle. The surface of the TIBIA that lies close under the skin.

shinbone the TIBIA.

shin splints a popular term for pain in the lower leg muscles and bones occurring in runners and football players and made worse by exertion.

shivering a rapid succession of contractions and relaxations of muscles and an important means of heat production in the body. The temperature rise in high fever is caused mainly by shivering.

shock 1 a syndrome featuring low blood pressure, a prejudiced blood supply to important organs such as the brain and heart, and low kidney output. Causes of shock include severe blood loss, burns, severe infection, allergy, heart damage from coronary thrombosis and head injury. Untreated shock may be rapidly fatal.
2 a temporary state of psychological overburdening from severe mental distress, often associated with stupefaction.

short-loop negative feedback direct controlling hormonal influence by the pituitary gland on the hypothalamus.

short-sightedness see MYOPIA.

shoulder-blade see SCAPULA.

shunt any bypassing or sidetracking of flow, especially of fluid such as blood or cerebrospinal fluid. A shunt may be the result of disease or may be surgically induced, or inserted as a prosthesis, to effect treatment.

SI units SI is *abbrev. for* Systeme Internationale. SI units are now almost universally used in medicine. They include the metre for length, the kilogram for weight, the mole for amount of substance in a solution, the joule for energy and the pascal for pressure. These units are qualified by decimal multipliers or divisors such as mega- (a million), kilo- (a thousand), deci- (a tenth), centi- (a hundredth), milli- (a thousandth), micro- (a millionth), nano- (a thousand millionth), pico- (a million millionth), and femto- (a thousand million millionth).

sial-, sialo- *combining form denoting* saliva or salivary gands.

sialagogue anything that increases the flow of saliva.

Siamese twins identical (monozygous) twins that have failed fully to separate after the first division of the ovum and remain partially joined together at birth. The junction is usually along the trunk or between the two heads. From the male twins, Chang and Eng, born in Siam in 1811.

sib a SIBLING.

sibilant 1 hissing.
2 a speech sound, such as 's', 'sh' or 'z'.
3 a sibilant consonant.

sibling any member of a group of related brothers or sisters.

sibling rivalry strong competition or feelings of resentment between SIBLINGS, especially between an older child and a new baby. Sibling rivalry may persist throughout life.

sick building syndrome a varied group of symptoms sometimes experienced by people working in a modern office building and attributed to the building. Symptoms include fatigue, headache, dryness and itching of the eyes, sore throat and dryness of the nose. No convincing explanation has been offered.

sick headache a headache accompanied by nausea.

sickle cell anaemia an inherited blood disease mainly affecting black people and those of Mediterranean origin. The basic defect is in the HAEMOGLOBIN molecule of the red blood cells, which is abnormal (haemoglobin S) and deforms in conditions of low oxygen tension causing the red cells to become sickle-shaped (sickling) and to rupture readily. The HETEROZYGOUS form is comparatively mild; the HOMOZYGOUS form is severe and dangerous and an incidental illness can bring on a sickling crisis calling for prompt and energetic treatment.

sickling crisis an acute episode in homozygous sickle cell disease in which a massive breakdown of red blood cells occurs with widespread severe pain from blockage of small blood vessels by sickled red cells. Some cases involve great enlargement of the liver or spleen and gross anaemia, others may affect the lungs with severe breathlessness and pleuritic pain, or the brain causing a stroke. Skilled and energetic management in hospital is mandatory for the sickling crisis.

sickling test the observation of a blood film, mixed with a solution of sodium metabisulphite, under the microscope. If haemoglobin S is present the red cells will assume a sickle shape within 20 minutes. See also SICKLE CELL ANAEMIA.

side effect any effect of a drug or other treatment additional to the required effect. Most side-effects are unwanted and some are dangerous.

sidero- *combining form denoting* iron.

siderosis any condition in which there is an excessive accumulation of iron in the body.

sievert the SI UNIT of equivalent absorbed dose of ionizing radiation. Compare RAD.

sigmoid or **sigmoidal** 1 S-shaped. 2 pertaining to the SIGMOID COLON.

sigmoid colon the S-shaped lower end of the colon extending down from about the level of the brim of the pelvis to the RECTUM. Sigma is the Greek letter S.

sign an objective indication of disease, perceptible by an external observer. Compare SYMPTOM.

signal transduction the common process by which the binding of a molecule to a receptor on a cell plasma membrane results in the transmission of a signal within the cell (second messenger) to trigger off a biochemical pathway in the cell.

silent mutation a change in DNA that has no effect.

silent site in a gene, one of the positions at which a mutation does not change the product.

silicone any polymeric (long-chain), organic compounds of silicon and oxygen in which each silicon atom is linked to an alkyl group. Silicones may be produced as oils, greases or rubbers. Silicone rubber (Silastic) is a valuable prosthetic surgical structural material as it is inert and permeable to oxygen and well tolerated by the tissues.

sinew a popular term for a TENDON.

single strand binding proteins proteins that keep single strands of DNA apart during replication. They do so by attaching to the separated strands at the point of separation (the replication fork) so as to seal them off from each other and prevent immediate re-linking.

single-unit smooth muscle smooth muscle tissue that contracts as a unit. The muscle cells are joined by gap junctions that allow electrical changes to pass from cell to cell.

singultus see HICCUP.

sinoatrial node a small area of specialized muscle cells in the upper right chamber of the heart (right atrium) that acts as the natural pacemaker of the heart, setting the rate in conjunction with other controlling influences, and transmitting impulses throughout the heart muscle.

sinus 1 one of the paired mucous membrane-lined air cavities in a bone, specifically the frontal sinuses in the forehead, the maxillary sinuses (antrums) in the cheek bones, the multicelled ethmoidal sinuses on either side of the upper part of the nose and the sphenoidal sinuses in the base of the skull.
2 any wide blood channel such as the venous sinuses in the MENINGES.
3 any tract or FISTULA leading from a deep infected area to a surface.

situs inversus an uncommon mirror-image reversal of the organs of the trunk. The heart points to the right, the LIVER and APPENDIX are on the left and the stomach and spleen on the right. Situs inversus is seldom of medical significance, but may confuse diagnosis until detected. See also DEXTROCARDIA.

skatole 3-methylindole, the substance in faeces that confers the smell. It is a breakdown product of bile and can be prepared by the action of potassium hydroxide on egg albumin. It attracts dung flies but, for evolutionary reasons connected with the bacterial content of faeces, repels humans. See also INDOLE.

skeletal pertaining to the SKELETON.

skeleton

The general shape of the body is conferred by the 206 bones of the skeleton which provide attachments for the muscles, and support and protection for the internal organs. The axis of the skeleton consists of the skull and the spine (vertebral column). This is a curved column of individual bones, called vertebrae, all of the same general shape but enlarging progressively in size and altering in proportion, from the top to the bottom. Each vertebra has a stout, roughly circular body in front and an arch behind that encloses a wide opening. The bones are secured neatly together by fibro-cartilaginous intervertebral discs and longitudinal ligaments, and the sequence of arches form a long flexible tube in which lies the spinal cord. There are seven vertebrae in the neck (cervical vertebrae), twelve in the chest (thoracic or dorsal vertebrae) and five in the lumbar region. The bottom end of the vertebral column is firmly attached to the central bone of the pelvis – the sacrum – and at the bottom of this is the tail bone, the coccyx. Both the sacrum and the coccyx are formed, evolutionarily, from fused vertebrae. The ring-shaped pelvis has a deep hollow, the acetabulum, on either side, in which articulates the almost spherical head of the thigh bone (femur).

The chest vertebrae provide attachment for the twelve pairs of ribs, and the front ends of most of these connect, by flexible cartilages, to the breast-bone (sternum) or to cartilages attached to the breast-bone. This forms the thoracic cage. Lying almost free over the upper ribs at the back are the two flat shoulder-blades (scapulae). Running outwards and upwards from the upper corners of the sternum lie the two collar bones (clavicles), the outer ends of which are attached by ligaments to bony processes on each scapula. The scapulae and clavicles form the shoulder girdle. The upper arm bone (humerus) articulates with a shallow hollow on the outer surface of the shoulder-blade.

There is an analogy between the bones of the arm and of the leg. The humerus of the upper arm corresponds to the femur of the thigh. The radius and ulna of the forearm correspond to the tibia and fibula of the lower leg. The eight tarsal bones of the wrist correspond to the seven carpal bones of the foot. The five metacarpals of the hand to the five metatarsals of the foot and the 14 phalanges of the fingers correspond to the 14 phalanges of the toes.

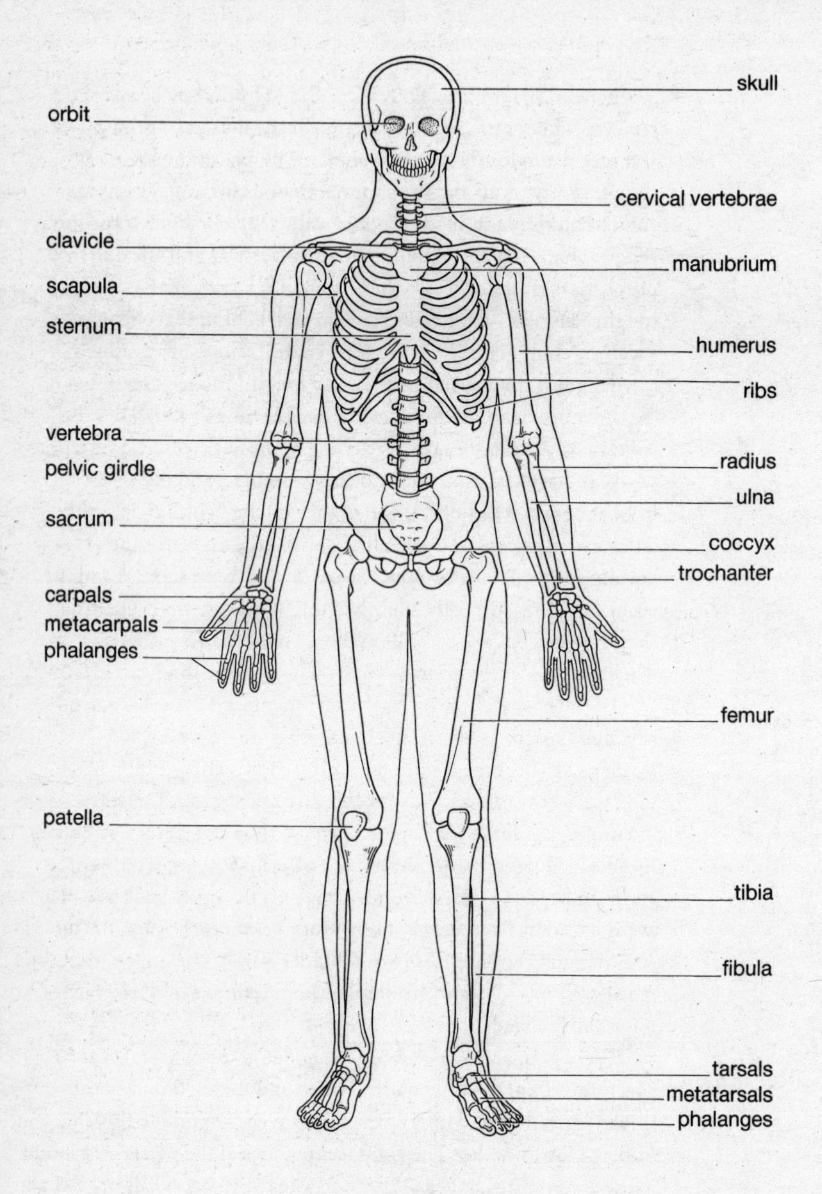

Fig. 16 **Skeleton**

Skull

The primary purpose of the skull is to accommodate and protect the most important part of the body – the brain – and to provide support and protection for the main sensory organs – the eyes and the ears. The floor of the interior of the skull is moulded to the exact shape of the surface of the brain and is provided with suitably-placed holes (foramina) through which the nerves emerging directly from the brain pass. The most conspicuous of these holes is the foramen magnum in the centre of the floor through which the spinal cord runs to enter the longitudinal channel in the vertebral column. The front part of the skull – the facial skeleton – contains the eye sockets (the orbits), and the double-sided air passage for the nose. It also contains four pairs of bony sinuses. These are lined hollows in the bone that lighten the skull and provide resonance to the voice. The lower jaw, or mandible, articulates with the skull high up on each side, just in front of the ear openings. The vault of the skull is made up of six double-layered, thin, curved bones fixed together by complex, jig-saw-like joints (sutures).

skewed of data that are not normally distributed. See PARAMETRIC.

skia- *combining form denoting* a shadow.

skin

The internal body environment is largely fluid. All the inner cells are bathed in tissue fluid and most of the body substance is infiltrated by fluid-containing blood vessels and lymphatics. So the body needs a waterproof outer covering to prevent loss of fluid. But the skin is much more than a simple sealant. It is a major organ with several important functions. It provides a remarkable degree of protection from the many hazards of the outside world and is self-renewing and self-repairing. It is the main effector of body temperature regulation. It is exquisitely sensitive to touch, pressure, pain, itching, heat and cold, and provides a major sensory interface between the body and the outer world, endlessly receiving environmental information and sending it to the brain.

The skin is 1.5–2 square metres in area and varies considerably in thickness, from the transparent fineness of the eyelids, at one extreme, to the protective solidity of the soles of the feet, at the other. It affords an excellent shield against various forms of radiation, especially light, heat and alpha particles. Solar radiation, otherwise severely damaging because of its high ultraviolet content, is largely absorbed by the skin pigment melanin, and exposure to increased solar radiation results in an automatic protective increase in melanin production (tanning). A healthy, intact skin offers remarkable resistance to bacterial attack, and accumulations of bacteria in the outer layer are disposed of by the constant process of shedding of this layer. Dry skin has high electrical resistance and does not readily conduct electricity. Considerable biochemical activity occurs in the cells of the deeper layers of the skin, a notable feature being the synthesis of vitamin D from sterols under the influence of sunlight. This is sufficient to prevent rickets or osteomalacia in people adequately exposed to the sun, even if the diet is deficient in the vitamin.

Skin structure

The skin has two layers, the outer epidermis which, as the name implies, lies upon the inner layer, the dermis. The epidermis has two main functions. It generates an outermost dead, horny layer, that is constantly being shed, and it synthesizes the pigment melanin. The epidermis is structurally simple and has no nerves or blood vessels. It acts as a rapidly replaceable surface, resistant to, or capable of tolerating, much abrasion and wear.

At the interface between the epidermis and the dermis is a collagen membrane called the basal lamina. Attached to the outer side of this is a layer of cells called the basal cell layer. This is the deepest layer of the epidermis and it is the basal cells that reproduce to form the cells nearer the surface – the prickle cell layer and the horny outer layer. It is the basal cells which grow abnormally in the common, light-induced skin cancer basal cell carcinoma or rodent ulcer. Most of the basal cells are called keratinocytes because they are the parents of the prickle cells which, as they approach the surface become condensed and flattened so that they consist of little more than flakes of the tough protein keratin. Some of the basal cells are different and are called melanocytes. These are the pigment-generating cells. The ratio of melanocytes to keratinocytes determines the colour of the skin. Dark-coloured people have a high proportion of melanocytes. Prickle cells are the cells which grow abnormally in common warts and melanocytes are the cells which, when affected by rare malignant change, form malignant melanomas. The pigment melanin is yellow to brown in colour and is made from the amino acid tyrosine linked to a protein.

The dermis, or 'true' skin (corium) is thicker than the epidermis and is less homogeneous. It has a connective tissue framework, largely made of collagen which enmeshes blood vessels, lymph

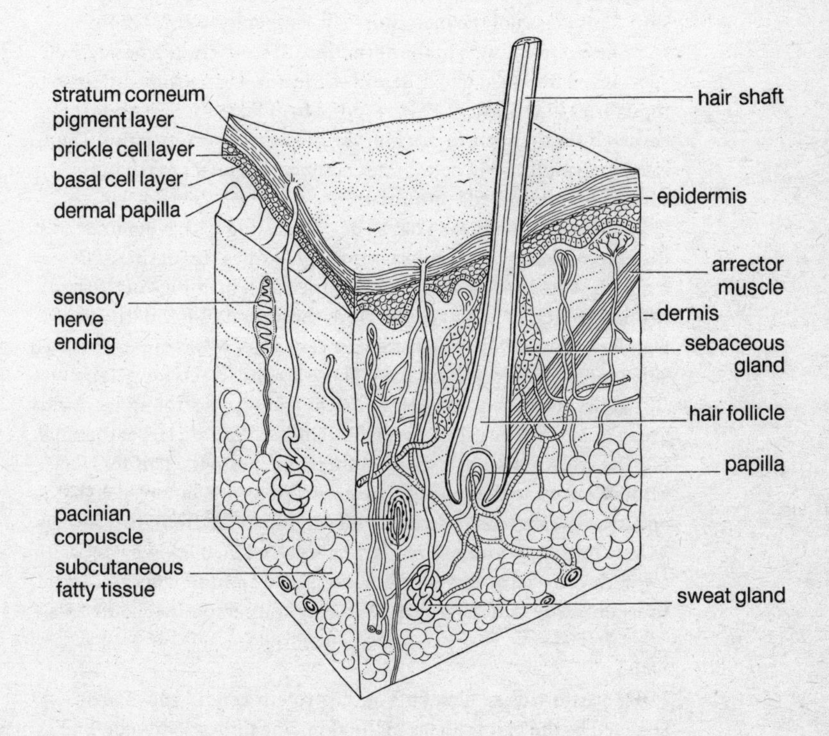
stratum corneum
pigment layer
prickle cell layer
basal cell layer
dermal papilla
sensory nerve ending
pacinian corpuscle
subcutaneous fatty tissue
hair shaft
epidermis
arrector muscle
dermis
sebaceous gland
hair follicle
papilla
sweat gland

Fig. 17 **Skin**

vessels, nerve endings, glands and various types of cells. The finest blood vessels are arranged in numerous finger-like processes (papillae) that push upwards, indenting the epidermis. A transverse cut made at about the level of the interface – as in skin grafting – shows countless bright red points where the papillae are cut through. The papillary vessels are the capillaries of the larger skin arterioles and the latter have an important layer of circular muscles in their walls which can greatly narrow them, so limiting the rate of blood flow through the skin. This is part of the mechanism of temperature regulation.

The nerve endings in the dermis are of a variety of types, specialized to respond to different stimuli and to provide different modalities of sensation. They respond to deformation, such as that caused by light touch, to pricking, to strong pressure, to painful and itching stimuli and to cold, and send nerve impulses to the brain which are interpreted appropriately.

The sweat glands each have a deep secreting part, lying under the skin, and a duct that runs up through both layers of the skin to reach the surface. The skin pores are the small openings in the skin through which the sweat passes from these ducts. Most of the sweat glands produce a fluid which is 99 per cent water with a dozen dissolved substances, mainly salt and urea. These are called eccrine glands. But in the armpits, groins, genital areas and nipples, the sweat glands are of a special type called apocrine glands. These are larger than the eccrine sweat glands and many of them open into the hair follicles rather than on to the surface. The main difference, however, is that apocrine sweat glands also throw off solid material from their linings. This is initially odourless, but is quickly acted on by skin bacteria to produce a characteristic odour. For social comfort, apocrine sweat must be washed off daily or inhibited by antiperspirant deodorants.

Hair

Hairs are threadlike filaments of the protein keratin which are secreted by the hair follicles in the skin. The follicles are tube-like structures widened at the bottom (the bulb) to accommodate the growing cell mass (the papilla). The papilla and the bulb are liberally supplied with blood vessels to bring in the enormous numbers of amino acids and the large quantity of energy fuel needed to synthesize hair protein. The papilla also contains nerve endings. Above the bulb, the tube, or sheath, of the follicle is lubricated by fatty sebaceous material (sebum) secreted into it by small adjoining sebaceous glands that open into the sheath. This material facilitates the slide of the hair outwards as it is continuously created at its lower end. In some areas, such as the outside of the nose, the hairs in the follicles are small in comparison with the sebaceous glands, so that the pore appears to be concerned solely with sebum production.

Each hair follicle has a tiny muscle, the erector pili muscle, attached to its outer side and anchored to the connective tissue. When this muscle contracts, as it does under the influence of intense emotion, the hair literally 'stands on end'.

Individual hairs each have an outer layer (the cuticle) of overlapping flat cells. Under this is the thick cortex, consisting of horny keratin. The inside of the hair is made of softer rectangular cells. This construction is the basis of the trichologists' claim that the hair is a tube, and of the unwarranted claim that the tube must be sealed after cutting. Different hair colour comes from different concentrations and depths of melanin secreted by melanocytes. This single pigment produces the whole spectrum of hair colour, from black to blond. Very blonde people have no melanocytes. When the melanocytes die, the hair turns grey or white. Stories of hair turning white overnight are apocryphal; existing hair cannot spontaneously lose its colour. Curly hair follicles have flatter cross-sections than straight hair. Curved follicles produce very curly hair.

For periods of two to five years, most of the cells of the bulb are actively reproducing and secreting the protein from which the hairs are made. This growth phase is followed by a resting period of about three months during which no growth occurs and the hairs separate from their papillae, form a clubbed end and remain in the follicle. When growth resumes, the old hairs are pushed out by the new growing hairs.

Nails

These are protective plates covering the vulnerable finger and toe ends. The nails are invaluable tools for many manipulative purposes, such as picking, scratching and unravelling. They also provide counter-pressure when feeling with the finger-pads. The nail is a curved plate of keratin – the same protein from which hair and horny skin is made – resting on the nail bed. Nails grow continuously from the nail matrix, which is the growing zone of the bed at the root of the nail, taking some four to five months, in the case of fingernails, to reach the tip of the finger. Toenails grow more slowly and take about three times as long. The matrix extends about one third of the distance from the root towards the tip of the nail. Damage to the matrix invariably distorts, or prevents growth of, the nail. The nail separates from the bed just short of the tip of the digit. The inturned skin edge around the nail is called the nail fold, and the cuticle is the free skin edge over the pale half moon (luna). The cause of the luna is still disputed.

The skin and temperature control

It is essential that the temperature of the body should be kept within narrow limits in spite of wide variations in the environmental temperature. Most body functions can continue normally up to about 40°C. but children commonly suffer convulsions at temperatures above 40.5°C. and permanent brain damage is likely in anyone whose temperature rises above 42°C. Lowered temperature is less dangerous, causing a general slowing of all metabolic processes, but the heart stops beating (cardiac arrest) at very low internal temperatures.

Blood temperature is monitored constantly by regulating centres in the underside of the brain, just above the pituitary gland (the hypothalamus). Any variations from normal (37°C.) automatically activates mechanisms to compensate. Extra heat is produced, when needed, by shivering – rapid repetitive muscle contraction, and heat loss is prevented by contraction of the muscles in the walls of the skin arterioles so that the vessels are narrowed and the flow of blood through the skin is minimized. In fever, loss of excessive heat occurs almost entirely via the skin. Under the control of the hypothalamus, the arteriolar wall muscles relax so that the blood flow through the skin increases and heat is lost by radiation. If this is insufficient, the sweat glands are stimulated to secrete. Evaporation of sweat from the skin surface results in considerable heat loss from the latent heat of evaporation. Evaporation of one gram of water requires 539 calories.

skin biopsy a portion of diseased skin removed for laboratory analysis, usually under local anaesthesia.

skin cancer one of a range of cancers of the skin that includes malignant melanoma, basal cell carcinoma, squamous cell carcinoma, Paget's disease of the nipple, mycosis fungoides and Kaposi's sarcoma.

skin disorders these include acne, basal cell carcinoma, boils, carbuncles, cellulitis, cold sores, cutaneous horns, dermatitis, eczema, epidermophytosis, haemangioma, impetigo, keratoacanthoma, keratosis, malignant melanoma, naevus, nappy rash, prickly heat, psoriasis, purpura, pustules, rosacea, scabies, scar, sebaceous cyst, shingles, squamous cell carcinoma, tinea, vitiligo, warts and xanthelasma.

skinfold thickness measurement a method of assessing the amount of fat under the skin by means of special calipers, sprung to exert a standard pressure and fitted with a scale. Skinfold thickness measurements may more accurately assess obesity than weighing.

skin graft the transference of an area of skin from one part of the body to another. A plastic surgical technique used to repair areas of deficient skin. Skin grafts may be split-skin or full-thickness or may have attached blood vessels that are rejoined by microsurgery to vessels at the new location.

skin peel a cosmetic procedure to improve the appearance of the facial skin by removing small wrinkles, scars, freckles and other blemishes. A paste containing carbolic acid is used.

skin tests investigations to determine allergic sensitivity to various substances by injecting small quantities into or under the skin or by applying the substance under patches (patch tests).

skull the bony skeleton of the head and the protective covering for the brain. The part of the skull that encloses the brain is called the cranium.

sleep the natural, regular, daily state of reduced consciousness and METABOLISM that occupies about one-third of the average person's life. Sleep requirements vary considerably in health, between about 4 and 10 hours in each 24 hour period. The purpose of sleep is unknown but prolonged deprivation is harmful, causing depression and mental disturbances, including hallucinations.

sleep apnoea repetitive periods without breathing, occurring during sleep and lasting for 10 seconds or longer. Most cases are due to over-relaxation of the muscles of the soft palate, in heavy snorers, which sag and obstruct the airway. Obesity is a common factor. Less commonly, the condition is due to disturbance of the brain mechanisms that maintain respiration.

sleep paralysis a feeling, experienced for a few seconds when falling asleep, of being unable to move. This is a feature of NARCOLEPSY but may affect healthy people.

sleep terror a childhood phenomenon featuring sudden screaming, an appearance of severe agitation, apparent inability to recognize faces or surroundings, return to sleep and no subsequent memory of the event. Sleep terror appears to be harmless and ceases in adolescence. Also known as night terror.

sleepwalking a state of dissociated sleeping and waking common in children, especially boys, and lasting usually for only a few minutes, in which the child gets out of bed and moves about. Sleepwalking in childhood is never purposeful and is of little importance so long as danger from falls is avoided. The child should be guided gently back to bed. Sleepwalking in adults usually has a hysterical basis.

slough 1 dead tissue cast off or separated from its original site.
2 the casting off of dead tissue.

Smad a family of eight proteins that participate in tumour suppression in conjunction with transforming growth factor-beta (TGF-β). Smad 1,2,3,5 and 8 are receptor-activated; Smad 4 is a co-mediator; and Smad 6 and 7 are inhibitory. The absence of Smad 3 is a feature of acute T cell lymphoblastic LEUKAEMIA. The term is derived from the proteins' homology to *Caenorhabditis elegans* Sma and drosophila MAD proteins.

small bowel see SMALL INTESTINE.

small intestine the longest, but narrowest part of the intestine. The part in which digestion and absorption of food is performed. The small intestine extends from the outlet of the stomach (the PYLORUS) to the CAECUM at the start of the large intestine (COLON), and consists of the DUODENUM, the JEJUNUM and the ILEUM.

smear a thin film of tissue, cells, blood or other material spread on a transparent slide for microscopic examination.

smegma accumulated, cheesy-white, sebaceous gland secretions occurring under the foreskin of an uncircumcised male with poor standards of personal hygiene. Smegma becomes infected, foul-smelling and irritating and can cause local inflammation. It has been said to be carcinogenic to sexual consorts but this has not been proved. Associated papillomaviruses may be the cause.

smell one of the five senses. Smell is mediated by airborne chemical particles that dissolve in the layer of mucus on the upper part of the nose lining and stimulate the endings of the olfactory nerve twigs. The olfactory system is capable of distinguishing a large number of distinct odours.

smooth muscle the unstriped involuntary muscle occurring in the walls of blood vessels, the intestines and the bladder, and controlled by the AUTONOMIC NERVOUS SYSTEM and by HORMONES.

sneezing a protective reflex initiated by irritation of the nose lining and resulting in a blast of air through the nose and mouth that may remove the cause. The vocal cords are tightly approximated, air in the chest is compressed and the cords suddenly separated.

Snellen's chart test a standard vision-testing chart of letters of diminishing size, used at the standard distance of 6 m. The eyes are tested one at a time and the result recorded in terms of the lowest line that can be correctly read. (Hermann Snellen, 1834–1908, Dutch ophthalmologist).

snoring a noise caused by vibration of the soft palate and other soft tissue in the upper airway by turbulent air flow. Snoring occurs during sleep usually when the mouth is open and is thus commonest when the snorer is lying on the back or when the nose is blocked. Snoring is never heard by the person causing the sound. Snoring may be treated by oral appliances to advance the mandible, by surgical palatoplasty or by hypnotherapy.

social anxiety disorder social phobia, a phobic disorder featuring disabling and distressing embarrassment, anxiety and humiliation experienced in social contexts, especially in public. The condition is commoner than was formerly supposed and the true prevalence has been masked by under-diagnosis because of the shame and reticence of the sufferers.

sodomy 1 anal copulation with a male. 2 anal or oral copulation with a woman. 3 copulation with an animal. Also known as buggery.

soft palate the mobile flap of muscle covered with mucous membrane that is attached to the rear edge of the hard palate. The soft palate seals off the cavity of the nose from the mouth during swallowing.

soiling inappropriate discharge of faeces after the age of about 3 or 4. Soiling is usually accidental. Compare ENCOPRESIS.

solar pertaining to the sun or to the SOLAR PLEXUS.

solar plexus a large network of autonomic nerves situated behind the stomach, around the coeliac artery. It incorporates branches of the VAGUS NERVE and the splanchnic nerves and sends branches to most of the abdominal organs. Also known as the coeliac plexus. The term derives from the sun-like appearance of the radiating branches.

solipsism the philosophic notion that the universe exists only in the perception of the person holding this view. This extreme position is seldom seriously proposed, however, except to illustrate the general point that none of us can be sure that this is not so.

somat-, somato- *combining form denoting* the body.

somatic 1 pertaining to the body (soma), as opposed to the mind (psyche). 2 pertaining to general body cells that divide by MITOSIS, as distinct from ova and spermatozoa that are formed by MEIOSIS. All the body cells except those in the ovaries and testes that produce ova and spermatozoa. 3 relating to the outer walls or framework of the body.

somatic gene therapy genetic treatment that affects only the SOMATIC cells and thus is limited in its effect to the individual treated. The alternative form – genetic treatment affecting the germ cells in the ovaries or testicles – may be perpetuated through succeeding generations. For this reason it is generally prohibited.

somatic mutation a mutation affecting SOMATIC cells that can affect only those cells and their offspring, so cannot be passed on to future generations. Such a mutation dies with the death of the individual.

somatization disorder the current term for hysteria, adopted as a euphemism.

somatopsychic psychosomatic. Pertaining to both body and mind.

somatostatins tetradecapeptides widely distributed in the body, that inhibits the secretion of many HORMONES and NEUROTRANSMITTERS. Somatostatins are secreted in the hypothalamus, elsewhere in the brain, the gastrointestinal tract, pancreas, retina, spinal cord and various endocrine glands. They have been suggested as antidotes to growth hormone in conditions such as ACROMEGALY but have a very short duration of action and their use results in rebound over-secretion.

somatotropin growth hormone produced by recombinant DNA techniques (genetic engineering) and used to treat growth defects.

somatotype the physical build of a person. A body type, claimed by some, with little evidence, to have a reliable correlation with personality. See also ECTOMORPH, ENDOMORPH, MESOMORPH.

-some *suffix denoting* body, as in chromosome.

somn-, somni- *prefix denoting* sleep.

somnambulism see SLEEPWALKING.

soporific 1 tending to induce sleep. 2 a sleep-inducing drug.

sore any local breakdown of a body surface (ULCER) or septic wound.

sorption the general term for the passive movement of liquid molecules, as by adsorption, absorption or persorption.

Southern blotting a method of identifying a fragment of DNA containing a specific sequence of bases. A mixture of fragments, produced by cutting DNA with restriction enzymes, is separated on a gel block by electrophoresis. The fragments are denatured to single strand DNA and transferred, by blotting, to a nitrocellulose sheet. The position of the desired fragment can then be shown by hybridizing it with a DNA probe labelled with radioactive phosphorous that will cause a black line on an X-ray film. The method was named after its developer Edward M. Southern. A similar technique for RNA analysis has been humorously called 'Northern blotting' and the play on words has been extended to include 'Western blotting'.

spacial summation the combined effect of input by dendrites at various points on the surface of a nerve cell body that produces a larger electrical charge than that caused by a single input.

spasm involuntary strong contraction of a muscle or muscle group. Spasms may be brief or sustained (cramps) and may result from minor muscle disorders, disease of the nervous system or habit (TICS).

spastic pertaining to spasms.

spasticity rigidity in muscles causing stiffness and restriction of movement. Spasticity may or may not be associated with paralysis or muscle weakness. Spasticity with paralysis is a feature of many cases of stroke. It occurs in spastic paralysis (cerebral palsy) and sometimes in multiple sclerosis.

spatulate shaped like a spatula.

specific ascending pathway a sensory input bundle of linked neurons dedicated to a single modality of sensation.

specific immune defence an immune response based on the recognition of a particular antigen. The immune system also provides general, non-specific defence.

specificity the characteristic of a binding site or receptor to be activated only by a single molecule or class of molecules.

SPECT *acronym for* Single Photon Emission Computed Tomography, a type of radionuclide scanning.

spectacles pairs of simple thin lenses, usually mounted in frames and used for the correction of short sight (MYOPIA), long sight (HYPERMETROPIA), ASTIGMATISM and PRESBYOPIA.

specular pertaining to a mirror, as in specular reflection.

speech

Speech is a remarkably complicated activity involving the operation of many parts of the brain. The use of language implies comprehension, the formulation of thoughts, the acquisition, storage and recall of words and their relationships, the selection of words (whether appropriate or not), their arrangement in meaningful sequences, and their articulation.

Verbal thinking and the articulation of speech, although closely connected, involve quite different neurological activities. Human speech differs qualitatively from the communication sounds made by the lower animals and involves considerable activity in the cortex of the brain . In the other primates, vocalization appears to be initiated and controlled by the limbic system rather than by the cortex. So it is probably correct to suggest that speech is one of the distinguishing attributes of the human being. Speech and language are much more than merely ways of immediate communication. With technological advance they have become the means of recording and consolidating individual and community experience and have become the single most important element in cultural expression.

Speech centres in the brain

There are five readily distinguishable brain areas concerned with language . The area in the frontal lobe, near the motor cortex (Broca's area), is concerned with the control of the many muscles in the face, tongue, throat and jaw involved in the articulation of speech. The other areas, further back along the temporal lobe of the brain (conveniently lumped together as Wernicke's area) are concerned with all the many sensory aspects of speech. Wernicke's area is connected to Broca's area by a thick tract of nerve fibres. Speech is believed to be formulated in Wernicke's area and the information passed to Broca's area where the program for the motor commands is put together. This is then passed to the motor cortex so that the appropriate muscles can be activated. Wernicke's area receives input from the visual cortex for written or printed language and from the auditory cortex for heard speech.

Phonation

The vocal cords are held apart during normal breathing. The production of voice sounds begins with the approximation of the cords during expiration. As they come together the pressure in the lungs rises. The degree of this pressure rise varies with the loudness of the sounds we wish to make. Air passing between the tightly-pressed cords forces them suddenly apart. This leads to a sudden drop of pressure so that the cords are able to come together again and the pressure under them rises once more. The effect is a rapidly repeated series of separations and closures so that a succession of compressions and rarefactions is imposed on the column of air in the throat, mouth and nose.

The pitch of the tones produced by the vocal cords depends on the frequency of vibration; the higher the frequency, the higher the pitch. This, in turn, depends on the tension on the vocal cords and on the length of cord allowed to vibrate. A short length produced a high pitch and the full length and lowest tension produces the deepest tone possible. The cords vibrate in a complex manner and different parts can vibrate simultaneously at different frequencies. In general, women and children have shorter vocal cords than men, so their voices are pitched higher. The range of the fundamental pitch of the voice extends from about 80 Hz in men to about 400 Hz in women. A soprano will easily produce a singing note of 1000 Hz.

Voice quality

The basic sound produced by the vocal cords has a wave-form described as a saw-tooth. This is the kind of wave produced by any double reed vibrating musical instrument such as an oboe or a bassoon. The saw-tooth waveform is quite different from the sine-wave form produced by a tuning fork or a flute. Pure sine-waves have a fundamental tone of a particular frequency (the recognizable pitch of the tone) and nothing else. Saw-tooth waves have a fundamental tone but also have a rich collection of harmonics. These are additional tones, or overtones, having frequencies which are simple multiples of the fundamental frequency. The harmonics are usually of lower amplitude than the fundamental and so are not heard as basic tones but as an alteration in quality. The quality of the voice, however has little to do with the richness and relative amplitude of the various harmonics produced by vocal cord vibration. Laryngeal tone, by itself, is thin and weak.

The final quality of the voice depends on the principle of sympathetic resonance. Hollow cavities have a natural frequency of vibration, as can be shown by blowing across the top of a bottle. There are many such cavities in the human head and neck – the throat cavity, nasal cavity, mouth cavity and the bony sinuses. Each of these reinforces, to a greater or lesser degree, the particular harmonics in the basic laryngeal tone that correspond to their natural resonant frequency. The effect of these cavities on voice quality varies with their varying volume, with the force of the expiration of the air and with the degree of communication between them. Speech quality is, for instance, greatly affected by alterations in the degree of swelling of the lining of the nose, as during a cold.

Formants

Resonances have an even more important function – to produce the vowel sounds of speech. Each vowel sound involves a combination of two frequencies to produce what is called a formant. These pairs of frequencies are common to all vowels, whether produced by men, women or children. Formant frequency has a bearing on the ease with which vowel sounds can be produced when singing high notes. Opera composers have long known empirically to avoid writing high notes for those vowels with a high upper formant frequency. They may not have known about formants, but errors in this matter would be quickly pointed out to them, in rehearsal, by the sopranos.

Articulation

Modulation of the basic pitch-varying tones to produce speech is primarily the function of the mouth cavity. This is done by ever-changing variations in its shape and volume and in the area and shape of its three openings – to the throat, to the nose and to the mouth. Quick changes in the mouth opening, in particular, allows a wide range of different sounds to be produced. The relationships of the tip of the tongue to the teeth, lips and the roof of the mouth, and of the lips to each other, are especially important in forming the outlet shape necessary to produce these sounds from the harmonic-rich vocal cord tones. Thus articulation of sound involves precise and accurately-timed contractions and relaxations of the muscles of the tongue, the lips, the soft palate and the face. All these movements have to be coordinated in the brain.

speech therapy treatment designed to help people with a communication difficulty arising from a disturbance of language, a disorder of articulation, difficulty in voice production or defective fluency of speech.

speedball a slang term for a dose of cocaine and heroin taken intravenously.

speed freak a slang term for a habitual amphetamine (amfetamine) user.

sperm 1 a spermatozoon.
2 semen (spermatic fluid).

spermat-, spermato- *combining form denoting* spermatozoa or semen.

spermatic pertaining to SPERM.

spermatic cord a cord-like structure consisting of the VAS DEFERENS surrounded by a dense plexus of veins and other blood vessels, lymphatic vessels and nerves. The spermatic cord runs upwards from the back of the testicle through the INGUINAL CANAL into the abdominal cavity where the vas leaves it to run into the PROSTATE GLAND.

spermatic fluid SEMINAL FLUID.

spermatids immature sperm cells formed in the testicle, having half the normal number of chromosomes (haploid), that develop into SPERMATOZOA without further division.

spermatocyte a cell of the seminiferous tubules of the testis that is converted by MEIOSIS into four SPERMATIDS.

spermatozoa microscopic cells about 0.05 mm long occurring in millions in seminal fluid. Spermatozoa are male GAMETES, carrying all the genetic contribution from the father and bearing either an X chromosome to produce a daughter, or a Y chromosome to produce a son.

spermaturia SPERMATOZOA in the urine.

sperm capacitation changes in spermatozoa, occurring in the female reproductive tract, that make them potentially capable of fertilization of an ovum.

sperm count a method of determining the concentration of SPERMATOZOA in a semen sample of known dilution. Counts are done on a slide engraved with squares of known size, using a microscope. Fertility is unlikely if the count is below 20,000,000 per ml.

sperm donation seminal fluid provided by a donor for the purposes of fertilization of women whose husbands or partners are sterile. Seminal fluid can be preserved indefinitely frozen in a glycerol cryoprotectant in phials, or plastic straws and kept in liquid nitrogen sperm banks.

spermicides contraceptive preparations designed to kill spermatozoa. In general, spermicides used alone are unreliable as contraceptives.

sperm injection IVF a method of in vitro fertilization in which an ovum is held steady by a suction device while a single sperm is injected directly into it through a very fine needle. The method, which was adopted to ensure fertilization in cases in which the father's sperm count is too low, has been criticized on the grounds that it may increase the likelihood of birth defects. It is suggested that it interferes with the natural selection process in which only the fittest sperm are able to penetrate the egg.

sphenoid wedge-shaped, or cuneiform. From Greek *sphenoidis*, a wedge.

sphenoid bone the wedge-shaped, bat-like central bone of the base of the skull.

sphincter a muscle ring, or local thickening of the muscle coat, surrounding a tubular passage or opening in the body. When a sphincter contracts it narrows or closes off the passageway.

sphincter of Oddi the smooth muscle ring that surrounds the common bile duct at the point at which it enters the DUODENUM. (Ruggero Oddi, Italian physician, 1864–1913).

sphygmomanometer a mercury manometer or aneroid instrument used to measure blood pressure. See KOROTKOFF SOUNDS.

spider naevus a common, tiny skin blemish consisting of a small, central, slightly raised, bright red area from which fine red lines, like spider legs, radiate. Numerous spider naevi occur in serious liver disease, such as cirrhosis and sometimes in pregnant women or those receiving hormone replacement therapy.

spin-, spino- *combining form denoting* spine, spinal or spinal cord.

spinal canal the tube-like space running the length of the VERTEBRAL COLUMN formed by the arches of the successive vertebrae through which the SPINAL CORD and its membranes pass.

spinal column see VERTEBRAL COLUMN.

spinal cord the downward continuation of the BRAINSTEM that lies within a canal in the spine (VERTEBRAL COLUMN). The cord is a cylinder of nerve tissue about 45 cm long containing bundles of nerve fibre tracts running up and down, to and from the brain. These tracts form SYNAPSES with the 62 spinal nerves that emerge in pairs from either side of the cord, between adjacent vertebrae, and carry nerve impulses to and from all parts of the trunk and the limbs.

spinal nerves the 31 pairs of combined MOTOR and sensory nerves that are connected to the spinal cord.

spinal reflex a reflex mediated by a nerve pathway with a sensory afferent to the spinal cord, linkages within the cord, and an efferent to muscle from the cord. Spinal reflexes commonly occur without brain influence.

spindle a term used adjectivally in anatomy and referring to any elongated cell or structure pointed at both ends. Fusiform.

spine see VERTEBRAL COLUMN.

spinothalamic tracts the nerve bundles running up the spinal cord carrying sensory impulses to the great sensory nucleus of the basal ganglia – the THALAMUS.

spinous process the rearward projection of a vertebra.

splanchnic pertaining to the internal organs (viscera).

spleen a solid, dark purplish organ, lying high on the left side of the abdomen between the stomach and the left kidney. The spleen is the largest collection of lymph tissue in the body and contains a mass of pulpy material consisting mainly of LYMPHOCYTES, PHAGOCYTES and red blood cells. The spleen is the main blood filter, removing the products of breakdown of red blood cells and other foreign and unwanted semisolid material. It is a source of lymphocytes and a major site of antibody formation.

splen-, spleno- *prefix denoting* spleen.

splenic pertaining to the spleen.

splenius one of two muscles at the back of the neck, running from the back and sides of the vertebrae to the OCCIPITAL bone of the skull. The splenius muscles rotate and extend the head.

splicing in genetics, the process in a DISCONTINUOUS GENE in MESSENGER RNA of removal of INTRONS and the joining together of exons.

splint a usually temporary support or reinforcement for an injured part, often used to minimize movement at the site of injury, especially in the case of a fracture of a bone.

split personality a rare condition in which the subject adopts, at different times, one of two or more distinct personas. The condition may be associated with epilepsy and there is often a history of abuse in childhood. It is not a feature of schizophrenia.

spondyl-, spondylo- *combining form denoting* VERTEBRA or vertebral.

spondylolisthesis the moving forwards of a vertebra relative to the one under it, most commonly of the 5th lumbar vertebra over the top of the SACRUM. This is due to a congenital weakness (SPONDYLOLYSIS) of the bony arch that bears the facets by which the vertebrae articulate together. Spondylolisthesis causes severe backache on standing and leads to nerve pressure effects. The condition may also affect vertebrae in the neck.

spondylolysis a symptomless congenital deficiency of bone in the arch of the 5th or 4th lumbar vertebra disorder of the spine. The arch is formed of soft fibrous tissue and there is a weak link with adjacent vertebrae so that the condition of SPONDYLOLISTHESIS may occur.

spongioblasts embryonic cells of the neural tube that give rise to the neural connective tissue (neuroglial) cells, the astrocytes and the oligodendrocytes.

spor-, sporo- *prefix denoting* spore.

sport-prohibited drugs a list of stimulants, narcotics, anabolic agents, diuretics and hormones prohibited by the Olympic Movement Anti-Doping Code. The list includes amphetamines (amfetamines), bromantan, caffeine (above 12 mcg/ml), carphedon, cocaine, ephedrine, certain beta agonists, diamorphine (heroin), morphine, methadone, pethidine, methandianone, nandrolone, stanozole, testosterone, clenbuterol, DHEA, androstenedione, 19-norandrostenediol, acetazolamide, frusemide (furosemide), hydrochlorothiazide, triamterene, mannitol, growth hormone, corticotrophin (corticotropin), chorionic gonadotrophin, pituitary and synthetic gonadotrophins,, erythropoietin, insulin, and all corresponding releasing factors and analogues.

sports medicine the branch of medicine concerned with the physiology of exercise and its application to the improvement of athletic performance and fitness, and with the prevention, diagnosis and treatment of medical conditions caused by, or related to, sporting activities of all kinds.

spot a popular term for any small lump or inflamed area on the skin such as a pustule, papule, comedone, cyst, macule, scab or vesicle.

sprain stretching or a minor tear of one of the ligaments that hold together the bone ends in a joint or of the fibres of a joint capsule.

sputum mucus, often mixed with PUS or blood, that is secreted by the goblet cells in the MUCOUS MEMBRANE lining of the respiratory tubes (BRONCHI and BRONCHIOLES). Excess sputum prompts the cough reflex. Also known as phlegm.

squamous 1 scaly. Covered with, or formed of, scales.

2 pertaining to, or resembling a scale or scales.

squamous cell a flat, scaly epithelial cell.

squamous epithelium an outer layer of a surface (epithelium) that is composed of flat, scaly cells.

SRY gene one of the few genes on the Y chromosome. This gene promotes the development of the testes in males.

stable 1 of an ill person, in a currently unchanging state, neither improving nor deteriorating.

2 of a personality, not liable to mental disturbances or abnormal behaviour.

stage a recognizable point or phase in the development of a progressive disease, particularly a cancer. In breast cancer, for instance, three recognizable stages might be: tumour confined to the breast tissue; tumour extended to the axillary lymph nodes; tumour widely metastasized.

staging determination of the stage to which a disease, especially a cancer, has progressed. Staging is important as an indication of the likely outcome (prognosis) and in deciding on the best form of treatment, as this may differ markedly at different stages.

staining the use of selected dyes to colour biological specimens such as cells, cell products, thin slices of tissues or microorganisms to assist in examination and identification under the microscope.

standard deviation a measure of dispersion widely used in statistics. Standard deviation is the square root of the arithmetic average of the squares of the deviations of the members of a sample from the mean.

Stanford-Binet test a type of intelligence test on which many current tests are based. (Alfred Binet, 1857–1911, French psychologist. Test adapted at Stanford University).

stapes the innermost of the three tiny linking

bones of the middle ear (auditory ossicles). The footplate of the stapes lies in the oval window in the wall of the inner ear and transmits vibration to the fluid in the COCHLEA.

starch a complex polysaccharide carbohydrate consisting of chains of linked glucose molecules. Amylose is a chain of 200–500 glucose units. Amylopectin consists of 20 cross-linked glucose molecules. Most natural starches are a mixture of these two. Starch, in the form of potatoes, rice and cereals forms an important part of the average diet and about 70 per cent of the world's food.

Starling and Bayliss Earnest Henry Starling (1866–1927) and William Maddock Bayliss (1860–1924) were two English physiologists who formed a lifelong intellectual alliance based on their interest in research. Starling was educated at Guy's Hospital and became head of the department of physiology there. In 1899 he became a Fellow of the Royal Society and the same year was appointed Professor of Physiology at University College, London. Bayliss started training as a doctor at University College but decided medicine was not for him and went to Wadham College, Oxford to do research in physiology. Then, possibly attracted by Starling's sister whom he married a few years later, he decided to return to University College to join forces with Starling.

In 1902 Starling and Bayliss were investigating the puzzling question of how it was that the pancreas invariably produced its digestive juice as soon as food passed from the stomach into the duodenum. Pavlov had claimed that this was the result of messages passing along nerves, but Starling and Bayliss proved that the pancreas continued to secrete on time even if all the nerves to it were cut. Careful research showed that stomach acid contact with the lining of the small intestine caused a substance, which they named 'secretin', to be produced. This substance passed to the pancreas by way of the blood-stream and prompted it to start secreting. Later, as it became clear that there were other chemical messengers of this kind, Starling suggested the general term 'hormones' for these substances, derived from the Greek *hormao* meaning 'I set in motion'.

Starling's *Principles of Human Physiology* was published in 1912, went into many editions and was read by generations of medical students. In 1903 Bayliss was elected a Fellow of the Royal Society and in 1912 became Professor of General Physiology at University College. He was knighted in 1922.

starvation long-term deprivation of food and its consequences. These are severe loss of body fat and muscle, changes in body chemistry with KETOSIS and constant hunger.

stasis a reduction or cessation of flow, as of blood or intestinal contents.

statutory rape sexual intercourse with a girl below the age of consent.

STD *abbrev. for* sexually-transmitted disease. See SEXUALLY TRANSMITTED INFECTIONS.

steato- *combining form denoting* fat.

stem cell a pluripotential progenitor cell from which a whole class of cells differentiate. A stem cell in the bone marrow, for instance, gives rise to the entire range of immune system blood cells (neutrophils, eosinophils, basophils, monocytes/macrophages, platelets, T cells and B cells) and the red blood cells (erythrocytes). Stem cells from umbilical cord blood have a considerable potential for medical treatment. A major research effort to produce stem cells artificially for medical purposes is under way. The genetic material from an ovum can be removed and a cumulus cell from the outside of an egg, or a fibroblast, can be inserted.

stenosis narrowing of a duct, orifice or tubular organ such as the intestinal canal or a blood vessel.

stent see STENTING.

stenting the use of a physical device, such as a tubular stainless steel or plastic mesh or coil of wire, to keep a body tube fully open. Stents are used in the CORONARY ARTERIES, the AORTA, the renal arteries, the FEMORAL ARTERIES, the intestine and elsewhere. In addition to the widespread use in arteries, self-expanding metallic stents have been successfully used to maintain patency in narrowing of urethras and bile ducts and for swallowing difficulties (dysphagia) caused by cancer of the OESOPHAGUS. Silicone rubber stents have been used in the TRACHEA and BRONCHI. Stents in arteries are liable to blockage by blood clotting. Anticoagulant treatment is required. Drug-eluting coronary stents designed to prevent restenosis are currently displacing bare-metal stents. Sirolimus appears to be the drug of choice at the time of writing.

stercorous pertaining to excrement.

stereo- *combining form denoting* solid or three-dimensional.

stereognosis the ability to identify the shape, size and texture of objects by touch, without the benefit of sight.

stereoisomerism mirror-image molecular asymmetry. Also known as chirality. In the case of amino acids, the two forms are represented on paper with the carboxyl group of the carbon chain at the top. In the Laevo (L) form, the functional groups connected to the central carbon or carbon chain are shown as projecting to the left and in the dextro (D) form they are shown projecting to the right.

stereopsis the normal ability to perceive objects as being solid. Stereoscopic vision.

sterile 1 free from bacteria or other microorganisms.
2 incapable of reproduction.

sterilization 1 the process of rendering anything free from living micro-organisms.
2 any procedure, such as hysterectomy, tying of the fallopian tubes, vasectomy or castration that deprives the individual of the ability to reproduce.

sternal pertaining to the STERNUM.

sternum the breastbone.

steroid 1 sterol-like.
2 any member of the class of fat-soluble organic compounds based on a structure of 17 carbon atoms arranged in three connected rings of six, six, and five carbons. The steroids include the adrenal cortex hormones, the sex hormones, progestogens, bile salts, sterols and a wide range of synthetic compounds produced for therapeutic purposes. Anabolic steroids are male sex hormones that stimulate the production of protein.

steroid drugs a large group of drugs that are derived from, resemble, or simulate the actions of, the natural corticosteroids or the male sex hormones of the body.

sterols a group of mainly unsaturated solid alcohols of the steroid group occurring in the fatty tissues of plants and animals. The sterols include cholesterol and ergosterol.

stertorous of breathing, a heavy, coarse snoring associated with a falling back of the tongue.

STI *abbrev. for* SEXUALLY-TRANSMITTED INFECTION.

sticky ends complementary single strands of DNA protruding from the ends of a DNA fragment as a result of the cleaving of each half of the double helix at different points near to each other. Sticky ends readily provide attachment points for other pieces of DNA or for further DNA synthesis.

stillbirth birth of a dead baby. The distinction from MISCARRIAGE is arbitrary and, in Britain, is set at 28 weeks of pregnancy. Stillbirths must be registered and the cause of death established before a certificate of stillbirth can be provided and burial may take place.

stimulus anything that causes a response, either in an excitable tissue or in an organism.

stitch a brief, sharp pain in the abdomen or flank caused by severe or unaccustomed exercise, especially running.

stom-, stomato- *combining form denoting* mouth.

stoma a mouth or orifice, especially one formed surgically, as in a COLOSTOMY or ILEOSTOMY.

stomach the bag-like organ lying under the DIAPHRAGM in the upper right part of the ABDOMEN into which swallowed food passes, by way of the OESOPHAGUS. The stomach has an average capacity of about 1.75 l and secretes hydrochloric acid and the protein-digesting enzyme PEPSIN.

stomach disorders these include peptic ulcer, pernicious anaemia, pyloric stenosis and stomach cancer.

stools faeces.

stop codon any one of three nucleotide triplets which marks the end of every gene and indicates that protein synthesis ends at that point. The three stop codons are UAG (the amber codon), UAA and UGA. U is uracil, A is adenine and G is guanine.

stop mutation any mutation that changes a codon that codes for an amino acid into a codon that codes for a 'stop'. Stop codons are UAG (amber codon), UAA (ochre codon) and UGA (opal codon). The effect of a stop mutation is that a protein is shortened so that an abnormal form is produced. CYSTIC FIBROSIS, MUSCULAR DYSTROPHY and other genetic disorders are caused by stop mutations. See also PTC124.

storage diseases a range of metabolic disorders in which various substances accumulate in abnormal amounts in certain body tissues or organs such as the liver.

stork bites a popular term for the small, harmless, pinkish skin blemishes that commonly occur around the eyes and on the back of the neck in new-born babies. Stork bites are benign tumours of small blood vessels (haemangiomas) and usually disappear within the first year of life.

storm a sudden worsening of the symptoms and other features of a disease. Used more often in the adjectival form 'stormy' or as the metaphor 'stormy passage'.

strabismus squint. The condition in which only one eye is aligned on the object of interest. The other eye may be directed too far inward (convergent strabismus), too far outward (divergent squint), or upward or downward (vertical squint). Squint in childhood, or any squint of recent onset, requires urgent treatment. Untreated childhood squint often leads to AMBLYOPIA. New squints in adults usually imply a disorder of the nervous system.

strain stretching or tearing of muscle fibres, usually in the course of athletic overactivity. There is swelling, pain, bruising and a tendency to muscle spasm.

strangulation constriction or compression of any passage or tube in the body, such as the jugular veins of the neck in manual strangulation, or the intestine in hernia. Strangulation may also result from twisting of a part as in volvulus or torsion of the testis.

strangury a frequent, painful but unproductive desire to empty the bladder. Strangury is a feature of bladder stones, bladder cancer, cystitis and prostatitis.

strapping the use of adhesive tape or firm bandages to maintain the desired relationship of parts of the body or to rest an injured or inflamed part.

stratified having a number of superimposed layers. Epithelia are commonly stratified and in the epidermis.

strepto- *combining form denoting* twisted, chain-like or coiled.

streptokinase a protein-splitting ENZYME used as a drug to dissolve blood clot in a coronary artery so as to minimize the degree of myocardial infarction during a heart attack. It is also used to treat pulmonary embolism.

stress any physical, social or psychological factor or combination of factors that acts on the individual so as to threaten his or her well-being and produce a physiological, often defensive, response. The response to stress may be beneficial, distressing or, occasionally, dangerous. Responses such as the production of ADRENALINE and CORTICOSTEROIDS, raised heart rate and blood pressure, increased muscle tension and raised blood sugar, are natural; but persistent civilized suppression of the natural physical concomitants (fight or flight) may be damaging. Most medical scientists view with scepticism the proposition that many human diseases are caused by stress. There is, however, no questioning the fact that overwhelming stress can cause physical and psychological damage.

stress genes a general term for genes that are induced to transcribe following exposure to DNA-damaging agents such as radiation or the action of oxygen free radicals.

stretch marks a popular name for STRIAE.

stretch receptor a sensory nerve ending activated by stretching of the part in which it is situated. An example is the muscle spindle stretch receptor.

striae broad, purplish, shiny or whitish lines of atrophy on the skin, most commonly affecting pregnant women, and occurring on the abdomen, breasts or thighs. Striae are due to altered COLLAGEN. Also known as stretch marks.

striated striped, grooved, or ridged.

striated muscle voluntary, skeletal muscle and heart (cardiac) muscle, characterized by microscopic transverse stripes or striation. Compare SMOOTH MUSCLE.

stricture narrowing of a body passage.

stridor noisy breathing caused by narrowing or partial obstruction of the LARYNX or TRACHEA.

stroke the effect of acute deprivation of blood to a part of the brain by narrowing or obstruction of an artery, usually by thrombosis (80 per cent), or of damage to the brain substance from bleeding into it (cerebral haemorrhage) (15 per cent). Subarachnoid haemorrhage is the cause in 5 per cent. The results of such damage are most obvious if they involve the nerve tracts concerned with movement, sensation, speech and vision. These are situated close together, in the internal capsule of the brain, and are often involved together. There may be paralysis and loss of sensation down one side of the body or of one side of the face, loss of corresponding halves of the fields of vision, a range of speech disturbances or various disorders of comprehension or expression. In most cases a degree of recovery, sometimes considerable, may be expected. Haemorrhage into the brainstem, where the centres for the control of the vital functions of breathing and heart-beat are situated, is the most immediately dangerous to life. Diagnosis of the type of stroke is important and this requires neuroimaging of the brain.

stroke in progression brain damage caused by an obstruction to the blood supply (ischaemia) that increases progressively over the course of hours, days or weeks. Has also been defined as 'a stroke in which the neurological deficit is still increasing in severity or distribution after the patient is admitted to observation'. The condition is associated with high concentrations of glutamate in the blood and cerebro-spinal fluid.

stroma the tissue forming the framework of an organ. Compare PARENCHYMA.

structural gene a GENE that codes, as most do, for the amino acid sequence of a protein rather than for a regulatory protein.

stupor a state of severely reduced consciousness, short of COMA, from which the affected person can be briefly aroused only by painful stimulation.

styloid pointed and slender.

styloid process any pointed bony protuberance as that on the TEMPORAL BONE, the FIBULA, the RADIUS or the ULNA.

styptic causing contraction of tissues or blood vessels and tending to check bleeding.

sub- *prefix denoting* under, less.

subacute intermediate in duration between ACUTE and CHRONIC.

subarachnoid space the space between the PIA MATER and the arachnoid mater, occasioned by the fact that the pia, the innermost layer of the meninges, closely invests the surface of the brain while the arachnoid, external to it, bridges over the grooves. The subarachnoid space contains blood vessels.

subclavian 1 situated below or under the CLAVICLE.
2 pertaining to the subclavian artery or vein.

subclavian artery a short length of the major artery that branches from the aorta on the left side and from the innominate artery on the right side and continues as the axillary artery to supply the arm. The subclavian arteries also supply the brain via their vertebral branches.

subclinical of a degree of mildness, or of such an early stage of development, as to produce no symptoms or signs.

subconscious 1 of mental processes and reactions occurring without conscious perception.
2 the large store of information of which only a small part is in consciousness at any time, but which may be accessed at will with varying degrees of success.
3 in psychoanalytic theory, a 'level' of the mind through which information passes on its way 'up' to full consciousness from the unconscious mind.

subcutaneous under the skin. Many injections are given subcutaneously. An alternative term is hypodermic.

subjectivity the quality or condition of perception of the mind or the emotions rather than of external objects or events (objectivity).

sublimation deflection of socially unacceptable drives into acceptable channels so that the necessity for repression is avoided. Sublimation is a psychoanalytic concept and is considered to be a healthy feature of a mature personality.

subliminal perception the reception of stimuli, often complex or verbal and usually visual, that are presented for such a short time as to be barely noticed or unnoticed. Such stimuli can, however, influence behaviour and present potential opportunities for abuse. The conclusion, now verified by physiological research, that consciousness and information transmission may involve different systems that can operate independently has major implications for psychology and philosophy and has aroused much controversy.

sublingual under the tongue.

subluxation partial or incomplete dislocation of a joint.

submaxillary pertaining to the lower jaw. Under the MAXILLA.

submicroscopic too small to be see under a visible light microscope.

subscapular below or on the underside of the shoulder blade (scapula).

substance a general term meaning any physical matter, or the nature of the matter of which something is made, which has, in recent years acquired a new sense. The term, in this restricted sense, is applied to any chemical, solid, liquid or gaseous, capable of affecting the state of the mind. A psychoactive material.

substance abuse a general term referring to the non-medical and 'recreational' use of drugs such as amphetamine (amfetamine), cannabis, cocaine, methylenedioxy-methamphetamine (ecstasy), heroin, lysergic acid diethylamide (LSD), organic solvents by inhalation, and so on. The term is also applied to an intake of alcohol that is likely to prove harmful. Oddly enough is not currently applied to a commonly-used substance more dangerous than most of these – tobacco.

substantia nigra a layer of grey matter (nerve cell bodies) containing pigmented nerve cells, that spreads throughout the white substance of the midbrain and receives fibres from the BASAL GANGLIA. DOPAMINE is produced in the substantia nigra, and loss of the pigment cells is a constant finding in PARKINSON'S DISEASE.

substrate the substance on which an ENZYME acts. Any reactant in a reaction that is catalysed by an enzyme.

subungual under a nail.

sucrose cane or beet sugar. A crystalline disaccharide carbohydrate present in many foodstuffs and widely used as a sweetener and preservative. During digestion, sucrose hydrolyses to glucose and fructose.

sudamen a tiny fluid-filled VESICLE formed at a sweat pore, as in PRICKLY HEAT.

sudden infant death syndrome cot death. The sudden, unexplained death of an apparently well baby. No apparent cause is established, even after a detailed postmortem examination. Many theories have been put forward and it seems likely that a range of causes is operating, including putting babies down to sleep in the prone position. Many sudden deaths in healthy babies can be explained.

sudoriferous sweat-producing or secreting.

suffocation oxygen deprivation by mechanical obstruction to the passage of air into the lungs, usually at the level of the nose, mouth, LARYNX or TRACHEA.

suicide intentional self-killing. Depression is the commonest cause of suicide and severely depressed people are always at risk. Suicide is also common among alcoholics, people with SCHIZOPHRENIA and people with severe personality disorders.

sulcus a narrow fissure or groove especially one of the furrows that separates adjacent convolutions (gyri) on the surface of the brain.

sulphur an element occurring in AMINO ACIDS and hence in many proteins, including COLLAGEN. Sulphur is often incorporated into ointments used in the treatment of various skin disorders such as ACNE, DANDRUFF and PSORIASIS.

sunburn the damaging effect of the ultraviolet component of sunlight on the skin. This varies from minor reddening to severe, disabling blistering and an increase in the risk of skin cancer.

sunscreens creams or other preparations used to protect the skin from the damaging effects of sunlight. Most contain para-aminobenzoic acid (PABA) which absorbs ultraviolet radiation.

super- *prefix denoting* above or excessive.

superciliary pertaining to the eyebrow. Literally, situated above the eyelashes.
supercoiling the secondary coiling of a DNA helix to form a coiled coil. In positive supercoiling, both strands of the double helix coil together in the same direction as the coiling of the strands.
superego a psychoanalytic term for the conscience. See also FREUDIAN THEORY.
superfecundation the fertilization of more than one ovum within a single menstrual cycle by separate acts of coitus, so that twins may be born with different fathers.
superfetation the rare occurrence of two or more fetuses of different ages in a womb that are the result of fertilizations occurring in different menstrual cycles.
superficial near the surface.
superior above, higher than, with reference to the upright body. Compare INFERIOR.
superiority complex an unrealistically exaggerated belief in one's own merits. Alfred Adler (1870–1937, Austrian psychologist) suggested that in some people a superiority complex is a response to feelings of inferiority.
superior vena cava the final venous channel that returns blood from the head, shoulders, upper limbs and thoracic structures other than the lungs to the upper chamber (atrium) of the right side of the heart.
superjacent lying immediately above or upon something.
supernatant floating on the surface.
supernumerary more than the normal NUMBER, as in supernumerary nipples or supernumerary teeth.
superovulation the production of more than one or two ova at one time. Superovulation is common when drugs are used to stimulate ovulation in the treatment of infertility.
superoxide dismutase a natural body enzyme that converts the superoxide free radical to hydrogen peroxide, which is then catalyzed to water. The gene for superoxide dismutase is on the long arm of chromosome 21 near the Alzheimer's locus. Brain tissue is highly susceptible to free radical damage. People with Alzheimer's disease have reduced levels of superoxide dismutase.
supinator one of the forearm muscles whose action is to rotate the hand into the palm-up position.
supine lying on the back with the face upwards.
supination the act of turning the body to a SUPINE position or of turning the horizontal forearm so that the palm of the hand faces upward. Compare PRONATION.
suppressor gene a GENE capable of suppressing the expression of a mutant gene at a different locus.
suppuration the production or discharge of PUS.
supra- *prefix denoting* above.
suprachiasmatic nuclei collections of nerve cells situated above the optic chiasma in the hypothalamus that act as a pacemaker for biorhythms.
supraorbital situated above the bony eye cavern (orbit), as supraorbital artery, vein and nerve.
suprapubic referring to the region on the centre of the front wall of the abdomen immediately above the pubic bone.
suprarenal glands a now obsolete term for the adrenal glands.
sural pertaining to the calf of the leg.
surface tension a property of a liquid surface, arising from unbalanced molecular cohesive forces, in which the surface behaves as if it were covered by a thin elastic membrane under tension and tends to adopt a spherical shape.
surfactants detergent-like substances that reduces surface tension and promotes wetting of surfaces. The lungs contain a surfactant to prevent collapse of the alveoli. Pulmonary surfactant is a complex mixture of proteins and lipids, and this may be deficient in premature babies leading to the respiratory distress syndrome (hyaline membrane disease). Surfactants can also be used to interfere with the motility of spermatozoa and so act as supplementary contraceptives with a spermicidal action.
surgeon a medical practitioner whose practice is limited to the diseases treated by surgical operation and who performs such treatment. General surgeons operate on almost all parts of the body but are now uncommon. The practice of specialist

surgeons includes cardiovascular surgery, neurosurgery, orthopaedic surgery, ophthalmic surgery, genitourinary surgery and ear, nose and throat surgery.

surgery 1 the treatment of disease, injury and deformity by physical, manual or instrumental interventions.
2 the diagnosis of conditions treated in this way.
3 the practice of operative treatment.
4 a room or suite used for medical consultation and treatment. From the Greek *cheirourgia*, hand work, as in *cheir*, hand and *ergon*, work.

surgical 1 pertaining to surgery or surgeons.
2 used in surgery.
3 pertaining to diseases treated by surgical rather than by medical means.

surgical drainage provision of an easy outflow route for infected or contaminated secretions or other unwanted fluid from an operation site or an area of infection or disease. Drains include soft rubber or plastic tubes or corrugated rubber sheeting.

surrogacy an agreement by a woman to undergo pregnancy so as to produce a child which will be surrendered to others. Fertilization may be by seminal fluid provided by the future adoptive father, or an ovum fertilized IN VITRO may be implanted in the surrogate mother. Surrogacy for gain is illegal in Britain and in some other countries.

suspensory ligament any LIGAMENT from which an organ or bodily part hangs, or by which it is supported.

sustentacular supporting.

susceptibility a more than normal tendency to contract an infection or other disease.

swayback an increase in the forward curvature of the spine in the lower back region (lumbar LORDOSIS) with a compensatory increase in the backward curvature in the chest region (dorsal KYPHOSIS).

sweat glands tiny, coiled tubular glands deep in the skin that open either directly on to the surface or into hair follicles and secrete a salty liquid. Apocrine glands occur only on hairy areas and open into hair follicles. Apocrine sweat contains organic matter that can be decomposed by skin bacteria and cause odours. Eccrine glands open on the surface, especially of the palms of the hands and the soles of the feet.

sweating see HYPERHIDROSIS.

swollen glands a misnomer. 'Swollen glands' are LYMPH NODES enlarged by inflammation or by infiltration with parasites or cancer. Inflammation of lymph nodes is called lymphadenitis.

sym-, syn- *prefix denoting* together, conjointly.

symbiosis a close association, of interdependence or mutual benefit, between two or more organisms, often of different species.

symbolism an important element in art and literature that was deemed by Freud, Jung and others to have major psychological implications, especially in the significance of the content of dreams. Interest in this line of speculation, once intense, has now waned.

sympathetic nervous system one of the two divisions of the AUTONOMIC NERVOUS SYSTEM, the other being the PARASYMPATHETIC. The sympathetic system causes constriction of blood vessels in the skin and intestines and widening (dilatation) of blood vessels in the muscles. It increases the heart rate, dilates the pupils, widens the bronchial air tubes, relaxes the bladder and reduces the activity of the bowel.

sympathomimetic of any agent that simulates or stimulates the sympathetic nervous system. Adrenergic. Sympathomimetic action can increase the heart output by 100 per cent over resting values. It can constrict arteries and increase the blood pressure. And it can widen the air tubes of the lungs allowing more free access of large volumes of air.

symphysis a joint in which the component bones are immovably held together by strong, fibrous cartilage. There is a symphysis between the two pubic bones at the front of the pelvis.

symptom a subjective perception suggesting bodily defect or malfunction. Symptoms are never perceptible by others. Objective indications of disease are called signs.

syn- *prefix denoting* together, conjointly or joined.

synaesthesia the phenomenon in which stimulations of one sense modality produces the effect of stimulation of another. Thus, a person may consistently experience a particular letter of the alphabet, or a musical tone, as a particular colour.

synapse the junctional area between two connected nerves, or between a nerve and the effector organ (a muscle fibre or a gland). Nerve impulses are transmitted across a synapse by means of a chemical NEUROTRANSMITTER such as ACETYLCHOLINE or NORADRENALINE. Synapses allow impulses to pass in one direction only and single brain cells may have more than 15,000 synapses with other cells. This complexity, allowing logical 'gate' operation, partly or wholly underlies the computational and storage abilities of the brain.

synaptic cleft the narrow, fluid-filled space between neurons or between a neuron and a muscle or gland cell across which neurotransmitters diffuse to convey a nerve impulse from one neuron to another or to promote contraction or secretion.

syncope fainting.

syndactyly fusion of two or more adjacent fingers or toes. Syndactyly is usually CONGENITAL. The fusion may involve skin only allowing easy surgical separation, but in more severe cases the bones may also be fused.

syndesmosis a joint in which the bones are connected by ligaments or fibrous sheets.

syndrome a unique combination of sometimes apparently unrelated symptoms or signs, forming a distinct clinical entity. Often the elements of a syndrome are merely distinct effects of a common cause, but sometimes the relationship is one of observed association and the causal link is not yet understood. Originally, the term was applied only to entities of unknown cause but many syndromes have now been elucidated and their names retained because of familiarity. From the Greek *syn*, together, and *dromos*, a course or race.

syndrome E a condition of desensitization to violence, loss of human sympathy, excessive arousal, obsessive ideas and compulsive repetition that enables affected individuals, nearly all men, to engage in repetitive killing of defenceless people, especially when this is done with the approval of authority. The syndrome has been observed repeatedly throughout human history. Critics have suggested that to call this entity a syndrome may be to excuse it.

synergism cooperative action, especially of groups of muscles, so as to achieve an end impossible by individual action.

syngamy the fusion of two GAMETES. Sexual reproduction.

syngenesis sexual reproduction.

synovial membrane the secretory membrane that lies within the capsule of a joint and produces the clear, sticky, lubricating synovial fluid without which smooth joint movement would be impossible. The synovial membrane covers all the internal structures of the joint except the bearing surfaces (the articular cartilages). Also known as the synovium.

synovium the SYNOVIAL MEMBRANE.

synthase an enzyme that catalyzes a reaction in which two chemical units are joined without the direct participation of ATP.

syring-, syringo- *combining form denoting* a tube or FISTULA.

system a group of related organs that act together to perform a common function. Body systems include the digestive system, the respiratory system, the cardiovascular system, the urinary system and the reproductive system.

systematic of each body system considered separately.

systemic 1 pertaining to something that affects the whole body rather than one part of it.
2 of the blood circulation supplying all parts of the body except the lungs.
3 of a drug taken by mouth or given by injection, as distinct from a drug applied externally.

systole the period during which the chambers of the heart (the atria and the ventricles) are contracting. Atrial systole, in which blood passes down into the ventricles, precedes the more powerful ventricular systole in which blood is driven into the arteries. Systole alternates with a relaxing period called DIASTOLE.

Tt

tachy- *combining form denoting* rapid or abnormally rapid.

tachycardia a rapid heart rate from any cause.

tachypnoea abnormally fast breathing, as from exercise, heart or lung disease or anxiety.

tactoreceptor a sensory nerve ending that responds to touch.

tailbone the COCCYX.

taliped clubfooted.

talipes clubfoot. A congenital deformity affecting the shape or position of one or both feet. In talipes cavus, there is exaggeration of the curvature of the longitudinal arch. In talipes equinovarus the ankle is extended and the heel and sole turned inwards.

talus the second largest bone of the foot that rests on top of the heel bone (calcaneus) and articulates with the TIBIA and FIBULA to form the ankle joint.

tampon a cylindrical mop of absorbent material placed in the VAGINA to absorb menstrual blood and allow freedom of activity during the menstrual period.

tapeworm a ribbon-like population, or colony, of joined flatworms, of the class *Cestoda*, derived from a common head (scolex) equipped with hooks or suckers by which it is attached to the lining of the intestine. Each segment, of which there may be a thousand, is called a proglottid and each contains both male and female reproductive organs.

tarsal 1 pertaining to the tarsal bones of the foot.

2 pertaining to the fibrous skeleton (tarsal plate) of the eyelid.

tarsal gland a MEIBOMIAN gland in the tarsal plate of the eyelid.

tarsus 1 the part of the foot between the leg and the metatarsal bones.

2 the seven bones of the tarsus

3 a fibrous plate that gives rigidity and shape to the eyelid.

taste and smell

The association of taste and smell is not arbitrary, as can at once be appreciated by anyone temporarily deprived of the sense of smell by a severe cold. Taste, by itself, is a very limited faculty, having some utilitarian value but few or no aesthetic possibilities. Taste, accompanied by a normal sense of smell, however, allows us to enjoy the whole range and subtlety of gastronomic experience. The reason for this seeming anomaly is that when we use the word 'taste' we are really referring to the simultaneous employment of both faculties.

Taste and smell are primitive functions present early in the evolutionary scale. These senses are comparatively of much greater importance to less highly evolved animals than to humans. This is especially true of the sense of smell which is no longer consciously regarded by most people as an important vehicle of information. We should not, however, underestimate the effect of olfaction as a means of conveying subtle and perhaps barely recognized intelligence between individuals. Flavour, as a means of identifying and enjoying substances is clearly important.

Taste buds

Taste buds are specialized nerve endings situated mostly on tiny mushroom-shaped protrusions on the tongue known as fungiform papillae. They also occur on the roof of the mouth, the throat and the upper third of the gullet (oesophagus). There are some 10,000 taste buds altogether and each has a life of seven to ten days, being replaced by new structures that differentiate from the surrounding epithelium. Although they cannot be distinguished from each other anatomically, it is commonly stated that taste buds are of four types. These are said to respond, respectively, to the four modalities of taste sensation – sweet, sour, salt and bitter. Using taste alone, these are the only taste sensations we can appreciate. Some authorities deny that there are four kinds of buds and point out that single taste buds can respond to substances in more than one of the four categories. The taste buds that respond to salt are distributed evenly on the tongue; those sensitive to sweet flavours are concentrated on and around the tip of the tongue. Sour tastes are detected mainly by buds situated on the sides of the tongue, and bitter flavours are experienced at the back of the tongue.

Taste buds are constructed like miniature oranges or bunches of bananas. Each contains some 50–57 slender curved cells with outer protrusions ending in small open pores that are bathed in the salivary fluid layer on the lining of the mouth and throat. The presence of this fluid is essential for their function, and the chemical substances that give rise to the sensation of taste cannot operate on the taste buds until they have been dissolved in the fluid. Once these dissolved substances enter the pores they are able to come into contact with the special taste receptors on the outer membrane of the receptor cells in the taste bud.

It seems likely that the membranes of single receptor cells carry different receptors for different substances. When the receptor cells are stimulated in this way they release neurotransmitters which diffuse across synapses and give rise to nerve impulses in nerve fibres running from the taste buds to the brain. These fibres run in the glossopharyngeal, lingual and vagus nerves.

As in other parts of the nervous system, intensity of sensation is coded as an increase in the frequency of repetitive nerve impulses (frequency modulation). A high frequency informs the brain of high intensity. For a particular nerve fibre, differences in the frequency of nerve impulses also appear to code for the different modalities of taste. Some fibres fire at a high rate in the presence of a sweet substance while others fire only slowly. Some respond only to one modality, most respond to two, some respond to three and some to all four. So it seems probable that perception of taste depends on the pattern of response of a group of nerve fibres. The nerve connections for taste pass up the brainstem to the sensory cell nucleus, the thalamus and end eventually in the sensory cortex of the brain in the region mapped for the mouth .

Olfactory receptors

Odour is conveyed to us by gaseous molecular particles conveyed in the air and these must reach the olfactory sense receptors in the nose and be dissolved in the layer of fluid mucus on the nose lining before they can be appreciated. The smell receptors lie in a small patch of mucous membrane, of area 2.5 square cm, lying immediately under the thin bony plates that form the roof of the nose. These plates are perforated with many small holes through which pass the nerve fibres of the olfactory nerves that run from the mucous membrane directly to the brain.

The olfactory nerve bodies are long and narrow and lie in the mucous membrane. Their lower ends are enlarged into knob-like processes from which six to twelve fine, hair-like fingers called cilia pass to the surface of the membrane so that their tips lie in the mucus layer. Mucus readily dissolves fine particles in the inspired air, especially if this is sniffed in through the nose. The dissolved odorous particles interact with receptors on the cell membranes, causing the cells to fire and despatch nerve impulses to the brain. Again, as in the case of taste, the intensity of the stimulus and the differentiation of olfactory quality is coded by differences in the frequency of these impulses in a given time. As in the case of the taste buds, no structural differences are apparent between the different olfactory nerve cells.

The olfactory nerves enter the olfactory bulbs – elongated, stem-like projections from the underside of the brain that lie on top of the perforated bone plates. From the bulbs, the nerve fibres pass to a part of the brain cortex lying on the underside of the frontal lobes, known as the limbic system . This system is a kind of primitive brain concerned with some of the more basic functions such as the satisfaction of the various appetites. It is also concerned with the hypothalamus and its regulatory functions.

Chemistry and the sense of smell
The olfactory system is very sensitive and is capable of detecting odours from material present in a concentration of only a few parts per million. Sensitivity to smell is said to be 100,000 times greater than sensitivity to taste. Because of this and because of the rapid diffusion of molecular particles in air we are able to detect odours coming from a considerable distance. The many thousands of different identifiable odours probably arise from combinations of a comparatively few different stimuli.

Odorous substances must be volatile enough to give off particles into the air. They must also be sufficiently soluble to go into solution in the nasal mucus. Many fat and water-soluble substances are strongly odorous. But volatility and solubility are not enough to explain olfaction. Most odorous substances are organic and it is possible to relate many chemical groups to particular kinds of smells. A wide range of molecules that are derived from the benzene ring have a pleasant aroma. This fact led chemists to describe the whole class of benzene ring compounds as the aromatic compounds. Many of the smells of plants derive from aromatic compounds such as the essential oils.

The relationship between chemistry and smell is not, however, an obvious one and very different molecules may have similar smells. On the other hand, different spacial arrangements of identical molecular structures (steroisomers) can produce very different smells. Clear distinctions can result from sometimes quite subtle changes in the molecular structure of odorous material. Many organic compounds can exist in two forms that are identical except that one molecule is a mirror image of the other. Such mirror-image forms are known as enantiomers. Even such a minor difference as this can affect the way in which the olfactory apparatus reacts. A case in point is the compound limonene. One enantiomer of limonene smells strongly of lemons; the other smells strongly of oranges.

Factors affecting the sense of smell
Volatility is influenced by temperature, which is one reason why perfumes are more effective on the body than in the bottle and why refrigerators often smell stale when switched off. Hunting dogs perform better in warm weather than in cold.

Any factor that interferes with ready access of air to the olfactory nerve endings will affect the sense of smell. When the nasal mucous membrane is swollen as a result of a cold or other inflammatory disorder (rhinitis) the sense of smell may be severely inhibited. An injury to the base of the skull that causes a fracture of the roof of the nose will commonly lead to permanent and total loss of the sense of smell (anosmia) from tearing of the olfactory nerve filaments. Increase in nasal sensitivity occurs when we are very hungry, and women are said to experience an increased sensitivity to substances related to the sex hormones at certain phases of the menstrual cycle.

Adaptation to persistent smells occurs within a matter of minutes. This useful aspect of physiology allows people to work comfortably in conditions that may seem appalling to the new entrant. Adaptation is not wholly a function of the olfactory nerve endings but is also the result of some form of inhibition occurring in the brain. It is one of the characteristics of the nervous system that it tends to ignore persistent stimuli and show the greatest sensitivity to change. To some extent, cross adaptation occurs between different strongly smelling substances. A nose adapted to one essential oil, such as oil of Wintergreen or eucalyptus, may have difficulty in detecting another.

Olfactory influences on behaviour

Pheromones are specific olfactory attractants that play a major part in determining the behaviour of lower animals. Many species have pheromone receptors for particular substances. Bees from a strange hive are identified by smell and are killed. The antennae of the male silk moth carries receptors that respond only to the pheromones given off by the female silk moth. The female does not have pheromone receptors. Stimulation of the pheromone receptors causes the males to fly in the direction of maximal pheromone concentration.

The behaviour of mammals, too, can be strongly influenced by olfactory stimuli. Attraction to sexually receptive females of many species is often mediated by smell, even by smell imparted to objects that have been touched. The secretion of reproductive hormones can be powerfully affected by olfactory stimuli, even to the extent of terminating a pregnancy when the recently mated female smells the odour of a strange male. It seems probable that some vestige of this kind of mechanism persists in humans. Musk, one of the central ingredients in the most provocative perfumes, is chemically related to the human sex hormones. It is derived from the sex glands of the musk deer.

taste bud one of the many spherical nests of cells containing specialized nerve endings distributed over the edges and base of the tongue. The taste buds respond to the crude flavours of substances dissolved in saliva. See also TASTE.

tattooing deliberate or accidental insertion of coloured material into the deeper layers of the skin. Tattooing commonly follows deep abrasions or nearby explosions. Brown or black particles buried deep in the skin cause a blue colouring. Decorative tattooing may be complicated by skin infection or allergy, hepatitis b, aids, psoriasis, lichen planus and discoid lupus erythematosus.

tau protein a major structural protein associated with the microtubules that form the cytoskeleton of nerve cells. An abnormal form with a shorter molecule is found in the helical filaments in senile neural plaques and in the insoluble neurofibrillary tangles of Alzheimer's disease. It is found in the cerebrospinal fluid of people with dementia but is present in higher concentration in Alzheimer's disease than in other forms of dementia. Abnormal tau protein binds strongly to the normal protein and the latter then become shortened by nerve cell proteolytic enzymes into the abnormal form. The latter is not split by these enzymes.

taurocholic acid one of the bile acids and an important constituent of bile. From the Greek *tauros*, a bull, from the bile of which the acid was first obtained.

tautomers structural ISOMERS that exist in equilibrium with each other. They have identical chemical formulae but their molecular structures differ slightly. A change in tautomeric form in the bases in DNA can cause a mutation as the alteration in the position of the atoms can interfere with the formation of HYDROGEN BONDS between the base pairs..

taxis movement of an organism toward or away from a stimulus.

taxon any group of organisms constituting one of the formal categories of classification, such as a phylum, class, order, family, genus, or species. See also TAXONOMY.

taxonomy the science or principles of biological classification and the assignment of appropriate names to species.

T-box gene family a family of genes that code for a range of transcription factors essential for normal organ development and body pattern development in both vertebrates and invertebrates. The family has a highly-conserved DNA binding feature known as a T-box. See also HOMEOBOX GENES.

T cell one of the two broad categories of LYMPHOCYTE, the other being the B cell group.

tears the secretion of the lacrimal glands. Tears consist of a solution of salt in water with a small quantity of an antibacterial substance called lysozyme. The tear film on the cornea also contains mucus and a thin film of oil.

technetium an artificial radioactive element that can be incorporated into various molecules for use in RADIONUCLIDE SCANNING.

tectorial membrane a structure in the organ of Corti in the cochlea overlying the bristles of the acoustic hair cells on the basilar membrane. As particular cells vibrate under the influence of sound waves the bristles wipe against the tectorial membrane thus activating those hair cells to send nerve impulses to the brain.

tectum any roof-like body structure such as the back (dorsal) part of the midbrain.

teeth the instruments of biting (incisors), tearing (canines) and grinding (molars) of food. There are 20 primary teeth and 32 permanent teeth, but it is common for one or more of the third molars, at the back (the 'wisdom teeth') to remain within the gum (unerupted) until well into adult life. The permanent teeth are numbered, 1–8, from the centre, in each quadrant. A dentist might thus refer to an 'upper right 3' meaning the patient's top right canine tooth.

teething the eruption of the primary teeth. This usually starts around the age of 6 or 7 months and all 20 primary teeth have usually erupted within 30 months. Teething often causes fretfulness but is not a cause of fever, convulsions, diarrhoea or loss of appetite.

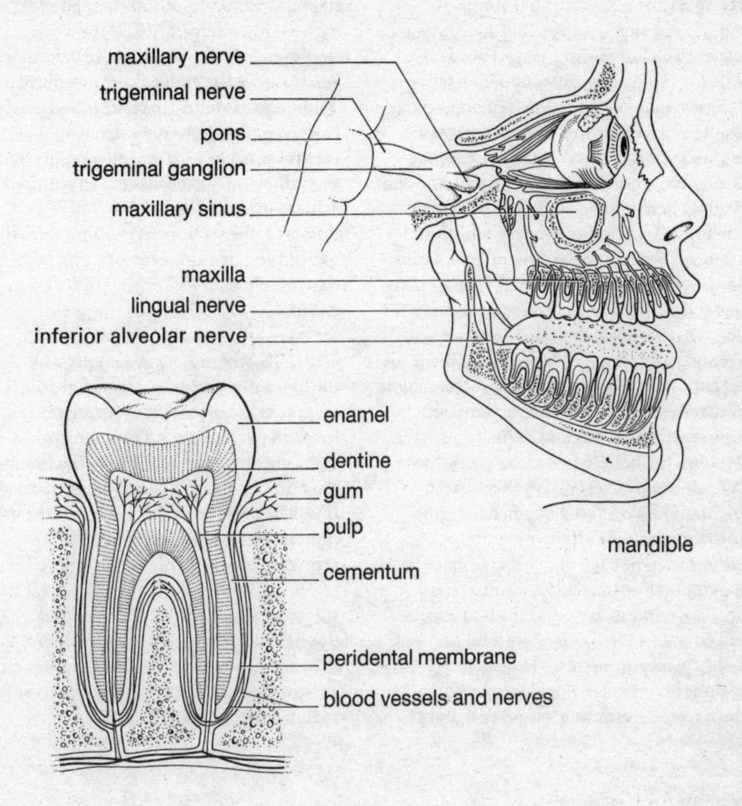
maxillary nerve
trigeminal nerve
pons
trigeminal ganglion
maxillary sinus
maxilla
lingual nerve
inferior alveolar nerve
mandible
enamel
dentine
gum
pulp
cementum
peridental membrane
blood vessels and nerves

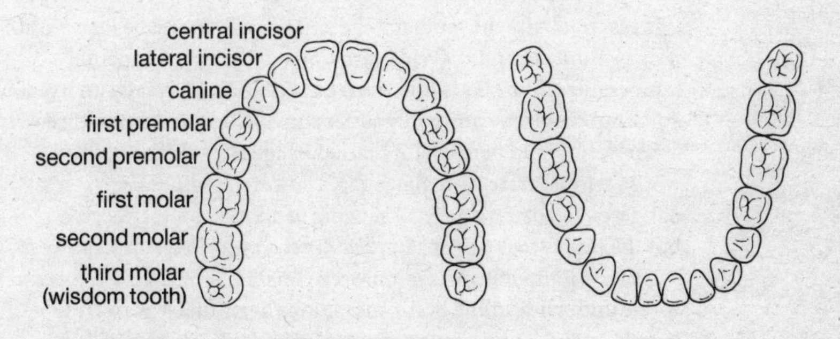
central incisor
lateral incisor
canine
first premolar
second premolar
first molar
second molar
third molar
(wisdom tooth)

Fig. 18 **Teeth**

teichopsia the scintillating scotoma of migraine. A slowly-expanding area of visual loss with a shimmering, jagged border that lasts for about 20 minutes and then fades away leaving normal vision. Teichopsia may or may not be followed by headache. Also known as 'fortification spectra', hence the Greek derivation from *teichos*, a wall, and *ops*, vision. There is anecdotal evidence that teichopsia is commoner in academics and intellectuals than in the general population.

telangiectasia a local or general increase in the size and number of small blood vessels in the skin. Often incorrectly called 'broken veins'. Telangiectasia may be present from birth but may be caused by undue exposure to sunlight or may be a feature of rosacea, psoriasis, lupus erythematosus and dermatomyositis.

teleology the believe that there is a purpose or design behind events or phenomena and that they are best explained in terms of their seeming purpose or end. It would, for instance, be a teleological argument to suggest that we have an immune system because, without it, we should all die from infections. Medical scientists tend to reject teleological arguments. Theological arguments are often teleological. From the Greek *tele* meaning 'far off or distant'.

telepathy communication claimed to occur without the interposition of the senses.

telomerase the enzyme that can reform the TELOMERES at the ends of chromosomes. Telomerase is found in cancers and is able to prevent the shortening that would otherwise occur with repeated replication, thus allowing cancerous cells in culture to achieve immortality.

telomeres the sections of DNA that form the natural end of a CHROMOSOME. The points that resist union with fragments of other chromosomes. Telomeres consist of repeated groups of the base sequence TTAGGG, where T, A and G represent the bases thymine, adenine and guanine, respectively. Formation of telomeres involves the shaving-off of some junk DNA at the chromosome end. There is a limit to the number of times that this can occur and this is believed to be the reason for the upper limit in the number of times cells can reproduce. Telomere length is a function of age but there is recent evidence that it may also be an X-linked familial trait.

telophase the last phase of MITOSIS, in which the chromosomes of the two daughter cells are grouped together at each separating pole to form new nuclei.

temperature regulation

It is essential that the temperature of the body should be kept within narrow limits in spite of wide variations in the environmental temperature. Most body functions can continue normally up to about 40°C. but children commonly suffer convulsions at temperatures above 40.5°C. and permanent brain damage is likely in anyone whose temperature rises above 42°C. Lowered temperature is less dangerous, causing a general slowing of all metabolic processes, but the heart stops beating (*cardiac arrest*) at very low temperatures.

Blood temperature is monitored constantly by regulating centres in the underside of the brain, just above the pituitary gland (the *hypothalamus*). Any variations from normal (37°C.) automatically activates mechanisms to compensate. Extra heat is produced, when needed, by shivering – rapid repetitive muscle contraction, and heat loss is prevented by contraction of the muscles in the walls of the skin arterioles so that the vessels are narrowed and the flow of blood through the skin is minimized. In fever, loss of excessive heat occurs almost entirely via the skin. Under the control of the hypothalamus, the arteriolar wall muscles relax so that the blood flow through the skin increases and heat is lost by radiation. If this is insufficient, the sweat glands are stimulated to secrete. Evaporation of sweat from the skin surface results in considerable heat loss from the latent heat of evaporation. Evaporation of one gram of water requires 539 calories.

temper tantrum an expression of frustration by a child who is prevented from demonstrating unconstrained action. The temper tantrum is an effective weapon in the war of independence and may involve screaming, floor-rolling, head-banging and breath-holding.

temporal 1 pertaining to the temples. 2 pertaining to time.

temporal bone one of two bones forming part of the sides and base of the skull and containing the hearing apparatus. Each temporal bone bears the mastoid process and articulates on its under surface with the head of the jaw bone (mandible).

temporal summation the build-up of an electrical charge on a membrane as a result of more than one small successive input.

temporomandibular joint the joint, immediately in front of the ear, between the head of the lower jaw bone (the mandible) and the under side of the temporal bone of the skull. Movement at this joint can often be seen through the skin if the mouth is opened widely.

ten-day rule the regulation, observed in all X-ray departments, that radiological examination of the lower abdomen in women of childbearing age should be restricted to the 10 days immediately following the first day of the last menstrual period. This is to avoid irradiating a young embryo.

tenderness pain elicited by touch or pressure.

tendon a strong band of COLLAGEN fibres that joins muscle to bone or cartilage and transmits the force of muscle contraction to cause movement. Tendons are often provided with sheaths in which they move smoothly, lubricated by a fluid secreted by the sheath lining. Tendons may become inflamed, or may be torn or cut.

tendon jerk a reflex contraction of the muscle to which a tendon is attached when the tendon is struck sharply so as to exert a sudden pull on the muscles. This reflex demonstrates the integrity of both the sensory and the motor nerve supply to the muscle. An exaggerated response may indicate the absence of higher nervous control on the reflex.

tendon of Achilles see ACHILLES' TENDON.

tenesmus a frequently recurring or continuous sense of wishing to empty the bowels. Tenesmus leads to ineffective straining and is a feature of the irritable bowel syndrome, haemorrhoids, ulcerative colitis, dysentery, polyps in the rectum, prolapse of the rectum and sometimes cancer of the rectum.

TENS *abbrev. for* Transcutaneous Electrical Nerve Stimulation. This is a method of treating long-persistent pain by passing small electric currents into the spinal cord or sensory nerves by means of electrodes applied to the skin. The necessary equipment is miniaturized and is readily portable.

tension muscle contraction as a reflection of anxiety. Most headaches are caused in this way. Tension, and associated symptoms, can often be relieved by formal relaxation procedures.

tensor a muscle that tenses a part.

terat-, terato- *combining form denoting* developmental abnormality.

teratogen any agent capable of causing a severe congenital bodily anomaly (monstrosity).

teratogenesis the production of a fetal congenital bodily abnormality.

teratology the study of the processes operating during early embryonic and fetal development that lead to major anomalies of body structure.

terminal care care of the dying.

termination codon see STOP CODON.

tertian recurring on alternate days (on the third day).

tertiary the ordinal that follows primary and secondary, used especially when the third stage of a disease has distinct characteristics, as in syphilis.

tertiary structure used mainly of a protein to refer to its three-dimensional folded shape in space. The primary structure is simply the order of the amino acids in the polypeptide chain; the secondary structure may be a coiling, the alpha helix, or a layering known as the beta-pleated sheet. The tertiary structure is determined by chemically active groups spaced along the chain that attract each other to form bonds at specific positions. The tertiary structure of proteins such as enzymes, is crucial to their function.

testicle see TESTIS.

testicular feminization syndrome a rare X-linked, genetically induced defect of the male sex hormone receptors on the surface of body cells in men. Male hormones are thus unable to act and normal male characteristics cannot develop. There is male PSEUDOHERMAPHRODITISM. The testicles are present, but undescended, the penis is rudimentary and there is a short, blind-ended vagina. The breasts may be well developed. It is usual to removed the testicles and give female sex hormones to promote full development of the nominal female sex.

testis one of the two male gonads, suspended in the scrotum by the spermatic cord. The testis, or testicle, contains the long, coiled seminiferous tubules in which the SPERMATOZOA are formed. Between the tubules are cells that secrete testosterone and other masculinizing steroid hormones and oestrogens.

testis disorders these include epididymitis, orchitis, torsion of testis and undescended testis.

testosterone the principal male sex hormone (androgen) produced in the INTERSTITIAL cells of the TESTIS and, to a lesser extent in the OVARY. Testosterone is ANABOLIC and stimulates bone and muscle growth and the growth of the sexual characteristics. It is also used as a drug to treat delayed puberty or some cases of infertility or to help to treat breast cancer in post-menopausal women. It may be given by mouth, depot injection or skin patch.

test tube baby a popular term for a baby derived from an ovum fertilized outside the body (in vitro fertilization).

tetany muscle spasm resulting from abnormally low levels of blood calcium. Low extracellular calcium on excitable cell membranes opens sodium channels leading to increased excitability. The drop in serum calcium can result from a reduction in blood acidity from deliberate or hysterical over-breathing (hyperventilation) or from underaction of the PARATHYROID GLANDS. Tetany affects mainly the hands and feet, causing a claw-like effect with extension of the nearer joints and bending of the outer joints (carpopedal spasm).

tetraplegia quadriplegia. Paralysis of all four limbs.

thalamus one of two masses of grey matter lying on either side of the midline in the lower part of the brain. It receives sensory nerve fibres from the spinal cord and connections from the midbrain, the eyes, the ears and the cerebral CORTEX. It sends fibres to the sensory part of the cerebral cortex. It is the collecting, coordinating and selecting centre for almost all sensory information, other than OLFACTORY, received by the body. Only part of the mass of information it receives is passed to the cortex. From the Greek *thalamos*, an inner chamber.

thanatology 1 the study of death.
2 the forensic study of the causes of death and their relation to postmortem appearances.

THC *abbrev. for* tetrahydrocannabinol, the active ingredient in marijuana.

theca a casing, outer covering or sheath.

thenar eminence the fleshy, muscular mass on the palm at the base of the thumb.

theology the study of the existence and nature of God and of man's relationship with God. Theological speculation and argument have led to the production of an immense body of writing encompassing descriptions of man's concept of God, philosophic studies, academic enquiry, doctrinal exposition and explanation, Canon law, formal dogma, moral studies, scriptural interpretation (exegesis) and so on.

therapeutic community a small local population of people of strong antisocial tendency, set up under the supervision of medical staff but in a non-clinical environment, to try to treat personality disorder. The object is to demonstrate the effects of such behaviour and to try to instil constructive patterns of conduct and improve social and interpersonal skills. Results have been promising.

therapist 1 a person providing or conducting any form of medical or psychological treatment.
2 a psychotherapist.

therapy treatment of disease or of conditions supposed to be diseases. The term is often qualified to limit its range, as in chemotherapy, physiotherapy, psychotherapy, radiotherapy, hydrotherapy and hypnotherapy.

therm-, thermo- *combining form denoting* heat.

thermogenesis heat production. Normal heat production results from the oxidation of food components; a sudden demand for heat causes heat production by shivering; abnormal thermogenesis from a breakdown of the heat-regulation mechanism (thermoregulation) is dangerous and sometimes fatal.

thermography a scanning technique in which temperature differences on the surface of the skin are represented as an image or as colour differences. Thermography readily demonstrates areas of inflammation and variations in blood supply, but has not proved particularly useful as an aid to diagnosis.

thermometer a device for registering body temperature. Thermometers may be analogue, as in the case of the common mercury expansion thermometer or colour-change devices, or may have a digital display.

thermoreceptor a sensory receptor that responds to heat.

thermoregulation the control of body temperature and its maintenance within narrow limits in spite of factors tending to change it.

thiamine vitamin B_1.

thick filaments the myosin filaments in muscle cells.

thin filaments the actin filaments in muscle cells.

third ventricle the small, slit-like midline, fluid-filled space in the centre of the brain which communicates with the two large lateral ventricles, one in each hemisphere, and with the FOURTH VENTRICLE behind.

thirst the strong desire to drink, arising from water shortage (dehydration) causing an increased concentration of substances dissolved in the blood. This change is monitored by nerve receptors in the HYPOTHALAMUS in the brain, and thirst is induced by a nerve reflex.

thorac-, thoraco- *combining form denoting* chest.

thoracic pertaining to the chest.

thoracic duct the main terminal duct of the lymphatic system that runs up alongside the spine from the abdomen to the root of the neck and then discharges into the large neck veins.

thoracic surgeon a surgeon specializing in operative treatment within the chest cavity, especially on the windpipe (trachea), the lungs, the heart and the gullet (oesophagus).

thorax the part of the trunk between the neck and the ABDOMEN. The thorax contains a central compartment, the MEDIASTINUM that contains the heart and separates the two lungs. The thorax also contains the TRACHEA, the OESOPHAGUS, a number of large arteries and veins connected to the heart. Its walls consist of the dorsal VERTEBRA, the breastbone (sternum) and the rib cage.

thought disorders a group of mental aberrations occurring in schizophrenia and various forms of dementia. They include false beliefs (DELUSIONS), memory loss, defects of attention and concentration, thought blocking and flight of ideas. Thought disorders may also be manifested by the use of word connections on the basis of phonetic resemblance rather than logic (clang associations) and by the use of meaningless or invented words (neologisms).

threonine one of the 20 AMINO ACIDS that form proteins and an essential ingredient in the diet. The substance is used as one of the ingredients in the externally-applied antibiotic preparation Cicatrin.

throat see PHARYNX.

thromb-, thrombo- *combining form denoting* blood clot or THROMBUS.

thrombin an enzyme in the blood that converts fibrinogen to fibrin, thus forming a blood clot.

thrombocyte a blood PLATELET, one of the numerous, non-nucleated fragments derived from the large cells, the megakaryocytes, and necessary for blood clotting. The term is ill-chosen because platelets are not cells.

thrombocytopenia an abnormally low number of PLATELETS in the blood. The lower limit of normality is about 150,000 per cubic millimetre.

thrombomodulin a receptor on ENDOTHELIUM that binds thrombin, inhibiting its clot-forming ability and causing it to activate protein C.

thromboplastin blood clotting factor III, an obsolete term referring to what is now known to be several blood clotting factors operating together.

thrombopoietin a hormone responsible for the growth of colonies of megakaryocytes and the production of platelets. The existence of thrombopoietin had been suspected for 30 years but was realized only in 1994. In 1997 the results of a trial of polyethylene-conjugated genetically-engineered (recombinant) thrombopoietin – described as recombinant human megakatyocyte growth and development factor (PEGrHuMGDF) – was published. It was found to have a powerful stimulatory effect on PLATELET production. Thrombopoietin is the ligand for the cytokine receptor Mpl.

thrombosis clotting of blood within an artery or vein so that the blood flow is reduced or impeded. If the vessel is supplying a vital part, such as the heart muscle or the brain, and the thrombosis cuts off the flow, the result may be fatal. Thrombosis of arteries supplying limbs or organs may lead to GANGRENE, heart attacks and strokes.

thromboxanes substances similar to the prostaglandins that promote blood clotting. Aspirin in small doses reduces the production of thromboxane by PLATELETS.

thrombus a blood clot forming especially on the wall of a blood vessel. This is commonly the result of local damage to the inner lining of the vessel (the endothelium).

thymic aplasia congenital absence of the THYMUS.

thymine a pyrimidine base, one of the four whose particular sequence forms the genetic code in the deoxyribonucleic acid (DNA) molecule.

thymus a small flat organ of the lymphatic system situated immediately behind the breastbone, that is apparent in children but inconspicuous after puberty. The thymus processes primitive LYMPHOCYTES so that they differentiate into the T cells of the immune system. It also differentiates T cells into T1 and T2 classes, a process that is influenced by the early environment of the individual.

thyro- *combining form denoting* thyroid.

thyroglobulin a large protein molecule, the storage form and precursor of the thyroid gland hormones.

thyroglossal duct a duct in the embryo that runs between the THYROID GLAND and the back of the tongue. Normally, the thyroglossal duct disappears before birth but part or all of it may persist.

thyroid cartilage the largest of the cartilages of the LARYNX, consisting of two backward-sloping, broad processes joined in front to form the protuberance in the neck known popularly as the Adam's apple.

thyroid gland an ENDOCRINE GLAND, situated in the neck like a bow tie across the front of the upper part of the windpipe (trachea). The gland secretes hormones that act directly on almost all the cells in the body to control the rate of their METABOLISM.

thyroid gland disorders these include cretinism, dyshormonogenesis, goitre hashimoto's thyroiditis, hyperthyroidism, hypothyroidism and thyrotoxicosis.

thyroid scanning a method of assessing the rate of uptake of iodine by, and the pattern of distribution in, the thyroid gland using a compound containing radioactive iodine and a GAMMA CAMERA.

thyroid-stimulating hormone the hormone, thyrotropin, produced by the PITUITARY GLAND that prompts secretion of thyroxine (levothyroxine) by the thyroid gland.

thyrotrophin releasing hormone a hormone produced in the HYPOTHALAMUS that passes to the PITUITARY GLAND where it prompts the secretion of thyroid stimulating hormone.

thyroxine the principal thyroid hormone. Thyroxine has four iodine atoms in the molecule and is often known as T_4. The sodium salt of thyroxide (levothyroxine) is sold as a drug used to treat thyroid deficiency disorders.

tibia the shin bone, the stronger of the two long bones in the lower leg. The front surface of the tibia lies immediately beneath the skin. Its upper end articulates with the femur (thigh bone) to form the knee joint and the lower end forms part of the ankle joint. Its companion bone, the fibula, lies on its outer side and is attached to it by ligaments.

tics repetitive, twitching or jerking movements of any part of the face or body and occurring at irregular intervals to release emotional tension. Children are commonly affected but tics seldom persist into adult life. They do not indicate organic disorder and can be controlled by an effort of will.

tidal volume the volume of air entering and leaving the lungs during a single respiratory effort, whether the rate is normal, high or low.

tight junctions junctions between cells the extracellular surface of whose cell membranes are joined. Tight junctions restrict molecular movement between cells.

tincture an alcoholic solution of a drug.

tinnitus any sound originating in the head and perceptible by the person concerned. Tinnitus may be a hissing, whistling, clicking or ringing sound, appearing to come from one or both ears, or from the centre of the head. It is usually associated with deafness and may be caused by anything that damages the hearing mechanism of the inner ear, such as loud noise, drugs toxic to the ear, Méni?re's disease, otosclerosis and PRESBYACUSIS.

tissue any aggregation of joined cells and their connections that perform a particular function. Body tissues include bone, muscle, nerve tissue, nerve supporting glial tissue, epithelium, fat, fibrous and elastic tissue.

tissue factor a protein found on the membrane of subendothelial cells that is necessary for the initiation of blood clotting. Tissue factor becomes accessible on wounding.

tissue fluid the fluid occupying the spaces between body cells by way of which oxygen, carbon dioxide and dissolved substances are passed to and from the cells. Also known as interstitial fluid.

tissue culture the artificial growth of sheets of human tissue in the laboratory. Tumour cells are readily cultured and some appear to be immortal. These are widely used for laboratory purposes. Normal skin cells (keratinocytes) can be cultured and used for grafting in the same person. Three-layered arteries have been grown as have sheets of urethral endothelium for purposes of urethral reconstitution in hypospadias. It has even been possible to grow a new ear around a mould of polymer mesh.

tissue plasminogen activator (TPA) a natural body ENZYME involved in the breakdown of blood clots. TPA is a small molecule protease that activates the conversion of plasminogen to plasmin. Plasmin is an enzyme that can convert fibrin strands in the blood clot to soluble products so that blood clot can be dissolved. TPA has been produced synthetically, as alteplase and is used in the treatment of conditions, such as heart attacks, caused by blockage of arteries by clotted blood.

tissue proteinase one of a range of enzymes that bring about controlled protein breakdown in a wide range of developmental, physiological and pathological processes. These enzymes are active in such processes as embryogenesis, angiogenesis, blood coagulation, fibrinolysis and inflammation. Over-expression of some of the proteinases has been implicated in a range of pathologies, especially connective tissue disorders such as rheumatoid arthritis in which collagen breakdown is an important factor.

tissue typing the identification of particular chemical groups present on the surface of all body cells and specific to the individual. These groups are called the histocompatibility antigens and include the human leukocyte antigens (HLAs) which are short chains of linked amino acids unique to each person except that they are shared by identical twins. Tissue typing is important in organ donation to reduce the risk of rejection. See also MAJOR HISTOCOMPATIBILITY COMPLEX.

titin a protein in muscles cells that extends from the Z line to the thick filaments and M line of the SARCOMERE.

titre a measure of the concentration and strength of a substance, such as an antiserum or antibody, in solution. Titre is estimated by the highest dilution that still allows a detectable effect.

tobacco dried leaves of the plant *Nicotina tabacum*. Tobacco contains the drug nicotine for the effects of which it is smoked, chewed or inhaled as a powder (snuff). All these activities are dangerous. Cigarette smoking, in particular, is responsible for a greatly increased risk of cancer of the lung, mouth, bladder and pancreas and for an increased likelihood of chronic bronchitis, emphysema, coronary artery disease and disease of the leg arteries. Smoking is also harmful during pregnancy, leading to smaller and less healthy babies.

tobacco-free social norm the concept that the avoidance of smoking should be generally regarded as natural and normal and that smoking is harmful and undesirable. The establishment of such a norm is especially important as a means of discouraging young people from smoking.

tocopherol one of the group of substances constituting vitamin E.

toe disorders these include bunion, hallux valgus, hammer toe, ingrown toenail.

-tomy *suffix denoting* a cutting or making an incision into.

tone the degree of tension maintained in a muscle when not actively contracting. In health, this is slight. Tone is abolished in certain forms of paralysis and greatly increased in others.

tongue the muscular, mucous membrane-covered, highly flexible organ that is attached to the lower jaw (mandible) and the HYOID BONE in the neck, and forms part of the floor of the mouth. The mucous membrane contains numerous small projections called papillae. On the edges and base of the tongue are many special nerve endings subserving taste and called taste buds.

tongue disorders these include glossitis, leukoplakia and tongue tie.

tongue tie a rare condition in which the mucous membrane partition under the tongue (the frenulum) is so tight as to limit tongue movement and even affect speech. The condition is easily corrected by snipping the frenulum.

tonic 1 of continuous activity.
2 a mythical remedy commonly prescribed by doctors as a PLACEBO.

tono- *combining form denoting* pressure or tension.

tonsil 1 an oval mass of lymphoid tissue, of variable size, situated on the back of the throat on either side of the soft palate.
2 any bodily structure resembling the palatine tonsil.

tonsil test for CJD prion the identification of rogue prion protein in a small biopsy sample of tonsil from a suspected victim of new-variant CJD. The protein is digested with an enzyme proteinase K and the fragments subjected to electrophoresis. The resulting banding pattern in new-variant CJD differs from that of other forms of CJD.

tooth see TEETH.

tooth decay local destruction of the enamel and underlying dentine of a tooth so that cavities form and infection can gain access to the pulp. Such infection can destroy the internal blood vessels and nerves and kill the tooth. Tooth decay is caused by acids formed by bacterial action on dental plaque and can be prevented by regular brushing and flossing. Also known as dental caries.

toothpaste a tooth-cleaning preparation containing a fine abrasive powder, such as chalk, a little soap or detergent, some flavouring, often peppermint, and some sweetening agent and, ideally, a fluoride salt. Many dentifrices also contain a chemical to coagulate protein in the tooth tubules and desensitize them to acids and temperature changes.

topical pertaining to something, usually medication, applied to a surface on or in the body, rather than taken internally or injected. Examples of topical applications are skin ointments or creams, eye and ear drops or ointments and vaginal pessaries.

topoisomerases enzymes that assist in the topological manipulation of DNA. Before DNA can replicate it must be unwound. This occurs in short segments and is achieved by a Type I topoisomerase that cuts one strand so that the free end can be rotated around the unbroken strand. Topoisomerase 1 then re-ligates the broken strand. Type II topoisomerases cut both strands. The term is pronounced 'topo – isomerase'. The derivation is Greek, *topos*, a place; *iso*, equal; *meros*, a part; with the -ase suffix denoting an enzyme.

topological isomers of DNA DNA molecules that are identical except for the number of times one strand of the double helix crosses over the other in a given length. Any change in this number (the linking number) requires that at least one strand must be broken. This process to change the linking number is catalyzed by a TOPOISOMERASE.

torsion twisting or rotation, especially of a part that hangs loosely on a narrow support. Torsion may affect a loop of bowel, the testicle or other organ and commonly results in dangerous obstruction to the blood supply of the part. Urgent surgical correction may be needed.

torticollis a permanent, intermittent or spasmodic twisting of the head or neck to one side. Causes of torticollis include birth injury to one of the long neck muscles, scarring and shortening of the neck skin, spasm of the neck muscles, whiplash injury, tics and vertical imbalance of the eye muscles. Treatment depends on the cause. Also known as wry neck.

totipotentiality totipotency, the ability of a cell to differentiate to form any kind of fully differentiated body cell. In stem cell research totipotentiality is taken to be a property of the cells of the embryo, the zygote and the immediate descendants of the first two cell divisions. Compare PLURIPOTENTIALITY and UNIPOTENTIALITY.

toxaemia the presence of bacterial or other poisons (toxins) in the blood. Compare SEPTICAEMIA.

toxi-, toxico- *combining form denoting* poison or toxin.

toxicity the quality or degree of poisonousness.

toxicology the study of the nature, properties and identification of poisons, of their biological effects on living organisms and of the treatment of these effects.

toxic psychosis a mental disorder caused by the effect on the brain of any poison or drug. Possible causes include alcohol, lead, mercury, cocaine, amphetamine (amfetamine), cannabis and hallucinogenic drugs.

toxin any substance produced by a living organism that is poisonous to other organisms. Bacterial disease is largely the result of poisoning by the toxins they produce. Some bacteria release soluble exotoxins that act remotely. Others produce only endotoxins which operate only locally. Some bacterial toxins are among the most poisonous substances known.

toxo- *combining form denoting* bow-shaped.

TPA *abbrev. for* TISSUE-PLASMINOGEN ACTIVATOR.

trabecula supporting strands of connective tissue constituting part of the framework of an organ.

trabeculation the formation of ridges on a surface.

trace elements dietary minerals required only in tiny amounts to maintain health. They include zinc, copper, chromium and selenium, and are rarely deficient except under unusual circumstances such as artificial feeding.

tracer 1 a biochemical that has been tagged with a radioactive atom so that its destination can be determined.
2 a length of nucleic acid tagged with a radioactive atom that can be used to find and identify samples of its complementary strand.

trach-, tracheo- *combining form denoting* trachea.

trachea the windpipe. A cylindrical tube of mucous membrane and muscle reinforced by rings of CARTILAGE, that extends downwards into the chest from the bottom of the LARYNX for about 10 cm. The trachea terminates when it branches into two main bronchi.

tracheostomy an operation to make an artificial opening through the front of the neck into the windpipe (trachea). A tube is then inserted to maintain the opening and allow breathing. Tracheostomy is necessary when life is threatened by obstruction to the airway or when breathing must be maintained artificially for long periods by an air pump.

trach- *combining form denoting* rough.

tract 1 an associated group of organs forming a pathway along which liquids, solids or gases are moved. Examples are the digestive tract, the urinary tract and the respiratory tract.
2 a bundle of myelinated nerve fibres with a common function.

traction the process of exerting a sustained pull on a part of the body, to achieve and maintain proper alignment of parts, as in the treatment of fractures. Spinal traction is used to reduce the tendency for soft tissue to be squeezed out of INTERVERTEBRAL DISCS.

tragus the small projection of skin-covered cartilage lying in front of the external ear opening (meatus).

training the inculcation of skills and abilities and of improved muscular bulk, power and performance by repetitive action in applying a force. Physical training alters muscle in several ways, some as subtle as mitochondrial changes, and improves the efficiency of the heart and the respiratory system. Other forms of training involve psychological or sensory modification.

trait 1 any inheritable characteristic.
2 a mild form of a recessive genetic disorder.

trance a state of reduced consciousness with diminished voluntary action. Trances may occur in some forms of epilepsy, in catalepsy, in hysteria and in hypnosis.

TRANCE *acronym for* tumour-necrosis-factor-related activation-induced cytokine. This cytokine stimulates osteoclast differentiation and offers the possibility of developing new control over bone loss in osteoporosis.

tranquillizer drugs drugs used to relieve anxiety or to treat psychotic illness.

trans in chemistry, of two groups having the configuration of being on opposite sides of a ring or a double bond.

trans- *prefix denoting* across, through or beyond. In stereochemistry, indicating that two groups are in the TRANS configuration. The prefix is usually italicized.

transactional analysis a psychological interpretation of behaviour based on a study of social interactions in which the individual is perceived as adopting one of three roles – 'adult', 'parent' or 'child'. Relationships are said to be satisfactory if the choice of role is appropriate and complementary, but are disruptive if the people concerned refuse to play the game. Transactional analysis explores the way people play these life games and identifies bad play which may damage the quality of life.

transaminase any enzyme that catalyses the transfer of amino groups in the metabolism of amino acids (transamination). Also known as aminotransferase.

transamination the reaction in which the amino group of an amino acid is transferred to a ketoacid, converting it into an amino acid.

transcriptase any enzyme that catalyses transcription of a molecule, as in the process of producing a copy of RNA from DNA. Reverse transcriptase is an enzyme found in viruses, such as HIV, that catalyses DNA from an RNA template – the reverse of the usual procedure.

transcription the synthesis of RNA on a DNA template.

transcription factors proteins that initiate the transcription of genes.

transduction 1 the conversion of energy in one form into energy in another.
2 the transfer of a gene from one bacterial host to another by means of a phage.
3 the transfer of a gene from one cell host to another by a retrovirus.

transferase any enzyme that catalyses the movement of atoms or groups of atoms from one molecule to another. A kinase adds a phosphate group to a substrate such as an amino acid.

transference the transfer of emotional wishes or thoughts experienced in relation to one person, to another person, especially a psychotherapist. Freud regarded transference in psychoanalysis as essential to success.

transfer factor a LYMPHOKINE released by T cells that activates MACROPHAGES and prompts them to attack fungi.

transferrin a blood protein (beta globulin) that can combine reversibly with iron and transport it to the cells.

transfer RNA a short-chain RIBONUCLEIC ACID molecule present in cells in at least 20 different varieties, each capable of combining with a specific amino acid and positioning it appropriately in a polypeptide chain that is being synthesized in a ribosome. Transfer RNA is a four-armed, clover-leaf-like structure. At the end of one arm is an ANTICODON, complementary to the codon for an amino acid in MESSENGER RNA. At the end of the opposite arm is a site to which the appropriate amino acid can be covalently linked. When a molecule of transfer RNA is linked to the amino acid corresponding to its anticodon it becomes aminoacyl-tRNA. The identity of the passenger on a particular tRNA molecule is determined by its anticodon rather than by its attached amino acid.

transferrin an iron-binding protein that carries iron in the blood plasma.

transfusion the replacement of lost blood by blood, or blood products, usually donated by another person. Blood transfusion is given in cases of shock or severe anaemia and is often life-saving. Blood must be of a compatible group and is invariably checked by cross-matching before being given.

transgenic animal an animal that has had a new DNA sequence introduced into its germ line by insertion into a fertilized egg or an early embryo. All subsequent offspring will carry the new genetic material in their genomes and will show its effects.

transient global amnesia a state, lasting for less than 24 hours, in which there is inability to form new memories and in which there is loss of memory for periods of up to years (retrograde amnesia) prior to the attack. After the attack there is permanent memory loss for the period of the attack. The condition may be brought on by emotional upset, physical exertion, sexual intercourse or the Valsalva manoeuvre.

transient ischaemic attacks disturbances of body function lasting for less than 24 hours and caused by localized nervous system defects. These occur because of temporary interruption or reduction of the blood supply to part of the brain. There may be visual loss, weakness in an arm or leg, numbness, speech difficulty or confusion. Transient ischaemic attacks are a warning of the danger of stroke and should always be investigated.

translation a final stage in the expression of a gene; the lining up of amino acids and synthesis of a polypeptide on a ribosome by means of transfer RNA. Translation requires several enzymes.

translocation 1 a form of chromosome mutation in which a detached part of a CHROMOSOME becomes attached to another chromosome, or parts of two chromosomes may be joined. Translocations may be inherited or acquired. In many cases they cause no effect on the body because all the normal chromosomal material is present. But if a translocation results in a deficiency or excess of chromosomal material the results are serious.

2 of a gene when a new copy of the gene appears at a location on the genome remote from the original location.

3 the movement of a RIBOSOME along a MESSENGER RNA molecule from one CODON to the next.

4 of the movement of a protein across a membrane.

transmembrane crossing a membrane.

transplacental passing across the PLACENTA.

transporter a membrane protein that acts as a conduit for the passage of molecules through the membrane.

transposition in genetics, the movement of a length of genetic material from one point in a DNA molecule to another.

transposition of the great vessels one of the types of congenital heart disease. In this condition, the AORTA, the main supply artery of the body, is wrongly connected to the right ventricle, which normally pumps blood returning from the body to the lungs. The pulmonary artery is connected to the left ventricle. As a result, the lungs are effectively bypassed and the blood supplied to the body is insufficiently oxygenated. There is blueness of the skin (cyanosis), breathlessness and failure to thrive. Heart surgery is usually necessary.

transposons discrete mobile sequences in the genome that can transport themselves directly from one part of the genome to another without the use of a vehicle such as a phage or plasmid DNA. They are able to move by making DNA copies of their RNA transcripts which are then incorporated into the genome at a new site. Sometimes called 'jumping genes'.

transsexualism a persistent conviction that the true gender is the opposite of the actual anatomical sex. This may cause depression and anxiety. Transsexualism is mainly experienced by men, and SEX-REASSIGNMENT SURGERY is often sought.

transudate 1 a fluid that has passed through a membrane.

2 a collection of fluid resulting from increased capillary pressure in capillary beds or decreased osmosis from reduced blood protein.

transverse colon the middle third of the colon that loops across the upper part of the ABDOMEN.

transverse process a bony protuberance on each side of a vertebra to which muscles are attached.

transvestism male desire to wear women's clothing often for reasons of sexual gratification. The reciprocal phenomenon is not normally referred to as transvestism as it is not usually related to sexual pleasure. Transvestites are not usually transsexuals.

trapezius muscle a large, triangular back muscle extending from the lower part of the back of the skull (occiput) almost to the lumbar region of the spine on each side. Each muscle extends outward from the rear processes of the vertebral column to the spine of the shoulder blade and the outer tip of the collar bone (clavicle). The trapezius muscle braces the shoulder blade and rotate it outwards when the arm is raised.

trauma 1 any injury caused by a mechanical or physical agent.

2 any event having an adverse psychological effect.

traumatology the study and practice of the management of patients suffering from recent physical injury, as from traffic or other accidents or assault.

travellers' diarrhoea a popular term for gastroenteritis usually caused by faecal contamination of food or water. The organisms most commonly involved are *Escherichia coli*, *Campylobacter jejuni*, *Salmonella* species, and *Shigella* species.

tremor rhythmical oscillation of any part of the body, especially the hands, the head, the jaw or the tongue. Tremor does not necessarily imply disease but is a feature of conditions such as cerebellar ataxia, encephalitis, essential-familial tremor, liver failure, mercury poisoning, multiple sclerosis, Parkinson's disease, thyrotoxicosis and Wilson's disease. It is also a side effect of many antipsychotic and other drugs.

TRH *abbrev. for* THYROTROPHIN RELEASING HORMONE.

triage a selection process, used in war or disaster, to divide casualties into three groups so as to maximize resources and avoid wastage of essential surgical skills on hopeless cases. In triage, an experienced surgeon sorts cases rapidly into those needing urgent treatment, those that will survive without immediate treatment, and those beyond hope of benefit from treatment. Triage is also used to assign treatment in the event of the appearance of a number of men suffering acute chest pain.

triceps muscle a three-headed muscle attached to the back up the upper arm bone (humerus) and to the outer edge of the shoulder-blade (scapula) and running down the back of the arm to be inserted by a strong TENDON into the curved process (olecranon process) on the back of one of the two forearm bones (the ulna). The triceps muscle straightens the elbow in opposition to the biceps muscle, which bends it.

trich-, tricho- *combining form denoting* hair or eyelash.

trichotillomania an apparent compulsion to pull out one's own hair, sometimes manifested by people with psychotic disorders or severe mental retardation. Anxious or frustrated children sometimes pull out hair.

tricuspid 1 having three cusps or points, as on a molar tooth.
2 pertaining to the three-cusped heart valve that separates the chambers on the right side.

tricuspid valve the valve lying between the upper and lower chambers of the right side of the heart.

trifocal 1 having three focal lengths.
2 distance-viewing spectacles having additional segments for intermediate and close ranges.

trigeminal nerve the 5th of the 12 pairs of cranial nerves and the sensory nerve of the face. Each trigeminal nerve divides into three main branches, the ophthalmic, the maxillary and the mandibular nerves, which then branch to supply the corresponding parts of the face.

triglycerides triacylglycerides. See FATS.

triploblastic of an embryo, having three primitive germ layers from each of which particular parts of the body develop.

triploid having three HAPLOID sets of chromosomes in each nucleus. The normal state of a body cell is DIPLOID.

trismus lockjaw. Tight closure of the mouth from uncontrollable spasm of the chewing muscles. Causes include tetanus, disorders of the jaw joint, tonsillitis, quinsy, mumps, tooth decay, Parkinson's disease or anorexia nervosa.

trisomy the occurrence of an extra chromosome in one of the 23 matched and identifiable pairs so that there are three, instead of two, of a particular chromosome. This anomaly may cause a wide range of structural abnormalities or even early death of the fetus. Trisomy of chromosome 21 (trisomy 21) is the cause of Down's syndrome.

trisomy 13 the presence of three copies of chromosome 13. Patau syndrome. There is microcephaly, cleft lip or palate, deafness, blindness, extra fingers and finger deformity. Such infants rarely survive for more than a few weeks

trisomy 18 syndrome the effect of TRISOMY of chromosome 18. Edwards syndrome. There is failure to thrive, severe mental deficiency, persistent contraction of muscles with clenching of the hands and anomalies of the face, hands, breastbone and pelvis. Survival beyond a few months is uncommon. This trisomy leads to spontaneous abortion in nine cases out of ten.

tritanopia unable to appreciate the colour blue. This is a rare defect of colour perception.

tRNA TRANSFER RNA.

trochanter one of two major bony processes, the greater and lesser trochanters, on the upper part of the FEMUR. Muscle tendons are attached to the trochanters.

trochlea a body structure that resembles a pulley, especially that for the tendon of the superior oblique eye muscle.

trochlear nerve the 4th of the 12 pairs of cranial nerves. The trochlear nerve supplies one of the muscles that moves the eye – the superior oblique muscle. Contraction of this muscle turns the eye downwards and outwards.

-trophic *combining form denoting* nutrition or nourishment.

tropho- *combining form denoting* nutrition or food.

trophoblast the outer layer of the BLASTOCYST.

-tropic *combining form denoting* moving, turning or changing in response to the stimulus specified.

tropical diseases these include amoebiasis, cholera, diphtheria, hookworm, leishmaniasis, malaria, malnutrition, onchocerciasis, plague, rabies, schistosomiasis, shigellosis, strongyloidiasis, tapeworm, tropical sprue, tropical ulcer, trypanosomiasis, tuberculosis, typhoid, typhus and yellow fever.

tropism an automatic movement made by an organism towards or away from a source of stimulation.

-tropism *combining form denoting* TROPISM in relation to the stated stimulus, as in phototropism (moving towards or away from light).

tropomyosin a regulatory protein in striated muscle concerned with the binding sites for the cross bridges on thin filaments in the SARCOMERE. Tropomyosin reversibly prevents bonding, thereby preventing muscles from being in a continuous state of contraction.

troponin a regulatory protein in striated muscle bound to TROPOMYOSIN and ACTIN in thin filaments. Troponin is the site of calcium binding that starts the contraction of the SARCOMERE.

truss a belt-like appliance with a pad that exerts pressure over the orifice of a HERNIA so as to prevent protrusion of the bowel. This is an unsatisfactory substitute for surgical repair.

trypsin one of the digestive enzymes secreted by the pancreas as the precursor trypsinogen, that breaks down protein to polypeptide fragments. These are then split further to amino acids by carboxypeptidase from the pancreas and aminopeptidase from the small intestine.

trypsinogen the precursor of TRYPSIN produced by the pancreas that is converted into the active form, trypsin, when acted upon by the enzyme enterokinase in the small intestine.

tubal pregnancy the commonest type of ECTOPIC pregnancy.

tubercle 1 a small nodular mass of tubercular tissue.
2 an informal term for TUBERCULOSIS.
3 any small, rounded protrusion on a bone.

tuberosity any prominence on a bone to which tendons are attached.

tubular reabsorption the recovery of needed substances from the contents of the tubule of the nephron into the peritubular capillaries.

tubular secretion the passage of unneeded substances from the peritubular capillaries into the tubule of the nephron and hence into the urine.

tubule a small tube or tubular structure.

tubulin a contractile protein that forms microtubules. These form the spindle fibres in MITOSIS that draw chromosomes apart is the course of cell division.

tum-, tume- *combining form denoting* swelling.

tumefacient producing or tending to produce swelling.

tumescence 1 a swelling or enlarging of a part.
2 a swollen condition.
3 a penile erection.

tumour a swelling. The term usually refers to any mass of cells resulting from abnormal degree of multiplication. Tumours may be BENIGN or MALIGNANT. Benign tumours enlarge locally, and are often enclosed in capsules, but do not invade tissue or spread remotely. Malignant tumours infiltrate locally and also seed off into lymphatic vessels and the bloodstream to establish secondary growths (metastases) elsewhere in the body.

tumour necrosis factor one of two related CYTOKINES capable of killing certain cancer cells and which also have regulatory functions on the immune system. Tumour necrosis factor is implicated in various inflammatory diseases including rheumatoid arthritis.

tumour-specific antigens substances produced by specific kinds of tumours that can be detected in the blood and may act as indicators or markers of the presence of tumours or of recurrences of tumours. The method is not wholly reliable because these substances can sometimes occur in the body apart from tumours.

tunica a covering or investing membrane or layer of tissue.

tunnel vision a narrowing of the extent to which peripheral visual perception is possible while looking straight ahead. Such loss of the visual fields may be caused by damage to the RETINAS, OPTIC NERVES, the nerve connections with the brain or to the visual cortex of the brain itself. Causes of such damage include glaucoma, retinitis pigmentosa, tumour of the pituitary gland, stroke, head injury or multiple sclerosis.

turbinate bones three pairs of small, curled bony processes, the upper two pairs being parts of the ETHMOID BONES and the lower pair of the MAXILLAS, that extends horizontally along the outer walls of the nasal passage. The turbinates are covered with mucous membrane which warms and moistens the incoming air. Also known as conchae.

turgid swollen and congested.

twins two offspring from a single pregnancy. Twins may be derived from a single fertilized egg that separates into two after the first division. Such monozygotic twins are genetically identical. Alternatively, twins may be derived from two eggs fertilized at the same time. Such twins are dizygotic and non-identical and in about half the cases are of different sex. The rate of monozygotic twinning in humans is fairly stable at 0.35 per cent, while the rate of dizygotic twinning varies more widely around an average of 1 per cent of pregnancies.

twitch a brief muscular contraction resulting from a sudden spontaneous impulse in a nerve supplying a group of muscle fibres. Twitching is common and is seldom of any medical significance.

Tylor, Edward Sir Edward Burnett Tylor (1832–1917) is generally considered the founder of British anthropology. As a wealthy young man travelling in America for the sake of his health, Tylor became fascinated with the ethnology and archaeology of the Mexicans. On his return to England he published *Anahuac: Mexico and the Mexicans, Ancient and Modern* (1861). This suggested that information about early stages of social evolution could be obtained from what he called 'survivals' – early traits that lingered on in more advanced cultures. Tylor was also much concerned with the tendency of early religions to endow natural or inanimate objects with a soul – a principle he described as animism.

In 1871 Tylor crystallized the important concept of 'culture' which he defined as 'That complex whole which includes knowledge, belief, art, morals, law, custom and any other capabilities and habits acquired by man as a member of society'. In spite of never having taken a University degree, Tylor became Professor of Anthropology at Oxford in 1896. His most important books are *Primitive Culture* (1871) and *Anthropology* (1881).

tympan-, tympano- *combining form denoting* eardrum or middle ear auditory bones, or both.

tympanic membrane the ear drum.

tympanum 1 the ear drum and the middle ear cavity.
2 the middle ear and its contents.

type genus any taxonomic genus selected as being representative of the family to which it belongs.

typh-, typho- *combining form denoting* typhus or typhoid.

typing a procedures to establish the group or classification of blood or tissues. See also BLOOD GROUPS and TISSUE TYPING.

tyrosinase a copper-containing enzyme that occurs in melanocytes and catalyses the production of MELANIN from TYROSINE. Also known as monophenol monooxygenase.

tyrosine the AMINO ACID 4-hydroxyphenylalanine. This is one of the 20 amino acids that are incorporated into protein.

Uu

ubiquitin a small 76-residue protein found in all animal cells and known to have altered minimally throughout evolutionary history. Ubiquitin is linked by covalent bonds to proteins destined for destruction by PROTEASOMES.

ulcer a local loss of surface covering (EPITHELIUM) and sometimes deeper tissue in skin or MUCOUS MEMBRANE. An open sore.

ulna one of the pair of forearm long bones. The ulna is on the little finger side. At its upper end it has a hook-like process, the olecranon, that fits into a hollow at the back of the lower end of the upper arm bone (the humerus) and prevents the elbow from over-extending. When the hand is turned on the long axis of the arm, the radius bone rotates around the ulna.

ulnar nerve one of the main nerves of the arm that supplies some of the muscles of the forearm and all the small muscles of the hand. It also provides sensation to the skin of the third of the hand on the little finger side. It is near the surface at the back of the elbow where it is liable to be painfully struck. This is the basis for the notion of the 'funny bone'.

ultra- *prefix denoting* beyond or on the other side of.

ultracentrifuge a device for rotating small containers at extremely high speed so as to expose the liquid contents to powerful centrifugal force, of the order of 100,000 g. Ultracentrifuges are used to separate particles of molecular size and determine molecular weights.

ultrafiltration filtration through a semipermeable membrane that allows only the passage of small molecules, such as those smaller than protein molecules.

ultramicroscopic too small to be resolved by an ordinary optical microscope but visible when illuminated by light from the side against a dark background.

ultrasonic pertaining to sound waves above the upper limit of audibility, that is above about 20,000 Hz.

ultrasonography the use of ultrasonic waves to image body structures for diagnostic purposes. See ULTRASOUND SCANNING.

ultrasound scanning a method of body imaging based on the reflectivity of sound. By using very high frequency (ultrasonic) sound the wavelengths are brought down to the necessary small dimensions. Piezo-electric transducers are used both to generate the waves and to pick up the reflected sound. Ultrasound scanning is believed to be completely safe and is widely used in obstetrics as well as in other disciplines.

ultraviolet light electromagnetic radiation of shorter wavelengths than visible light but longer wavelengths than X-rays. Ultraviolet light is divided into three zones – UVA with wavelengths from 380–320 nanometres (billionth of a metre), UVB from 320 nm down to 290 nm and UVC from 290 nm down to one tenth of a nanometre. UVC and most of UVB are absorbed by the ozone layer in the earth's stratosphere. Ultraviolet light causes sunburning and damages the skin's elastic protein, collagen. It is also a major factor in the development of the skin cancers rodent ulcer (basal cell carcinoma), malignant melanoma and squamous cell carcinoma. It causes pinguecula and pterygium in the eyes.

umbilical cord the nutritional, hormonal and immunological link between the mother and the fetus during pregnancy. The umbilical cord arises from the PLACENTA and enters the fetus at the site of the future navel. It carries two arteries and a vein that connect to the fetal circulation.

umbilicus the scar formed by the healing at the exit site of the UMBILICAL CORD after this has been tied and cut and the tissues have died and dropped off.

uncinate hooked (in the literal sense). Hook-shaped.

uncircumcized in possession of a FORESKIN (prepuce). Anatomically complete.

unconditioned response an automatic or instinctive response produced by a stimulus without any prior learning or conditioning process.

unconditioned stimulus the stimulus that evokes an UNCONDITIONED RESPONSE.

unconscious mind a term with more than one definition. In general, it is taken to be that division of the mental process that proceeds without immediate awareness of the fact on the part of the possessor. Clearly, this is an essential arrangement. Were the whole content of the mind (or memory) in consciousness at all times, we would be overwhelmed, so some form of selectivity is necessary. Access to the unconscious component is of variable difficulty. Ease of access seems to depend largely on associative links with the currently conscious part, such as mnemonics, and with the length of time since last the same information was accessed. Freudian psychoanalysts have their own definition. To them, the unconscious mind, or at least a part of it, is a domain into which are repressed those id functions too unpalatable to be constantly presented to consciousness. The unresolved conflicts within such material are, they claim, the source of all our psychological troubles.

unconsciousness a state of unrousability caused by brain damage and associated with reduced activity in part of the BRAINSTEM called the reticular formation. Unconsciousness varies in depth from a light state, in which the unconscious person responds to stimuli by moving or protesting, to a state of profound coma in which even the strongest stimuli evoke no response. Causes include head injury, inadequate blood supply to the brain, fainting, asphyxia, poisoning, near drowning, starvation and low blood sugar (hypoglycaemia).

uncus a hook-shaped part near the front of the temporal lobe of the brain that is concerned with the senses of smell and taste. Disease in this area may produce hallucinations of foul smells.

undescended testicle a testicle that has remained in the abdomen, where testicles normally develop, or in the inguinal canal by which they normally descend to the SCROTUM. Undescended testicles do not become fertile and are more than normally prone to cancer later in life. For these reasons it is wise to have an undescended testicle brought down in childhood.

ungual pertaining to a fingernail or toenail. Subungual means 'under a nail'.

uni- *prefix denoting* single or one.

uniarticular pertaining to only one joint. Monoarticular.

unicellular consisting of a single cell, as in a unicellular organism.

unilateral on or affecting one side only. One-sided. From Latin *unus*, one and *latus*, a side or flank.

uniovular originating from one egg, as in the case of monozygotic twins.

unipara a woman who has had only one baby. Primipara.

unipolar having one pole or process as in the case of a nerve cell with one AXON. As applied to an electrode, the term is something of a misnomer. A second electrode is needed to complete the circuit, but this is often attached at a remote point from the point of application of the electrode.

unipotentiality the full limitation of the ability of cells to differentiate. For example, basal epidermal cells in the skin cannot differentiate to form other types of cells; they can only produce keratinised squames. See also PLURIPOTENTIALITY and TOTIPOTENTIALITY.

unmedullated see UNMYELINATED.

unmyelinated of a nerve fibre, lacking a MYELIN SHEATH.

unsaturated pertaining to a compound, especially of carbon, in which atoms are linked by double or triple valence bonds. A saturated compound has only single bonds. In general, unsaturated compounds are less stable than saturated compounds and can undergo a wider variety of reactions.

unsaturated fatty acid a fatty acid with one double bond between carbons.

upstream in genetics, at a stage in the sequence of processes in the expression of a gene that is further away from the final protein product. The term is also used to mean in the direction of the 5'-end of a chain of bases in DNA (see PHOSPHODIESTER BOND). In both cases the opposite sense is called 'downstream'.

urachus a primitive structure in the embryo from which the bladder develops. Later it is represented by a fibrous cord that extends from the top of the bladder to the UMBILICUS.

uracil one of the four bases that form the nucleotide code in RNA.

urate a salt of uric acid.

ur-defence a belief, such as a conviction of personal immortality, or of the inherent goodness of man, considered by some to be essential to psychic well-being.

urea a substance formed in the liver from the excess of nitrogenous material derived from amino acids and excreted in solution in the urine. Urea can be used as an osmotic diuretic and as a cream for ICTHYOSIS and other hyperkeratotic skin disorders.

urease an enzyme that breaks down urea to ammonia and carbon dioxide.

-uresis *combining form denoting* excreted in the urine or the excretion of urine.

ureter a tube that carries urine downwards from each kidney to the urinary bladder for temporary storage. The ureters have muscular walls that can contract to assist in the propulsion of the urine.

ureteric pertaining to the URETER.

urethra the tube that carries urine from the bladder to the exterior. In the male, the urethra runs along the penis and opens at the tip. In addition to urine it carries seminal fluid during ejaculation. In the female, the urethra is shorter and runs directly downwards from the bladder in front of the VAGINA, opening between the vaginal orifice and the CLITORIS.

urethral pertaining to the URETHRA.

urethro- *combining form denoting* URETHRA.

-uria *combining form denoting* the presence in the urine of a specified substance or specifying the state or quantity of the urine, usually implying abnormality. Examples are HAEMATURIA and OLIGURIA.

uric acid the main end product of PURINE metabolism. Uric acid is derived from ADENINE and GUANINE, two of the purines in DNA and RNA (nucleic acids). An excess of uric acid salts in the body can cause gout and kidney stones.

uridine the nucleoside of URACIL found in RNA and other NUCELOTIDES.

urinary bladder the muscular bag for the temporary storage of urine situated in the midline of the pelvis at the lowest point in the abdomen, immediately behind the pubic bone. The bladder wall relaxes at intervals to allow filling but as the internal pressure rises the intervals become shorter and the urgency to empty the bladder becomes more frequent and then continuous. Unless emptied voluntarily, the bladder will eventually empty spontaneously.

urinary bladder disorders these include bladder stones, bladder cancer, cystitis and incontinence.

urinary catheterization the passage of a blunt-ended, rubber or plastic tube along the URETHRA into the bladder so as to release urine in cases of obstruction to outflow or inability to pass urine voluntarily for other reasons (urinary retention).

urinary system

Like any other chemical engineering plant, the human body continuously produces unwanted waste material which must be disposed of. The effluent from carbohydrate consumption consists of carbon dioxide (CO_2) and water. CO_2 is the major waste product of metabolism and is disposed of, along with water vapour, with every expired breath. The amount of water lost in the breath is roughly equal to that produced by metabolism. The breakdown of the protein components, amino acids, to form carbohydrates or fats involves a process known as deamination – removal of amino chemical groups containing ammonia (NH_3). In any quantity, this ammonia would be dangerous, so it is converted in the liver to a stable and inert substance called urea. About 30 g of urea is produced each day. This is the main waste product arising from protein breakdown, and a principal function of the body's excretory system – the urinary system – is to dispose of urea. Surprisingly, very little of the waste disposed directly from the intestinal canal derives either from the food or from body metabolic processes.

The two kidneys – reddish brown, bean-shaped structures, each about 11 cm long – lie in pads of fat high up on the inside of the back wall of the abdomen, one on each side of the spine. Kidney function, in disposing of urea and in maintaining correct levels of other substances in the blood, is essential to life. Its importance is reflected in the location of the kidneys close to and on either side of the main artery and vein of the body – the aorta and the inferior vena cava. The kidneys are connected to these large blood vessels by short wide arteries and veins – the renal vessels – and receive blood from the aorta under high pressure. Over a litre of blood is filtered by the kidneys every minute – some 1640 litres per day – and the kidneys receive about one fifth of the entire heart output.

Kidney function

The functional unit of the kidney is called the nephron. This is a microscopic structure of which there are about 1 million in each kidney. The nephron consists of a filtering unit and a complex system of tubules. The filtering unit is simply a small tuft of blood capillaries, the glomerulus, surrounded by a cup-shaped, double-walled structure known as Bowman's capsule. Bowman's capsule is best regarded as a thin-walled hollow sphere pushed in (invaginated) at one side by the glomerulus. The hollow interior of Bowman's capsule is continuous with the start of the tubule system of the nephron. From the capsule the tubule takes a few twists and turns then runs straight for a time before looping back to the region of Bowman's capsule where it has a few more twists before ending in a urine collecting duct. The tiny blood vessels that form the glomerulus are derived from branches of the renal artery. On exiting from the glomerulus they form a dense network of capillaries surrounding the entire tubular system of the nephron. The capillaries drain into tiny veins that join to form the draining veins of the kidney and empty into the inferior vena cava.

All the blood entering each nephron thus passes first to the glomerular capillary tuft inside Bowman's capsule, then passes on to come into intimate contact with the whole of the nephron tubule system, and finally returns to the circulation by way of the renal veins. The formation of urine starts with a massive filtration of blood plasma (blood less the cells) through the glomerulus into Bowman's capsule. This filtrate, however, contains a great deal of water and much low-molecular-weight material essential to the body. Large-molecule substances, such as proteins and fats, do not pass into the capsule. If all the filtrate were lost as urine we would be dead in a matter of hours. The final composition of the urine is determined by the remarkable powers of selective reabsorption and secretion of the tubule system. As the crude filtrate passes down the tubules from Bowman's capsule much of the water, together with any substances needed by the body, is reabsorbed into the blood. At the same time, certain unwanted substances that did not pass into Bowman's capsule are actively secreted into the urine by the tubule. Sodium, potassium, calcium, chloride, bicarbonate, phosphate, glucose, amino acids, vitamins and many other substances are returned to the blood and conserved. Selective reabsorption is a highly complicated process under the control of various hormones such as aldosterone from the adrenal gland, the antidiuretic hormone from the pituitary gland and parathyroid hormone from the parathyroid glands.

Urine

Urine is a sterile solution containing water, urea, uric acids and various inorganic salts in varying concentrations. By adjusting the amount of water and acidic substances reabsorbed, and hence the composition of the urine, the kidneys are largely responsible for ensuring that the body contains the right amount of water and that the blood is of the correct degree of acidity. Most of us drink more water than is strictly necessary, so the kidneys regularly excrete dilute urine to maintain the body's water balance. On average, about 1 ml of urine is formed per minute – a daily output of 1200–2000 ml. The volume of urine produced varies greatly, however, with variations in fluid intake and the amount of fluid lost in the sweat. The urine is usually acid and contains creatinine and various products of blood cell breakdown. The yellow colour comes from a pigment urochrome. Most drugs or their breakdown products are eliminated through the kidney. Many diseases, but especially those of the kidneys, cause characteristic

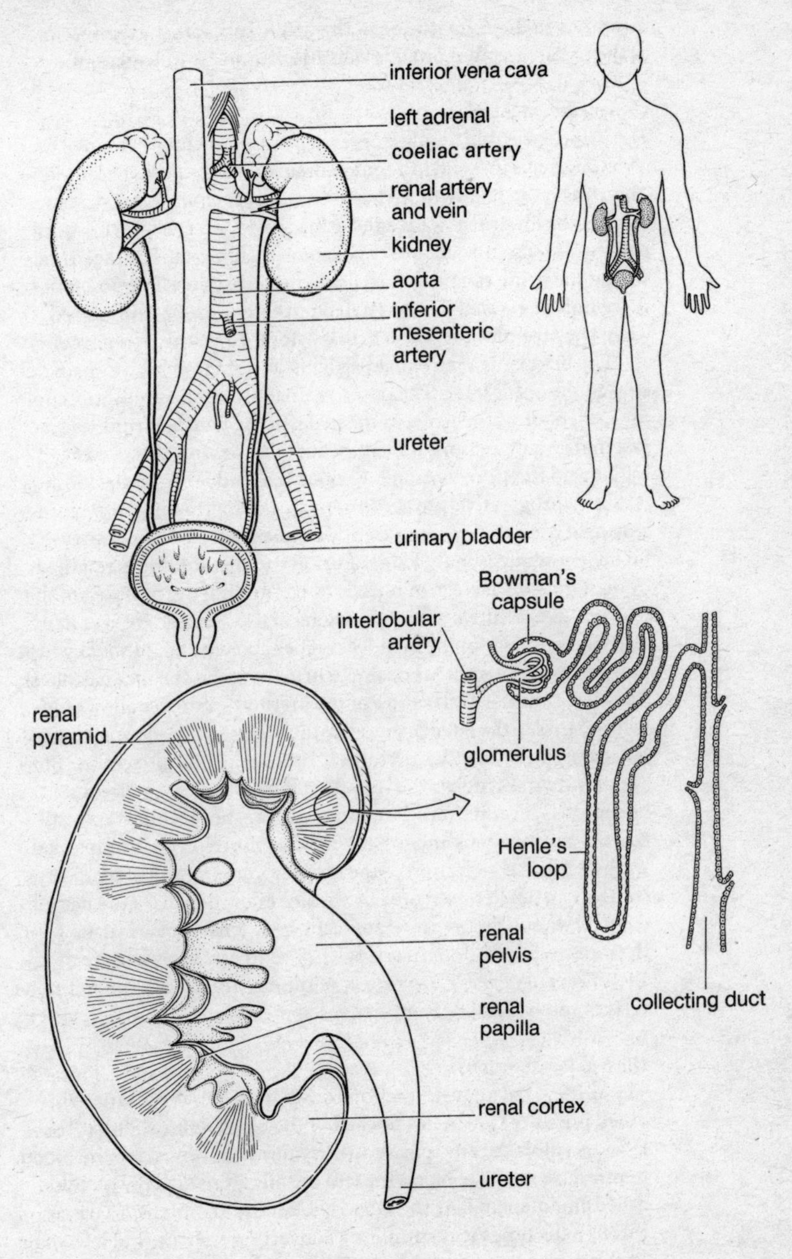

Fig. 19 **Urinary system**

variations in the constitution of the urine. Laboratory examination of the urine can often provide valuable diagnostic information.

Urinary drainage system

The urine collecting tubules form a separate branching system of ever-widening ducts that end in a conical drain called the pelvis of the kidney. This runs into a hollow tube, the ureter, 40–45 cm long which descends to enter the lower rear wall of the urinary bladder. The ureter is not a passive tube. Its wall contains circular muscle fibres capable of peristalsis so that the urine can be milked down to the bladder. The power of these muscles is well demonstrated by the agonizing pain caused when they attempt to move a urinary stone downwards (renal colic).

The bladder is a muscular bag lying low in the pelvis, immediately behind the pubic bone. The outlet drainage tube, the urethra, occupies the underside of the penis in the male. In the female it runs down just in front of the front wall of the vagina, opening between the vagina and the clitoris. As urine enters the bladder from the kidneys it relaxes progressively to accommodate the increasing volume, and storing it until it may conveniently be disposed of. When full, the bladder contains about 350 ml of urine and this volume is retained by the tight contraction of muscles surrounding the first part of the urethra. In health, we are unconscious of the accumulation of urine until the pressure in the bladder reaches about 20 cm of water. At this point there is a conscious desire to urinate, and if the inclination is gratified, voluntary relaxation of the urethral sphincter allows the urine to pass to the outside by way of the urethra. If the desire to urinate is repressed, the bladder relaxes a little and the impulse ceases. But as pressure rises higher the impulse returns. This sequence may be repeated several times, but occurs at ever shorter intervals until there is a continuous and urgent demand for release, accompanied by rhythmical contraction of the bladder muscle. At this extremity, control can be exercised only by a major effort of will on the part of the now wholly preoccupied subject who is tensely contracting all the muscles in the floor of the pelvis. Eventually, at a pressure of about 100 cm of water, regardless of the circumstances, the sphincter relaxes and the bladder empties spontaneously.

Other kidney functions

The kidneys have some functions other than excretion. When the blood pressure falls below normal or the blood volume drops, the kidneys automatically release an enzyme called renin into the blood. Renin acts on a large blood protein to split off a small polypeptide called angiotensin I. As the blood containing angiotensin I passes through the lungs it encounters a converting enzyme which changes angiotensin I to angiotensin II. The latter hormone acts on the muscle in the walls of blood vessels causing them to constrict. This at once raises the blood pressure. Angiotensin II is also a powerful stimulator of the hormone aldosterone from the adrenal glands. Aldosterone acts on the kidney tubules, causing them to reabsorb sodium and retain water. This kind of automatic feedback control mechanism (servomechanism) is typical of physiology. An important group of blood pressure control drugs, the angiotensin converting enzyme inhibitors (ACE inhibitors) apply this knowledge.

The kidneys also produce a substance, erythropoietin, which stimulates the rate of formation of blood cells in the bone marrow.

urine the fluid excretion of the kidneys, a solution in water of organic and inorganic substances, most of which are waste products of METABOLISM. Normal urine is clear, of varying colour, of specific gravity between 1.017 and 1.020 and slightly acid. It contains UREA, URIC ACID, creatinine, ammonia, sodium, chloride, calcium, potassium, phosphates and sulphates.

uro- *combining form denoting* urine or the urinary system.

urobilinogen a pigment formed from BILIRUBIN in the intestine by bacterial action that is absorbed into the bloodstream and excreted in the urine. Excess is found in the urine in haemolytic anaemia and certain liver disorders. None is found in complete obstruction to the the outflow of bile from the liver.

urodynamics functioning of the urinary bladder, urethral sphincter and pelvic floor muscles. Urodynamic studies, which may be made in various ways, involve measurement, over a period, of such parameters as urine flow rates; total bladder capacity; bladder urine volume before voiding; residual urine volume; bladder pressure before and during voiding; bladder contractability; urethral sphincter pressure; patient's perception of bladder fullness; and ability to inhibit voiding. Urodynamics has been described as the 'gold standard investigation' in the management of all forms of urinary incontinence.

urogenital, urinogenital pertaining to both urinary and genital structures or functions.

urologist a doctor who specializes in the diagnosis and treatment of disorders of the KIDNEYS, the URETERS, the URINARY BLADDER and the URETHRA.

urology the scientific study of the disorders of the kidneys and the urine drainage system.

uterine 1 pertaining to the UTERUS.
2 having the same mother but not the same father.

uterus the female organ in which the fetus grows and is nourished until birth. The uterus is a hollow, muscular organ, about 8 cm long in the non-pregnant state, situated at the upper end of the VAGINA and lying behind and above the URINARY BLADDER and in front of the RECTUM. It is suspended by LIGAMENTS from the walls of the pelvis. The lining of the uterus is called the endometrium. Under the influence of hormones from the ovaries this thickens progressively until shed during menstruation. In pregnancy, the uterus expands considerably with the growth of the fetus until it rises almost to the top of the abdominal cavity.

uterus disorders these include cervical erosion, cervicitis, endometriosis, fibroids and prolapse. Retroverted uterus is not considered to be a disorder.

utricle a small sac or pocket.

uvea the coat of the eye lying immediately under the outer SCLERA (the CHOROID), together with its continuum, the CILIARY BODY and the IRIS. The uvea contains many blood vessels and a variable quantity of pigment. From the Greek word uvea, a grape, because of the resemblance of the uvea to a peeled black grape). See also UVEITIS.

uvula the small fleshy protuberance that hangs from the middle of the free edge of the soft PALATE. Like the rest of the soft palate, the uvula is composed of muscle and connective tissue covered by MUCOUS MEMBRANE.

Vv

vaccination see IMMUNIZATION.
vaccinal pertaining to VACCINE or IMMUNIZATION.
vaccine a suspension of microorganisms of one particular type that have been killed or modified so as to be safe, given to promote the production of specific ANTIBODIES to the organism for purposes of future protection against infection.
vacuole a small, clear region in the CYTOPLASM of a cell, sometimes surrounded by a membrane. Vacuoles may be used to store cell products or may serve an excretory function.
vacuum extraction a method of assisting childbirth used as an alternative to FORCEPS DELIVERY. A cup-like device is applied to the baby's scalp and firmly secured by suction. Traction can then be applied via a short chain and handle. The equipment is known as a ventouse.
vagal pertaining to the VAGUS NERVE.
vagina literally a sheath. In the female it acts as a receptacle for the penis in coitus and as the birth canal. The vagina is a fibromuscular tube, 8–10 cm long lying behind the URINARY BLADDER and URETHRA and in front of the RECTUM. The cervix of the UTERUS projects into its upper part. The vagina is highly elastic and has a thickened and folded mucous membrane lining that can stretch readily.
vaginal pertaining to the VAGINA or to a sheath.
vaginal disorders these include gonorrhoea, thrush, trichomoniasis, vaginal discharge, vaginismus, vaginitis and vulvovaginitis.
vaginismus apparently involuntary rejection, by a woman, of attempted sexual intercourse or gynaecological examination. The legs are straightened, the thighs pressed together and the muscles of the pelvic floor, that surround the vagina, tighten. Sexual desire may appear normal until penetration is tried. Treatment involves full explanation and instruction in the insertion of vaginal dilators of gradually increasing size.
vagus nerves the 10th of the 12 pairs of cranial nerves that arise directly from the brain. The vagus nerves arise from the sides of the MEDULLA OBLONGATA and pass down the neck to supply and control the throat, LARYNX, BRONCHI, lungs, OESOPHAGUS and heart. They then enter the abdomen on the front and back of the oesophagus and supply the stomach and the intestines as far as the descending COLON. The vagus is an important part of the AUTONOMIC NERVOUS SYSTEM. From the Latin *vagus*, wandering or straying.
valence, valency the property of an atom or group of atoms to combine with other atoms or groups of atoms in specific proportions. The number of atoms of hydrogen with which an atom can combine or displace in forming compounds. An atom may be monovalent, divalent, trivalent or tetravalent.
valetudinarian 1 a person constantly suffering from one illness or another, especially one deeply preoccupied with ill health.
2 a HYPOCHONDRIAC.
valgus, valgum, valga abnormal displacement of a part in a direction away from the midline of the body. Hallux valgus is the condition in which the big toe is bent outwards so as to point towards the little toe. Compare VARUS.

valine one of the ESSENTIAL AMINO ACIDS, a constituent of protein.

vallate surrounded by a rimmed depression.

vallecula a small hollow, groove or depression on the surface of an organ.

valsalva manoeuvre the effort to breathe out forcibly while the mouth and nose are firmly closed or the vocal cords pressed together. The valsalva manoeuvre is employed while straining at stool and in other circumstances. It causes a rise in blood pressure followed by a sharp drop and then a second sharp rise in blood pressure. This may be dangerous in people with heart disease and should be avoided. (Antonio Maria Valsalva, 1666–1723, Italian anatomist).

valve a structure that allows movement in a predetermined direction only. There are valves in the heart, the veins, the lymphatics, the urethra and elsewhere.

valvular pertaining to, or possessing, a VALVE or valves.

vampirism a product of Slavic folklore inspired by the blood-sucking vampire bat and now done to death by uninspired writers for television. The vampire operates only at night, rising from its grave to suck the jugular blood from its victims. These, in turn, become vampires after death, and can be restored to respectable mortality only if a sharpened wooden stake is driven through their hearts.

van der Waals forces weak attractions between non-polar parts of molecules. (Johannes D. van der Waals, Dutch physicist and Nobel laureate, 1837–1923).

variable region the part of an antibody at the tip of each arm (N-terminal region) that varies considerably in its amino acid sequence from one antibody to another. The remaining parts of the structure of antibodies are fixed and almost identical. This region is coded for by the V gene.

varices varicosities. Swollen, twisted and distorted lengths of vessels, usually veins. Veins affected by varices are called varicose veins and these are commonest in the legs. Oesophageal varices are the varicosities of the veins at the lower end of the OESOPHAGUS that occur when the portal vein drainage through the liver is impeded by cirrhosis. They are liable to cause dangerous bleeding. The singular form of the word is varix.

varico- *combining form denoting* VARICES or varicosity.

varicose Pertaining to VARICES.

varix see VARICES.

varus, varum, vara displaced or angulated towards the midline of the body. Coxa vara is a deformity at the upper end of the thigh bone (femur) in which the angle between the neck and the shaft is decreased.

vas a vessel or channel conveying fluid. See VAS DEFERENS.

vascular endothelial growth factor (VEGF) a naturally-occurring POLYPEPTIDE of between 121 and 206 amino acids. It occurs in four forms, the commonest of which has 165 amino acids. VEGF promotes the production of tiny new blood vessels by stimulating the growth, migration and proliferation of endothelial cells. It increases permeability of existing vessels and causes widening of blood vessels through the mediation of NITRIC OXIDE. Local shortage of blood supply (ischaemia) increases the gene expression of VEGF. Chronic lymphocytic leukaemia B cells resist apoptosis by secreting and binding VEGF. VEGF in retinal ischaemia is believed to be the basis of one kind of age-related macular degeneration (ARMD).

vascularization the process of forming new blood vessels.

vasculature any system of blood vessels supplying an organ, an area of the body or the whole body.

vasculitis widespread inflammation of blood vessels occurring as the principal feature of a range of conditions including erythema nodosum, polyarteritis nodosa, some forms of purpura, rheumatoid arthritis, temporal arteritis and thromboangiitis obliterans.

vas deferens the fine tube that runs up in the SPERMATIC CORD on each side from the EPIDIDYMIS of the TESTICLE, over the pubic bone and alongside the bladder to end by joining the seminal vesicle near its entry to the PROSTATE GLAND. The vas deferens conveys spermatozoa from the testicle to the seminal vesicle. See also VASECTOMY.

vasectomy the common operation for male sterilization. The VAS DEFERENS is exposed on each side through a short incision, just below the root of the penis, and is cut through and the ends tied off and secured well apart. Following this no newly produced spermatozoa can reach the exterior. The operation can be reversed but fertility is not always restored.

vaso- *combining form denoting* a vessel, especially a blood vessel.

vasoactive affecting blood vessels.

vasoconstriction active narrowing of small arteries as a result of contraction of the circular smooth muscle fibres in their walls. This severely reduces the flow of blood through them. Compare VASODILATATION.

vasodilatation widening of blood vessels as a result of relaxation of the muscles in the walls. This allows a greater volume of blood to pass through in a given time. Compare VASOCONSTRICTION.

vasodilation see VASODILATATION.

vasomotor pertaining to the control of the muscles in the walls of blood vessels and hence the rate of blood flow.

vasopressin a hormone secreted in the HYPOTHALAMUS and stored in and released from the PITUITARY GLAND. Vasopressin controls water retention by the kidneys and thus the water content of the body. Deficiency of vasopressin causes diabetes insipidus.

vasospasm tightening or spasm of blood vessels.

vasovagal pertaining to the VAGUS NERVE and to its effects on blood vessels.

VD *abbrev. for* venereal disease, now, for reasons of political correctness, largely obsolete. See SEXUALLY TRANSMITTED DISEASES.

vector an animal such as an insect, capable of transmitting an infectious disease from one person to another. The disease organism develops and multiplies in the vector and may pass through various stages, or may even be transmitted through one or more generations of the vector, before being passed on to a human host. From the Latin *vectus*, one who carries.

vegetarianism the policy of deliberate exclusion of animal muscle protein and sometimes of other animal products, such as eggs and milk, from the diet. Strict vegetarians, such as vegans (who avoid eggs and dairy products), must exercise care to ensure that all essential elements are present in the diet.

vegetative state see PERSISTENT VEGETATIVE STATE.

VEGF *abbrev. for* VASCULAR ENDOTHELIAL GROWTH FACTOR.

vein thin-walled blood vessel containing blood at low pressure which is being returned to the heart from tissues that have been perfused by arteries.

vein disorders these include haemorrhoids, phlebitis, thrombophlebitis, varices, varicocele and varicose veins.

vena cavae the largest veins in the body. The superior vena cava drains blood from all parts of the body above the level of the heart, the inferior vena cava from all parts below the heart. The two veins empty into the right atrium of the heart.

venene a mixture of snake venoms used to produce a general antidote (antivenin).

venereal pertaining to love. The term is seldom, if ever, used in its strict sense and has come to be indissolubly associated with disease spread during coitus. Today, however, even in this context the word has a very old-fashioned, even archaic, ring and has been almost universally replaced by the less emotive 'sexually-transmitted'. From the Latin *Venus*, the Roman goddess of love.

venereal diseases see SEXUALLY TRANSMITTED DISEASES.

venereology the medical speciality concerned with the sexually transmitted diseases. Now usually known, for purposes of euphemism, as genitourinary medicine (GUM).

venepuncture entry into a vein, usually with a hollow needle so as to gain access to the bloodstream for the purpose of obtaining a sample of blood or giving an injection directly into it.

venesection cutting of a vein for the purposes of removing blood. This is done to obtain blood for transfusion, or, rarely, to treat conditions, such as polycythaemia, haemochromatosis and porphyria.

venom poison produced by scorpions, some jellyfish, some fish, a few snakes, some toads, the Gila monster, some spiders and a few insects such as bees, wasps or hornets. Venoms act in various ways and may affect either the nervous system, to cause paralysis, or the blood to cause either widespread clotting or bleeding. Venoms are seldom fatal except in very young or debilitated people.

ventilator a mechanical air or oxygen pump used to maintain breathing in a paralysed, deeply anaesthetized or brain-damaged person unable to breathe spontaneously. Ventilators provide an intermittent flow of air or oxygen under pressure and are connected to the patient by a tube inserted into the windpipe (trachea) either through the mouth or nose or through an opening in the neck (a tracheostomy).

ventouse the suction equipment used for assisting in childbirth. See VACUUM EXTRACTION.

ventral pertaining to the front of the body. From the Latin *venter*, the belly. Compare DORSAL.

ventricle a cavity or chamber filled with fluid, especially the two lower pumping chambers of the heart and the four fluid-filled spaces in the brain.

ventricular pertaining to a VENTRICLE.

venule a very small VEIN.

vergence 1 movement of one or both eyes so that the visual axes converge or diverge. 2 the effect caused on a parallel beam of light by a convex (converging) or concave (diverging) lens.

vermi- *combining form denoting* worm.

vermiform appendix see APPENDIX.

verminous infected with worms or infested with ectoparasites (vermin).

vernix an abbreviation of vernix caseosa, a layer of greasy material, skin scales and fine hairs with which fetuses and new-born babies are covered. Vernix is easily washed off after birth.

verruca a WART on any part of the skin.

verrucose, verrucous covered with warts or wart-like protrusions.

versicolor of various colours.

version 1 a procedure in obstetrics to turn the fetus in the womb into a position more suitable for delivery.
2 rotation of both eyes simultaneously in the same direction.

vertebra one of the 24 bones of the VERTEBRAL COLUMN.

vertebral column the bony spine. A curved column of bones, called vertebrae, of the same general shape but increasing progressively in size from the top of the column to the bottom. Each vertebra has a stout, roughly circular body behind which is an arch that encloses an opening to accommodate the spinal cord. The arch bears bony protuberances on either side and at the back and facets for articulation with the vertebrae above and below. The vertebral bodies are fixed together by cushioning intervertebral discs and strong longitudinal ligaments. There are 7 neck vertebrae, 12 in the back and 5 in the lumbar region. The 5th lumbar vertebra sits on top of the sacrum, which is formed from the fusion of five vertebrae. The coccyx, hanging from the lower tip of the sacrum, is the fused remnant of the tail.

vertebral column disorders these include ankylosing spondylitis, kyphosis, lordosis, osteoporosis, scoliosis, slipped disc, spina bifida, spondylolisthesis and spondylolysis.

vertex 1 the top of the head.
2 any apex or highest point on a body structure.

vertical transmission transmission, as of a hereditary characteristic or of a disease, from parent to offspring. Horizontal transmission is transmission between contemporaries or individuals of the same generation.

vertigo the illusion that the environment, or the body, is rotating. Severe vertigo causes the sufferer to fall. It may be due to TRAVEL SICKNESS, fear of heights, anxiety, alcohol, drugs or HYPERVENTILATION. Some cases of the most severe and persistent vertigo may be caused by disorders of the balancing mechanisms in the inner ears, such as Méniere's disease or labyrinthitis, or to disease of the cerebellum or its connections from insufficient vertebrobasilar blood supply, tumour or multiple sclerosis.

Vesalius, Andreas Andreas Vesalius (1514–64) the foremost 16th century anatomist and professor at the University of Padua, did not, at first, dare to challenge the writings of the great authority, Galen. But when artists, such as Leonardo da Vinci, who sketched accurately from human dissections and had no interest in medical dogma, showed that Galen was frequently wrong, Vesalius took heart. In 1543 he published his masterpiece, *De fabrica corporis humani*, a beautifully illustrated text on human anatomy based on meticulous dissections and observations of dead bodies. This great work directly confronted much of the error in Galen's teachings and showed that many of Galen's ideas had been derived from observations on animals and applied, without reservation, to humans.

Vesalius's book contained exact and carefully accurate descriptions of the gross anatomy of the whole body – the skeleton, the muscles, the heart and blood vessels, the nervous system and the organs. Quickly, the book became famous and its reputation widespread. Ambroise Paré, the outstanding surgeon of the 16th century, incorporated Vesalius's work into the anatomical section of his classic 1564 textbook on surgery. Paré was so enthusiastic about Vesalius's anatomical writings that he translated much of the book into French, greatly extending its influence. By the end of the century Vesalius had a monopoly on human anatomy. But his influence was much wider. Although his book contained nothing that was not already accessible to any patient dissector, it marked a radical change in method. From then on, speculative philosophy, dogmatic assertion and traditional authority had to give way to empirical observation.

Vesalius was succeeded at Padua and elsewhere by a number of distinguished men whose contributions to human anatomy and physiology remain essential to this day. Among these was Gabriel Fallopius who enlarged on Vesalius' work expanded much detail, and is remembered eponymously for his description of the uterine tube, and Hieronymus Fabricius who published a detailed description of veins including their valves. Fabricius contented himself with pure description and drew no conclusions from his finding, but his work was to have momentous consequences at the hands of the great William Harvey.

vesical pertaining to a bladder, especially the URINARY BLADDER.

vesication blistering.

vesicle 1 a small blister.
2 any small pouch, as the SEMINAL VESICLES, the small bladders which store semen.
From the Latin *vesiculum*, the diminutive of vesica, a bladder or bag.

vessel any closed channel for conveying fluid.

vestibular apparatus the system of semi-circular canals, utricle and saccule in the inner ear, lying within the temporal bone, that provides the functions of hearing and balance.

vestibule a space or cavity forming the entrance to another cavity.

vestige a body structure with no current apparent function which appears to have had a function at a previous evolutionary stage.

vestigial pertaining to a VESTIGE.

V gene a gene that codes for the main part of the VARIABLE REGION of an antibody.

vibration-induced disorders a range of disorders that includes deafness, vibration-induced neuropathy, Raynaud's phenomenon ('white finger'), and, in the case of whole body vibration, motion sickness, low-back pain, visual disturbances and insomnia. The hand-arm vibration syndrome (HAVS) features spasm of blood vessels with white fingers, sensory and motor nerve damage and even muscle, bone and joint changes. This syndrome was recognized by international agreement in 1985.

vibrator an electrically driven reciprocating device used to apply low-frequency repetitive force for the purpose of massaging any part of the body.

villous 1 pertaining to villi (see VILLUS).
2 featuring numerous fingerlike processes.

villous atrophy flattening and disappearance of the finger-like absorptive processes of the small intestine that is a feature of coeliac disease. Villous atrophy is associated with an increased density of LYMPHOCYTES in the bowel lining (intraepithelial lymphocytes), but whether they cause it is uncertain.

villi small finger-like processes on a surface, as in the small intestine, the PLACENTA, the tongue and the CHOROID PLEXUSES of the brain.

villin an actin-binding, severing and bundling protein found in the MICROVILLI of the apical membranes of certain cells, including the canalicular microvilli of hepatocytes. Abnormalities in the expression of the gene for villin, resulting in the absence of this protein are thought to be a cause of such liver disorders as biliary atresia.

vinculum a slender connecting band.

violaceous of a violet or purple colour.

viraginity 1 male-like psychology in a woman. 2 of a violent, ill-tempered personality in a woman (a virago).

viraemia the presence of viruses in the blood.

viral pneumonia inflammation of the lung (PNEUMONIA) caused by a virus infection.

virgin a person who has never had sexual intercourse. From the Latin *virgo*, a maiden, so should, strictly, apply only to a female. The claimed physical sign of virginity, an intact HYMEN, cannot always be relied upon because the part is subject to wide anatomical variations and to trauma. Both the term and the importance placed on an intact hymen are residua of a male-dominated past.

virgin birth conception and child-bearing by a woman who has never had sexual intercourse, as a result of artificial insemination with donated semen.

viricidal capable of killing VIRUSES.

virilism masculinization in the female. This may occur in tumours of the adrenal glands or ovaries which secrete abnormal quantities of the male hormones, the androsterones, or in the condition of congenital adrenal hyperplasia. There is increased growth of body hair, balding at the temples, acne, absence of menstruation, enlargement of the clitoris, increased muscular development and deepening of the voice. The treatment involves removal of the cause.

virility the quality of sexual maleness, strength, vigour and energy.

virion a complete virus particle, as found outside cells, and consisting of the genetic material and the surrounding capsid.

virology the study of viruses of medical importance and the diseases they produce.

virotherapy a new treatment modality in which viruses are used to detect cancerous cells, invade them, and replicate in them in such a way as to kill them. Viruses that bind only to proteins found on tumour cells may be used, or they may be modified so that their genes will transcribe only in tumour cells. Virotherapy may also involve the transport of cytotoxic drugs selectively into tumour cells. It has produced hopeful results in animals and clinical trials on humans have begun.

virtual patient a highly complex software model consisting of representations of the hundreds of different metabolic or immunological states that occur in the organs of human patients with a particular disease. In theory, and perhaps soon in practice, the method can be used to test new drugs by applying the detailed technical description of the drug to the program and observing its effect. Simulation of diabetes has already been achieved and tested for its validity.

virucidal-anhidrotics drugs used to remove warts. An example is glutaraldehyde (Glutarol).

virulence the capacity of any infective organism to cause disease and to injure or kill a susceptible host.

viruses infectious agents of very small size and structural simplicity, all of which are smaller than the smallest bacterium. They consist of a core of nucleic acid, either DNA or RNA encased in a protein shell. Viruses can maintain a life-cycle and reproduce only by entering a living cell and taking over part of the cell function. All living cells are believed to be susceptible to virus infection. The most important virus diseases are AIDS, arthropod-borne fevers, aseptic meningitis, Burkitt's lymphoma, chickenpox, cold sores, the common cold, cytomegalovirus inclusion disease, epidemic keratoconjunctivitis, equine encephalitis, some forms of gastroenteritis, glandular fever, influenza, Lassa fever, measles, molluscum contagiosum, mumps, ORF, parainfluenza, poliomyelitis, progressive multifocal leucoencephalopathy, rabies, SARS, shingles, vaccinia, warts and yellow fever.

virus interference protection of cells against virus infection, as a result of prior virus infection of neighbouring cells. This stimulates the production of proteins, called INTERFERONS, which become attached to the membranes of other cells and prompt them to produce enzymes which interfere with replication of subsequent viral invaders. Interferons also stimulate killer LYMPHOCYTES to attack and destroy cells which have been invaded with viruses.

viscera organs within a body cavity, especially digestive organs. The singular form of the word is viscus.

visceral leishmaniasis see KALA AZAR.

visceroptosis downward displacement or sagging of an organ or organs within a body cavity, especially the intestines, due to loss of support.

viscous of a liquid substance, thick and sticky so that there is resistance to flow.

vision

Very little remains to be discovered about the way in which the eye works. We also know a great deal about the neurological connections between the eye and the brain – the optical pathways – and about the mapping of the fields of vision in the cortex of the brain. But as to the physiological events that underlie the actual experience of vision, we know almost nothing at all.

The eye and the brain

Between the object of perception and the experience of vision lies a remarkable pathway in which information alters its form and nature several times. We see by virtue rays of light of varying intensity and direction, reflected from the surface of, or transmitted through, external objects. Some of this light enters the eye and is focused on the retina where it stimulates millions of photoreceptor cells to produce nerve impulses. These are coordinated in the retina and pass out of the back of the eye along the million or so separate fibres of the optic nerve. Some of these fibres from each eye remain on the same side, some cross over to the other side. All of them end in one of half a dozen layers in a pair of junction boxes on the underside of the brain, known as the geniculate bodies.

In each geniculate body, some layers are connected to one eye, some to the other. These layers consist of synapses between the nerve fibres (axons) arising in the retina and a new set of nerve cells whose axons run in great sweeping curves through the substance of the brain, right to the back. There, in the outer layer (the occipital cortex) lie a further set of nerve cells that are, in some way, concerned with the processes of visual experience. The link between the synapses in the geniculate body layers are made as a result of visual experience in childhood. If one eye is covered for the first few months of life, the layers, on each side, corresponding to that eye do not develop properly and the eye remains permanently blind.

Binocular vision requires that the two eyes should work together, aligning themselves with high precision on the object of interest so that the image of the object formed on each retina should fall upon corresponding areas of the two retinas. This is a demanding requirement and it is achieved under exacting brain control, by means of which the contractions of the six tiny muscles that move each eye are coordinated by a computing nerve network in the brainstem. The eyeballs can rotate freely as they are embedded in pads of fat within the orbits – bony caverns in the skull which provide protection from injury.

Like a closed-circuit TV camera, the eye contains two main parts – an optical, image-forming and focusing lens system consisting of the cornea, the iris diaphragm and the internal crystalline lens – and a transducer, the retina, which converts the images falling on it into patterns of frequency-modulated electrical signals. These pass back to the rear part of the brain by way of the visual pathways – the optic nerves, tracts and radiations.

Focusing is effected by an automatic process known as accommodation – an adjustment of the optical power of the crystalline lens lying immediately behind the pupil. This lens is elastic in young people and, if allowed to relax, assumes an almost spherical shape. It is, however, stretched into a flatter shape by delicate protein strands around its equator that are pulled radically to a muscle ring, the ciliary muscle, that surrounds the lens. When this muscle ring contracts, the tension on the supporting strands is released and the lens surfaces become more curved by its own elasticity. This allows the lens to converge the rays from a near object, which are more divergent than rays from a distant object. The stimulus for accommodation arises from the detection by the retina of the vergence of the rays passing through it.

Retina

The images that are formed on the retina by the optical system are inverted and are bit-mapped. As a result of the almost one-to-one correspondence between points on the retina and points on the visual cortex of the brain, the image can be considered as being represented there, also, as a bit-mapped image. The retina, however, is not simply a passive light-to-nerve impulse transducer. Its photocells, the rods and cones, are interconnected in such a way as to form a kind of computer that increases the range of contrast sensitivity and codes the output signals.

The rods are more sensitive to light than the cones but are colour-blind and are most concentrated at the edge of the retina. The cones are more numerous at the centre of the retina and are colour-sensitive. They are of three types, each type giving its maximum nerve impulse output when one of the three primary colours falls upon it. In this way, colour information is coded in the patterns of nerve impulses passing along the optic nerve fibres. The procedure is analogous, in reverse, to the way a colour image is produced in a TV tube. The concentration of photocells is greatest at the centre of the retina and only here, at the macula lutea, is high resolution vision possible. But the visual 'fixation' reflexes ensure that the eyes are always aligned accurately on objects we wish to see and in this way the image always falls on the macula. It is impossible to read normal print if the gaze is steadily fixed even a few degrees to one side of it.

Eye optical defects

Optical defects arise in various ways. The commonest result from a disparity between the focal length of the lens system and the axial length of the eye. If the lens is relatively too powerful the image of distant objects will tend to form in front of the retina. Only the light rays from near objects are sufficiently divergent to focus on the retina. This condition is called short-sightedness, or myopia, and is usually due to excessive curvature of the cornea or to excessive axial length of the eyeball. If the lens is relatively too flat or the eyeball too short, the eye is hypermetropic and extra focusing is needed to see clearly. If the cornea has a greater degree of curvature in any one meridian than in other meridia, the result is astigmatism. The effect of this is that the eye cannot sharply focus image lines lying in all orientations. If, for instance, vertical lines are sharp, horizontal lines will be blurred. The corneal meridia of maximal and minimal curvature need not be vertical and horizontal, however; they may lie in any two mutually perpendicular orientations.

The internal crystalline lens continues to secrete fibres throughout life and, in consequence, becomes 'tighter' and less elastic with increasing years. This results in a progressive loss of focusing power, usually manifesting itself by an inability to read comfortably around the middle forties. This is called presbyopia (Greek, *presbos*, an old man).

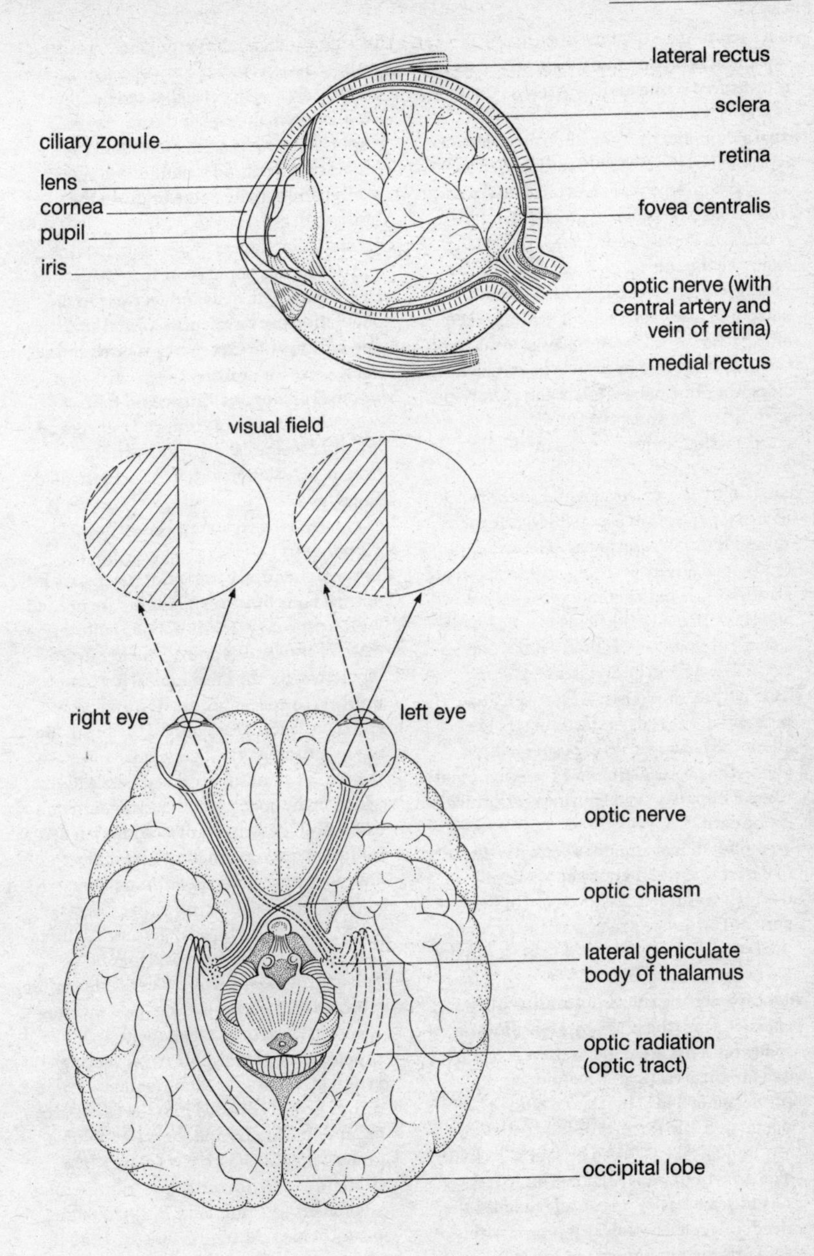

Fig. 20 **Vision**

visual acuity the extent to which an eye is capable of resolving fine detail. Visual acuity is measured by means of a SNELLEN'S CHART TEST.

visual aids optical devices, often of a telescopic type, used by the visually handicapped to assist vision. Also known as low visual aids.

visual evoked responses a method of modifying the ELECTROENCEPHALOGRAM by exposing the subject to visual stimuli. The most useful is a reversing chessboard pattern in black and white. The time taken after each reversal for the change to occur on the EEG is affected by defects of conduction along the optic nerves, tracts and radiations, such as may be caused by MULTIPLE SCLEROSIS. Unlike visual acuity tests, the method is purely objective.

visual fields the area over which some form of visual perception is possible while the subject looks straight ahead. The visual fields normally extend outwards to about 90° on either side but are more resticted below and above. Parts of the fields may be lost as a result of GLAUCOMA, retinal damage, optic nerve disease and brain disease.

visual purple rhodopsin. A light-sensitive pigment in the rods of the retina that is chemically changed on exposure to light and in the process stimulates production of a nerve impulse. Visual purple is reformed in the dark.

visuscope an instrument used in ORTHOPTICS to determine whether or not a patient is using the central macular region of the RETINA. The device projects an image on to the retina and the relationship of this to the FOVEA can be observed.

vital capacity the volume of air that can be expelled from the lungs by a full effort following a maximal inspiration.

vitalism the now largely abandoned philosophical idea that some kind of postulated 'life force' is necessary before any biological entity can be said to be living. The growth of molecular cell biology in recent years has progressively reduced the need for such a notion as it progressively explains the phenomena of life in physicochemical terms.

vital signs indications that a person is still alive. Vital signs include breathing, sounds of the heart beat, a pulse that can be felt, a reduction in the size of the pupils in response to bright light, movement in response to a painful stimulus and signs of electrical activity in the brain on the ELECTROENCEPHALOGRAM.

vital statistics figures of births, marriages and deaths in a population, from which the rate of natural increase or decrease in the population can be calculated. Vital statistics also indicate life expectancy at birth and the main causes of death.

vitamin D analogues drugs used to treat nutritional malabsorption and rickets and osteomalacia resistant to vitamin D treatment. Examples are alfacalcidol (Alfad , One Alpha), calcitriol (Calcijex, Rocaltrol), calcipotriol (Dovonex) and tacalcitol (Curatoderm).

vitamins chemical compounds necessary for normal body function. Vitamins are needed for the proper synthesis of body building material, HORMONES and other chemical regulators; for the biochemical processes involved in energy production and nerve and muscle function; and for the breakdown of waste products and toxic substances. The B group of vitamins are COENZYMES without which many body ENZYMES cannot function normally. The amount of vitamins needed for health are very small and are almost always present in adequate amounts in normal, well-balanced diets. Excess intake of vitamins A and D is dangerous. Vitamins C and E are antioxidants and may be valuable, in doses many times the minimum requirement, in combatting the damaging effect of FREE RADICALS. Folic acid supplements are valuable in preventing NEURAL TUBE DEFECTS. Vitamins are conventionally divided into the fat-soluble group A,D,E and K, and the water-soluble group, vitamin C (ascorbic acid) and the B vitamins – B_1 (thiamine), B_2 (riboflavine, riboflavin), nicotinic acid, B_6 (pyridoxine), pantothenic acid, biotin, folic acid and B_{12}. The term was derived from the mistaken belief that vitamins were 'vital amines'.

vitreous body the transparent gel that occupies the main cavity of the eye between the back of the CRYSTALLINE LENS and the RETINA.

vitreous detachment separation of the rear part of the VITREOUS BODY from the retina as a result of the natural shrinkage that occurs in the elderly. Perception of floating specks or moving clouds may be a conspicuous, but often temporary, feature of the process.

vocal cords a pair of pearly-white shelves of thin mucous membrane stretched across the interior of the LARYNX and capable of being tensioned to a widely varying degree by small laryngeal muscles. The vocal cords are caused to vibrate by the outwards passage of air from the lungs and the sound so produced is modulated by changes in the shape and volume of the mouth cavity to produce speech and song.

vocal cord disorders these include laryngitis, recurrent laryngeal nerves and singer's nodes.

voltage-gated channels protein channels in cell membranes that open or close under the influence of changes in the electrical charge on the membrane.

voluntary muscle striped muscle normally contracted in the course of volitional activity.

volar pertaining to the palm of the hand or the sole of the foot.

volvulus twisting of a loop of intestine. Volvulus causes obstruction to the flow of contents and threatens occlusion of the supplying blood vessels. This will inevitably lead to GANGRENE of the affected segment of bowel unless quickly relieved by surgery. This may involve removing the affected loop of bowel and joining up the free ends.

vomer a thin, flat plate of bone that forms the rear part of the partition of the nose (the nasal septum).

vomiting involuntary upward expulsion of the stomach contents. Vomiting is prompted by the presence of vomit-stimulating substances in the blood and effected by sudden, forceful downward movements of the DIAPHRAGM and inward movement of the abdominal wall. The stomach wall muscle plays no part.

vomitus material vomited.

von Willebrand factor a plasma protein secreted by endothelial cells that helps platelets to adhere to the damaged lining of blood vessels. (E. A. von Willebrand, Swedish physician 1870–1949).

vorticose veins four veins in the CHOROID coat of the eye, the branches of which have a whorled appearance.

voyeurism covertly observing people undressing or engaging in sexual intercourse, so as to obtain sexual stimulation. Voyeurism is a male activity, engaged in by the lonely and the socially inadequate, and is usually accompanied by MASTURBATION.

vulva the female external genitalia, comprising the mons pubis, the two pairs of LABIA, the area between the labia minora, and the entrance to the VAGINA.

vulval disorders these include bartholinitis, chancre, genital herpes, kraurosis vulvae, sexually transmitted diseases, thrush, trichomoniasis and vulvovaginitis.

Ww

waist the part of the trunk between the lower ribs and the pelvis.

waist circumference a measurement that has been shown to be a valid index identifying people who need weight management if they are to avoid a significant risk of heart attacks. Waist circumferences of more than 94 cm in men and more than 80 cm in women indicate danger.

walking aids supports for people with muscle weakness, joint disease or balancing problems. They include plain walking sticks, sticks with three or four small feet, light alloy Zimmer frame 'walkers', elbow crutches and walking calipers.

Wallace, Alfred Russel the Welsh amateur naturalist Alfred Russel Wallace (1823–1913) was born in Usk, Monmouthshire. He was not a professional scientist but was so intensely interested in natural history that he went on expeditions to the Amazon and to Malaya to collect specimens. His published accounts of these voyages were excellently done and were widely read. He recorded how struck he was by the sharp differences between the features of identical species that were separated by long stretches of water. He compared the characteristics of various species in Australia with those of the same species in Asia. Like Darwin, Wallace was influenced by the ideas of Malthus on the control of populations, and in 1858 he wrote a paper outlining the principle of evolution by natural selection and sent it to Darwin.

Darwin was a man of character and great generosity. When he read Wallace's paper that so closely embodied his own ideas, he made no attempt to publish *The Origin of Species* quickly, but magnanimously circulated Wallace's paper to other interested scientists. He even collaborated with Wallace in the first presentation of the theory to the Linnaean Society. It was Darwin's book, however that caught the public's attention and made Darwin famous – some, at the time said infamous – and Wallace's contribution, which certainly ranked with Darwin's, was largely ignored.

Wallace believed in spiritualism and was interested in psychic research. Although he fully accepted evolution by natural selection, he could never bring himself to believe that the higher intellectual functions of the human being had developed by evolution from earlier animal ancestors. He was convinced that these had occurred by some miraculous process.

wall-eyed having a divergent squint (STRABISMUS) or a large, white scar on the cornea.

wan unnaturally pale.

warts verrucas, non-malignant, localized skin excrescences caused by different strains of more than 130 human papilloma viruses. These stimulate overgrowth of the prickle cell layer at the base of the EPIDERMIS of the skin, resulting in excessive local production of the horny material KERATIN. Warts on the soles (plantar warts) are forced into the skin by the weight of the body.

water the oxide of hydrogen. Water is essential for life and provides about 70 per cent of the body weight in lean people and about 50 per cent in the obese. The body of the average 70 kg man contains about 40 l of water. Just over half the total body water is within the cells and the remainder is outside, partly in the blood, but mainly in the tissue spaces surrounding the cells. Water molecules are very small and move freely across cell membranes. Water is lost from the body in the urine, in evaporation from the skin, in the expired air and in the faeces. Losses are reduced automatically if there is reduced intake. Restricting water intake is dangerous especially in hot conditions.

waterbrash sudden, unexpected secretion of a quantity of saliva into the mouth as a reflex response to symptoms of dyspepsia.

water intoxication the effect of excessive water retention in the brain in the course of any disorder causing general OEDEMA. The condition features headache, dizziness, confusion, nausea and sometimes seizures and coma. Treatment is the correction of the cause and measures to withdraw water from the brain into the blood.

Watson-Crick model the double helix concept of the DNA molecule, proposed by two researchers in Cambridge in 1953, which triggered off a revolution in biology and medicine and led to an explosive succession of advances in genetics. (James D. Watson, b. 1928, American molecular biologist; and Francis H. C. Crick, 1916–2004, English biochemist and neurophysiologist).

Watson, James Dewey James Dewey Watson was born in 1928 in Chicago and enrolled in the University of Chicago at the early age of 15. He graduated in 1947 and proceeded to postgraduate work on viruses at Indiana University, where he was awarded a PhD in 1950. His initial interests had been in ornithology, but from 1950–51 he continued working on virus research at the University of Copenhagen. While there, he became convinced that the key to the understanding of heredity lay in the chemical structure of DNA, and he resolved to work on the problem. Accordingly, at the age of 23, he moved to the Cavendish Laboratory at Cambridge, where he began to study the possible structure of DNA. His work with Francis Crick in establishing the structure made him one of the two most important figures in the entire history of biology.

Soon after the publication of the *Nature* paper in 1953 which revealed the structure of DNA to the world, Watson moved to the California Institute of Technology, where he worked until 1955. He then moved to Harvard where he did research on molecular biology and where, in 1961, he was appointed Professor of Biology. In 1965 his book *Molecular Biology of the Gene* was published, and in 1968 he published *The Double Helix*, a personal account of the discovery of the structure of DNA and of the other people involved.

In 1968 Watson became Director of the Cold Spring Harbor Laboratory of Quantitative Biology, New York, where the work was mainly concentrated on cancer research. In 1981 he published *The DNA Story*. In 1988 Watson became Director of the Human Genome project at the National Institutes of Health, Washington – the largest single endeavour in biological science – which, early in the 21st century, was to complete the sequencing of the entire genetic chromosomal structure of the human being. He remained in that appointment until the Spring of 1992.

WBC *abbrev. for* white blood cell.

weak acid an acid with molecules of which that do not completely ionize to produce hydrogen ions when dissolved in water.

weakness a state of debility caused by prolonged bed rest, muscle disease or wasting, severe infection, anaemia, starvation or psychological disorder with loss of motivation. Once causes have been removed, the only cure for weakness is activity.

weaning substitution of solid foods for milk in an infant's diet.

webbing edge-to edge joining of the fingers or toes by flaps of skin. This is a common congenital abnormality and is easily corrected by surgery.

web-fingered see WEBBING.

weight loss the effect of an absorbed calorie intake that is smaller than the calorie expenditure. Weight loss is a feature of anorexia nervosa, cancer, depression, diabetes, persistent diarrhoea, deliberate dieting, malabsorption, starvation, thyrotoxicosis, tuberculosis and persistent vomiting. Unexplained weight loss is a warning sign of possible serious disease and should never be disregarded.

well-man a healthy man who attends a clinic or surgery to ensure that his general health, lifestyle, and sexual performance are satisfactory.

Wernicke, Carl Carl Wernicke (1848–1905) was a young professor of anatomy at Breslau. Stimulated by Paul Broca's demonstration of a case of speech defect arising from temporal lobe brain damage, Wernicke decided to investigate the temporal lobe of the left side of the brain by careful post-mortem study of the brains of people with a known history of acquired speech defects. During the 1870s he developed a theory of speech function that is generally accepted today and that has had an important influence on ideas of brain function generally.

Broca had already shown that there was an area near the front of the temporal lobe, for speech movements. Wernicke now showed that there were areas further back, connected to Broca's area, which were concerned with the more mental aspects of speech. Wernicke's published work includes a full description of the effects of brain damage in these areas – effects known as 'receptive' or 'sensory' aphasia. Such damage affected the understanding of spoken language or the ability to recognize or to understand the meaning of printed or written words (alexia). Alternatively, it might destroy the ability to produce meaningful writing (agraphia). In some cases, while speech remained fluent, the effect of the damage was to deprive it of meaning. All this was in marked contrast to Broca's aphasia in which comprehension was normal but the affected person was unable to speak.

Wernicke came to the conclusion that speech, and other higher functions, could not be assigned to definite single areas of the brain cortex as in the case of vision, hearing, smell and tactile sensation. These higher functions depended on the complex links between all these areas and between them and various parts of the temporal lobe. Wernicke's ideas have been supported by much subsequent research, and the areas he described and the aphasia resulting from their damage now both bear his name.

Western blotting a method of detecting very small quantities of a protein of interest in a cell or body fluid. The sample is spread by ELECTROPHORESIS on a block of polyacrylamide gel. The separated proteins are then transferred, by blotting, to a thin plastic sheet to make them more accessible for reaction with an antibody specific for the protein of interest. The antibody-antigen complex can then be detected by rinsing the sheet with a second, radioactive, antibody that recognizes the first. The sheet will now produce a dark band on X-ray film. See also SOUTHERN BLOTTING.

wet dream a popular term for an erotic dream culminating in a spontaneous orgasm and ejaculation of semen.

wet nurse a woman who breast feeds another woman's child.

whiff test a semi-humourous term applied to the test for *Gardnerella vaginalis* organisms. A drop of 10 per cent potassium hydroxide is added to a drop of vaginal discharge on a microscope slide. If the test is positive a distinct fishy smell, from the production of amines, becomes apparent. The test is not completely specific but is quick and of practical value.

whiplash injury a neck injury caused by the application of sudden accelerative or decelerative forces to the body so that the neck bends acutely in a direction opposite to the direction of the force. This results in immediate reflex contraction of the stretched muscles so that the head is jerked in the other direction. Neck ligaments may be stretched or torn or a neck VERTEBRAE may be fractured. There is pain and disability, often for weeks, and an orthopaedic collar may be needed.

white blood cell see LEUKOCYTE.

white matter those parts of the central nervous system that appear white on section because they consist mainly of myelinated nerve fibres. Compare grey matter which consists mainly of nerve cell bodies.

WHO *abbrev. for* WORLD HEALTH ORGANIZATION.

whole blood blood, usually for transfusion, from which no constituent has been removed.

whole body MRI the scanning of the entire body as a single event by magnetic resonance imaging. The technique has been found valuable for detecting skeletal and other cancer metastasis, as a means of whole body fat measurement and as an acceptable alternative to conventional autopsy. Critics have referred to the anxieties caused by false positive findings.

wild relating to an entity, such as a virus, bacterium or gene that arises naturally or that comes from a natural environment, rather than that originates in a laboratory or as a result of artificial circumstances.

Wilkins, Maurice Maurice Wilkins (1916–2004) was born in New Zealand and studied at St John's College, Cambridge. He was an intensely private and self-effacing man who was much loved by colleagues and students. Working at King's College, London in 1950, he was the first to use X-ray diffraction techniques to try to elucidate the chemical structure of DNA. The images produced by his group – which included the brilliant experimentalist Rosalind Franklin – were of unprecedented clarity and played an essential part in the recognition by Francis Crick and James Watson at Cambridge that the DNA molecule was a double helix. The roles of Wilkins and Franklin, which were crucial, have not always been fully acknowledged outside the scientific community. Wilkins was awarded the Nobel Prize in 1962 with Crick and Watson. His autobiography, *The Third Man Of The Double Helix*, was published in 2003.

wind a popular term for the result of air swallowing by greedy babies. Air swallowed along with a feed becomes compressed by PERISTALSIS and may cause colic and much crying. Slower feeding, dill water and silicone polymer oils, to reduce surface tension and form froth, are helpful.

windburn inflammation of the skin caused by exposure to hot dry wind. The condition can be prevented by covering exposed areas of skin.

wind chill the cooling effect of wind at low temperatures. This is greater than the effect of ambient cold alone and may cause rapid heat loss, adding to the risk of HYPOTHERMIA.

windpipe see TRACHEA.

wisdom tooth a popular term for the rearmost tooth in each of the four quadrants of the jaws. The third molar. Usually, the third molars do not erupt until the ages of 17–21, but often one or more is unable to emerge fully from the gum because of overcrowding. This is called an impacted wisdom tooth.

witches' milk brief milk production from the breasts of newborn babies of either sex due to the presence of the hormone PROLACTIN in the mother's blood. This passes through the placenta to the fetus before birth. The effect is harmless and soon wears off.

withdrawal bleeding bleeding from the lining of the womb (UTERUS) caused by withdrawal of the female sex hormones progesterone or oestrogen. This occurs naturally in menstruation, but is a feature of cessation of any treatment with these hormones, for any purpose, including contraception.

withdrawal syndrome the complex of symptoms experienced on withdrawal of a drug on which a person is physically dependent. Symptoms of heroin withdrawal include craving for the drug, restlessness, depression, running nose, yawning, pain in the abdomen, vomiting, diarrhoea, loss of appetite, sweating and gooseflesh ('cold turkey'). Those caused by withdrawal of other narcotic drugs are similar but less intense.

Wolffian duct a part of the embryo that develops into the ductal organs of the reproductive system in the male, but that degenerate in the female. (Kaspar F. Wolff, German embryologist, 1733–94).

womb see UTERUS.

word blindness see DYSLEXIA.

worried well people who do not need medical treatment but who visit the doctor to be reassured.

wound any injury involving a break in the surface of the skin or an organ by any means including surgical incision.

wrist the complex, many-boned joint between the hand and the arm. The eight wrist bones, or carpals, are arranged in two rows, the nearer row, which articulates with the forearm bones, containing the scaphoid, lunate, triquetral, and pisiform bones, and the farther row the trapezium, trapezoid, capitate, and hamate. These are connected
to the bones of the palm, the metacarpals. Many tendons, connecting forearm muscles to the fingers and thumb, run through the wrist. These pass under ligamentous straps (retinacula) which prevents them from springing away from the wrist. Arteries and nerves also pass through the wrist.

wrist disorders these include carpal tunnel syndrome, colles' fracture, scaphoid fracture and tenosynovitis.

writer's cramp a psychological disorder causing spasm of the muscles involved in holding a pen or pencil so writing becomes impossible. Other activities using the same muscles are usually unaffected. The condition is probably due to a mistaken vocation. Writers using word processors may develop a REPETITIVE STRAIN INJURY.

xanth-, xantho- *combining form denoting* yellow.

xanthelasma cholesterol deposits in the eyelid skin, near the inner corner of the eye, appearing as unsightly, raised, yellow plaques that enlarge slowly. Xanthelasma does not necessarily imply raised blood cholesterol, but this should be checked to eliminate the dangerous condition of familial hypercholesterolaemia, of which it is a feature. Plaques of xanthelasma can easily be removed but tend to recur.

xanthoma a yellowish or orange mass of fat-filled cells occurring in the skin of people with various disorders of fat metabolism. Also known as generalized XANTHELASMA.

xanthomatosis a condition occurring in various, often hereditary, disorders of fat metabolism, in which cholesterol-containing fatty nodules (xanthomas) occur in different parts of the body, including the tendons, arteries, the skin, the corneas, the crystalline lenses, the internal organs and the brain. The effects depend on the site of the deposits and may include mental deficit.

xanthopsia yellow vision. This sometimes occurs in JAUNDICE or in poisoning with digitalis.

X chromosome the CHROMOSOME which, with the Y chromosome, determines the sex of the individual. About 50 per cent of sperm carry an X chromosome and 50 per cent a Y. The sex of the future child is determined by whether an X-carrying or a Y-carrying sperm happens to fertilize the ovum. The ovum carries only an X chromosome. Females have two X chromosomes in each body cell, males have one X and one Y. The X chromosome is large and contains about 6 per cent of the genomic DNA. The Y is about half the size. Well over a hundred disorders are known to be determined by genes on the X chromosome. These are called X-linked conditions. See also X-INACTIVATION.

xeno- *combining form denoting* foreign or strange.

xenogeneic pertaining to the genetic differences between different species.

xeno-oestrogens substances with oestrogenic properties derived from other than biological sources. A number of organochloride compounds, for instance, have been found to be sufficiently oestrogenic to cause a rise in the incidence of breast cancer in groups of women exposed to them. The insecticide dieldrin is a case in point.

Xenopsylla a genus of fleas of which the species *X. cheopis*, the oriental rat flea, is the vector of plague and murine typhus.

xero- *combining form denoting* dry.

xeroderma dryness of the skin.

xerophthalmia dryness of the eyes with thickening of the CONJUNCTIVA, occurring in vitamin A deficiency, pemphigus and autoimmune disorders such as Sjogren's syndrome. Artificial tears must be used constantly to maintain the essential film of water over the cornea.

xerosis dryness, especially of the eyes, mouth, vagina or skin.

xerostomia dry mouth.

X-inactivation the normal failure of expression of one of the two X chromosomes in females. Early in development some cells switch off the paternal X chromosome, other cells switch off the maternal one. Inactivated chromosomes remain so in all subsequent daughter cells. Most women have a mixture of two different cells populations each expressing a different X chromosome. The inactivated chromosome is visible microscopically as the Barr body. This effect, sometimes called Lyonization after the British geneticism Mary Frances Lyon (1925–), who proposed it in 1961, accounts for a number of observed phenomena in genetics.

xiphisternum the flat, leaf-like process hanging down from the lower end of the breastbone (sternum). The xiphisternum is cartilaginous in childhood, partly bony (ossified) in adult life and wholly ossified and fused to the sternum in old age. Also known as the xiphoid process.

X-linked pertaining to genes, or to the effect of genes, situated on the X CHROMOSOME. X-linked disorders are those caused by mutated genes on the X chromosome. They include agammaglobulinaemia, albinism, Alport syndrome, Charcot-Marie-Tooth peroneal muscular atrophy, colour blindness, diabetes insipidus, ectodermal dysplasia, glucose-6-phosphate dehydrogenase deficiency, Fabry disease, glycogen storage disease VIII, gonadal dysgenesis, haemophilia a, one form of hydrocephalus, hypophosphataemia, ichthyosis, Turner's syndrome, one form of mental retardation, Becker and Duchesse muscular dystrophy, one form of retinitis pigmentosa and the testicular feminization syndrome.

X-linked recessive pertaining to a gene situated on an X chromosome which is expressed if the chromosome is carried on both X chromosomes in a female (which is necessarily rare). In males, however, the Y chromosome carries little or no genetic material and does not contain the normal ALLELE, so the gene on the X chromosome will always manifest itself. An X-linked recessive condition will thus usually occur only in males (who have one X and one Y chromosome) but cannot be transmitted by a father to his son because the son receives only the Y chromosome. The characteristic is, however, transmitted via the daughters, who are carriers. Their sons have a 50/50 chance of acquiring the X chromosome and manifesting the characteristic.

XO configuration the state of the CHROMOSOMES when only one sex chromosome, an X, is present. This is the sex chromosome abnormality most commonly found in Turner's syndrome. Also known as monosomy X.

X-ray a form of electromagnetic radiation produced when a beam of high-speed electrons, accelerated by a high voltage, strikes a metal, such as copper or tungsten. X-radiation penetrates matter to a degree depending on the voltage used to produce it and the density of the matter. It acts on normal photographic film in much the same way as does visible light, but can also produce an image on a fluorescing screen. These properties make X-radiation valuable in medical diagnosis. X-rays are damaging to tissue, especially rapidly reproducing tissues, and can be used to treat various cancers .

XXX configuration the state of the body cells when an additional X CHROMOSOME is present. Females possessing this chromosome configuration are often mentally retarded but their children, if any, are usually normal.

XYY configuration a male chromosome abnormality in which an additional male sex chromosome is present in every body cell. This configuration has been found in normal men, but is often associated with mental retardation and criminal tendencies.

Yy

yawning an involuntary and often infectious act of slow, deep inspiration accompanied by an almost uncontrollable desire to open the mouth widely. The purpose of yawning remains a matter of speculation but it stretches and opens the air sacs of the lungs and helps to improve the return of blood to the heart, reducing blood stagnation and increasing its oxygen content.

year-and-a-day rule the long-established legal principle that a person cannot be held criminally liable for a death that occurs more than a year and a day after the commission of the act alleged to be its cause.

yeasts single-celled nucleated fungi that produce enzymes capable of fermenting carbohydrates. The yeasts of chief medical interest are those of the *Candida* and *Monilia* species which cause thrush. Yeasts are rich in B vitamins.

yellow spot the MACULA LUTEA of the RETINA.

yin and yang the opposite but complementary principles of Chinese philosophy incorporated into traditional Chinese medicine. Yin is feminine, dark and negative, Yang masculine, bright and positive. Their interaction and balance is claimed to maintain the harmony of the body.

yoga one of the six orthodox systems of Indian philosophy. In Hatha Yoga the emphasis is on physical preparation for spiritual development. It incorporates a series of poses, known as asanas, by which one may retain youthful flexibility and control of the body and achieve relaxation and peace of mind.

yolk sac a tiny bag attached to the embryo that provides early nourishment before the PLACENTA is formed.

yoni a symbol for the VULVA, as a source of pleasure, in Indian religion.

Young, Thomas Thomas Young (1773–1828) studied medicine at London, Edinburgh, Cambridge and Gottingen and set up in practice in London in 1799. His lively interest in all branches of science, however, made him restless, and within two years he had abandoned medicine to devote himself to scientific research. Although he distinguished himself in many different fields including physics, physiology, medicine and Egyptology, some of his most important discoveries related to light and to the eye. In 1793 Young read a paper entitled Observations on Vision to the Royal Society. In this, he was able to show how accommodation of the eye for near vision was the result of changes in the curvature of the crystalline lens. For a time it was rumoured that Young had stolen the idea from the surgeon John Hunter, having picked it up at a dinner party attended by Boswell, Sir Josua Reynolds and others. Young's whole way of life was, however, inconsistent with plagiarism, the slur was rejected, and this paper led to his election to the Fellowship of the Royal Society.

In 1800 Young read a remarkable paper, *The Mechanism of the Eye* to the Royal Society. In this, among other original findings, he described his invention, the 'optometer', by which he had been able to measure the radius of curvature of the cornea and to show that astigmatism was caused by non-spherical curvatures. In 1801 Young was appointed Professor of Natural Philosophy at the new Royal Institution. His thrice-weekly lectures there proved to be much too difficult for a popular audience and little of what he said was understood. The lectures were, however, published in 1807 and are now recognized as containing more original contributions to science than any comparable work.

Young formulated the currently accepted three-colour theory of colour vision; he worked out the wave theory of light – probably the greatest contribution he made to theoretical physics; he demonstrated chromatic aberration in the eye; and he calculated the wave-lengths of the seven colours of the spectrum. Rightly, Young has been called the father of physiological optics. In other fields he was equally distinguished. He established the physics of elasticity; he wrote a *Dictionary of the Ancient Egyptian Language*; he contributed extensively to the first edition of the *Encyclopaedia Britannica*; he wrote on a wide range of subjects including actuarial science, hydraulics, tides, annuities, bridge, carpentry, integrals, weights and measures and languages; and he translated the Rosetta stone, thereby establishing the long-forgotten hieroglyphic text.

Zz

ZAP-70 a tyrosine kinase normally expressed in T cells and NK cells.

Z DNA an uncommon configuration for DNA. It is a left-hand helix with about 12 bases per turn.

zinc a metallic element required in small quantities for health. Deficiency is rare but may occur in people with certain malabsorption conditions, with anorexia nervosa, diabetes, severe burns, prolonged feverish illness, severe malnutrition in childhood and in alcoholics. Zinc deficiency is associated with atrophy of the thymus gland and depressed cell-mediated immunity, skin atrophy, poor wound healing, loss of appetite, persistent diarrhoea, apathy and loss of hair. A normal diet contains plenty of zinc but a small zinc supplement is said to shorten the duration of the common cold.

zinc oxide a white powder with mild astringent properties used as a dusting powder or incorporated into creams or ointments and used as a bland skin application. Mixed with oil of cloves, zinc oxide forms an effective and pain-relieving temporary dressing for a tooth cavity. Zinc oxide is an ingredient in numerous proprietary medical preparations.

Z line the structure delimiting each end of the sarcomere in striated muscle. It gives anchorage to titin and to the outer ends of the thin filaments.

zombification the purported conversion by magic means to a state in which the awareness of an individual is retained by a sorcerer in a bottle or jar while the body, lacking will or agency, becomes the slave of the sorcerer. Zombification is deemed to be murder in Haiti even if the victim is manifestly still alive. The state is either induced by various poisons or by strong suggestion in a context of powerful superstitious belief. Some zombies have been found to be schizophrenics.

zonule the delicate suspensory ligament of the CRYSTALLINE LENS of the eye.

zoo blot the use of SOUTHERN BLOTTING to check whether a DNA probe from one species can hybridize with the DNA of other species.

zoonoses diseases of animals that can affect people. The zoonoses do not include human diseases transmitted from person to person by animal vectors. The zoonoses include anthrax from cattle, brucellosis and Q fever from goats and sheep, glanders from horses, leptospirosis and plague from rats, psittacosis from birds, rabies from any mammal, Rocky Mountain spotted fever from small mammals, toxocariasis from dogs, toxoplasmosis from cats, tuberculosis from cows and yellow fever from monkeys.

zoophobia irrational fear of animals.

zwitterions the dipolar ions that form from amino acids when they are in solution at a neutral pH.

zygoma the cheek bone.

zygomatic pertaining to the ZYGOMA.

zygomatic arch the bony arch extending below and to the outer side of the eye socket (the orbit) and forming the prominence of the cheek.

zygomatic process any of the three processes that make up the ZYGOMATIC ARCH.

zygote an egg (ovum) that has been fertilized but has not yet undergone the first cleavage division. A zygote contains a complete (DIPLOID) set of chromosomes, half from ovum and half from the fertilizing sperm, and thus all the genetic code for a new individual.

zymogen the inactive precursor of an enzyme.

Human Biology Resources on the Internet

This is a selection of addresses for websites relating to human biology, grouped under the main subdivisions of anatomy, physiology, biochemistry, genetics, human embryology, molecular cell biology, human evolution and human ecology, with a few general entries under human biology. The quantity of material available on the internet on these topics is virtually limitless, but it is hoped that the samples provided here may encourage students and other interested people to browse further. All addresses were current in 2006. Finding reliable and free data should not be difficult although a few points need to be borne in mind. The material should be up to date. Small organizations and departments within academic institutions sometimes encounter funding difficulties and are unable to continue with their researches. Make sure to look at the 'Last updated' section of the main website page before using any data. Try clicking on the links to make sure that they have been maintained properly and do not result in error messages. Ideally information should be obtained from websites run by universities, research institutes, and other reputable organizations. Websites maintained by individuals may not be up to date and comprehensive. It is also possible that the prejudices of those maintaining the websites will be reflected in the content and list of links.

Human Biology

Annals of Human Biology

www.tandf.co.uk/journals/titles/03014460.asp

An international journal of the Society for the Study of Human Biology with papers concerned with human population including genetics, ecology, epidemiology and ageing.

BBC Education – AS Guru Biology

www.bbc.co.uk/education/asguru/biology/intro.shtml

Produced in consultation with the chief examiner of AS Level Biology and teachers, this website focuses specifically on the difficult parts of the core curriculum.

The Biology Project: Human Biology

www.biology.arizona.edu/human_bio/human_bio.html

Apply your knowledge of Mendelian genetics to humans, and learn about constructing pedigrees. Also take the quiz on sexual reproduction.

Anatomy

Instant Anatomy

www.instantanatomy.net/

A specialized website with anatomical illustrations of the human body to help with learning this topic.

An@tomy.tv

www.anatomy.tv/

The most detailed and accurate 3D model of human anatomy. This is a wonderful educational tool as models can be rotated while layers of anatomy can be added or deleted.

Cyberanatomy

http://anatome.ncl.ac.uk/tutorials/

Anatomy tutorials by Dr Donald Shanahan, University of Newcastle, England.

Introductory Anatomy: Digestive System

www.leeds.ac.uk/chb/lectures/anatomy8.html

A complete look at the this system including the mouth, swallowing, oesophagus, stomach, duodenum, small and large intestines, salivary glands, etc.

Anatomy for Beginners
www.channel4.com/science/microsites/A/anatomy/
Channel 4's website is the first broadcast showing a real demonstration of human anatomy. Dr Gunther von Hagens dissects the bodies while the pathologist, Professor John Lee, explains how both healthy and diseased bodies work.

The Anatomy of the Immune System
www-micro.msb.le.ac.uk/MBChB/2b.html
An in-depth look at the immune system and all its separate parts.

Journal of Anatomy
www.blackwellpublishing.com/journal.asp?ref=0021-8782
Publication of the Anatomical Society of Great Britain and Ireland containing original papers, invited review articles and book reviews. Abstracts available free electronically.

Edinburgh Human Developmental Anatomy
www.ana.ed.ac.uk/anatomy/database/humat/
Database on the ontology of human developmental anatomy.

In Our Time
www.bbc.co.uk/radio4/history/inourtime/inourtime_20020214.shtml
BBC programme on the history of anatomy over 2,000 years including Vesalius and the infamous bodysnatchers, Burke and Hare.

Instant Anatomy
www.instantanatomy.net/index.html
Instant anatomy is a specialized website dealing with the anatomy of the human body as a companion to the CD-ROM, *Instant Anatomy*.

Introductory Anatomy: Respiratory System
www.leeds.ac.uk/chb/lectures/anatomy7.html
Lectures from the Faculty of Biological Sciences, University of Leeds

PATTS – Anatomy: Significant Body Systems
www.webschoolsolutions.com/patts/systems/anatomy.htm
A companion series for the PATTS Certificate Program with interactive learning tools.

BUBL LINK: Human anatomy
www.bubl.ac.uk/link/h/humananatomy.htm
An academic website for the best internet resources on this subject. BUBL is an information service provided by the Digital Library Team based at the University of Strathclyde, UK.

History of Ophthalmology
www.mrcophth.com/Historyofophthalmology/anatomy.htm
Ophthalmic anatomy, ophthalmology, eye, anatomy, medical history.

BUBL LINK: Medical Imaging
http://bubl.ac.uk/Link/m/medicalimaging.htm
An academic website for the best internet resources on this subject. BUBL is an information service provided by the Digital Library Team based at the University of Strathclyde, UK.

BBC: Science & Nature – Human Body and Mind
www.bbc.co.uk/science/humanbody/body/factfiles/skeleton_anatomy.shtml
Anatomical diagram showing a frontal view of a human skeleton with label links to detailed information on parts of the body.

Physiology

Human Physiology images
www.fleshandbones.com/imagebank/bytitle.cfm
A website for medical students and instructors containing digital photographs and professional illustrations. There are many physiology diagrams at
www.fleshandbones.com/physiology/davies/

GCSE Human Physiology and Health
www.icslearn.co.uk/gcse-A-level/science/gcse-human-physiology-health/
The UK distance learning GCSE Human Physiology and Health Course.

GCSE Human Physiology and Health
www.aqa.org.uk/qual/gcse/hph.html
The Assessment and Qualifications Alliance (AQA) guidance to teaching this subject.

National Extension College (NEC) Courses: Human Physiology and Health GCSE
www.nec.ac.uk/courses/product?product_id=889&category_id=526
Discover how the human body works and the varied aspects of development. This site also has practical experiments to help develop analytical skills.

BioMed Central (BMC)
www.biomedcentral.com/bmcphysiol/
BMC Physiology publishes original research articles in all functional and developmental processes of physiological processes. Abstracts and full text are available free of charge but can also be searched on MEDLINE and BIOSIS.

Physiology Online
www.physoc.org/
The electronic information service of the Physiological Society (UK) including the *Journal of Physiology, Experimental Physiology etc.*

Physiology Homepage
www.members.aol.com/Bio50/
Lecture notes, study guides, anatomical illustrations, genetic code and model of DNA listed by University of Texas, University of Pittsburgh and Trinity College, Perth, Australia.

Human Physiology
cwx.prenhall.com/bookbind/pubbooks/silverthorn2/
Studies of human physiology by Dee Unglaub Silverthorn, Ph.D. University of Texas, with illustrations, quizzes and current research links.

Experimental Physiology
www.journals.cambridge.org/action/displayJournal?jid=EPH
A bimonthly journal publishing original work on all facets of physiology. Only abstracts are free, full text is available for subscription fee.

Visual Physiology
www.lifesci.sussex.ac.uk/home/George_Mather/Linked%20Pages/Physiol/
A visual physiology interactive tutorial which includes the paths of visual processing from the eyes to the brain cortex.

Biochemistry

Biochemical Society
www.biochemistry.org/
Details of the activities of the Society, and contains a good listing of useful links.

BioMed Central (BMC) Biochemistry
www.biomedcentral.com/bmcbiochem/
BMC Biochemistry publishes original research articles on all aspects of biochemistry, including metabolic pathways, enzyme functions, and small molecular components of cells.

Biotechnology* and *Applied Biochemistry
www.babonline.org/
Both journals publish papers on new technology including recombinant engineering, antibodies and genes, and stem-cell therapeutics. Now available in EESI-View.

BUBL LINK: Biochemistry
www.bubl.ac.uk/link/b/biochemistry.htm
An academic website for the best internet resources on this subject. BUBL is an information service provided by the Digital Library Team based at the University of Strathclyde, UK.

FEBS Journal
www.febsjournal.org/
Federation of European Biochemical Societies (FEBS) publishes papers on bioinformatics, genomics, nanoscience and many other subjects. This website also has a Journal prize for young scientists.

International Union of Biochemistry and Molecular Biology
www.chem.qmul.ac.uk/iubmb/
Recommendations on biochemical and organic nomenclature, symbols and terminology, etc from the Department of Chemistry, Queen Mary University of London.

The Medical Biochemistry Page
www.dentistry.leeds.ac.uk/biochem/thcme/home.html
A site designed principally as a teaching tool for students of medical biochemistry. However, any student taking biochemistry will benefit.

The Association for Clinical Biochemistry
www.acb.org.uk/
Organization dedicated to promoting the advancement of clinical biochemistry in the United Kingdom and the Republic of Ireland.

European Journal of Biochemistry
www.ingentaconnect.com/content/bsc/ejb
Only abstracts of the papers published in this journal are available free of charge online. For access to full text, a subscription fee is charged or a password can be obtained from your academic institution.

Glossary of Biochemistry and Molecular Biology
www.portlandpress.com/pp/books/online/glick/search.htm
Very useful searchable electronic glossary of biochemical and molecular terminology by David M. Glick.

Structural Biochemistry@Edinburgh
www.bch.ed.ac.uk/
The Structural Biochemistry Group is based at the University of Edinburgh but has an international mix of personnel and expertise in protein expression and purification. Protein structures discovered by the researchers can be browsed on this site.

Archives of Physiology and Biochemistry
www.tandf.co.uk/journals/titles/13813455.asp
Abstracts are available online for free on subjects such as molecular, biochemical and cellular aspects of metabolic diseases but for full text articles, a subscription fee is required.

National Institute for Medical Research (NIMR)
www.nimr.mrc.ac.uk/physbiochem/
The NIMR has four major research groups – Genetics and Development, Infections and Immunity, Neurosciences, Structural biology.

Genetics

National Coalition for Health Professional Education in Genetics (NCHPEG)
www.nchpeg.org/
The NCHPEG promotes access to information on recent genetic research.

Behavioural Genetics Interactive Modules
statgen.iop.kcl.ac.uk/bgim/index2.html
An online introduction to behavioural genetics for anyone wishing to pursue this subject.

BUBL LINK: Human Genetics
bubl.ac.uk/link/h/humangenetics.htm
An academic website for the best internet resources on this subject. BUBL is an information service provided by the Digital Library Team based at the University of Strathclyde, UK. Also includes information on the ethics of human genetics.

Human Genetics Commission
www.hgc.gov.uk/
A governmental advisory body on how new developments in human genetics will impact on people and on health care.

Genetic Interest Group (GIG)
www.gig.org.uk/
GIG provides information to raise awareness of genetic disorders and also tries to improve services and treatment for people affected by disorders.

GenePool
http://libraries.nelh.nhs.uk/genepool/
This is a specialist library for clinical genetics containing information on syndromes and diseases.

Genetics
www.super-memory.com/sml/colls/genetics.htm
Basic facts about DNA, chromosomes, genetic engineering, transgenics, genetics and genetic diseases.

British Society for Human Genetics
www.bshg.org.uk/
An independent body representing professionals working in research groups such as Clinical Genetics Society and Cancer Genetics Group.

Human Genetics Alert
www.hgalert.org/
An independent watchdog group.

Wellcome Trust Centre for Human Genetics
www.well.ox.ac.uk/
Research into the genetic basis of common diseases such as hypertension, diabetes, heart disease, multiple sclerosis, rheumatoid arthritis, etc.

Medical Research Council (MRC): Human Genetics Unit
www.hgu.mrc.ac.uk/
This research establishment aims to advance the understanding of genetic factors implicated in human disease, and normal and abnormal development.

Annals of Human Genetics
www.journals.cambridge.org/action/displayJournal?jid=HGE
This journal publishes original research directly concerned with human genetics or the application of scientific principles and techniques to any aspect of human inheritance. Only online abstracts are free.

Twin Research and Human Genetics
www.ingentaconnect.com/content/aap/twg
Issues of this journal are available online with free access to abstracts but a subscription fee is required for access to full article.

Genetics and Human Behaviour: the Ethical Context
www.nuffieldbioethics.org/go/ourwork/behaviouralgenetics/introduction
This report by the Nuffield Council on Bioethics can be downloaded or browsed online.

Human Genetics
www.srtp.org.uk/humgen.shtml
The Society, Religion and Technology Group's website on human genetics containing substantial work in areas such as gene therapy and embryo selection for designer babies.

The Sanger Institute: Human Genetics
www.sanger.ac.uk/genetics/humgen.shtml
The Human Genome Project formed the core of human genetic research during the early years of the Sanger Institute, culminating in the Human GeneMap, chromosome maps.

The Human Genome Project (HGP)
http://www.genome.gov/10001772
The human genome sequence was completed in April 2003. Now, scientists have access to an amazing international research tool – a database which maps all the human genes, the genetic blueprint of a human being.

Excite UK
www.excite.co.uk/directory/Science/Biology/Genetics/Eukaryotic/Animal/Mammal/Human
International forum in human molecular genetics; it has a mailing list.

Human embryology

Databases from the Biomedical School of the University of Edinburgh
www.ana.ed.ac.uk/anatomy/database/
Online publication of the human embryo taxonomy and other useful databases.

Human Fertilization and Embryology Authority
www.hfea.gov.uk/Home
The HFEA licenses and monitors all human embryo research.

Reproductive Biomedicine Online
www.rbmonline.com
An international journal available online. It publishes papers concerned with biomedical research and ethical issues surrounding conception of the human embryo.

Review of the Human Fertilization and Embryology Act 1990
www.dh.gov.uk/PublicationsAndStatistics/Publications/PublicationsPolicyAndGuidance/PublicationsPolicyAndGuidanceArticle/fs/en?CONTENT_ID=4123774&chk=556b/v
The UK Government's consultation paper on the above act, can be downloaded free of charge.

OMNI: Embryology
www.omni.ac.uk/browse/mesh/D004626.html
OMNI is a free catalogue of hand-selected and evaluated internet resources in health and medicine. Human embryology page has specimens representing 10 stages of development.

Cloning Human Embryos for Spare Cells and Tissues: an ethical dilemma
www.srtp.org.uk/clonembr.htm
The Society, Religion and Technology Group of the Church of Scotland examine the ethics of technology for a New Millennium.

Association of Medical Research Charities (AMRC)
www.amrc.org.uk/index.asp?id=15036
Discussion on the current restrictions on embryo research and the Human Fertilization and Embryology Authority (HFEA).

Molecular cell biology

Medical Research Council (MRC): Laboratory for Molecular Cell Biology
www.ucl.ac.uk/LMCB/
Information on the research, aims of the unit and events for the above organization.

Molecular Cell Online
www.molecule.org/
This journal publishes reports of novel results that are of unusual significance and of interest to researchers in the field.

Dictionary of Cell Biology
www.mblab.gla.ac.uk/~julian/Dict.html
An online dictionary for cell and molecular biology.

Molecular Cell Biology Research Communications
www.ingentaconnect.com/content/ap/rm
This journal has some issues available electronically but free online access is limited to abstracts only. For full text articles, a subscription fee is required.

Dr Chromo's school: the biology of a cell
www.rothamsted.bbsrc.ac.uk/notebook/courses/guide/cell.htm
An online package for the hands-on teaching of molecular biology.

Molecular Cell Biology
www.palgrave.com/science/lifesciences/lodish/index.asp
This textbook by Lodish, H. *et al.* has three chapters freely available online.

BUBL LINK: Cell biology
www.bubl.ac.uk/link/c/cellbiology.htm
An academic website for the best internet resources on this subject. BUBL is an information service provided by the Digital Library Team based at the University of Strathclyde, UK. Subjects: cell biology, genetics research, molecular biology.

Genes to Cells
www.blackwellpublishing.com/journal.asp?ref=1356-9597
This journal is published on behalf of the Molecular Biology Society of Japan. Free online access to full articles is available within institutions in the developing world.

The Institute of Cancer Research (IRC): Molecular Cell Biology Team
http://dynamic.icr.ac.uk/icontactdb/galleries/teams/Molecular_Cell_Biology_gallery.shtml
Information on the staff and research of the Molecular Cell Biology Team at the IRC.

Human evolution

BBC Science & Nature: the Evolution of Man
www.bbc.co.uk/sn/prehistoric_life/human/human_evolution/index.shtml
An educational essay on human beginnings and evolution from the BBC website.

Human Biology, Lecture notes from the University of Leeds
www.leeds.ac.uk/chb/humbmods.html
Evolutionary developmental biology and introductory anatomy.

Journal of Human Evolution
www.ingentaconnect.com/content/ap/hu
Many issues of this journal are available online with free access to abstracts but a subscription fee is required for access to the full article.

Natural Environment Research Council (NERC)
www.nerc.ac.uk/funding/thematics/efched/
Information on the new NERC science programme entitled 'Environmental factors in the chronology of human evolution and dispersal'.

Leverhulme Centre for Human Evolutionary Studies and Duckworth Laboratory
www.human-evol.cam.ac.uk/Centre/centre.htm
Both the Centre for Human Evolutionary Studies and Duckworth Laboratory are devoted to the study of human evolution and variation at Cambridge University.

Human Evolution on Yahoo
http://uk.dir.yahoo.com/Science/Biology/Evolution/Human_Evolution/
Websites and categories on human evolution from the Yahoo search engine including *Evolution of Modern Humans: A Survey of the Biological and Cultural Evolution*, and also, *Human Evolution: Evolution and the Structure of Health and Disease.*

Natural Selection
http://130.88.13.174/
Natural Selection is a subject gateway of evaluated internet resources to the natural world and new sources are added weekly.

Bradshaw Foundation
www.bradshawfoundation.com/migration.html
Human evolution, migration and a genetic map are all available on this website.

Monkeys and Human Evolution
www-biol.paisley.ac.uk/courses/Tatner/biomedia/units/monk1.htm
An interactive section of the Biological Sciences website hosted at Paisley University, UK.

Antiquity of Man
www.antiquityofman.com/Human_evolution_articles.html
Articles discussing human evolution and civilization including, *Recent developments in Human Biological and Cultural Evolution*, by Professor J. Desmond Clark.

Why Y?
www.ucl.ac.uk/tcga/ScienceSpectra-pages/SciSpect-14-98.html
In this paper, Neil Bradman and Mark Thomas look at the Y chromosome in the study of human evolution, migration and prehistory.

Human Ecology

Human Ecology Research Group (HERG) University College London
www.ucl.ac.uk/herg/
An overview of HERG which specializes in sub-Saharan Africa. The Group focuses on developing new understandings and narratives of people's interaction with the environment.

Centre for Human Ecology
www.che.ac.uk/index.php/
The Centre for Human Ecology exists to stimulate and support fundamental change towards ecological sustainability and social justice.

What is human ecology?
www.homepages.which.net/~gk.sherman/b.htm
Online papers by Grant K Sherman and also, *Accounting for environmental decision-making*, which was his MSc Thesis, Warwick University.

Human Ecology
www.ingentaconnect.com/content/klu/huec
This journal has many issues available electronically but free online access is limited to abstracts only. For full text articles, a subscription fee is required.

Environmental Change Institute (ECI) University of Oxford
www.eci.ox.ac.uk/humaneco/
The ECI's Human Ecology research programme and information on publications.

Friends of Human Ecology Forum
http://uk.dir.yahoo.com/Social_Science/Sociology/Human_Ecology/
Yahoo forum for human ecologists, students, academics, professors, anyone else interested. Works with individuals, communities, and organizations to act for ecological sustainability.

Glasgow ePrints Service: Human ecology. Anthropogeography
http://eprints.gla.ac.uk/view/subjects/GF.html
Published and peer-reviewed papers from Glasgow University Library

Cultural Anthropology Methods Journal (CAM)
http://lucy.ukc.ac.uk/cam.html
This journal publishes articles on the real qualitative and quantitative research method with reviews of software of interest to all social sciences researchers.